Corynebacteria
Genomics and Molecular Biology

Edited by

Andreas Burkovski

Lehrstuhl für Mikrobiologie,
Friedrich-Alexander Universität Erlangen-Nürnberg,
91058 Erlangen, Germany

Caister Academic Press
Norfolk UK

www.caister.com

British Library Cataloguing-in-Publication Data
A catalogue record for this book is available from the British Library
ISBN: 978-1-904455-30-1

Cover Greenfluorescent *Corynebacterium glutamicum* cells expressing the *gfpuv* gene (kindly provided by N. Jeßberger, Erlangen.

Printed and bound in Great Britain

Other Books of Interest

Plasmids: Current Research and Future Trends	2008
Pasteurellaceae: Biology, Genomics and Molecular Aspects	2008
Vibrio cholerae: Genomics and Molecular Biology	2008
Pathogenic Fungi: Insights in Molecular Biology	2008
Helicobacter pylori: Molecular Genetics and Cellular Biology	2008
Staphylococcus: Molecular Genetics	2008
Leishmania: After The Genome	2008
Archaea: New Models for Prokaryotic Biology	2008
RNA and the Regulation of Gene Expression	2008
Legionella Molecular Microbiology	2008
Molecular Oral Microbiology	2008
Epigenetics	2008
Animal Viruses: Molecular Biology	2008
Segmented Double-Stranded RNA Viruses	2008
Acinetobacter Molecular Microbiology	2008
Pseudomonas: Genomics and Molecular Biology	2008
Microbial Biodegradation: Genomics and Molecular Biology	2008
The Cyanobacteria: Molecular Biology, Genomics and Evolution	2008
Coronaviruses: Molecular and Cellular Biology	2007
Real-Time PCR in Microbiology: From Diagnosis to Characterisation	2007
Bacteriophage: Genetics and Molecular Biology	2007
Candida: Comparative and Functional Genomics	2007
Bacillus: Cellular and Molecular Biology	2007
AIDS Vaccine Development: Challenges and Opportunities	2007
Alpha Herpesviruses: Molecular and Cellular Biology	2007
Pathogenic *Treponema*: Molecular and Cellular Biology	2007
PCR Troubleshooting: The Essential Guide	2006
Influenza Virology: Current Topics	2006
Microbial Subversion of Immunity: Current Topics	2006
Probiotics and Prebiotics: Scientific Aspects	2005
Foodborne Pathogens: Microbiology and Molecular Biology	2005

Contents

Contributors v

Preface vii

1 The Discovery of *Corynebacterium glutamicum* and Birth of Amino Acid Fermentation Industry in Japan 1
Shigezo Udaka

2 Genomics of Industrially and Medically Relevant Corynebacteria 7
Andreas Tauch

3 DNA Microarray-based Transcriptome Analysis in *C. glutamicum* 33
Volker F. Wendisch

4 Proteomics of *Corynebacterium glutamicum* and Other Corynebacteria 55
Jörn Kalinowski, Dirk Wolters, and Ansgar Poetsch

5 Metabolic Network Analysis and Design in *Corynebacterium glutamicum* 79
Christoph Wittmann and Elmar Heinzle

6 Plasmids and Promoters in Corynebacteria and Their Applications 113
J. Nešvera and M. Pátek

7 Regulation of Carbon Metabolism in *Corynebacterium glutamicum* 155
Annette Arndt and Bernhard J. Eikmanns

8 Molecular Mechanisms of Nitrogen Control in Corynebacteria 183
Eva Hänßler and Andreas Burkovski

9 Phosphorus Metabolism and its Regulation 203
Volker F. Wendisch and Michael Bott

10 Sulfur Metabolism in *Corynebacterium glutamicum* 217
Christian Rückert and Jörn Kalinowski

11 Regulation of Iron Homeostasis in *Corynebacterium glutamicum* 241
Julia Frunzke and Michael Bott

12 Structure and Synthesis of the Cell Wall 267
Lothar Eggeling, Gurdyal S. Besra and Luke Alderwick

13 General and Regulatory Proteolysis in Corynebacteria 295
Johannes Amon, Alja Lüdke, Fritz Titgemeyer, and Andreas Burkovski

14 Environmental Stress Response of *Corynebacterium glutamicum* 313
Susanne Morbach and Reinhard Krämer

Index 335

Contributors

Luke Alderwick
School of Biosciences
University of Birmingham
Edgbaston
Birmingham B15 2TT
United Kingdom

Johannes Amon
Lehrstuhl für Mikrobiologie
Friedrich-Alexander-Universität Erlangen-Nürnberg
Staudtstr. 5
91058 Erlangen
Germany

Annette Arndt
Institute of Microbiology and Biotechnology
University of Ulm
89069 Ulm
Germany

Gurdyal S. Besra
School of Biosciences
University of Birmingham
Edgbaston
Birmingham B15 2TT
United Kingdom

Michael Bott
Institut für Biotechnologie 1
Forschungszentrum Jülich
D-52425 Jülich
Germany

Andreas Burkovski
Lehrstuhl für Mikrobiologie
Friedrich-Alexander-Universität Erlangen-Nürnberg
Staudtstr. 5
91058 Erlangen
Germany

Lothar Eggeling
Research Centre Juelich
D-52425 Juelich
Germany

Bernhard J. Eikmanns
Institute of Microbiology and Biotechnology
University of Ulm
89069 Ulm
Germany

Julia Frunzke
Institut für Biotechnologie 1
Forschungszentrum Jülich
D-52425 Jülich
Germany

Eva Hänßler
Lehrstuhl für Mikrobiologie
Friedrich-Alexander-Universität Erlangen-Nürnberg
Staudtstr. 5
91058 Erlangen
Germany

Elmar Heinzle
Biochemical Engineering Institute
Saarland University
Campus A1.5
66123 Saarbrücken
Germany

Jörn Kalinowski
Institut für Genomforschung und Systembiologie
Centrum für Biotechnologie
Universität Bielefeld
Universitätsstr. 25
33615 Bielefeld
Germany

Reinhard Krämer
Institut für Biochemie der Universität zu Köln
Zülpicher Str. 47
50674 Cologne
Germany

Alja Lüdke
Lehrstuhl für Mikrobiologie
Friedrich-Alexander-Universität Erlangen-Nürnberg
Staudtstr. 5
91058 Erlangen
Germany

Susanne Morbach
Institut für Biochemie der Universität zu Köln
Zülpicher Str. 47
50674 Cologne
Germany

Jan Nešvera
Institute of Microbiology
Academy of Sciences of the Czech Republic
Víděnská 1083
14220 Prague 4
Czech Republic

Miroslav Pátek
Institute of Microbiology
Academy of Sciences of the Czech Republic
Víděnská 1083
14220 Prague 4
Czech Republic

Ansgar Poetsch
Lehrstuhl für Biochemie der Pflanzen
Universitätsstr. 150
Ruhr-Universität Bochum
44801 Bochum
Germany

Christian Rückert
Institut für Genomforschung und Systembiologie
Centrum für Biotechnologie
Universität Bielefeld
Universitätsstraße 25
33615 Bielefeld
Germany

Andreas Tauch
Institut für Genomforschung und Systembiologie
Centrum für Biotechnologie
Universität Bielefeld
Universitätsstraße 25
D-33615 Bielefeld
Germany

Fritz Titgemeyer
Lehrstuhl für Mikrobiologie
Friedrich-Alexander-Universität Erlangen-Nürnberg
Staudtstr. 5
91058 Erlangen
Germany

Shigezo Udaka
Grand Maison C-1112
Hoshigaoka-Yamate 401
Chikusa-ku
Nagoya 464-0808
Japan

Volker F. Wendisch
Institute for Molecular Microbiology and Biotechnology
Westfalian Wilhelms University Muenster
Corrensstr. 3
D-48149 Muenster
Germany

Christoph Wittmann
Lehrstuhl für Biotechnologie
Institut für Biochemie. WWU Münster
Wilhelm-Klemm-Strasse 2
48149 Münster
Germany

Dirk Wolters
Analytische Chemie
Universitätsstr. 150
Ruhr-Universität Bochum
44801 Bochum
Germany

Preface

Corynebacteria are Gram-positive microorganisms closely related to mycobacteria and inhabiting diverse ecological niches such as soil, vegetables, sewage, skin, and cheese smear. The genus *Corynebacterium* was originally defined two centuries ago when Lehmann and Neumann introduced the diphtheroid bacillus *Corynebacterium diphtheriae* as the type strain of a new genus (Lehmann and Neumann, 1896), comprising nonmotile, nonsporulating bacteria which typically form straight or slightly curved rods with tapered or sometimes clubbed ends (coryne: Greek for club) and V-forming cell pairs or palisades of several cells due to the corynebacterial snapping mechanism of cell division.

Although *C. diphtheriae* had an immense medical importance and contributed significantly to child death in times when antibiotics and vaccines were not available, which is still a problem in African and Asian countries, today *Corynebacterium glutamicum* is the best investigated species of this genus. When *C. glutamicum* was isolated about half a century ago in a screening program for L-glutamate-producing bacteria from a soil sample collected at Ueno Zoo in Tokyo, this was the beginning of a tremendous success story as described in an introductory essay in this book. Today *C. glutamicum* is one of the biotechnologically most important bacterial species with an annual production of more than two million tons of amino acids, mainly L-glutamate and L-lysine (Leuchtenberger et al., 2005). Due to this immense industrial importance, *C. glutamicum* has been studied extensively starting already shortly after its discovery. Although obviously successful, former investigations had the disadvantage that they were typically focused on distinct transporters, enzymes, or regulators; a global analysis without prior assumptions and expectations was impossible in the 1990s. This situation changed when the *C. glutamicum* genome was sequenced and published independently by different industry-supported groups (Ikeda and Nakagawa, 2003; Kalinowski et al., 2003). The knowledge of this genome sequence, which was the first available for corynebacteria, gave a fresh impetus to *C. glutamicum* research, since it allowed the development of new molecular biology tools such as DNA microarrays for transcriptome analyses as well as proteome analyses by two-dimensional polyacrylamide gel electrophoresis and mass spectrometry.

This book describes the molecular biology toolbox available today for *Corynebacterium* research, including global analyses techniques such as comparative genomics, transcriptome, proteome and metabolome analysis as well as the most recent knowledge on *Corynebacterium* promoter structures and vector systems. Furthermore, topics such as regulatory networks controlling carbon, nitrogen, phosphorus, sulphur and iron metabolism, cell wall structure, proteolysis and environmental stress response are covered. The fascinating advances made in the last few years unequivocally demonstrate that especially *C. glutamicum* has become one of the best investigated and understood microorganisms. Clearly, the high G+C Gram positive corynebacteria are significantly different from low G+C Gram positive bacteria such as *Bacillus subtilis*, and the application of modern molecular biology analysis techniques has made *C. glutamicum* not only an excellent model organism for *Corynebacterinae*, but also a model organism for systems biology (Wendisch et al., 2006).

Andreas Burkovski

References

Ikeda, M., and Nakagawa, S. (2003). The *Corynebacterium glutamicum* genome: features and impacts on biotechnological processes. Appl. Microbiol. Biotechnol. *62*, 99-109.

Kalinowski, J., Bathe, B., Bartels, D., Bischoff, N., Bott, M., Burkovski, A., Dusch, N., Eggeling, L., Eikmanns, B.J., Gaigalat, L., Goesmann, A., Hartmann, M., et al. (2003). The complete *Corynebacterium glutamicum* ATCC 13032 genome sequence and its impact on the production of L-aspartate-derived amino acids and vitamins. J. Biotechnol. *104*, 5-25.

Lehmann, K.B., and Neumann, R. (1896). Atlas und Grundriss der Bakteriologie und Lehrbuch der speziellen bakteriologischen Diagnostik. J.F. Lehmann, Munich, Germany.

Leuchtenberger, W., Huthmacher, K., and Drauz, K. (2005). Biotechnological production of amino acids and derivatives: current status and prospects. Appl. Microbiol. Biotechnol. *69*, 1-8.

Wendisch, V.F., Bott, M., Kalinowski. J., Oldiges. M., and Wiechert, W. (2006). Emerging *Corynebacterium glutamicum* systems biology. J. Biotechnol. *124*, 74-92.

From: Corynebacteria: Genomics and Molecular Biology. Edited by: Andreas Burkovski.

The Discovery of *Corynebacterium glutamicum* and Birth of Amino Acid Fermentation Industry in Japan

1

Shigezo Udaka

Abstract

Amino acids were once prepared by decomposition of proteins and considered as precious natural products of limited availability. Nowadays, amino acids such as L-glutamate and L-lysine are produced on a scale of million of tons by fermentation. The prosperity of amino acid fermentation industry started about half a century ago with my discovery of a glutamic acid producing bacterium. This chapter summarizes the story of the isolation of *Corynebacterium glutamicum*, the first approaches, the successful screening procedure and the prerequisites that were crucial for the discovery of this amino acid-producing bacterium.

1. The discovery of *Corynebacterium glutamicum*

Monosodium glutamate (MSG), whose "umami" (well-tasting) was discovered by Professor K. Ikeda in early 20th century, enhances the flavor of food and was traditionally used in Japanese cuisine in the form of the kelp preparation kombu and MSG from protein hydrolysate. After World War II, the shortage of food, especially protein-rich meat was very serious in Japan and the idea was to use MSG for seasoning of very tasteless food. However, since L-glutamate was prepared from the chemical decomposition of soybean and wheat proteins, it was too expensive. So, several Japanese researchers in the field of Applied Microbiology started to try to produce glutamic acid by fermentation processes. Most of them planned to synthesize α-ketoglutaric acid, the precursor of L-glutamate, by fermentation, and subsequently convert α-ketoglutaric acid to L-glutamate. The fermentative production of organic acids, such as citric acid, had by that time a long history of research and industrialization. Japanese specialists contributed extensively to the fermentative production of α-ketoglutaric acid and reported a high yield of about 50% on sugar basis with the use of some bacteria such as *Pseudomonas* species. Several methods have been presented to prepare L-glutamic acid from α-ketoglutaric acid. Amination by transaminases may be the simplest method, but this requires the addition of some other amino acid as amino group donor. As an alternative, an enzymatic method which uses glutamic acid dehydrogenase and NAD(P)H in the presence of ammonium salts was proposed. The latter method required the regeneration of the oxidized coenzyme. For this reduction process, the use of living cells was attempted at that time. In any case, these methods are composed of two stages: the production of α-ketoglutaric acid and its transformation into L-glutamic acid. A direct synthesis of L-glutamate from glucose and ammonia in a one-step process would have been superior to the two step procedure, however, researchers at that time believed that glutamate is essential for cell growth and that the extracellular production of large amounts of glutamate was most unlikely. This was the state of research, when I started to work on glutamate production.

At the end of 1955, I came back from study abroad in the U.S.A. to the Research Laboratory of Kyowa Hakko Kogyo Co. and was asked to survey and examine "fermentation processes and related papers" without giving me any particular research theme. I was permitted to study freely for some time and since I was young and wanted to create something new, I chose to develop a new and efficient production of L-glutamate by using microorganisms. At that time, solvents such as ethanol and organic acids like citric acid were known to be efficiently produced through catabolic pathways; however, there was no report on overproduction of a cell-building metabolite such as glutamate by anabolism. It was also known that amino acids are either metabolized very rapidly to other compounds or used directly to form proteins.

To avoid these problems, my first idea was the production of polymerized glutamate and its subsequent hydrolysis to obtain the monomeric form. At that time, a report appeared showing that a highly viscous mucoidal material produced by *Bacillus subtilis* was mainly a polymer of glutamate (Thorne et al., 1954). This polymer is an important part of Natto, a common Japanese food produced by *B. subtilis* var. *natto*. I collected the viscous material from Natto and hydrolyzed it by hydrochloric acid. However, unexpectedly the hydrolysate did not contain much L-glutamate.

My second idea was to use a highly concentrated suspension of living cells. When large amounts of glucose and ammonium sulfate were added to a non-growing cell suspension I expected that the cellular metabolism may give rise to L-glutamate, because of the 20 natural amino acids, glutamate is known to be most actively and rapidly synthesized. Since there were no sign of success, I abandoned these trial experiments quickly.

The third and finally successful idea was to produce L-glutamate by culturing microorganism and suppressing growth so that utilization of glutamate for cellular proteins would be prevented. Nobody seemed to be working in this way. Since such an organism might be quite rare if it exists at all, I developed an efficient method for the screening of a glutamate-producer within a few weeks (Udaka, 1960). This method is illustrated in Figure 1.1. In short, bacteria isolated from various environmental samples were grown

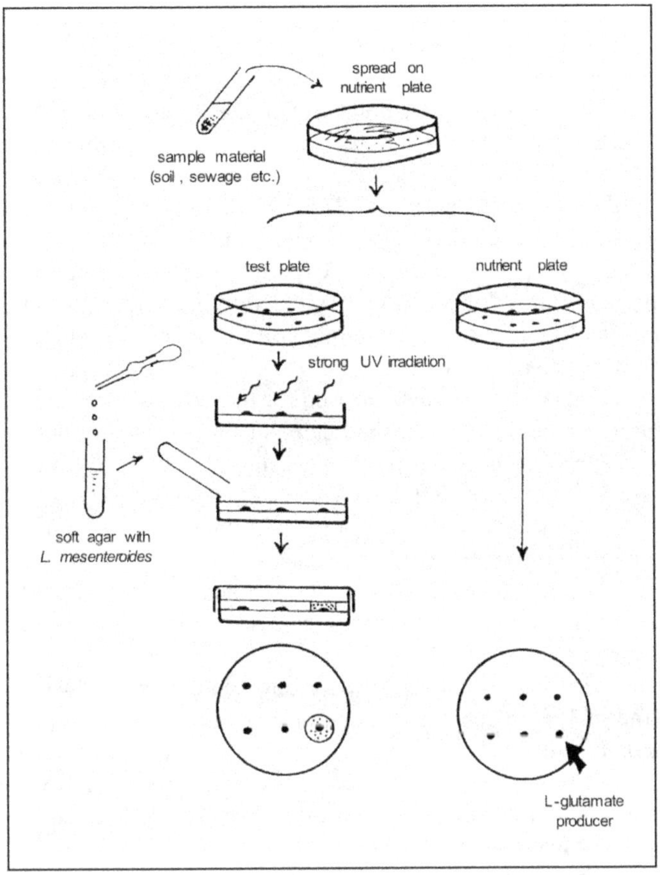

Figure 1.1. Schematic representation of the successful screening procedure leading to isolate No. 534, the first identified colony of *Corynebacterium glutamicum*. Diverse samples from different environments were spread on agar plates. Colonies appearing on these plates were killed by strong UV irradiation. Then, the plates were overlaid with a basal agar medium for glutamic acid bioassay containing *Leuconostoc mesenteroides*, an L-glutamic acid auxotrophic bacterium. Upon further incubation, the bacteria which produced a halo (growth of the assay organism) on their surrounding areas of the colonies were selected from a nutrient plate inoculated in parallel (Udaka, 1960).

on agar plates. Colonies appearing on these plates were killed by strong UV irradiation. Then, the plates were overlaid with a basal agar medium containing *Leuconostoc mesenteroides* strain P-60, a lactic acid bacterium, which requires L-glutamic acid for growth, as bioassay organism for glutamate production. Upon further incubation, the bacteria which produced a halo (growth of the assay organism) on their surrounding areas of the colonies were selected. In my initial test, about 500 bacterial isolates were examined. As the results, it was observed that the culture medium of one bacterial isolate, namely No. 534, which had given a dense halo, contained considerable amounts of glutamic acid. A filter paper, on which the culture supernatant had been developed, was sprayed with ninhydrin solution and heated. A deep purple spot appeared at the point corresponding to glutamic acid! More than a dozen bacteria showed a faint spot due to glutamic acid and about ten others a dense spot at the point of alanine. In addition, the presence of a glutamic acid- and alanine-producing isolates and a valine-producing one were confirmed (Kinoshita et al., 1957). No spots corresponding to any other amino acids were detected. Isolate No. 534 accumulated the largest amount of L-glutamate as supposed from the density of the halo and ninhydrin staining. In subsequently carried out experiments using a liquid test medium, the purified culture of this organism yielded to my surprise about 10 g of L-glutamate per liter of medium containing 50 g of glucose after 4 days, namely, 20% yield on sugar basis. I separated the crystalline amino acid from the culture supernatant by the method used for the separation of glutamate from wheat gluten hydrolysate and the chemical properties and optical rotatory activity of the crystalline sample were identical to the authentic L-glutamic acid. Thus, it became evident that a bacterium can produce L-glutamate very efficiently from a sugar source and ammonia. This was the birth of the amino acid fermentation process not only in Japan, but also in the world!

Strain No. 534 isolated by me was initially described as *Micrococcus glutamicus*, later it was identified as a member of the genus *Corynebacterium* and re-named *Corynebacterium glutamicum*.

I was very lucky to find an excellent glutamate producer within the short time of less than three months of research. When I think about why I succeeded, four reasons seem to be of critical importance for my success. If even any one of the following conditions was not met, development of microbial production process of amino acids would have been delayed by 5 to 20 years, or even indefinitely.

The first reason for success was the use of an efficient screening method for the identification of a glutamic acid hyperproducer. With the established screening, it became possible without too much difficulty to test thousands of microbial isolates and the effects of many kinds of cultural conditions. The second crucial point was the suitability of test culture media. When I undertook this pioneering study, the composition of culture media suitable to glutamic acid production was entirely unknown. Taking the biosynthetic pathway of glutamic acid into consideration, I prepared culture media with a carbon/nitrogen ratio corresponding to that of glutamic acid by adding high concentration of ammonium source to the medium used to fermentative production of α-ketoglutaric acid. The third important reason for success was the wide screening for glutamic acid-producing bacteria carried out with samples from various natural environments. I thought that in any fermentation production process, the production efficiency depends on the genetic properties and on the culture conditions (environment) of the microbes, and that the issue of a new fermentation will be strongly influenced by the specificity of the microbes themselves. Although I was inexperienced in screening, yet I noticed intuitively that there are innumerable microbes in nature, and that the kind of microbes differs from environment to environment. I collected the samples for microbe screening from different natural environments. They were, for example, soils, polluted waters from sewage treatment plants and animal feces from zoological gardens. In fact, strain No. 534 was isolated from a soil sample collected at Ueno Zoo in Tokyo, which was contaminated with avian feces. The generation of my idea might be unconsciously affected by the famous phrase of late Prof. Kin-ichiro Sakaguchi, "Ask to microbes, and they never deceive us", and also by the tradition of Applied Microbiology in Japan such that screening procedures were regarded as very important. At that time I thought already of a possible use of mutants for glutamic acid production, if glutamic acid-producing bacteria could not be isolated from natural sources. Since then I always keep this idea without much modification. The fourth crucial point for the discovery of *C. glutamicum* was the ease of glutamic acid analysis. I can think of many reasons which permitted the realization of glutamic acid fermentation and subsequently different amino acid fermentation. As one of the major reasons, I would mention progress of analytical method. Amino acids, important constituents of living body and foods, comprise various kinds of compounds. Analysis of each amino acid was a difficult task for a long time. After World War II, however, the situation changed completely by development of microbiological determination (microbioassay) and of paper chromatography. An expert of microbioassay was by my side in the laboratory where I did experiments for my graduation thesis and I practiced paper chromatography when I started my job in industry. From such experiences, I applied easily a bioassay system for screening of glutamic acid-producing bacteria and determining culture products by paper chromatography without any difficulties.

2. Later research milestones

After discovery of *C. glutamicum*, I focused my research further on amino acid production. When I cultured a *C. glutamicum* mutant strain, I found that glutamate production had disappeared and instead the mutant produced an amino acid, which showed the spot position of lysine on paper chromatogram after the reaction with ninhydrin. However, further analyses showed that the spot was L-ornithine, which is an intermediate in the biosynthetic pathway from glutamate to arginine. The enzyme converting ornithine to citrulline was blocked due to a mutation in the corresponding gene and as a result accumulation of large amounts of ornithine occurred. Although this mutant could not grow without addition of arginine and biotin to the culture medium, the amounts produced of glutamate and ornithine varied as a function of the amount of the nutrients added (Udaka and Kinoshita, 1958a). With limiting amounts of arginine and sufficient biotin, an efficient production of ornithine was obtained. I found and characterized particular enzymes pertinent to the biosynthetic pathway of ornithine and then demonstrated the control of ornithine production by the metabolic regulatory mechanism for the first time (Udaka and Kinoshita, 1958b). By proving the importance of the metabolic regulation in amino acid fermentation, I pioneered a new way to metabolically controlled fermentation. Because of these studies, I was invited to attend to a memorable Cold Spring Harbor Symposium on the cellular regulatory mechanism (Umbarger, 1961) in 1961, which is known as a landmark in Molecular Biology. It led me to a new research field, the investigation of Molecular Biology of protein synthesis and secretion (Udaka, 1970; Yamagata et al., 1989).

3. Final remarks

For me, the beginning of research on amino acid metabolism and fermentation is very vivid and I am still excited by the "discovery of this new continent". More than 50 years after my isolation of the L-glutamate producer *C. glutamicum*, not only glutamate but also other amino acids are produced by the bacteria. Additionally, not only *C. glutamicum* but also other bacteria such as *Escherichia coli* and *B. subtilis* were found to be able to produce amino acids, vitamins and nucleotides. Within a few decades, an industry producing more than 1.5 million tons of L-glutamate and 650,000 tons of L-lysine arose from the finding of the first colony of isolate No. 534. Today, new exciting techniques as described in this book are available and are broadly applied for further strain development. However, one should keep in mind that this tremendous success story was only possible due to a new way of thinking and some luck, as all major breakthroughs in research are.

References

Kinoshita, S., Udaka, S., and Shimono, M. (1957). Studies on the amino acid fermentation part.1. Production of L-glutamic acid by various microorganisms. J. Gen. Appl. Microbiol. *3*, 193-205.

Thorne, C.B., Gomez, C.G., Noyes, H.E., and Housewright, R.D. (1954). Production of glutamylpolypeptide by *Bacillus subtilis*. J. Bacteriol. *68*, 307-315.

Udaka, S. (1960). Screening method for microorganisms accumulating metabolites and its use in isolation of *Micrococcus glutamicus*. J. Bacteriol. *79*, 754-755.

Udaka, S. and Kinoshita, S. (1958a). Studies on L-ornithine fermentation I. The biosynthetic pathway of L-ornithine in *Micrococcus glutamicus*. J. Gen. Appl. Microbiol. *4*, 272-282.

Udaka, S., and Kinoshita, S. (1958b). Studies on L-ornithine fermentation II. The change of fermentation product by a feedback type mechanism. J. Gen. Appl. Microbiol. *4*, 283-288.

Umbarger, H.E. (1961). Feedback control by end product inhibition. Cold Spring Harbor Symposia on Quantitative Biology 26, 301-312.

Udaka, S. (1970). Isolation of arginine repressor in *Escherichia coli*. Nature *228*, 336-338.

Yamagata, H., Nakahama, K., Suzuki, Y., Kakinuma, A., Tsukagoshi, N., and Udaka, S. (1989). Use of *Bacillus brevis* for efficient synthesis and secretion of human epidermal growth factor. Proc. Natl. Acad. Sci. USA *86*, 3589-3593.

From: Corynebacteria: Genomics and Molecular Biology. Edited by: Andreas Burkovski.

Genomics of Industrially and Medically Relevant Corynebacteria

2

Andreas Tauch

Abstract

The genus *Corynebacterium* is a diverse collection of Gram-positive microorganisms with a DNA of high G+C content. Many species of this genus were originally isolated from human clinical samples, whereas others were detected in animals and diverse environmental habitats. The apparent heterogeneity of the genus *Corynebacterium* poses challenges and opportunities when characterizing individual species at the genomic level. To date, the complete genome sequences of four species have been published, and sequencing of at least two more species is ongoing. These genomic data dramatically improved our understanding of the corynebacterial genome architecture, metabolic processes, species-specific traits, industrial capabilities, and potential roles in pathogenicity of humans and animals. This article describes the current status of *Corynebacterium* genome projects and summarizes the major findings of functional and comparative genomics studies.

1. Introduction

The genus *Corynebacterium* was originally defined to accommodate bacterial species that showed morphological similarities to the diphtheroid bacilli (Liebl, 2005). Thereafter, the genus *Corynebacterium* comprised, for several decades, an extremely diverse collection of morphologically similar Gram-positive microorganisms, including both pathogenic and non-pathogenic species. During the last decades, the use of chemotaxonomic markers and phylogenetic approaches based on systematic 16S rDNA sequence analysis not only helped to define the taxonomic borderline of the genus *Corynebacterium*, but also clearly demonstrated that the members of this genus form a monophyletic group (Pascual et al., 1995). However, corynebacterial species exhibit considerable heterogeneity, for instance in terms of mycolic acid content (Funke et al., 1997) and in DNA base composition that covers the wide range of 46 to 74% G+C (Funke et al., 1995), although most species have between 51 and

68% G+C (Liebl, 2005). Meanwhile, over 70 validly published species were taxonomically classified into the genus *Corynebacterium*. Many new species were described relatively recently and were isolated from clinical samples (Otsuka et al., 2005; Riegel et al., 2006; Yassin, 2007) or from wild animals (Goyache et al., 2003; Fernandez-Garayzabal et al., 2004), while others were obtained from saline soil (Chen et al., 2004) or the surface of smear-ripened cheese (Brennan et al., 2001). The large number of newly identified species apparently reflects both the growing industrial and medical interest in corynebacteria. The complete genomes of four *Corynebacterium* species have been sequenced and published so far, including two species of biotechnological relevance (*Corynebacterium glutamicum* and *Corynebacterium efficiens*) and two human pathogens (*Corynebacterium diphtheriae* and *Corynebacterium jeikeium*). The genomes online database GOLD (Bernal et al., 2001) lists two additional corynebacterial species (*Corynebacterium urealyticum* and *Corynebacterium pseudotuberculosis*) whose genomes are currently being sequenced. Below information about the respective genomes is summarized, addressing the individual sequencing projects, the deduced gene content of genomic and phage islands, prominent biological features, the architecture and stability of the genomes, and the core gene content of corynebacteria. General features of the completely sequenced corynebacterial genomes are listed in Table 2.1 and Table 2.2.

2. Genomics of *Corynebacterium glutamicum*

2.1 The genomes of *C. glutamicum* ATCC 13032 and *C. glutamicum* R

C. glutamicum is a fast-growing, saprophytic microorganism that can be isolated from soil, soil contaminated with bird feces, sewage, manure, vegetables, and fruits (Liebl, 2005). The sequenced type strain of this species, *C. glutamicum* ATCC 13032, was originally isolated from a soil sample of the Ueno Zoo in Tokyo (Japan) due to its unique ability to secrete significant amounts of glutamic acid under suitable conditions (Udaka, 1960). Nowadays, *C. glutamicum* is applied in the fermentative production of amino acids, in particular of L-glutamic acid and L-lysine, which play important roles in human and animal nutrition, respectively (Leuchtenberger et al., 2005). Both glutamic acid and lysine are the most relevant products among the amino acids in terms of market volume and are produced on a large scale by specifically developed production strains of *C. glutamicum*. Due to the importance of *C. glutamicum* for the biotechnological industry, the genome of the type strain ATCC 13032 was sequenced independently by Bielefeld University (Germany) and Kitasato University (Japan), using the ordered-clone approach and the whole-genome shotgun strategy, respectively, and revealing two slightly different variants of this strain (Tauch et al., 2002; Kalinowski et al., 2003; Ikeda and Nakagawa, 2003).

The ordered-clone approach was applied by Bielefeld University and Degussa to determine the complete nucleotide sequence of the type

Table 2.1. General features of completely sequenced corynebacterial genomes

Feature	*Corynebacterium*					
	***C. glutamicum* ATCC 13032**	***C. glutamicum* ATCC 13032**	***C. glutamicum* R**	***C. efficiens* YS-314**	***C. diphtheriae* NCTC 13129**	***C. jeikeium* K411**
Reason of interest	industrial amino acid production	industrial amino acid production	industrial amino acid production	potential amino acid producer	human pathogen	nosocomial pathogen
Genome size	3,282,708 bp	3,309,401 bp	3,314,179 bp	3,147,090 bp	2,488,635 bp	2,462,499 bp
G+C content	53.8%	53.8%	54.1%	63.4%	53.5%	61.4%
No. of CDS	3002	3099/2993[a]	2990	2950	2320	2104
Average gene length	952 bp	933 bp	957 bp	979 bp	962 bp	1030 bp
Coding density	87%	86.8%	86.3%	90.4%	87.9%	89.2%
No. of rRNA operons	6	6	6	5	5	3
No. of tRNAs	60	60	57	56	54	50
Episomes	0	0	1 (PCGR1)	2 (pEC2 and pEC3)	0	1 (pKW4)
Refseq	NC_006958	NC_003450	NC_009342	NC_004369	NC_002935	NC_007164
Reference	Kalinowski et al., 2003	Ikeda and Nakagawa, 2003	Yukawa et al., 2007	Nishio et al., 2003	Cerdeno-Tarraga et al., 2003	Tauch et al., 2005

[a] The higher number of predicted coding regions derived from the original genome sequence submission GenBank BA000036 (Ikeda and Nakagawa, 2003), the lower number from the sequence annotation deposited with Refseq NC_003450.

Table 2.2. General features of episomes sequenced in the course of corynebacterial genome projects

Corynebacterium	Episome name	Episome size	G+C content	No. of CDS	Coding density	Prominent features	Refseq	Reference
C. glutamicum R	PCgR1	49,120 bp	53.9%	28	37%	cryptic; genes with weak similarity to phages	NC_009343	Yukawa et al., 2007
C. efficiens YS-314	pEC2	23,743 bp	54.4%	15	73%	cryptic	NC_004319	Nishio et al., 2003
C. efficiens YS-314	pEC3	48,672 bp	56.4%	41	73%	genes involved in metal transport/homeostasis	NC_004320	Nishio et al., 2003
C. jeikeium K411	pKW4	14,323 bp	53.8%	16	71%	genes involved in bacteriocin synthesis	NC_003080	Tauch et al., 2005

strain *C. glutamicum* ATCC 13032 (Tauch et al., 2002). This approach is generally based on large-insert libraries to construct a map of overlapping clones covering the whole genome and on subsequent shotgun sequencing of selected clones to obtain the whole-genome sequence. The first step towards an ordered clone library of *C. glutamicum* ATCC 13032 was made by determining the genome size and establishing a combined physical and genetic map of the chromosome (Bathe et al., 1996). The genome size of *C. glutamicum* ATCC 13032 was calculated to be approximately 3.1 Mbp by separating, with pulsed-field gel electrophoresis, large DNA fragments generated with the meganucleases *Swa*I, *Pac*I and *Pme*I. The *Swa*I restriction fragments were additionally arranged to a circular chromosomal map by Southern hybridizations, using labeled probes of a *C. glutamicum* ATCC 13032 cosmid library for linking (Bathe et al., 1996). The cosmid library and an additional bacterial artificial chromosome (BAC) library were used to define and select ordered minimal sets of large-insert clones for sequencing of the *C. glutamicum* ATCC 13032 genome (Tauch et al., 2002). Ninety-five cosmid clones were arranged in 18 contiguous genomic segments and subsequently sequenced by the shotgun approach, providing a genome coverage of 86.6%. An additional set of 21 BACs represented linking clones between the sequenced cosmid contigs and was used to complete the genome sequence of *C. glutamicum* ATCC 13032. A complete coverage of the genome with large-insert clones was obtained by mapping of 44 unique BACs, which revealed an average overlap of 22 kb (Tauch et al., 2002). The nucleotide sequences of the ordered cosmid and BAC clones were finally assembled to a whole-genome sequence of 3,282,708 bp with a G+C content of 53.8%, containing 3002 predicted coding regions (Kalinowski et al., 2003).

The whole-genome shotgun strategy was applied by Kitasato University and Kyowa Hakko Kogyo to determine the genome sequence of *C. glutamicum* ATCC 13032 (Ikeda and Nakagawa, 2003). The direct shotgun strategy does not require preliminary data (such as a map or an ordered clone library) before the sequencing phase. However, a cosmid library was used as a scaffold to link the contig ends that were obtained from shotgun sequencing. Finally, a total genome sequence of 3,309,401 bp was determined, including the prediction of 3099 genes on the genome (Ikeda and Nakagawa, 2003). It is quite evident that the genome sequences of *C. glutamicum* ATCC 13032 determined by Bielefeld University and Kitasato University were not identical, when simply considering the different genome sizes. In particular, the number of insertion sequences and the presence of a putative prophage island (CGP4) in the larger genome represent the major differences between the genomic sequences (Kalinowski et al., 2003). The *C. glutamicum* ATCC 13032 genome sequence determined by Bielefeld University revealed the presence of three additional insertion sequences of the IS*Cg1* and IS*Cg2* type. The additional copy of IS*Cg1* disrupted by transpositional integration the chaperone gene *groEL1* (Barreiro et al., 2005), whereas the additional IS*Cg2* elements are located in distinct intergenic regions of the chromosome (Quast et al., 1999).

Table 2.3. Genomic islands and putative prophages localized in the *C. glutamicum* ATCC 13032 and *C. efficiens* YS-314 genome sequences

Genomic island	CDS at left end	CDS at right end	Size	No. of genes	G+C content	Prominent features	Reference
***C. glutamicum* ATCC 13032**							
LCG1	*cg0414*	*cg0443*	26.9 kb	25	45.6%	Low GC island; *murA3* genes and genes involved in surface polysaccharide biosynthesis	Kalinowski, 2005
HGC1	*cg3267*	*cg3295*	20.6 kb	29	62.6%	High GC island; genes for cation-transporting P-type ATPase, two-component system and multicopper oxidase; IS*Cg5* at 5′ junction, IS*Cg19* at 3′ junction	Kalinowski, 2005
CGP1	*cg1507*	*cg1524*	12.7 kb	20	47.2%	Putative prophage island; integrase gene at 5′ junction, $tRNA^{Leu}$ at 3′ junction	Kalinowski, 2005
CGP2	*cg1746*	*cg1752*	3.9 kb	6	47.6%	Putative prophage island; integrase gene at 3′ junction	Kalinowski, 2005
CGP3	*cg1890*	*cg2071*	185.8 kb	175	48.4%	Putative prophage island; phage-related genes, DNA restriction-modification systems; $tRNA^{Val}$ at 5′ junction, integrase gene at 3′ junction	Kalinowski, 2005; Zhang and Zhang, 2004
CGP4[a]	*ncgl1775*	*ncgl1806*	23.5 kb	32	50.8%	Putative prophage island integrated in CGP3; integrase gene at 3′ junction, 4.4-kb sequence duplication at integration site	Kalinowski, 2005; Zhang and Zhang, 2004
***C. efficiens* YS-314**							
CEGI-1	*ce0211*	*ce0269*	70.5 kb	59	59.5%	Low GC island; *ask-asd* genes of the lysine biosynthesis pathway, *leuA* gene of the leucine biosynthesis pathway; *recR* gene at 5′ junction, $tRNA^{Pro}$ at 3′ junction	Zhang and Zhang, 2005
CEGI-2	*ce0340*	*ce0390*	54.6 kb	51	55.5%	Low GC island; genes involved in surface polysaccharide biosynthesis; transposase gene at 5′ junction	Zhang and Zhang, 2005
CEGI-3	*ce0816*	*ce0865*	40.4 kb	50	59.5%	Putative prophage island; phage-related genes; integrase gene at 5′ junction, 26-bp direct repeat element at 5′ and 3′ junctions	Zhang and Zhang, 2005
CEGI-4	*ce1212*	*ce1245*	32.3 kb	34	56.6%	Low GC island; transposase and IS-related genes; transposase gene at 3′ junction	Zhang and Zhang, 2005

[a] Genomic island CGP4 is only present in the *C. glutamicum* ATCC 13032 genome sequenced by Kitasato University (Ikeda and Nakagawa, 2003) and was detected together with CGP3 in the genomic island CGGI-1 (Zhang and Zhang, 2004).

Moreover, genome sequencing of the wild-type strain *C. glutamicum* R, isolated in Japan from a meadow soil sample, has been reported very recently (Yukawa et al., 2007). The genome was sequenced by the whole-genome shotgun strategy, and gaps between the contigs were closed by using a phage library. The *C. glutamicum* R chromosome has a size of 3,314,179 bp and contains 2990 predicted coding regions. The G+C content of the chromosome is 54.1%. A detailed comparison of the gene content of *C. glutamicum* R and *C. glutamicum* ATCC 13032 revealed that only 160 and 189 genes, respectively, are strain-specific (Yukawa et al., 2007). A remarkable difference between *C. glutamicum* R and *C. glutamicum* ATCC 13032 is the presence of 11 strain-specific islands larger than 10 kb in the chromosome of strain R (Suzuki et al., 2005a). Furthermore, *C. glutamicum* R habors the extrachromosomal element PCgR1 exhibiting a G+C content of 53.9%, which is very similar to that of the chromosome of strain R. PCgR1 encodes only 28 putative proteins and therefore has a relatively low coding density (Yukawa et al., 2007).

2.2 Genomic islands of the *C. glutamicum* ATCC 13032 genome

Genomic islands of the *C. glutamicum* ATCC 13032 genome were defined by a window-based GC profile method (Kalinowski et al., 2003). The genomic island LCG1 was detected due to the relatively low G+C content of the respective DNA region (Table 2.3). The LCG1 island has a size of 26.9 kb and comprises 25 predicted coding regions with varying G+C content, ranging from 41 to 50%. Prominent genes of this genomic island are *murA* and *murB* that are both involved in murein formation, whereas other genes of the LCG1 island may contribute to the biosynthesis of surface polysaccharides. The *C. glutamicum* ATCC 13032 genome contains additional copies of the *murAB* genes (*murA2* and *murB2*), exhibiting a G+C content of the coding regions that is typical for this strain. Notably, the LCG1 island is absent from the *C. glutamicum* R genome, suggesting that differences could exist between the cell walls of the two strains (Yukawa et al., 2007).

The genomic island HGC1 has a size of 20.6 kb and deviates significantly to a high G+C content of the DNA, since the genes of this genomic region have G+C contents up to 69% (Table 2.3). The HGC1 island is flanked at the 5′ junction by a defective copy of insertion sequence IS*Cg5*, whereas the nonfunctional IS element IS*Cg19* is located at the 3′ junction. Genes encoding a putative cation-transporting P-type ATPase, a two-component system and a putative multicopper oxidase are prominent features of the central part of the HGC1 island. DNA regions with up to 99% nucleotide sequence identity were detected on strain-specific island 3 of *C. glutamicum* R (Yukawa et al., 2007) and on plasmid pLEW279b from *Corynebacterium* sp. L2-79-05 (Williams et al., 2006), suggesting the plasmid-mediated transfer of the HGC1 island to the *C. glutamicum* chromosome.

In addition to the low and high GC genomic islands, four putative prophage islands were detected in the *C. glutamicum* ATCC 13032 genome (Table 2.3), when not only considering deviations in the nucleotide composition but also the presence of phage integrase genes (Zhang and Zhang, 2004; Kalinowski,

2005). In principle, phage integrases are enzymes that mediate the unidirectional site-specific recombination between two recognition sequences, the phage attachment site and the bacterial attachment site, resulting in the site-specific integration of the phage into the bacterial chromosome (Groth and Calos, 2004). Each of the four detected prophage islands of *C. glutamicum* contains an integrase gene or a remnant thereof at the 5′ or 3′ junction (Table 2.3). The putative prophage islands found in the *C. glutamicum* genome are diverse in size. Prophage island CGP1 has a length of 12.7 kb and is flanked by a $tRNA^{Leu}$ gene at the 3′ junction, indicating that the $tRNA^{Leu}$ locus contains the integration site for this corynephage. Prophage island CGP2 codes for six genes and has a size of only 3.9 kb, suggesting that this genomic island is a highly degenerated remnant of a formerly functional prophage genome (Kalinowski, 2005). CGP3 is the largest prophage island of *C. glutamicum* with a size of 185.8 kb and located adjacent to a cluster of seven tRNA genes. The genomic island has a G+C content of 48.4% and comprises 175 predicted coding regions. The vast majority of the genes encode hypothetical proteins, with a few exceptions, including the *cglIM*, *cglIR* and *cglIIR* genes of the *C. glutamicum* DNA restriction-modification system. The *cglIR* and *cglIIR* genes encode a stress-sensitive restriction system of *C. glutamicum* ATCC 13032 that represents the major barrier for efficient gene transfer into this strain (Schäfer et al., 1994), whereas the *cglIM* gene codes for the cognate 5-cytosine methyltransferase (Schäfer et al., 1997). The *cgl* gene region is unique to *C. glutamicum* ATCC 13032 and apparently absent from other *C. glutamicum* strains, including *C. glutamicum* R (Schäfer et al., 1994; Yukawa et al., 2007). The putative prophage island CGP3 therefore contributes to the evolution of the *C. glutamicum* ATCC 13032 genome in such a way that it couples the efficiency of horizontal gene transfer to the presence of unfavorable environmental conditions. The acquired foreign genetic information might enhance the capability of the *C. glutamicum* cell to deal with the particular environmental requirement.

The putative prophage island CGP4 has a size of 23.5 kb and is flanked by a duplicated DNA region of approximately 4.4 kb in size (Kalinowski, 2005). Since the CGP4 region is located within the genomic island CGP3, the 4.4-kb sequence duplication is apparently related to the integration process of the phage genome, as is the generation of an imperfect direct repeat element of approximately 500 bp at the 5′ junction of CGP3 and the 3′ junction of CGP4 (Zhang and Zhang, 2004). The CGP4 island represents one of the major differences between the *C. glutamicum* ATCC 13032 genome sequences determined by Bielefeld University and Kitasato University. The genomic content of prophages apparently contributes to the rapid divergence of the *C. glutamicum* ATCC 13032 genomes.

2.3 Genomic islands of the *C. glutamicum* R genome

Comparison of the genomic information of the wild-type strains *C. glutamicum* ATCC 13032 and *C. glutamicum* R revealed the presence of 11 strain-specific islands (SSIs) larger than 10 kb in the genome of strain R (Suzuki et al., 2005a). These SSIs exist in the highly conserved common backbone of both genomes, whose conservation rate is approximately 97% between the strains ATCC 13032 and R. The G+C content of the SSIs ranges from 45.7 to 60.7%. Notably, none of the SSIs is flanked by obvious DNA repeats (Yukawa et al., 2007). The SSIs mainly contain mobile elements and genes of unknown function, although several proteins of a diverse functional context are also encoded on the islands (Yukawa et al., 2007). For instance, SSI 8 codes for proteins with similarity to gene products of the lytic corynephage BFK20 (Bukovska et al., 2006).

By means of the Cre/*loxP* recombination system, the 11 SSIs were individually excized from the *C. glutamicum* R chromosome, thereby deleting a total of 250 kb containing 233 predicted coding regions (Suzuki et al., 2005a; Suzuki et al., 2005b). The resulting *C. glutamicum* R mutants revealed a normal growth behavior under standard laboratory conditions, indicating that the SSIs were dispensable for cell survival. Multiple deletions of SSIs were generated with genetically modified Cre/*loxP* systems, resulting in the loss of three and eight SSIs, respectively (Suzuki et al., 2005c; Suzuki et al., 2005d). The largest genomic deletion, coding for a total number of 188 predicted proteins, comprised 190 kb of the *C. glutamicum* R chromosome. Despite the loss of numerous SSIs, the *C. glutamicum* R mutants exhibited normal growth under laboratory conditions. The apparent dispensability for growth and survival of the SSIs, either individually or in combination, supported the view that the respective genes only marginally contribute to the fitness of *C. glutamicum* R and were acquired by recent horizontal transfer.

3. Genomics of *Corynebacterium* efficiens

3.1 The genome of *C. efficiens* YS-314

C. efficiens (formerly named '*C. thermoaminogenes*') was originally isolated from onion bulbs and soil samples (Fudou et al., 2002). Three strains of this species were obtained by incubation of samples at 45°C in the course of a systematic search for new glutamic acid-producing bacteria that are more heat-tolerant than *C. glutamicum* and can grow at an elevated temperature. In principle, such strains can be applied in amino acid production processes at elevated temperature, with the potential advantage of reducing the costs needed for cooling during the fermentation. Comparative 16S rDNA sequence analysis revealed 95.3% identity of *C. efficiens* with the 16S rDNA gene of *C. glutamicum*, which is the closest taxonomic relative (Fudou et al., 2002).

The direct shotgun sequencing strategy was used to determine the genome sequence of the type strain *C. efficiens* YS-314 (Nishio et al., 2003) that was isolated from a soil sample collected at Kanagawa, Japan (Fudou et al., 2002). The DNA sequence of the *C. efficiens* YS-314 chromosome displayed a total size of 3,147,090 bp with a G+C content of 63.4%, and contains 2950 predicted coding regions. *C. efficiens* YS-314 harbors two small plasmids, named pEC2 and pEC3, with a size of 23,743 bp and 48,672 bp, respectively. Both plasmid sequences showed a relatively low G+C content (54.4% and 56.4%), indicating the recent acquisition of these plasmids by *C. efficiens* YS-314.

3.2 Genomic islands of the *C. efficiens* YS-314 genome

To detect genomic islands in the *C. efficiens* YS-314 genome sequence, a cumulative GC profile method was applied to display the G+C content distribution (Zhang and Zhang, 2005). Since the cumulative GC profile is a windowless method, it provides a high resolution to compute the distribution of the genomic G+C content. By evaluating the resulting *z'* curve of the cumulative GC profile method, four low GC genomic islands were defined in the genome sequence of *C. efficiens* (Table 2.3). The genomic islands vary in size from 32.3 kb to 70.5 kb, with CEGI-1 being the largest island. A gene coding for a recombinational DNA repair protein of the RecF pathway is located at the 5′ junction of CEGI-1, whereas a $tRNA^{Pro}$ locus is located at the 3′ junction of this element, suggesting that this tRNA locus is the integration site. The CEGI-1 island encodes various enzymes, such as aspartate kinase, aspartate-semialdehyde dehydrogenase and isopropylmalate synthase. No further genes encoding these amino acid biosynthesis enzymes are present in the *C. efficiens* YS-314 genome (Nishio et al., 2003). A remarkable feature of the aspartate kinase of *C. efficiens* is the low thermal stability of the enzyme when compared with the aspartate kinase of *C. glutamicum*, whereas other amino acid biosynthesis enzymes and enzymes of the central metabolism revealed a high thermal stability in the heat-tolerant species *C. efficiens* (Nishio et al., 2003). The presence of the *ask* (*lysC*) gene on genomic island CEGI-1 provides an explanation for the low thermal stability of the enzyme in such a way that the adaptive mutations necessary for thermostability have not occurred extensively due to the recent acquisition of CEGI-1 by horizontal gene transfer.

Genomic island CEGI-2 of *C. efficiens* YS-314 has a size of 54.6 kb and contains genes that may be involved in the biosynthesis of surface polysaccharides, whereas the CEGI-4 island harbors several transposase genes (Table 2.3). Both low GC islands of *C. efficiens* YS-314 are flanked by insertion sequences, either at the 5′ junction or the 3′ junction.

Genomic island CEGI-3 of *C. efficiens* YS-314 has a size of 40.4 kb and contains 50 genes (Table 2.3). The CEGI-3 region is characterized by a typical structure of genomic islands, since it is flanked by a 26-bp direct repeat element and encodes an integrase at the 5′ junction (Zhang and Zhang, 2005). Most of the predicted coding regions of CEGI-3 encode hypothetical proteins, with the exception of seven genes whose deduced proteins revealed similarity to gene products of bacteriophage BFK20 (Bukovska et al., 2006). The high content of genes coding for hypothetical proteins and the similarity to the phage genome of BFK20 are strong indications that the genomic island CEGI-3 has a foreign origin and represents a putative prophage island.

3.3 Thermotolerance of *C. efficiens* YS-314

C. efficiens has an average G+C content of the DNA that is 10% higher than that of *C. glutamicum*, with which it shares a conserved order of orthologous genes (Nishio et al., 2003). However, the *C. efficiens* genome lacks the mutator gene *mutT*, whereas *C. glutamicum* retains it (Nakamura et al., 2003). Defective mutation of the *mutT* gene in *Escherichia coli* apparently enhanced the frequency of unidirectional transversions from an A-T to a C-G pair, and thus increased the G+C content of the *E. coli* genome (Cox and Yanofsky, 1967). Accordingly, the higher G+C content of *C. efficiens* may be due to the absence of a *mutT* gene (Nakamura et al., 2003).

The difference in G+C content between *C. efficiens* and *C. glutamicum* is largely reflected in codon usage and nucleotide substitutions (Nishio et al., 2003). The codon usage of *C. efficiens* genes is much more biased than that of *C. glutamicum*. Among the ten most frequently used codons in *C. efficiens*, all have G or C nucleotides in the third position, whereas none of the ten rarely used codons contains G or C in the third position. A comparison between the amino acid frequencies in the predicted proteins of *C. efficiens* and *C. glutamicum* revealed a tremendous bias in amino acid substitutions in all orthologous proteins of *C. efficiens* (Nishio et al., 2003). The analysis of synonymous codon replacements indicated that among the 30 most frequent synonymous substitutions, 26 represent changes in the third letter from A/T in *C. glutamicum* to G/C in *C. efficiens*. Likewise, among the 30 most frequent nonsynonymous substitutions in *C. efficiens*, 27 increased the G+C content of the *C. efficiens* chromosome. In particular, the high frequency of arginine and glycine in *C. efficiens* proteins concurs with the view that the frequency of these amino acid residues increased parallel to the raise of the G+C content.

Moreover, the thermal stability of 13 pairs of enzymes belonging to the glutamate or lysine biosynthesis pathways in *C. efficiens* and *C. glutamicum* were compared on the basis of the enzymatic activities remaining after heat treatment of protein crude extracts (Nishio et al., 2003). This comparative approach suggested a significant correlation between specific amino acid substitutions and the thermal stability of *C. efficiens* proteins. The analysis of amino acid substitutions in the orthologous proteins revealed that three asymmetric substitutions from *C. glutamicum* to *C. efficiens* are important for thermostability in *C. efficiens*: from lysine to arginine, serine to alanine, and serine to threonine. Thus, *C. efficiens* seems to have acquired thermostability through the accumulation of specific amino acid substitutions, which in part are represented by the increase in the G+C content of the chromosome that has occurred during the course of evolution (Nishio et al., 2003).

4. Genomics of *Corynebacterium diphtheriae*

4.1 The genome of *C. diphtheriae* NCTC 13129

C. diphtheriae is the etiological agent of the acute, communicable disease diphtheria that is generally characterized by local growth of the bacterium in the pharynx along with the formation of an inflammatory pseudomembrane. (Hadfield et al., 2000). The major virulence factor of *C. diphtheriae*, diphtheria toxin, is a potent exotoxin that inhibits eucaryotic protein synthesis *via* ADP-ribosylation of elongation factor 2 and kills susceptible host cells (Holmes, 2000). The gene encoding diphtheria toxin (*tox*) is present in the genome of some closely related corynephages. Diphtheria toxin is therefore only synthesized by *C. diphtheriae* strains that habor *tox*$^+$ phages in their chromosome, and expression is greatest under iron-limiting conditions (Holmes,

Table 2.4. Genomic islands localized in the *C. diphtheriae* NCTC 13129 and *C. jeikeium* K411 chromosomes

Genomic island	CDS at left end	CDS at right end	Size	No. of genes	G+C content	Prominent features	Reference
C. diphtheriae NCTC 13129							
CDGI-1	*dip0180*	*dip0222*	36.5 kb	43	52.2%	*tox*$^+$ corynephage	Cerdeno-Tarraga et al., 2003
CDGI-2	*dip0223*	*dip0244*	17.5 kb	22	56.9%	Fimbrial associated sortase-like proteins	Cerdeno-Tarraga et al., 2003
CDGI-3	*dip0282*	*dip0287*	6.6 kb	6	55.5%	Putative iron transport system; two-component regulatory system	Cerdeno-Tarraga et al., 2003
CDGI-4	*dip0334*	*dip0357*	18.3 kb	24	51.2%	Phage-related genes; putative alkylated DNA repair protein	Cerdeno-Tarraga et al., 2003
CDGI-5	*dip0438*	*dip0445*	8.7 kb	8	52.2%	Putative surface-anchored proteins	Cerdeno-Tarraga et al., 2003
CDGI-6	*dip0752*	*dip0766*	15 kb	15	50.9%	Putative lantibiotic modifying enzyme; lantibiotic transport system	Cerdeno-Tarraga et al., 2003
CDGI-7	*dip0795*	*dip0820*	19 kb	26	59.7%	Phage-related genes	Cerdeno-Tarraga et al., 2003
CDGI-8	*dip1645*	*dip1663*	20.2 kb	19	50.4%	Putative membrane-anchored protein	Cerdeno-Tarraga et al., 2003
CDGI-9	*dip1817*	*dip1837*	14.6 kb	21	55.1%	Phage-related genes	Cerdeno-Tarraga et al., 2003
CDGI-10	*dip2010*	*dip2015*	11.1 kb	6	49.4%	Fimbrial associated sortase-like protein	Cerdeno-Tarraga et al., 2003
CDGI-11	*dip2066*	*dip2093*	30.1 kb	28	52.3%	Surface-anchored fimbrial associated protein; Sdr-family adhesin	Cerdeno-Tarraga et al., 2003
CDGI-12	*dip2148*	*dip2168*	30 kb	21	58.2%	Phage-related genes; siderophore biosynthesis and iron transport system	Cerdeno-Tarraga et al., 2003
CDGI-13	*dip2208*	*dip2234*	25.9 kb	27	50.7%	Phage-related genes; fimbrial associated sortase-like proteins	Cerdeno-Tarraga et al., 2003

C. jeikeium K411							
JKGI-1	*jk0522*	*jk0530*	15.6 kb	9	54.9%	DNA restriction-modification system	Tauch et al., 2005
JKGI-2	*jk1226*	*jk1264*	44.9 kb	39	55.2%	DNA restriction-modification systems, *mrr*-like restriction system	Tauch et al., 2005
JKGI-3	*jk1579*	*jk1602*	24.3 kb	24	55.8%	Molybdate transport system and molybdenum cofactor biosynthesis	Tauch et al., 2005
JKGI-4	*jk1767*	*jk1791*	33.6 kb	25	60.1%	Siderophore biosynthesis and iron transport system	Tauch et al., 2005
JKGI-5	*jk1803*	*jk1823*	30.7 kb	21	60.3%	Siderophore biosynthesis and iron transport systems	Tauch et al., 2005

2000). After infection, *C. diphtheriae* can colonize, for instance, the upper repiratory tract of humans, where it releases diphtheria toxin to the host cells, thereby causing the local symptoms of the disease (Hadfield et al., 2000).

The direct shotgun sequencing strategy has been applied to determine the genome sequence of *C. diphtheriae* NCTC 13129 that was isolated from a pharyngeal membrane of a patient with clinical diphtheria (Cerdeno-Tarraga et al., 2003). The genome of *C. diphtheriae* NCTC 13129 has a size of 2,488,635 bp with a G+C content of 53.5%, and contains 2320 predicted coding regions, of which 45 were annotated as pseudogenes. The genome sequence derived from two genomic shotgun libraries and terminal sequences from a large-insert BAC library that was used for generating a scaffold.

4.2 Genomic islands of *C. diphtheriae* NCTC 13129

Genomic islands in the *C. diphtheriae* NCTC 13129 genome sequence were defined by examining local anomalies in the nucleotide composition of the DNA, such as G+C content, GC skew and/or dinucleotide frequency deviations that can be indicative of the recent acquisition of DNA regions by horizontal gene transfer (Cerdeno-Tarraga et al., 2003). The most prominent genomic island comprises the complete nucleotide sequence of the *tox*$^+$ corynephage of *C. diphtheriae* NCTC 13129. The prophage genome has a size of 36,566 bp with a G+C content of 52.2% and encodes 43 predicted proteins. Sequence similarities on the amino acid level were detected to structural proteins of bacteriophage BFK20 from '*Brevibacterium flavum*' (Bukovska et al., 2006). Both ends of the prophage genome showed a significantly decreased G+C content and are flanked by tRNA^{Arg} sequences. The tRNA^{Arg} locus of *C. diphtheriae* provides the conserved box 1 sequence for the recombination process at the *attB-attP* (*attL*) site (Ratti et al., 1997). The *tox* gene, encoding the major virulence factor diphtheria toxin, is located at the right end of the prophage genome, adjacent to the attachment site and within the region of low G+C content. This specific location of the *tox* gene is indicative of a bacterial gene that was acquired from a previous host and is dispensible for the life cycle of the phage, but may affect the phenotype or fitness of the lysogenic bacterium (Brüssow et al., 2004).

In addition to the *tox*$^+$ corynephage, twelve genomic regions with local anomalies in the nucleotide composition were detected in *C. diphtheriae* NCTC 13129, of which five islands contain phage-related genes (Table 2.4), indicating that additional remnants of prophages are present in the genome (Cerdeno-Tarraga et al., 2003). It is noteworthy that several genes potentially involved in pathogenicity of *C. diphtheriae* NCTC 13129 are located on the genomic islands. These putative pathogenicity islands encode, for instance, a siderophore biosynthesis and export system, a putative lantibiotic biosynthesis system, and three sortase-related fimbrial systems. It is therefore likely that *C. diphtheriae* has recently acquired by horizontal transfer specialized genes that may be involved in the pathogenic lifestyle by encoding fimbrial structures for the adherence of the bacterium to the host cell surfaces.

4.3 The fimbrial structures of *C. diphtheriae* NCTC 13129

Analysis of the gene clusters encoding fimbrial proteins of *C. diphtheriae* revealed that the respective genes are involved in the synthesis of three morphologically distinct surface organelles, contributing largely to the structural design of the cell surface (Ton-That and Schneewind, 2003; Gaspar and Ton-That, 2006; Swierczynski and Ton-That, 2006). Each gene cluster is composed of at least one pilin-specific sortase gene (*srtA-E*) and three sortase-mediated pilus assembly genes (*spaA-I*). Pilin-specific sortases are involved in pilus assembly by catalyzing ordered cross-linking reactions, whereby the sortase enzymes cleave pilin precursor proteins at the carboxy-terminal sorting signals (LPXTG) and involve the side-chain amino groups of pilin motif sequences to generate covalent linkages between pilin subunits (Ton-That et al., 2004). Morphological and genetic studies clearly demonstrated that each pilus of *C. diphtheriae* is composed of three pilin subunits and that the pilus assembly process requires pilin-specific sortases. In the case of the SpaA-type pilus, SpaA represents the major pilin that is distributed uniformly along the pilus shaft, whereas SpaB is observed only at regular intervals. The SpaC protein seems to be located at the tip of the pilus structure (Ton-That and Schneewind, 2003). Both minor pilins, SpaB and SpaC, but not the major pilin SpaA, act as adhesin to pharyngeal epithelial cells (Mandlik et al., 2007). The minor pilins SpaB and SpaC are also components of the cell wall of *C. diphtheriae*, probably enabling an intimate attachment of the bacterium to pharyngeal epithelial cells. This tight contact might facilitate the delivery of diphtheria toxin to the host cell, thereby creating a safe niche for the pathogen after damaging the host cell through the inhibition of protein synthesis (Mandlik et al., 2007).

5. Genomics of *Corynebacterium jeikeium*

5.1 The genome of *C. jeikeium* K411

C. jeikeium was initially isolated from human blood cultures and was associated with bacterial endocarditis following cardiac surgery (Jackman et al., 1987). Subsequent medical reports established that *C. jeikeium* is part of the normal human skin flora and is implicated in a variety of nosocomial infections (Funke et al., 1997). *C. jeikeium* is referred to as lipophilic or lipid-requiring species, since visible growth depends on the addition of lipids to culture media (Funke et al., 1997). The sequenced clinical isolate *C. jeikeium* K411 was originally recovered from the axilla of a bone marrow transplant patient who received immunosuppressive therapy and broad-spectrum antibiotics (Kerry-Williams and Noble, 1986).

The direct shotgun sequencing strategy was used to determine the genome sequence of *C. jeikeium* K411 (Tauch et al., 2005). The chromosome of *C. jeikeium* K411 consists of 2,462,499 bp with a G+C content of 61.4%, and comprises 2104 predicted coding regions, of which 68 apparently represent pseudogenes. Assembly of the shotgun sequences was facilitated by using large-insert clones from fosmid and BAC libraries as a scaffold. *C. jeikeium* K411 harbors the bacteriocin-producing plasmid pKW4 (Kerry-

Williams and Noble, 1986). The plasmid genome displayed a size of 14,323 bp and a G+C content of 53.8%.

5.2 Genomic islands of the *C. jeikeium* K411 genome

Evaluation of the cumulative GC profile for the *C. jeikeium* K411 chromosome indicated the presence of five genomic islands, ranging in size from 15.6 to 44.9 kb (Table 2.4). The genomic islands JKGI-1 and JKGI-2 most likely contribute to the high restriction barrier of *C. jeikeium* K411 for incoming foreign DNA, since they encode an *mrr*-like DNA restriction system and different types of restriction-modification systems, belonging to classes I and II (Bickle and Krüger, 1993). Genomic island JKGI-3 encodes several proteins that may be involved in molybdate uptake and molybdenum cofactor biosynthesis (Table 2.4). The genomic islands JKGI-4 and JKGI-5 are located in immediate vicinity and are only separated by a small DNA region encoding ribosomal proteins. Both genomic islands are apparently involved in iron uptake of *C. jeikeium* K411, since they contain genes for siderophore biosynthesis systems and iron ABC transporters (Table 2.4). The genomic islands of *C. jeikeium* K411 are flanked by insertion sequences at the 5′ and 3′ junctions, indicating that IS-mediated transpositional recombination played a major role in integrating the islands into the chromosomal backbone. Genomic island JKGI-4 is flanked by defective copies of IS*3503* at both junctions. The presence of target site duplications at the integration site of the genomic island suggests that JKGI-4 derived from a formerly active composite transposon (Tauch et al., 2005).

5.3 Lipophilism and pathogenicity of *C. jeikeium* K411

C. jeikeium belongs to the group of lipophilic corynebacteria whose growth is generally enhanced by the addition of lipids to the culture medium (Funke et al., 1997). Functional annotation and deduced metabolic analysis of the *C. jeikeium* K411 genome sequence indicated that the lipophilic phenotype originates from the absence of a fatty acid synthase gene and represents a lipid auxotrophy (Tauch et al., 2005). A prominent feature of the *C. jeikeium* K411 genome is the presence of several genes encoding a complete β-oxidation pathway. The respective genes are involved in the activation of exogenous fatty acids of various chain lengths concomitant with transport into the cell and subsequent fatty acid degradation. Degradation of fatty acids results in their gradual conversion to acetyl-CoA and in the production of one equivalent each of $FADH_2$ and NADH, which may be used in ATP generation (Hiltunen and Qin, 2000). The predominant recovery of *C. jeikeium* isolates from the axillary, inguinal and perineal areas of the human skin reflects a relationship between the natural habitats and the lipophilic phenotype of the bacterium (Tauch et al., 2005). The respective areas are intertriginous sites of the human body that are characterized by an elevated moisture of the skin in conjunction with a substantial formation of hydrolipid films composed of triglycerides, fatty acids, ceramides, cholesterol, and cholesterol esters (Stewart, 1992). Thus, *C. jeikeium* seems to colonize preferably those sites of the human body that provide an appropriate amount of exogenous fatty acids for growth.

It is furthermore likely that some of the predicted virulence factors of *C. jeikeium* are involved in generating free fatty acids by damaging the plasma membrane of host cells. The *C. jeikeium* K411 genome codes for a putative alkaline ceramidase with homology to the respective enzyme from *Pseudomonas aeruginosa* (Tauch et al., 2005). Ceramidases are lipolytic enzymes that hydrolyze the amide bond in ceramides to yield sphingosine and free fatty acids. *C. jeikeium* also encodes a cholesterol esterase that apparently catalyzes the hydrolysis of long-chain fatty acid esters from cholesterol and may thus be involved in fatty acid metabolism (Tauch et al., 2005). Furthermore, an acid phosphatase with similarity to the AcpA protein of *Francisella tularensis* is encoded by the *C. jeikeium* K411 genome. Purified acid phosphatase from *F. tularensis* revealed a broad substrate specificity, hydrolyzed phosphatidylcholine and thereby released diglycerides (Reilly et al., 1996). Consequently, the analysis of the genome sequence of *C. jeikeium* K411 sheds light on its adaptation to the lipophilic lifestyle in such a way that a complete β-oxidation pathway enables the bacterium to utilize exogenous fatty acids present on the human skin and that enzymatic functions of the predicted virulence factors are closely linked to fatty acid metabolism by releasing additional fatty acid substrates for growth (Tauch et al., 2005).

6. Genomics of *Corynebacterium urealyticum*

C. urealyticum (formerly named *Corynebacterium* group D2) is frequently isolated from human urine samples and involved in the etiology of urinary tract infections in patients with underlying predisposing factors, such as urinary tract instrumentation (Soriano et al., 1990). Urinary tract infections caused by *C. urealyticum* are significantly associated with struvite (ammonium magnesium phosphate) stone formation (Soriano et al., 1986). The type strain of the species, *C. urealyticum* DSM7109, was originally isolated from an inpatient diagnosed with alkaline-encrusted cystitis (Soriano et al., 1985). A recently performed pyrosequencing approach provided first information on the genome of *C. urealyticum* DSM7109 (Tauch et al., 2006).

Ultrafast sequencing of *C. urealyticum* DSM7109 was performed with the Genome Sequencer 20 System (Margulies et al., 2005), and yielded 657,410 sequence reads that were finally used for *de novo* genome assembly (Tauch et al., 2006). A total number of 2,294,755 bases were assembled into 69 contigs with an average contig size of 33,257 bp. The largest assembled contig had a size of 175,964 bp. In addition, 154 small contigs (≤ 500 bp) with an average length of 144 bp were assembled, producing a total of 22,211 bp of genetic information. The contiguous sequences of the *C. urealyticum* DSM7109 genome were automatically annotated, resulting in the prediction of 2027 protein-coding sequences that were functionally classified into COGs (Tauch et al., 2006). Approximately three-quarters of the deduced proteins (1589) revealed significant similarity to proteins encoded by *C. jeikeium* K411 and are apparently shared between both species, whereas 438 proteins were non-homologous between both corynebacteria. This result is consistent with the close taxonomic relationship of *C. urealyticum* and *C. jeikeium* (Pascual et al., 1995). It may soon become possible to gain deeper insights into the lifestyle and pathophysiology of *C. urealyticum* from in-depth analysis of the assembled sequences.

7. Genomics of *Corynebacterium pseudotuberculosis*

C. pseudotuberculosis is an important animal pathogen and is the causative agent of a disease that is commonly called caseous lymphadenitis (Dorella et al., 2006a). This disease is found in the major sheep and goat production areas worldwide and causes significant economic losses. Despite the importance of *C. pseudotuberculosis* for animal health, there is little information about the pathogenesis and the facultative intracellular lifestyle of this bacterium. Only three virulence factors have been identified in *C. pseudotuberculosis* so far (Dorella et al., 2006a). The most prominent one is phospholipase D, a sphingomyelin-degrading exotoxin (McKean et al., 2007).

The strain selected for genome sequencing, *C. pseudotuberculosis* 1002, was originally isolated from goat caseous granulomas in Bahia state (Brazil), and has been licensed as a live attenuated vaccine strain (Dorella et al., 2006a). Currently, only 215 genomic servey sequences (GSSs) obtained from random samples of a bacterial artificial chromosome library of *C. pseudotuberculosis* 1002 are publicly available (Dorella et al., 2006b). This BAC library contains about 18,000 clones with inserts ranging in size from 25 to 120 kb and representing a 390-fold coverage of the *C. pseudotuberculosis* genome, when considering a genome size between 2.5 and 3.1 Mbp. Many GSSs (80.4%) revealed high similarity to sequences from *C. diphtheriae*, confirming the close taxonomic releationship of both species (Dorella et al., 2006b). Once the complete genome sequence of *C. pseudotuberculosis* 1002 is established, it will be possible to better understand the molecular and genetic basis of virulence of this bacterium.

8. Structural features of corynebacterial genomes

Assembly and scaffolding of the corynebacterial genome sequences demonstrated that the genetic information is stored in the form of circular chromosomes. The genome sizes of the soil bacteria *C. glutamicum* and *C. efficiens* are significantly larger than those of the genomes of the pathogenic species *C. diphtheriae* and *C. jeikeium*. Since the coding density of bacterial chromosomes does not vary much (approximately one gene per kb of DNA), the genome size is directly proportional to the number of genes (Bentley and Parkhill, 2004). The higher number of predicted coding regions in the genomes of the saprophytic corynebacteria might reflect the greater metabolic diversity that is necessary to survive and adapt to the changing environmental conditions of a soil-based habitat, whereas fewer metabolic pathways and functions are apparently required in the specialized ecologial niche of the human-adapted pathogens. A comparison of the genetic content of pathogenic and non-pathogenic corynebacteria by the reciprocal best-hit method with the BLAST algorithm revealed that gene loss has played a major role in the evolution of the *C. diphtheriae* genome (Nishio et al., 2004).

The analysis of the GC skew of microbial chromosomes is generally applicable to define the leading and lagging strands, since the nucleotide composition of the chromosome changes at two points that coincide with the origin of replication and the replication terminus (Lobry and Louarn, 2003). GC skew profiles of the corynebacterial chromosomes indicated a bidirectional replication mechanism (Kalinowski et al., 2003; Cerdeno-Tarraga et al., 2003; Tauch et al., 2005), though the GC skew plot of *C. efficiens* YS-314 is a bit ambiguous (Nishio et al., 2003). The putative origin of replication (*oriC*) is located close to the *dnaA* gene, and a potential *dif* locus involved in replication termination is present at a position of 180° from *oriC*. The location of the cognate *dif* locus of *C. jeikeium* K411 appears to be skewed from the expected position of 180° from *oriC* (Tauch et al., 2005). Whether this predicted structural asymmetry indeed results in a 244-kb size difference between the replichores of the *C. jeikeium* K411 chromosome remains to be elucidated.

The G+C content of the sequenced corynebacterial chromosomes varied between 53.5 and 63.4% (Table 2.1). This broad variation of the G+C content is a typical feature of the genus *Corynebacterium* and reflects its genetic diversity (Liebl, 2005). The DNA base composition of corynebacteria covers the wide range of 46 to 74% G+C, but most species have a G+C content between 51 and 68% (Funke et al., 1995; Liebl, 2005). The DNA base composition of the sequenced corynebacterial species is thus in the prevalent range. A remarkable structural feature of the *C. diphtheriae* NCTC 13129 genome sequence is that the G+C content is not constant across the chromosome (Cerdeno-Tarraga et al., 2003). In particular, a genomic region of about 740 kb in size has a significantly lower G+C content (49.99%) than the remainder of the genome (54.96%). This genomic region encompasses the replication terminus, but has no clearly defined boundaries, suggesting a gradual transition from a higher G+C content in the region near the origin of replication to a lower content in the region around the replication terminus. Many bacterial chromosomes are structured in this way, with a tendency toward an A+T-enrichment near the replication terminus (Daubin and Perriere, 2003). Several hypotheses may explain the pattern of decreasing G+C content near the replication terminus, considering for instance a higher frequency of horizontal gene transfer in this region or structural constraints due to the resolution of chromosome dimers at the end of the replication process (Daubin and Perriere, 2003). However, the skew of Rag (RGNAGGGS) motifs, a family of polarized, G-rich octamers (Lobry and Louarn, 2003), is maintained in the *C. diphtheriae* NCTC 13129 chromosome, even in the genomic region with low G+C content surrounding the replication terminus (Cerdeno-Tarraga et al., 2003).

Another structural feature of the *C. efficiens*, *C. diphtheriae* and *C. jeikeium* genome sequences is the presence of Clustered Regularly Interspaced Short Palindromic Repeats (CRISPRs) (Jansen et al., 2002; Edgar, 2007). CRISPRs are a family of direct repeats of 21 to 37 bp that are separated by regularly sized, non-repetitive spacer sequences. Members of the CRISPR family were detected in many prokaryotic genomes and are commonly associated with *cas* genes (Haft et al., 2005). CRISPRs have been implicated, for instance, in large-scale chromosome rearrangements and DNA condensation, but a role in acquired resistance against bacteriophages has been demonstrated very recently (Barrangou et al., 2007). The *C. jeikeium* K411 genome contains a CRISPR family sequence with 61 repeats (Tauch et al., 2005). These highly conserved DNA sequences have a size of 29 bp, contain an imperfect inverted repeat of 7 bp and are separated by a variable nucleotide sequence of 32 bp. The CRISPRs are part of a small genomic region with low G+C content that is flanked by a $tRNA^{Met}$ gene at the 5′ junction. Highly similar repetitive sequences were detected in the genome of *Nocardia farcinica* IFM 10152 (Ishikawa et al., 2004) and in the kanamycin biosynthesis gene cluster of *Streptomyces kanamyceticus* 12-6 (Yanai et al., 2006). The CRISPR signature of *C. jeikeium* K411 is furthermore similar to that of *E. coli* K-12 located downstream of the *iap* coding region (Nakata et al., 1989). The CRISPR of *C. efficiens* YS-314 has a length of 36 bp and is present in 22 copies, whereas the chromosome of *C. diphtheriae* NCTC 13129 contains two types of CRISPRs with length 28 and 36, respectively (Edgar, 2007).

9. Core genes and essential genes in corynebacteria

Although the corynebacterial genomes differ in size and gene content, they must contain all the information to allow the cell to perform the essential functions that are necessary to maintain metabolic homeostasis and to reproduce and evolve. Bacterial cells usually can import metabolites, but not functional proteins. Accordingly, they have to rely on their own gene products to perform such important cellular functions. To determine the core set of corynebacterial genes, the deduced proteins of the established genome sequences were compared among each other by reciprocal best BLASTP matches against the COG database (Yukawa et al., 2007). This computational method detected 835 genes that were considered orthologous in *C. glutamicum*, *C. efficiens*, *C. diphtheriae*, and *C. jeikeium* and apparently represent the conserved chromosomal backbone of the four species. Analysis of differences between the genomes of saprophytic and pathogenic species by the reciprocal best BLASTP method revealed that up to 762 genes are limited to the saprophytes, whereas only 100 non-core genes are present in both of the pathogenic species (Yukawa et al., 2007). On the other hand, *C. diphtheriae* and *C. jeikeium* have 502 and 378 specific genes, respectively, reflecting their adaptation to specific ecological niches on the human body.In another computational approach, the common orthologous genes in *C. glutamicum*, *C. efficiens*, *C. diphtheriae*, *Mycobacterium tuberculosis*, and *Streptomyces coelicolor* were defined by the reciprocal best-hit method with BLAST, using four species and excluding one corynebacterial genome (Nishio et al., 2004). This calculation revealed 748 orthologous genes in the five actinobacterial genomes, 768 when excluding *C. glutamicum*, and 773 when excluding *C. efficiens*, but 831 when excluding *C. diphtheriae*. This results implies that *C. diphtheriae* has lost many orthologous genes that were found in the four other bacteria after the divergence from the common ancestor of the corynebacteria (Nishio et al., 2004). Moreover, a high-throughput transposon mutagenesis strategy has been applied to reveal *C. glutamicum* R genes that are essential under normal laboratory conditions (Suzuki et al., 2006). Two mini-transposons based on IS*31831* or Tn*5* were used for systematic mutagenesis of the *C. glutamicum* R chromosome. This combination of two transposable elements resulted in the disruption of 2332 genes, representing 78.1% of the predicted coding regions of *C. glutamicum* R. Transposon insertions were not observed in 658 coding regions that may have been missed by chance or because the gene disruption was lethal. The respective genes therefore represent candidate essential genes. The number of essential genes might be significantly lower, since experimental strategies based on transposon mutagenesis systems tend to overestimate this set of genes (Gil et al., 2004). A comparison between *C. glutamicum* R genes and essential genes of *E. coli* and *Bacillus subtilis* indicated that 251 genes of strain R have strict orthologs among *E. coli* or *B. subtilis* essential genes. Interestingly, 221 of these are included in the list of candidate essential genes of *C. glutamicum* R (Suzuki et al., 2006).

10. Genome stability and genome rearrangements of corynebacteria

Bioinformatics comparisons of the order of orthologous genes between the sequenced corynebacteria revealed a high degree of synteny over the entire length of the *C. glutamicum*, *C. efficiens* and *C. diphtheriae* chromosomes. A remarkable difference in the overall synteny is only caused by the presence of genomic islands CGP3 and CGP4 in the *C. glutamicum* ATCC 13032 chromosome (Kalinowski, 2005; Ikeda and Nakagawa, 2003). A further exception is the *C. jeikeium* K411 chromosome that is characterized by the presence of ten apparent breakpoints of synteny, indicating the presence of few genome rearrangments in this species and including the insertion, translocation and inversion of distinct DNA segments (Tauch et al., 2005). A comparison of the order of orthologous genes in closely related *Mycobacterium* species showed, on the other hand, that extensive rearrangements have occurred in these species throughout the course of evolution, whereas the order of orthologous genes between mycobacteria and corynebacteria is apparently disrupted (Nakamura et al., 2003). These differences in synteny between the taxonomically related genera indicated that corynebacteria have exceptionally stable genome structures with respect to the order of orthologous genes and have maintained an ancestral genome organization even after the divergence of corynebacteria and mycobacteria. A comparison of the content of recombinational repair systems among corynebacterial and mycobacterial species suggested that the absence of the *recBCD* genes in corynebacteria is responsible for the suppression of genome shuffling in these species (Nakamura et al., 2003). Likewise, mutations of the *recBCD* pathway of *E. coli* suppressed the occurrence of chromosomal inversions in this organism (Kowalczykowski et al., 1994). The apparent genome stability of *Corynebacterium* species indicated that speciation has not taken place by genome shuffling, but particularly occurred by gene gain and gene loss, horizontal gene transfer, and nucleotide substitutions in the genome sequence.

References

Barrangou, R., Fremaux, C., Deveau, H., Richards, M., Boyaval, P., Moineau, S., Romero, D.A., and Horvath, P. (2007). CRISPR provides acquired resistance against viruses in prokaryotes. Science *315*, 1709-1712.

Barreiro, C., Gonzalez-Lavado, E., Brand, S., Tauch, A., and Martin, J.F. (2005). Heat shock proteome analysis of wild-type *Corynebacterium glutamicum* ATCC 13032 and a spontaneous mutant lacking GroEL1, a dispensable chaperone. J. Bacteriol. *187*, 884-889.

Bathe, B., Kalinowski, J., and Pühler. A. (1996). A physical and genetic map of the *Corynebacterium glutamicum* ATCC 13032 chromosome. Mol. Gen. Genet. *252*, 255-265.

Bentley, S.D., and Parkhill, J. (2004). Comparative genomic structure of prokaryotes. Annu. Rev. Genet. *38*, 771-792.

Bernal, A., Ear, U., and Kyrpides, N. (2001). Genomes OnLine Database (GOLD): a monitor of genome projects world-wide. Nucleic Acids Res. *29*, 126-127.

Bickle, T.A., and Krüger, D.H. (1993). Biology of DNA restriction. Microbiol. Rev. *57*, 434-450.

Brennan, N.M., Brown, R., Goodfellow, M., Ward, A.C., Beresford, T.P., Simpson, P.J., Fox, P.F., and Cogan, T.M. (2001). *Corynebacterium mooreparkense* sp. nov. and *Corynebacterium casei* sp. nov., isolated from the surface of a smear-ripened cheese. Int. J. Syst. Evol. Microbiol. *51*, 843-852.

Brüssow, H., Canchaya, C., and Hardt, W.D. (2004). Phages and the evolution of bacterial pathogens: from genomic rearrangements to lysogenic conversion. Microbiol. Mol. Biol. Rev. *68*, 560-602.

Bukovska, G., Klucar, L., Vlcek, C., Adamovic, J., Turna, J., and Timko, J. (2006). Complete nucleotide sequence and genome analysis of bacteriophage BFK20 - a lytic phage of the industrial producer *Brevibacterium flavum*. Virology *348*, 57-71.

Cerdeno-Tarraga, A.M., Efstratiou, A., Dover, L.G., Holden, M.T., Pallen, M., Bentley, S.D., Besra, G.S., Churcher, C., James, K.D., De Zoysa, A., Chillingworth, T., Cronin, A., Dowd, L., Feltwell, T., Hamlin, N., Holroyd, S., Jagels, K., Moule, S., Quail, M.A., Rabbinowitsch, E., Rutherford, K.M., Thomson, N.R., Unwin, L., Whitehead, S., Barrell, B.G., and Parkhill, J. (2003). The complete genome sequence and analysis of *Corynebacterium diphtheriae* NCTC13129. Nucleic Acids Res. *31*, 6516-6523.

Chen, H.H., Li, W.J., Tang, S.K., Kroppenstedt, R.M., Stackebrandt, E., Xu, L.H., and Jiang, C.L. (2004). *Corynebacterium halotolerans* sp. nov., isolated from saline soil in the west of China. Int. J. Syst. Evol. Microbiol. *54*, 779-782.

Cox, E.C., and Yanofsky, C. (1967). Altered base ratios in the DNA of an *Escherichia coli* mutator strain. Proc. Natl. Acad. Sci. USA *58*, 1895-1902.

Daubin, V., and Perriere, G. (2003). G+C3 structuring along the genome: a common feature in prokaryotes. Mol. Biol. Evol. *20*, 471-483.

Dorella, F.A., Pacheco, L.G., Oliveira, S.C., Miyoshi, A., and Azevedo, V. (2006a). *Corynebacterium pseudotuberculosis*: microbiology, biochemical properties, pathogenesis and molecular studies of virulence. Vet. Res. *37*, 201-218.

Dorella, F.A., Fachin, M.S., Billault, A., Dias Neto, E., Soravito, C., Oliveira, S.C., Meyer, R., Miyoshi, A., and Azevedo, V. (2006b). Construction and partial characterization of a *Corynebacterium pseudotuberculosis* bacterial artificial chromosome library through genomic survey sequencing. Genet. Mol. Res. *5*, 653-663.

Edgar, R.C. (2007). PILER-CR: fast and accurate identification of CRISPR repeats. BMC Bioinformatics *8*, 18.

Fernandez-Garayzabal, J.F., Vela, A.I., Egido, R., Hutson, R.A., Lanzarot, M.P., Fernandez-Garcia, M., and Collins, M.D. (2004). *Corynebacterium ciconiae* sp. nov., isolated from the trachea of black storks (*Ciconia nigra*). Int. J. Syst. Evol. Microbiol. *54*, 2191-2195.

Fudou, R., Jojima, Y., Seto, A., Yamada, K., Kimura, E., Nakamatsu, T., Hiraishi, A., and Yamanaka, S. (2002). *Corynebacterium efficiens* sp. nov., a glutamic-acid-producing species from soil and vegetables. Int. J. Syst. Evol. Microbiol. *52*, 1127-1131.

Funke, G., Lawson, P.A., and Collins, M.D. (1995). Heterogeneity within human-derived centers for disease control and prevention (CDC) coryneform group ANF-1-like bacteria and description of *Corynebacterium auris* sp. nov. Int. J. Syst. Bacteriol. *45*, 735-739.

Funke, G., von Graevenitz, A., Clarridge, J.E. 3rd, and Bernard, K.A. (1997). Microbiology of coryneform bacteria. Clin. Microbiol. Rev. *10*, 125-159.

Gaspar, A.H., and Ton-That, H. (2006). Assembly of distinct pilus structures on the surface of *Corynebacterium diphtheriae*. J. Bacteriol. *188*, 1526-1533.

Gil, R., Silva, F.J., Pereto, J., and Moya, A. (2004). Determination of the core of a minimal bacterial gene set. Microbiol. Mol. Biol. Rev. *68*, 518-537.

Goyache, J., Ballesteros, C., Vela, A.I., Collins, M.D., Briones, V., Hutson, R.A., Potti, J., Garcia-Borboroglu, P., Dominguez, L., and Fernandez-Garayzabal, J.F. (2003). *Corynebacterium sphenisci* sp. nov., isolated from wild penguins. Int. J. Syst. Evol. Microbiol. *53*, 1009-1012.

Groth, A.C., and Calos, M.P. (2004). Phage integrases: biology and applications. J. Mol. Biol. *335*, 667-678.

Hadfield, T.L., McEvoy, P., Polotsky, Y., Tzinserling, V.A., and Yakovlev, A.A. (2000). The pathology of diphtheria. J. Infect. Dis. *181*, S116-S120.

Haft, D.H., Selengut, J., Mongodin, E.F., and Nelson, K.E. (2005). A guild of 45 CRISPR-associated (Cas) protein families and multiple CRISPR/Cas subtypes exist in prokaryotic genomes. PLoS Comput. Biol. *1*, e60.

Hiltunen, J.K., and Qin, Y. (2000). β-oxidation - strategies for the metabolism of a wide variety of acyl-CoA esters. Biochim. Biophys. Acta *1484*, 117-128.

Holmes, R.K. (2000). Biology and molecular epidemiology of diphtheria toxin and the *tox* gene. J. Infect. Dis. *181*, S156-S167.

Ikeda, M., and Nakagawa, S. (2003). The *Corynebacterium glutamicum* genome: features and impacts on biotechnological processes. Appl. Microbiol. Biotechnol. *62*, 99-109.

Ishikawa, J., Yamashita, A., Mikami, Y., Hoshino, Y., Kurita, H., Hotta, K., Shiba, T., and Hattori, M. (2004). The complete genomic sequence of *Nocardia farcinica* IFM 10152. Proc. Natl. Acad. Sci. USA *101*, 14925-14930.

Jackman, P.J.H., Pitcher, D.G., Pelczynska, S., and Borman, P. (1987). Classification of corynebacteria associated with endocarditis (group JK) as *Corynebacterium jeikeium* sp. nov. System. Appl. Microbiol. *9*, 83-90.

Jansen, R., Embden, J.D., Gaastra, W., and Schouls, L.M. (2002). Identification of genes that are associated with DNA repeats in prokaryotes. Mol. Microbiol. *43*, 1565-1575.

Kalinowski, J. (2005). The genomes of amino acid-producing Corynebacteria. In Handbook of *Corynebacterium glutamicum*., L. Eggeling, and M. Bott, eds. (Boca Raton, FL: CRC Press), pp. 37-56.

Kalinowski, J., Bathe, B., Bartels, D., Bischoff, N., Bott, M., Burkovski, A., Dusch, N., Eggeling, L., Eikmanns, B.J., Gaigalat, L., Goesmann, A., Hartmann, M., Huthmacher, K., Krämer, R., Linke, B., McHardy, A.C., Meyer, F., Möckel, B., Pfefferle, W., Pühler, A., Rey, D.A., Rückert, C., Rupp, O., Sahm, H., Wendisch, V.F., Wiegräbe, I., and Tauch, A. (2003). The complete *Corynebacterium glutamicum* ATCC 13032 genome sequence and its impact on the production of L-aspartate-derived amino acids and vitamins. J. Biotechnol. *104*, 5-25.

Kerry-Williams, S.M., and Noble, W.C. (1986). Plasmids in group JK coryneform bacteria isolated in a single hospital. J. Hyg. *97*, 255-263.

Kowalczykowski, S.C., Dixon, D.A., Eggleston, A.K., Lauder, S.D., and Rehrauer, W.M. (1994). Biochemistry of homologous recombination in *Escherichia coli*. Microbiol. Rev. *58*, 401-465.

Leuchtenberger, W., Huthmacher, K., and Drauz, K. (2005). Biotechnological production of amino acids and derivatives: current status and prospects. Appl. Microbiol. Biotechnol. *69*, 1-8.

Liebl, W. (2005). *Corynebacterium* taxonomy. In Handbook of *Corynebacterium glutamicum*., L. Eggeling, and M. Bott, eds. (Boca Raton, FL: CRC Press), pp. 9-34.

Lobry, J.R., and Louarn, J.M. (2003). Polarisation of prokaryotic chromosomes. Curr. Opin. Microbiol. *6*, 101-108.

Mandlik, A., Swierczynski, A., Das, A., and Ton-That, H. (2007). *Corynebacterium diphtheriae* employs specific minor pilins to target human pharyngeal epithelial cells. Mol. Microbiol. *64*, 111-124.

Margulies, M., Egholm, M., Altman, W.E., Attiya, S., Bader, J.S., Bemben, L.A., Berka, J., Braverman, M.S., Chen, Y.J., Chen, Z., Dewell, S.B., Du, L., Fierro, J.M., Gomes, X.V., Godwin, B.C., He, W., Helgesen, S., Ho, C.H., Irzyk, G.P., Jando, S.C., Alenquer, M.L., Jarvie, T.P., Jirage, K.B., Kim, J.B., Knight, J.R., Lanza, J.R., Leamon, J.H., Lefkowitz, S.M., Lei, M., Li, J., Lohman, K.L., Lu, H., Makhijani, V.B., McDade, K.E., McKenna, M.P., Myers, E.W., Nickerson, E., Nobile, J.R., Plant, R., Puc, B.P., Ronan, M.T., Roth, G.T., Sarkis, G.J., Simons, J.F., Simpson, J.W., Srinivasan, M., Tartaro, K.R., Tomasz, A., Vogt, K.A., Volkmer, G.A., Wang, S.H., Wang, Y., Weiner, M.P., Yu, P., Begley, R.F., and Rothberg, J.M. (2005). Genome sequencing in microfabricated high-density picolitre reactors. Nature *437*, 376-380.

McKean, S.C., Davies, J.K., and Moore, R.J. (2007). Expression of phospholipase D, the major virulence factor of *Corynebacterium pseudotuberculosis*, is regulated by multiple environmental factors and plays a role in macrophage death. Microbiology *153*, 2203-2211.

Nakamura, Y., Nishio, Y., Ikeo, K., and Gojobori, T. (2003). The genome stability in *Corynebacterium* species due to lack of the recombinational repair system. Gene *317*, 149-155.

Nakata, A., Amemura, M., and Makino, K. (1989). Unusual nucleotide arrangement with repeated sequences in the *Escherichia coli* K-12 chromosome. J. Bacteriol. *171*, 3553-3556.

Nishio, Y., Nakamura, Y., Kawarabayasi, Y., Usuda, Y., Kimura, E., Sugimoto, S., Matsui, K., Yamagishi, A., Kikuchi, H., Ikeo, K., and Gojobori, T. (2003). Comparative complete genome sequence analysis of the amino acid replacements responsible for the thermostability of *Corynebacterium efficiens*. Genome Res. *13*, 1572-1579.

Nishio, Y., Nakamura, Y., Usuda, Y., Sugimoto, S., Matsui, K., Kawarabayasi, Y., Kikuchi, H., Gojobori, T., and Ikeo, K. (2004). Evolutionary process of amino acid biosynthesis in *Corynebacterium* at the whole genome level. Mol. Biol. Evol. *21*, 1683-1691.

Otsuka, Y., Kawamura, Y., Koyama, T., Iihara, H., Ohkusu, K., and Ezaki, T. (2005). *Corynebacterium resistens* sp. nov., a new multidrug-resistant coryneform bacterium isolated from human infections. J. Clin. Microbiol. *43*, 3713-3717.

Pascual, C., Lawson, P.A., Farrow, J.A., Gimenez, M.N., and Collins, M.D. (1995). Phylogenetic analysis of the genus *Corynebacterium* based on 16S rRNA gene sequences. Int. J. Syst. Bacteriol. *45*, 724-728.

Quast, K., Bathe, B., Pühler, A., and Kalinowski, J. (1999). The *Corynebacterium glutamicum* insertion sequence IS*Cg2* prefers conserved target sequences located adjacent to genes involved in aspartate and glutamate metabolism. Mol. Gen. Genet. *262*, 568-578.

Ratti, G., Covacci, A., and Rappuoli, R. (1997). A tRNA(2Arg) gene of *Corynebacterium diphtheriae* is the chromosomal integration site for toxinogenic bacteriophages. Mol. Microbiol. *25*, 1179-1181.

Reilly, T.J., Baron, G.S., Nano, F.E., and Kuhlenschmidt, M.S. (1996). Characterization and sequencing of a respiratory burst-inhibiting acid phosphatase from *Francisella tularensis*. J. Biol. Chem. *271*, 10973-10983.

Riegel, P., Creti, R., Mattei, R., Nieri, A., and von Hunolstein, C. (2006). Isolation of *Corynebacterium tuscaniae* sp. nov. from blood cultures of a patient with endocarditis. J. Clin. Microbiol. *44*, 307-312.

Schäfer, A., Schwarzer, A., Kalinowski, J., and Pühler, A. (1994). Cloning and characterization of a DNA region encoding a stress-sensitive restriction system from *Corynebacterium glutamicum* ATCC 13032 and analysis of its role in intergeneric conjugation with *Escherichia coli*. J. Bacteriol. *176*, 7309-7319.

Schäfer, A., Tauch, A., Droste, N., Pühler, A., and Kalinowski, J. (1997). The *Corynebacterium glutamicum cglIM* gene encoding a 5-cytosine -methyltransferase enzyme confers a specific DNA methylation pattern in an McrBC-deficient *Escherichia coli* strain. Gene *203*, 95-101.

Soriano, F., Ponte, C., Santamaria, M., Aguado, J.M., Wilhelmi, I., Vela, R., and Delatte, L.C. (1985). *Corynebacterium* group D2 as a cause of alkaline-encrusted cystitis: report of four cases and characterization of the organisms. J. Clin. Microbiol. *21*, 788-792.

Soriano, F., Ponte, C., Santamaria, M., Castilla, C., and Fernandez Roblas, R. (1986). *In vitro* and *in vivo* study of stone formation by *Corynebacterium* group D2 (*Corynebacterium urealyticum*). J. Clin. Microbiol. *23*, 691-694.

Soriano, F., Aguado, J.M., Ponte, C., Fernandez-Roblas, R., and Rodriguez-Tudela, J.L. (1990). Urinary tract infection caused by *Corynebacterium* group D2: report of 82 cases and review. Rev. Infect. Dis. *12*, 1019-1034.

Stewart, M.E. (1992). Sebaceous gland lipids. Semin. Dermatol. *11*, 100-105.

Suzuki, N., Tsuge, Y., Inui, M., and Yukawa, H. (2005a). Cre/*loxP*-mediated deletion system for large genome rearrangements in *Corynebacterium glutamicum*. Appl. Microbiol. Biotechnol. *67*, 225-233.

Suzuki, N., Okayama, S., Nonaka, H., Tsuge, Y., Inui, M., and Yukawa, H. (2005b). Large-scale engineering of the *Corynebacterium glutamicum* genome. Appl. Environ. Microbiol. *71*, 3369-3372.

Suzuki, N., Nonaka, H., Tsuge, Y., Okayama, S., Inui, M., and Yukawa, H. (2005c). Multiple large segment deletion method for *Corynebacterium glutamicum*. Appl. Microbiol. Biotechnol. *69*, 151-161.

Suzuki, N., Nonaka, H., Tsuge, Y., Inui, M., and Yukawa, H. (2005d). New multiple-deletion method for the *Corynebacterium glutamicum* genome, using a mutant *lox* sequence. Appl. Environ. Microbiol. *71*, 8472-8480.

Suzuki, N., Okai, N., Nonaka, H., Tsuge, Y., Inui, M., and Yukawa, H. (2006). High-throughput transposon mutagenesis of *Corynebacterium glutamicum* and construction of a single-gene disruptant mutant library. Appl. Environ. Microbiol. *72*, 3750-3755.

Swierczynski, A., and Ton-That, H. (2006). Type III pilus of corynebacteria: pilus length is determined by the level of its major pilin subunit. J. Bacteriol. *188*, 6318-6325.

Tauch, A., Homann, I., Mormann, S., Rüberg, S., Billault, A., Bathe, B., Brand, S., Brockmann-Gretza, O., Rückert, C., Schischka, N., Wrenger, C., Hoheisel, J., Möckel, B., Huthmacher, K., Pfefferle, W., Pühler, A., and Kalinowski, J. (2002). Strategy to sequence the genome of *Corynebacterium glutamicum* ATCC 13032: use of a cosmid and a bacterial artificial chromosome library. J. Biotechnol. *95*, 25-38.

Tauch, A., Kaiser, O., Hain, T., Goesmann, A., Weisshaar, B., Albersmeier, A., Bekel, T., Bischoff, N., Brune, I., Chakraborty, T., Kalinowski, J., Meyer, F., Rupp, O., Schneiker, S., Viehoever, P., and Pühler, A. (2005). Complete genome sequence and analysis of the multiresistant nosocomial pathogen *Corynebacterium jeikeium* K411, a lipid-requiring bacterium of the human skin flora. J. Bacteriol. *187*, 4671-4682.

Tauch, A., Trost, E., Bekel, T., Goesmann, A., Ludewig, U., and Pühler, A. (2006). Ultrafast *de novo* sequencing of *Corynebacterium urealyticum* using the Genome Sequencer 20 System. Biochemica *4*, 4-6.

Ton-That, H., and Schneewind, O. (2003). Assembly of pili on the surface of *Corynebacterium diphtheriae*. Mol. Microbiol. *50*, 1429-1438.

Ton-That, H., Marraffini, L.A., and Schneewind, O. (2004). Protein sorting to the cell wall envelope of Gram-positive bacteria. Biochim. Biophys. Acta *1694*, 269-278.

Udaka, S. (1960). Screening method for microorganisms accumulating metabolites and its use in the isolation of *Micrococcus glutamicus*. J. Bacteriol. *79*, 754-755.

Williams, L.E., Detter, C., Barry, K., Lapidus, A., and Summers, A.O. (2006). Facile recovery of individual high-molecular-weight, low-copy-number natural plasmids for genomic sequencing. Appl. Environ. Microbiol. *72*, 4899-4906.

Yanai, K., Murakami, T., and Bibb, M. (2006). Amplification of the entire kanamycin biosynthetic gene cluster during empirical strain improvement of *Streptomyces kanamyceticus*. Proc. Natl. Acad. Sci. USA *103*, 9661-9666.

Yassin, A.F. (2007). *Corynebacterium ureicelerivorans* sp. nov., a lipophilic bacterium isolated from blood culture. Int. J. Syst. Evol. Microbiol. *57*, 1200-1203.

Yukawa, H., Omumasaba, C.A., Nonaka, H., Kos, P., Okai, N., Suzuki, N., Suda, M., Tsuge, Y., Watanabe, J., Ikeda, Y., Vertes, A.A., and Inui, M. (2007). Comparative analysis of the *Corynebacterium glutamicum* group and complete genome sequence of strain R. Microbiology *153*, 1042-1058.

Zhang, R., and Zhang, C.T. (2004). A systematic method to identify genomic islands and its applications in analyzing the genomes of *Corynebacterium glutamicum* and *Vibrio vulnificus* CMCP6 chromosome I. Bioinformatics *20*, 612-622.

Zhang, R., and Zhang, C.T. (2005). Genomic islands in the *Corynebacterium efficiens* genome. Appl. Environ. Microbiol. *71*, 3126-3130.

From: Corynebacteria: Genomics and Molecular Biology. Edited by: Andreas Burkovski.

DNA Microarray-based Transcriptome Analysis in *C. glutamicum*

3

Volker F. Wendisch

Abstract

Since the availability of DNA microarray technology for *Corynebacterium glutamicum* and completion of the genome sequence, transcriptome analyses have immensely contributed to further our understanding of this bacterium, a non-pathogenic actinomycete of biotechnological importance. This chapter describes the use of DNA microarray-based transcriptome analyses to elucidate global gene expression changes in response to environmental changes (stimulons), to determine targets of transcriptional regulators (regulons), to map undirected mutations and to improve amino acid production by *C. glutamicum*.

1. Introduction

Biotechnology, like many fields of biology, is experiencing a shift towards systems-level characterizations. Systems biology seeks to determine the complete repertoire of genes (genome), mRNAs (transcriptome), proteins (proteome), metabolites (metabolome), fluxes of an organism or cell as well as their interactions across all levels based on global analysis techniques. These data are integrated into predictive models, which are tested experimentally and are further refined by global experimental analysis and modeling. Genome sequencing provided the genetic blueprint of the *Corynebacterium glutamicum* cell (Ikeda and Nakagawa, 2003; Kalinowski et al., 2003; Yukawa et al., 2007). Of the global analysis techniques, only transcriptome analysis yields near-complete data sets and therefore is one of the cornerstones of systems biology approaches. Whole-genome DNA microarray technology has been developed for *C. glutamicum* (for reviews see Polen and Wendisch, 2004; Wendisch, 2003) and has since then been used to study fundamental aspects of *C. glutamicum* physiology on the one side and has been applied to improve amino acid production by this bacterium on the other side.

DNA microarrays can be manufactured in three formats based on either dsDNA fragments immobilized onto functionalized surfaces, on long oligonucleotides immobilized onto functionalized surfaces, or on short oligonucleotides synthesized *in situ*. The last format has been commercialized by Affymetrix and is suitable for detection of single nucleotide polymorphisms under certain circumstances. Production of DNA microarrays of the other two formats is more flexible and allows easily changing the microarray layout by inclusion of additional or exclusion of unwanted DNA probes. Whole-genome DNA microarrays for *C. glutamicum* based on dsDNA fragments immobilized onto poly-L-lysine functionalized glass surfaces became available first (reviewed in Wendisch, 2003) and meanwhile also Affymetrix *C. glutamicum* Gene-Chips (Mitsuhashi et al., 2006) are in use, but are not freely available. Several labs develop DNA microarrays based on long oligonucleotides immobilized onto functionalized surfaces. Current DNA microarray layouts cover the protein coding genes, but probes to sample intergenic regions have not yet been included. All methods for RNA isolation, fluorescence labeling of RNA, hybridization, scanning, raw data and statistical analysis, and follow-up bioinformatics are in place or can be applied to *C. glutamicum* without further adjustments (for reviews see Khodursky et al., 2003; Polen and Wendisch, 2004; Rhodius et al., 2002; Wendisch, 2003).

Rather than summarizing all *C. glutamicum* transcriptomics work performed to date, this chapter describes selected examples of whole-genome DNA microarray based transcriptome analyses. These were performed for the following objectives: deducing physiological strategies to cope with an environmental change (stimulon analyses), unraveling transcriptional regulatory network topology (regulons), mapping undirected mutants and improving strains for amino acid production.

2. Characterization of stimulons of *C. glutamicum*

Bacteria are exposed to a constantly changing environment. In order to maintain homeostasis of e.g. the intracellular pH, concentrations of key metabolites and several other parameters, bacteria respond to environmental changes in several different ways. These responses may comprise metabolic regulation e.g. by enzyme activity control as well as by genetic control mechanisms. Transcriptome analysis has proven to be ideally suited to determine the genetic response of a bacterium to changes of environmental parameters.

From the characterization of a stimulon, i.e. the group of genes that change expression in response to an imposed stimulus such as a nutrient limitation or a pH change, it is possible to deduce the physiological strategy of an organism to cope with that environmental change. Transcriptome analyses were employed widely to determine stimulons of *C. glutamicum* (Table 3.1). In such an approach, the phosphate starvation stimulon of *C. glutamicum* as well as the kinetics of this gene expression response were characterized (Ishige et al., 2003). Of the genes of the phosphate starvation stimulon the *pstSCAB* operon coding for an ABC-type phosphate uptake system is induced first as soon as phosphate becomes scarce. The *ugpABCE* operon encoding an

Table 3.1. Stimulons of *C. glutamicum* characterized by transcriptome analysis

Category	Stimulon	Reference
C metabolism	Acetate	(Gerstmeir et al., 2003; Hayashi et al., 2002)
	Ethanol	(Arndt et al., 2008)
	Propionate	(Hüser et al., 2003)
	L-lactate	(Jakob et al., 2007; Stansen et al., 2005)
	Citrate	(Polen et al., 2007)
	Ribose	(Wendisch, 2003)
	Inositol	(Krings et al., 2006)
	Serine	(Netzer et al., 2004b)
P metabolism	Phosphate starvation	(Ishige et al., 2003)
N metabolism	Ammonium starvation	(Silberbach et al., 2005a; Silberbach et al., 2005b)
other	$MgCl_2$	(Pallerla et al., 2005)
	Iron	(Krug et al., 2005)
	Valine	(Lange et al., 2003)

ABC type sn-glycerol 3-phosphate transport system and the gene for glycerophosphoryldiester phosphodiesterase (*glpQ1*) are induced shortly afterwards enabling uptake of sn-glycerol 3-phosphate as alternative source of phosphorus. The genes *ushA* (5′-nucleotidase) (Rittmann et al., 2005), *nucH* (extracellular nuclease), *phoH1* (ATPase of unknown function) and cg0689 are induced subsequently. Finally, a number of further genes mainly coding for proteins of unknown function are induced. Thus, a three fold strategy allows *C. glutamicum* to cope with phosphate limitation and includes induction of genes for (a) a high-affinity system for phosphate uptake, (b) a system for uptake and utilization of the organophosphate sn-glycerol 3-phosphate and (c) systems allowing to access inorganic phosphate present in non-transportable organophosphates (Ishige et al., 2003).

Transcriptome analyses were also used to characterize gene expression changes of *C. glutamicum* specific for utilization of different carbon sources. Usually global gene expression is compared between utilization of two carbon sources, e. g. glucose/acetate, glucose/ribose or glucose/inositol (Gerstmeir et al., 2003; Krings et al., 2006; Wendisch, 2003). In these comparative analyses, it is however difficult to assign an observed expression change *a priori* to either the presence of carbon source 1 or to the absence of carbon source 2. For example, genes showing higher expression on citrate than on glucose might be subject either to glucose repression or to citrate induction (Polen et al., 2007). To circumvent this problem, global gene expression can be compared during growth on a complex medium such as LB in the presence or absence of e.g. citrate (Polen et al., 2007). Citrate specific induction of the genes for citrate uptake (*citM* and *tctABC*) was detected in both citrate/glucose and LB+citrate/LB transcriptome comparisons (Polen et al., 2007). On the other hand, increased transcript levels of *acn*, which encodes the TCA cycle enzyme aconitase, in the presence of citrate were observed only in the citrate/glucose comparison, but not in the LB+citrate/LB comparison suggesting *acn* control by glucose repression rather than by citrate induction (Polen et al., 2007).

L-lactate can be used as a carbon source for growth of *C. glutamicum* and can be secreted under conditions of oxygen deprivation and as a by-product during lysine and glutamate production. L-lactate may impose a pH stress on *C. glutamicum* cells (Seletzky et al., 2006). Transcriptome analysis have been performed to characterize global gene expression (a) with respect to L-lactate formation during temperature-triggered glutamate production (Stansen et al., 2005) (b) with respect to utilization of L-lactate as sole carbon and energy source (Stansen et al., 2005) and (c) in response to long-term adaptation to low pH due to L-lactic acid (Jakob et al., 2007).

L-lactate secretion by *C. glutamicum* occurs under oxygen deprivation (Okino et al., 2005) and requires reduction of pyruvate by NAD-dependent L-lactate dehydrogenase (EC 1.1.1.27; LdhA) (Inui et al., 2004). Addition of bicarbonate shifts acid production under oxygen deprivation from L-lactate formation to succinate production (Okino et al., 2005). A transcriptome comparison of the glutamate-producing *C. glutamicum* strain 2262 and its non-producing mutant during chemostat cultivation revealed a single difference: higher expression of the NCgl2816-*lldD* operon in *C. glutamicum* 2262 than in the non-producing strain 2262NP (Stansen et al., 2005). Although *C. glutamicum* 2262 produced L-lactate, while 2262NP did not, *ldhA* expression did not differ between the strains. It was shown that the NCgl2816-*lldD* operon encodes a putative permease as well as quinone-dependent L-lactate dehydrogenase (EC 1.1.2.3; LldD) and that its induction during temperature-triggered glutamate production by *C. glutamicum* 2262 ensures re-utilization of L-lactate produced earlier (Stansen et al., 2005).

When the transcriptome of *C. glutamicum* wild type grown on L-lactate was compared to growth on pyruvate or on glucose, expression differences for more than 30 genes were observed. With L-lactate as the carbon source, expression of 22 genes was more than fourfold higher, and expression of 10 genes was more than fourfold lower, than for pyruvate-grown cultures. However, clearly the most prominent difference was a more than 15 fold increased expression of the NCgl2816-*lldD* operon ((Stansen et al., 2005) Polen & Wendisch, unpublished). Thus, as expected from the fact that pyruvate and L-lactate utilization differ only by the uptake systems and by quinone-dependent L-lactate dehydrogenase, the operon required for L-lactate utilization showed the highest transcript level increase. Although it is now clear that the NCgl2816-*lldD* operon is induced in the presence of L-lactate, the underlying regulatory mechanism and possible further control mechanisms remain to be characterized.

L-lactic acid imposes a stress on *C. glutamicum* as e.g. initial L-lactic-acid concentrations from 10 to 30 g/l led to a decrease of the growth rate and the biomass substrate yield without effecting the oxygen substrate conversion likely due to changes in pH and osmolality (Seletzky et al., 2006). Long-term adaptation to L-lactic acid was accompanied by many gene expression changes, as the global gene expression profile of *C. glutamicum* adapted to pH 5.7 with lactic acid in a continuous culture showed increased expression for 116 genes and decreased expression for 90 genes as compared to pH 7.5 without lactic acid (Jakob et al., 2007). Most (45%) of the up-regulated genes code for hypothetical proteins, while the remaining genes encode transcriptional regulators, proteins responsible for ex- and import of solutes, metabolism and several proteins of unknown function. Although continuous cultivation in a chemostat avoided nutrient gradients typical for batch cultivations, overlapping effects due to a changed pH, changed osmolality and the presence of the carbon and energy source L-lactate precluded deducing a clear strategy of *C. glutamicum* to cope with these stimuli from the global gene expression data. Accordingly, of 36 gene disruption mutants constructed for up-regulated genes only 2 mutants exhibited an acid sensitive growth phenotype: the mutants lacking functional genes for the RNA polymerase sigma factors E and B (Jakob et al., 2007).

Recently, the stimulons for the two alcohols ethanol and myo-inositol have been characterized (Arndt et al., 2008; Krings et al., 2006). Thirty one genes responded to myo-inositol utilization and 21 of those were localized in two clusters on the *C. glutamicum* genome (Krings et al., 2006). Functional genomics revealed that cluster I is essential for myo-inositol utilization, while cluster II encodes redundant functions. The gene *iolD*, which encodes a putative thiamine pyrophosphate-requiring enzyme of myo-inositol catabolism and is localized on cluster I, was required for myo-inositol utilization. The functionally characterized myo-inositol uptake systems IolT1 and IolT2, which are encoded on cluster I and II, respectively, could replace each other, while a mutant lacking both IolT1 and IolT2 could not grow with myo-inositol as sole carbon source. Transcriptome analysis of ethanol-grown *C. glutamicum* cells with glucose- or acetate-grown cells revealed differential expression of 95 genes, most notably increased expression of genes coding for TCA cycle enzymes, for Zn^{2+}/Mn^{2+} uptake systems and for a putative acetaldehyde dehydrogenase and decreased expression of genes for iron uptake (Arndt et al., 2008). The ethanol stimulon comprised all genes relevant for activation of ethanol via acetate to acetyl-CoA and its conversion in the TCA and glyxoxylate cycles. Enzyme measurements and growth experiments showed that glucose represses ethanol utilization thus resulting in diauxic growth on glucose-ethanol mixtures and sequential utilization of glucose before ethanol (Arndt et al., 2008).

3. Determination of regulons of *C. glutamicum*

To determine the group of genes and operons coordinately regulated by a transcriptional regulator, i.e. the regulon, it is important to keep the principle of gratuity in mind. As described for the characterization of the NtrC regulon of *E. coli* (Zimmer et al., 2000) by Magasanik (Magasanik, 2000), the effects of the absence of a transcriptional regulator need to be determined under a condition where ideally the expression of the regulated genes does not affect the physiology of the cells. As this ideal can rarely be achieved in practice, transcriptome comparisons between the wild-type strain and a strain carrying a deletion of the transcriptional regulator gene or a plasmid for its overexpression performed to reveal those genes responsive to the concentration of the transcriptional regulator are often combined with experiments assaying binding of the transcriptional regulator to target DNA, e.g. by band-shift assays. While transcriptome analysis does not allow distinguishing direct regulatory events from indirectly propagated effects, the so-called chromatin immunoprecipitation (ChIP)-to-chip analysis offers the possibility to score transcriptional regulator binding sites in a genome-wide manner (Wendisch, 2006). DNA-protein interaction indicates direct transcriptional regulation. A number of regulons in *C. glutamicum* has already been characterized (Table 3.2).

Targets of the response regulator MtrA of the two-component signal transduction system MtrA-MtrB have been determined using both transcriptome and ChIP-to-chip analyses (Brocker and Bott, 2006; Möker et al., 2004). Deletion of *mtrAB* resulted in morphological changes (i.e. cells were segmented and threefold longer), increased sensitivity to penicillin, vancomycin and lysozyme but higher resistance to ethambutol (Möker et al., 2004). Transcript levels of the genes *proP* and *betP*, encoding carriers for proline and the osmocompatible solute betaine, respectively, were reduced and transcript levels of the cell-wall peptidase genes *mepA, ppmA* and *lpqB* were increased in the *mtrAB* deletion mutant. Before and after a hyperosmotic shock, transcript levels of *proP*, *betP* and *lcoP* were lower in the *mtrAB* deletion mutant than in the wild type. From the transcriptome analysis three hypothesis were deduced: (a) MtrA directly regulates both genes of peptidoglycan metabolism (*mepA, ppmA* and *lpqB*) and of osmoprotection (*betP, lcoP* and *proP*), (b) only regulation of *mepA, ppmA* and *nlpQ* expression by MtrA is direct, while differential expression of *proP*, *betP* and *lcoP* results as an indirect consequence; (c) MtrA directly regulates expression of *proP*, *betP* and *lcoP*, while the changed transcript levels of *mepA, ppmA* and *nlpQ* are an indirect consequence. In the ChIP-to-chip analysis to find direct targets of MtrA, first a rabbit anti-MtrA antiserum was used to precipitate protein-DNA complexes crosslinked by formaladehyde treatment of live *C. glutamicum* cells. After reversal of the crosslinks, DNA co-precipitated by the rabbit anti-MtrA antiserum was purified, amplified and labelled with fluorescent desoxynucleotides and then hybridized to a whole-genome DNA microarray (Brocker and Bott, 2006). DNA co-immunoprecipitated by the rabbit anti-MtrA antiserum using an *mtrA* deletion mutant served as a negative control. This experiment indicated that MtrA binds to the promoter region of *nlpC* (NCgl2108), the region covering NCgl0102 and NCgl0103 and the intergenic region between the divergently transcribed genes *nrdH* (NCgl2445) and *nadE* (NCgl2446). DNA affinity chromatography experiments corroborated binding of MtrA to the promoter or intergenic regions of *nlpC* and *nrdh/nadE*,

Table 3.2. Transcriptome analysis used for characterization of regulons of *C. glutamicum*

Category	Stimulon	Reference
C metabolism	AcnR; Aconitase regulator	(Krug et al., 2005)
	SugR; regulator of sugar PTS genes	(Engels and Wendisch, 2007)
P metabolism	PhoR; response regulator of the two-component system PhoS-PhoR involved in adaptation to phosphate starvation	(Kocan et al., 2006; Schaaf and Bott, 2007)
N metabolism	AmtR; nitrogen control repressor	(Beckers et al., 2005)
S metabolism	McbR; methionine and cysteine biosynthesis repressor	(Rey et al., 2005)
	SsuR; sulphonate sulphur utilization regulator	(Koch et al., 2005)
other	FarR;	(Hänßler et al., 2007)
	SigB; alternative RNA polymerase sigma factor B	(Larisch et al., 2007)
	SigM; ECF-type RNA polymerase sigma factor M	(Nakunst et al., 2007)
	ClgR; Clp protease gene regulator	(Engels et al., 2005)
	LtbR; leucine and tryptophan biosynthesis regulator	(Brune et al., 2007)
	DtxR; regulator of iron metabolism	(Brune et al., 2006; Wennerhold and Bott, 2006)
	RipA; regulator of iron protein genes	(Wennerhold et al., 2005)
	MtrA; response regulator of the two-component system MtrB-MtrA for control of compatible solute carrier and cell wall peptidase genes	(Brocker and Bott, 2006; Möker et al., 2004)

but not to NCgl0102/0103, and also showed that MtrA binds to the upstream regions of *proP* and *betP* (Brocker and Bott, 2006). Thus, the results from the transcriptome, ChIP-to-chip and DNA affinity chromatography experiments are commensurate with the view that MtrA activates genes for putative cell-wall peptidases (*mepA, nlpC*) and represses genes encoding carriers for compatible solutes (*betP, proP*).

4. Mutant mapping

DNA microarray experiments have been used to compare genomes of related strains as well as to identify mutations resulting in gene expression changes. In comparative genome analysis, genomic DNA from two related strains is fluorescently labeled and hybridized to a whole-genome DNA microarray. In such a genomotyping experiment comparing the glutamate-producing *C. glutamicum* strain 2262 with the type strain ATCC 13032, hybridization signals were observed for about 87% of the genes with genomic DNA from both strains using DNA microarrays designed for *C. glutamicum* ATCC 13032. This comparative genome analysis also revealed that 333 genes in 26 loci, which are present in *C. glutamicum* ATCC 13032, are lacking in *C. glutamicum* 2262 (Figure 3.1). As the DNA microarray is designed according to the genome of the reference strain, this kind of analysis only detects genes absent from the query strain, but present in the reference strain. In addition, copy number variations of genes present in the reference strain can also be detected, e.g. gene amplifications in the query strand. The resolution and coverage of these analyses is a direct consequence of the probe length and spacing. *C. glutamicum* DNA microarrays based on 1 PCR product of about 500 bp per gene (Wendisch, 2003) allow detection of gene deletions larger than 500-2000 bp.

Besides direct comparative hybridization of genomic DNA, DNA microarrays can also be used in transcriptomics experiments for detection of mutations. A prerequisite for this kind of analysis is that the mutation results in altered transcript levels in *cis* or in *trans*. The first example of using transcriptomics to map a spontaneous mutation in *C. glutamicum* was the characterization of a suppressor mutant of a pyruvate kinase gene deletion mutant (Netzer et al., 2004a). In the absence of pyruvate kinase *C. glutamicum* cannot grow on acetate or citrate which indicated that pyruvate generation from PEP by pyruvate kinase is the only source of pyruvate under these conditions. Upon prolonged incubation suppressor mutants able to grow on acetate or citrate in the absence of pyruvate kinase could be isolated (Netzer et al., 2004a). The suppressor mutant still did not show any pyruvate kinase activity, thus, the mutation did neither result in activation of a cryptic pyruvate kinase gene nor in appearance of a side pyruvate kinase activity of a related enzyme. Transcriptome analysis showed that during growth on acetate or citrate the malic enzyme gene was expressed in the suppressor mutant, but hardly in the wild type. As malic enzyme decarboxylates malate to pyruvate, a new source of pyruvate for growth on acetate or citrate could be revealed and corroborated by plasmid-borne overexpression of *malE*. As revealed by transcriptional fusion analysis, a point mutation identified upstream of *malE* was responsible for expression of *malE* in the suppressor mutant during growth on acetate or citrate (Netzer et al., 2004a). Besides *cis* mutations leading to expression changes like that observed for *malE* also *trans* mutations can be identified. Recently, we observed that a spontaneous mutant able to degrade isocitric acid expressed genes for uptake of citrate and isocitrate in the absence of the presumed inducer citrate. The mutant possessed a point mutation in the open reading frame of the gene *citA* leading to an amino acid exchange in the kinase domain of the sensor kinase CitA of the two-component signal transduction system CitA-CitB (Youn & Wendisch, unpublished).

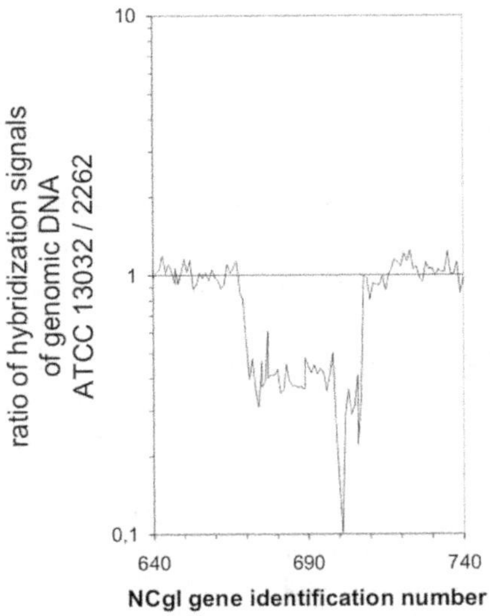

Figure 3.1. Comparative hybridization analysis of genomic DNA from *C. glutamicum* 2262 and ATCC 13032 using whole-genome DNA microarrays based on the genome sequence of *C. glutamicum* ATCC 13032. The ratio of the hybridization signals is plotted against the gene identification numbers, which reflect the consecutive gene order in the genome.

5. Global gene expression analysis of glutamate producing *C. glutamicum*

Since its isolation as glutamate producing bacterium *C. glutamicum* strains have been developed continuously to improve production of L-glutamic acid and its monosodium, potassium, magnesium, calcium and ammonium salts, which are used as flavoring agents in the food industry (Shimizu and Hirasawa, 2007). Glutamate production needs to be triggered by biotin limitation, temperature shift or addition of detergents like Tween, antibiotics such as penicillin or cerelunin and chemotherapeutics like ethambutol (Shimizu and Hirasawa, 2007). Alteration of the fatty acid and/or mycolic acid composition as a consequence of biotin limitation (reduced levels of the biotinylated acetyl-CoA carboxylase, a key enzyme of fatty acid synthesis), addition of cerulenin (a fatty acid synthase inhibitor) or detergents and cell wall alterations due to addition of β-lactam antibiotics or ethambutol (an inhibitor of the cell wall arabinosyltransferase) are thought to be one key element of glutamate production. Glutamate production requires reduction of the 2-oxoglutarate dehydrogenase complex activity. Transcriptomics analysis of glutamate production (Kataoka et al., 2006; Radmacher et al., 2005; Stansen et al., 2005) did not reveal significantly changed expression of the genes for the subunits of the 2-oxoglutarate dehydrogenase complex, although metabolic flux analysis indicated a low flux via 2-oxoglutarate dehydrogenase complex under the tested glutamate production conditions. This can now be explained as the signal transduction pathway leading to inhibition of the 2-oxoglutarate dehydrogenase complex has recently been unraveled (Niebisch et al., 2006; Schultz et al., 2007). Three proteins are involved: the 15-kDa protein OdhI, the serine/threonine protein kinase G (PknG) and the cg0062 gene product, a phospho-Ser/Thr protein phosphatase. In its unphosphorylated state, OdhI binds to the E1 subunit (OdhA) of the 2-oxoglutarate dehydrogenase complex and inhibits its activity (Niebisch et al., 2006).

Phosphorylation of OdhI at threonine-14 by PknG leads to relieve of 2-oxoglutarate dehydrogenase complex inhibition (Niebisch et al., 2006). Under conditions requiring little 2-oxoglutarate dehydrogenase complex activity, e.g. during glutamate production, phosphorylated OdhI is dephosphorylated by the phospho-Ser/Thr protein phosphatase (Schultz et al., 2007). Control of 2-oxoglutarate dehydrogenase complex flux by protein-protein interaction and phosphorylation thus is of major importance, while expression control of the respective 2-oxoglutarate dehydrogenase complex genes appears to be negligible.

In one study reduced expression of almost all genes for enzymes in glycolysis, the pentose phosphate pathway and the TCA cycle (including the 2-oxoglutarate dehydrogenase complex genes) was observed under glutamate production conditions (biotin limitation, addition of Tween 40 or penicillin). However, these changes were not specific to glutamate producing conditions as they also occurred when Tween 80 was added, a condition not triggering glutamate production (Kataoka et al., 2006). It is likely that most of these changes are due to slowed growth resulting from biotin limitation, addition of penicillin Tween 40 or Tween 80 as compared to the untreated control. Similarly, reduced expression of the 2-oxoglutarate dehydrogenase complex gene *odhA, lpd* as well as the F_1 ATPase and respiratory chain genes was observed late during ethambutol-triggered glutamate production, when growth ceased (Radmacher et al., 2005). Ethambutol addition led to restructuring of the cell wall, which was also reflected by sustained expression changes of cell wall-related genes, e.g. encoding the secreted metalloendopeptidase MepA (Möker et al., 2004) or the cell-division ATP binding protein FtsE. Immediately after ethambutol addition, expression of *yggB* (NCgl1221) increased (Radmacher et al., 2005). In the absence of YggB, which contains an N-terminal region similar to the *E. coli* mechanosensitive channel YggB and a C-terminal region unlike any known sequence, the betaine efflux rate was reduced in hypoosmotic shock experiments (Nottebrock et al., 2003). Indirect evidence suggests that YggB may be involved in glutamate excretion as overexpression of *yggB* increased glutamate production and *yggB* deletion reduced glutamate production (Nakamura et al., 2007). However, it has not yet been determined biochemically whether YggB enables glutamate efflux in addition to betaine efflux or whether indirect (regulatory) effects of deletion or overexpression of *yggB* result in altered glutamate production from glucose. Although expression of *yggB* increased immediately after ethambutol addition (Radmacher et al., 2005), other triggers of glutamate production (biotin limitation, addition of penicillin or Tween 40, temperature shift) were not reported to result in induction of *yggB*.

Temperature-shift induced glutamate production by *C. glutamicum* 2262 can be analysed under the balanced conditions of chemostat cultivation (Delaunay et al., 1999; Uy et al., 2003). However, the process was found to be instable and a non-producing mutant, *C. glutamicum* 2262NP, could be isolated (Uy et al., 2003). When these strains were exposed to a temperature-shock in continuous cultivations, *C. glutamicum* 2262 produced glutamate and 2262NP did not (Stansen et al., 2005; Uy et al., 2003). Under the defined conditions of chemostat cultivation, global gene expression hardly differed between the producing and the non-producing strain (Stansen et al., 2005). Only two genes, *lldD* and NCgl2816, showed higher mRNA levels in the producing strain than in the non-producing mutant. The putative permease gene NCgl2816 is organized in an operon with *lldD*, which was shown to be a quinone-dependent L-lactate dehydrogenase required for L-lactate utilization (Stansen et al., 2005). Increased mRNA levels of the L-lactate-inducible NCgl2816-*lldD* operon in *C. glutamicum* 2262 were due to formation of L-lactate as a by-product of glutamate production. L-lactate formation can be avoided by deletion of *ldhA*, the gene for the L-lactate-forming NAD-dependent L-lactate dehydrogenase (Inui et al., 2004). However, as L-lactate formed as a by-product can eventually be re-utilized, constitutive expression of NCgl2816-*lldD* offers an alternative strategy to improve the temperature-triggered glutamate production process.

6. DNA microarray-based strain improvement for lysine production

Transcriptome analysis has been employed to characterize lysine production and to identify new targets for the optimization of lysine producing strains. *C. glutamicum* possesses a split lysine biosynthesis pathway, which requires pyruvate and oxaloacetate as carbon precursors and the reducing power of NADPH (Wittmann and Becker, 2007). A feedback-resistant version of the key enzyme aspartokinase, which is encoded by $lysC^{FBR}$ (Schrumpf et al., 1992; Thierbach et al., 1990), is key to sustain a high flux into the split lysine biosynthesis pathway. Formation of other aspartate-family amino acids was reduced by restricting the homoserine dehydrogenase flux through *hom* mutations (Ohnishi et al., 2002). Carbon precursor supply for lysine biosynthesis was improved by enhancing anaplerosis through overexpression of the pyruvate carboxylase gene *pyc* or by reducing the opposite C_4 decarboxylation through deletion of the PEP carboxykinase gene *pckA* (Peters-Wendisch et al., 1998; Peters-Wendisch et al., 2001; Riedel et al., 2001). The NADPH supply could be improved by forcing entry of glucose-6-phosphate into the pentose phosphate pathway NADPH forming either by deletion of the phosphoglucoisomerase gene *pgi* (Marx et al., 2003), or by expression of mutant alleles of the glucose-6-phosphate dehydrogenase gene *zwf* or of the 6-phosphogluconate dehydrogenase gene *gnd* (Ando et al., 2002; Marx et al., 2003; Ohnishi et al., 2005). Increasing the pentose phosphate pathway flux improved lysine production on sucrose (Georgi et al., 2005) and, in some strains, on fructose (Becker et al., 2005) and was achieved by overexpression of *fbp*, which encodes fructose-1,6-bisphosphatase (Rittmann et al., 2003).

By transcriptome analysis gene expression differences between the classically obtained strain B-6 and the wild type ATCC 13032 were determined. As a general feature of this strain it could be shown that besides the aspartokinase gene and *dapD* of lysine biosynthesis many additional genes for biosynthetic enzymes of other amino acids showed higher mRNA levels in strain B-6 than in the wild type (Hayashi et al., 2006b). However, global gene expression in a genetically defined lysine-producing strain obtained by introducing certain *lysC*^{FBR}, *hom* and *pyc* alleles into the wild type genome did not differ from the wild type (Hayashi et al., 2006b). Also, while addition of serine hydroxamate, a trigger of the stringent response (Tauch et al., 2001), to *C. glutamicum* wild type resulted in increased expression of many amino acid biosynthesis genes, the pattern observed differed from that in the comparison between the lysine-producing strain B-6 and the wild type (Hayashi et al., 2006b). It was concluded that some unknown global regulatory mechanism governing amino acid biosynthesis needs to be identified and subsequently manipulated to optimize lysine production by *C. glutamicum.*

Lysine production by a different classically obtained lysine-producing strain, *C. glutamicum* ATCC 13287, was characterized during 9 h of batch cultivation with respect to fluxes, metabolite concentrations and mRNA levels (Krömer et al., 2004). In this strain gene expression did not change much under the tested conditions, only for the genes of gluconeogenetic PEP carboxykinase, ATPase α-chain, the glyoxylate cycle enzyme isocitrate lyase and the lysine exporter expression changes of three or more were observed for only a single time point (Krömer et al., 2004). High mRNA levels of the lysine exporter gene *lysE* during lysine production can be explained as its regulation is known to depend on the transcriptional regulator LysG, which activates *lysE* expression at elevated concentrations of the co-inducers lysine or arginine (Bellmann et al., 2001).

From the transcriptome analysis of the leucine-auxotrophic classically obtained *C. glutamicum* strain MH20-22B, which also possesses feedback-resistant aspartokinase, novel candidate genes to improve lysine production were deduced and subsequently tested experimentally (Sindelar and Wendisch, 2007). The transcriptome comparisons were also performed for genetically defined intermediate strains differing in one or more of the following characteristics: aspartokinase allele (*lysCWT* or *lysC*^{FBR}), leucine prototrophy or auxotrophy and wild-type or classically obtained strain background. Expression of 7 genes/operons was altered only in the classically obtained strain background irrespective of the aspartokinase allele or leucine auxotrophy (Sindelar and Wendisch, 2007). Plasmid-borne overexpression of the gene NCgl0855, putatively encoding a methyltransferase, and the operon *amtA-ocd-soxA*, encoding an ammonium uptake system, a putative ornithine cyclodeaminase and an uncharacterized enzyme, improved lysine production by MH20-22B by about 40% (Sindelar and Wendisch, 2007). The targets NCgl0855 and *amtA-ocd-soxA* have not been identified previously in classical approaches involving mutagenesis and screening/selection or by metabolic engineering approaches guided by carbon flux analysis or biochemical knowledge. This report was also the first demonstration that novel target genes for the improvement of lysine production could be identified in a discovery-driven approach based on global gene expression analysis. However, currently the mechanistic basis for improvement of lysine production through overexpression of NCgl0855 or *amtA-ocd-soxA* is not known.

As already mentioned above, NCgl0855 encodes a putative methyltransferase and as lysine biosynthesis in *C. glutamicum* does not include methylation reactions, an indirect beneficial effect of NCgl0855 on lysine production by this bacterium has to be invoked. The first gene of the *amtA-ocd-soxA* operon has been shown to encode the membrane-potential-driven uptake system for ammonium AmtA (Siewe et al., 1996), while the functions of the *ocd* and *soxA* gene products are unknown. The *ocd* gene product is similar to bacterial ornithine cyclodeaminases that liberate ammonium from ornithine during catabolism of arginine and ornithine e.g. in agrobacteria, clostridia, and pseudomonads (Glansdorff and Xu, 2007; Tricot et al., 1991). The *soxA* gene product shows sequence similarity to the N-terminal domain of monomeric sarcosine oxidases. Sarcosine oxidases demethylate sarcosine (N-methylglycine) to glycine, formaldehyde, and hydrogen peroxide. Subsequently, glycine can be converted to pyruvate by serine hydroxymethyltransferase and serine deaminase, which are present in *C. glutamicum* (Netzer et al., 2004b; Simic et al., 2002). Increased ammonium uptake due to higher AmtA levels might improve lysine production. However, it remains to be established whether overexpression of *amtA* alone improves lysine production by *C. glutamicum* MH20-22B or whether *ocd* and/or *soxA* are also required. *C. glutamicum* MH20-22B also exhibited high expression of genes of the terminal oxidase cytochrome *bd* and an ABC transporter required for the formation of an active cytochrome *bd* oxidase (*cydD* and *cydB*; Bott and Niebisch, 2005), which is commensurate with the finding that this strain contains higher levels of cytochrome *bd* than *C. glutamicum* wild type (Kabus et al., 2007). However, the energetic efficiency of the cytochrome *bc1-aa3* supercomplex is higher than that of cytochrome *bd* as transport of 6 and 2 protons, respectively, is coupled to electron transfer from menaquinol to oxygen (Bott and Niebisch, 2003). Consequently, deletion of the genes for the cytochrome *bd* oxidase in *C. glutamicum* MH20-22B increased lysine production significantly (Kabus et al., 2007). Therefore, the high *cydB* and *cydD* mRNA levels and the high cytochrome *bd* level exemplify that classically obtained lysine producer strains such as MH20-22B show traits detrimental to lysine production. Thus, as differentially expressed genes identified by transcriptome analysis of lysine producing strains can either be advantageous or disadvantageous for lysine production, a combination of overexpression and deletion strategies might prove useful for target validation.

7. Transcriptome analysis for the characterization of the metabolism of other amino acids and of pantothenate

Transcriptome analysis was also used to characterize and improve production of the branched-chain amino acid valine by *C. glutamicum* (Lange et al., 2003). Overexpression of the branched-chain amino acid biosynthesis genes *ilvBNC* combined with deletion of *ilvA*, encoding the first enzyme of isoleucine biosynthesis, and of the pantothenate biosynthesis genes *panBC* in *C. glutamicum* wild type was sufficient to enable production of valine (Radmacher et al., 2002). Growth of the valine producing strain, but not of the wild type, was inhibited by valine added to the growth medium. DNA microarray analysis identified gene expression changes only occurring in the valine producing strain in response to added valine (Lange et al., 2003). The high expression of *ilvBNC* and of an isoleucine tRNA synthetase gene indicated intracellular isoleucine limitation in the presence of valine in the medium. It was subsequently shown that this isoleucine limitation was linked to the *ilvA* deletion in the isoleucine auxotrophic valine producing strain (Lange et al., 2003). As isoleucine limitation and growth inhibition by valine could be overcome by addition of high isoleucine concentrations or of low concentrations of the dipeptide isoleucylisoleucine, competition of isoleucine uptake by the branched-chain amino acid uptake system BrnQ (Tauch et al., 1998) caused valine inhibition. Besides the identification of the molecular cause of the growth impairment of the producing strain by valine, this DNA microarray analysis also revealed a target for strain improvement. Because externally added valine increased *ilvBN* expression and stimulated valine production by the valine producing strain, the activity of the *ilvBN* encoded acetohydroxyacid synthase activity may still be a limiting factor for valine production in this valine producing strain (Lange et al., 2003). Feedback-resistant acetohydroxyacid synthase due to expression of the mutant $ilvN^{M13}$ allele improved valine production by this strain (Elisakova et al., 2005). In the production strain intracellular pyruvate availability correlated with the valine production rate (Ruklisha et al., 2007). Metabolic engineering to improve the supply of pyruvate as precursor for valine biosynthesis by deletion of the *aceE* gene, which encodes the E1p enzyme of the pyruvate dehydrogenase complex, improved valine production by *C. glutamicum* to a yield of 0.4 g g^{-1} and a volumetric productivity of 1.17 g l^{-1} h^{-1} (Blombach et al., 2007).

Transcriptome analysis was also used to characterize leucine biosynthesis as *C. glutamicum* strains overproducing the branched-chain amino acid leucine are available (Patek, 2007). The leucine and tryptophan biosynthesis regulator LbtR represses transcription of the *leuB* and *leuCD* genes encoding isopropylmalate dehydratase and isopropylmalate dehydrogenase of the leucine biosynthesis pathway as well as the tryptophan biosynthesis operon *trpEGDCFBA* (Brune et al., 2007). LbtR binds to conserved 12-bp motifs, CA(T/C)ATAGTG(A/G)GA, downstream of the -10 promoter regions of *leuB, leuC*, and *trpEGDCFBA*. Transcriptome analysis, which helped to define the LbtR regulon (Brune et al., 2007), has not yet been used to characterize leucine production, while the effect of limited leucine biosynthesis on global gene expression has been determined (Hayashi et al., 2006a). A partial leucine auxotrophy due to a mutant isopropylmalate dehydratase large subunit (encoded by $leuC^{G456D}$), which is beneficial for lysine production by *C. glutamicum*, caused high expression of many amino acid biosynthesis genes (Hayashi et al., 2006a). As these effects were also observed when the *rel* gene encoding bifunctional guanosine pentaphosphate synthetase (Wehmeier et al., 1998) was deleted, the stringent response could be ruled out as the primary cause (Hayashi et al., 2006a). The stringent response can be elicited artificially by addition of the serine analogue serine hydroxamate, which prevents charging of the seryl-tRNA by competitively binding to the seryl-tRNA synthetase. The global gene expression changes caused by the addition of serine hydroxamate (Brockmann-Gretza and Kalinowski, 2006; Hayashi et al., 2006a) differ from those arising from high intracellular serine concentrations due to addition of the tripeptide Ser-Ser-Ser (Netzer et al., 2004b). Comparative transcriptome analysis of *rel*-proficient and *rel* mutant strains with/without addition of serine hydroxamate identified the expected set of gene expression changes due to the stringent response (Brockmann-Gretza and Kalinowski, 2006). Unlike in *E. coli*, nitrogen metabolism genes also showed *rel*-dependent control. In addition, the set of genes showing *rel*-independent expression changes did not comprise genes encoding ribosomal proteins, a fundamental difference to the *E. coli* stringent response (Brockmann-Gretza and Kalinowski, 2006).

Pantothenate overproducing strains have been developed based on valine producing strains (Sahm and Eggeling, 1999) as the immediate precursor of valine, ketoisovalerate, is the substrate of the first enzyme in pantothenate biosynthesis. A promoter down-mutation of the *ilvE* gene for the enzyme transaminating ketoisovalerate to valine and overexpression of the pantothenate biosynthesis genes *panB, panC* and *panD* in a valine producer enabled pantothenate production (Hüser et al., 2005; Sahm and Eggeling, 1999). Transcriptome comparison of the pantothenate producing strain with a control strain revealed differential expression of genes involved in leucine, serine, and glycine biosynthesis, in regeneration of methylenetetrahydrofolate, in nicotinic acid mononucleotide biosynthesis, and in acyl coenzyme A conversion during batch cultivations. Of these, 21 genes showed differential expression between the producing and control strain. Obviously, these comprised the genes of valine and pantothenate biosynthesis overexpressed in the producer, but additionally genes encoding two putative acetyltransferases and the alternative sigma factor C as well as 12 uncharacterized proteins showed higher expression in the producer than in the control (Hüser et al., 2005). Although none of the putative targets was tested experimentally for its effect on pantothenate production, these findings provide a basis for further strain development of pantothenate producers.

References

Ando, S., Ochiai, K., Yokoi, H., Hashimoto, S., and Yonetani, Y. (2002). Novel glucose-6-phosphate dehydrogenase. Patent WO0198472 *(2002-01-02).*

Arndt, A., Auchter, M., Ishige, T., Wendisch, V.F., and Eikmanns, B. J. (2008). Ethanol catabolism in *Corynebacterium glutamicum*. J. Mol. Microbiol. Biotechnol, *in press.*

Becker, J., Klopprogge, C., Zelder, O., Heinzle, E., and Wittmann, C. (2005). Amplified expression of fructose 1,6-bisphosphatase in *Corynebacterium glutamicum* increases *in vivo* flux through the pentose phosphate pathway and lysine production on different carbon sources. Appl Environ. Microbiol. *71*, 8587-8596.

Beckers, G., Strösser, J., Hildebrandt, U., Kalinowski, J., Farwick, M., Krämer, R., and Burkovski, A. (2005). Regulation of AmtR-controlled gene expression in *Corynebacterium glutamicum*: mechanism and characterization of the AmtR regulon. Mol. Microbiol. *58*, 580-595.

Bellmann, A., Vrljic, M., Patek, M., Sahm, H., Krämer, R., and Eggeling, L. (2001). Expression control and specificity of the basic amino acid exporter LysE of *Corynebacterium glutamicum*. Microbiology *147*, 1765-1774.

Blombach, B., Schreiner, M.E., Moch, M., Oldiges, M., and Eikmanns, B.J. (2007). Effect of pyruvate dehydrogenase complex deficiency on L-lysine production with *Corynebacterium glutamicum*. Appl Microbiol. Biotechnol, *76,* 615-623.

Bott, M., and Niebisch, A. (2003). The respiratory chain of *Corynebacterium glutamicum*. J. Biotechnol. *104*, 129-153.

Bott, M., and Niebisch, A. (2005). Respiratory energy metabolism. In Handbook of *Corynebacterium glutamicum* (Eggeling L, Bott M, eds), CRC Press, Boca Raton, USA, pp 305-332.

Brocker, M., and Bott, M. (2006). Evidence for activator and repressor functions of the response regulator MtrA from *Corynebacterium glutamicum*. FEMS Microbiol. Lett. *264*, 205-212.

Brockmann-Gretza, O., and Kalinowski, J. (2006). Global gene expression during stringent response in *Corynebacterium glutamicum* in presence and absence of the *rel* gene encoding (p)ppGpp synthase. BMC Genomics *7*, 230.

Brune, I., Werner, H., Hüser, A.T., Kalinowski, J., Pühler, A., and Tauch, A. (2006). The DtxR protein acting as dual transcriptional regulator directs a global regulatory network involved in iron metabolism of *Corynebacterium glutamicum*. BMC Genomics *7*, 21.

Brune, I., Jochmann, N., Brinkrolf, K., Hüser, A. T., Gerstmeir, R., Eikmanns, B.J., Kalinowski, J., Pühler, A., and Tauch, A. (2007). The IclR-type transcriptional repressor LtbR regulates the expression of leucine and tryptophan biosynthesis genes in the amino acid producer *Corynebacterium glutamicum*. J. Bacteriol. *189*, 2720-2733.

Delaunay, S., Gourdon, P., Lapujade, P., Mailly, E., Oriol, E., Engasser, J.M., Lindley, N.L., and Goergen, J.L. (1999). An improved temperature triggered process for glutamate production with *Corynebacterium glutamicum*. Enz Microb Biotechnol. *25*, 762-768.

Elisakova, V., Patek, M., Holatko, J., Nesvera, J., Leyval, D., Goergen, J.L., and Delaunay, S. (2005). Feedback-resistant acetohydroxy acid synthase increases valine production in *Corynebacterium glutamicum*. Appl Environ. Microbiol. *71*, 207-213.

Engels, S., Ludwig, C., Schweitzer, J.E., Mack, C., Bott, M., and Schaffer, S. (2005). The transcriptional activator ClgR controls transcription of genes involved in proteolysis and DNA repair in *Corynebacterium glutamicum*. Mol. Microbiol. *57*, 576-591.

Engels, V., and Wendisch, V.F. (2007). The DeoR-type regulator SugR represses expression of *ptsG* in *Corynebacterium glutamicum*. J. Bacteriol. *189*, 2955-2966.

Georgi, T., Rittmann, D., and Wendisch, V.F. (2005). Lysine and glutamate production by *Corynebacterium glutamicum* on glucose, fructose and sucrose: Roles of malic enzyme and fructose-1,6-bisphosphatase. Metab Eng. *7*, 291-301.

Gerstmeir, R., Wendisch, V.F., Schnicke, S., Ruan, H., Farwick, M., Reinscheid, D., and Eikmanns, B.J. (2003). Acetate metabolism and its regulation in *Corynebacterium glutamicum*. J. Biotechnol. *104*, 99-122.

Glansdorff, N., and Xu, Y. (2007). Microbial Arginine Biosynthesis: Pathway, Regulation and Industrial Production. *In* Amino Acid Biosynthesis – Pathways, Regulation and Metabolic Engineering (Wendisch VF, ed), Springer, Heidelberg, Germany.

Hänßler, E., Müller, T., Jessberger, N., Völzke, A., Plassmeier, J., Kalinowski, J., Krämer, R., and Burkovski, A. (2007). FarR, a putative regulator of amino acid metabolism in *Corynebacterium glutamicum*. Appl Microbiol. Biotechnol, DOI 10.1007/s00253-007-0929-5.

Hayashi, M., Mizoguchi, H., Shiraishi, N., Obayashi, M., Nakagawa, S., Imai, J., Watanabe, S., Ota, T., and Ikeda, M. (2002). Transcriptome analysis of acetate metabolism in *Corynebacterium glutamicum* using a newly developed metabolic array. Biosci Biotechnol. Biochem. *66*, 1337-1344.

Hayashi, M., Mizoguchi, H., Ohnishi, J., Mitsuhashi, S., Yonetani, Y., Hashimoto, S., and Ikeda, M. (2006a). A *leuC* mutation leading to increased L-lysine production and *rel*-independent global expression changes in *Corynebacterium glutamicum*. Appl Microbiol. Biotechnol. *72*, 783-789.

Hayashi, M., Ohnishi, J., Mitsuhashi, S., Yonetani, Y., Hashimoto, S., and Ikeda, M. (2006b). Transcriptome analysis reveals global expression changes in an industrial L-lysine producer of *Corynebacterium glutamicum*. Biosci Biotechnol. Biochem. *70*, 546-550.

Hüser, A.T., Becker, A., Brune, I., Dondrup, M., Kalinowski, J., Plassmeier, J., Pühler, A., Wiegräbe, I., and Tauch, A. (2003). Development of a *Corynebacterium glutamicum* DNA microarray and validation by genome-wide expression profiling during growth with propionate as carbon source. J. Biotechnol. *106*, 269-286.

Hüser, A.T., Chassagnole, C., Lindley, N. D., Merkamm, M., Guyonvarch, A., Elisakova, V., Patek, M., Kalinowski, J., Brune, I., Pühler, A., and Tauch, A. (2005). Rational design of a *Corynebacterium glutamicum* pantothenate production strain and its characterization by metabolic flux analysis and genome-wide transcriptional profiling. Appl Environ. Microbiol. *71*, 3255-3268.

Ikeda, M., and Nakagawa, S. (2003). The *Corynebacterium glutamicum* genome: features and impacts on biotechnological processes. Appl Microbiol. Biotechnol. *62*, 99-109.

Inui, M., Kawaguchi, H., Murakami, S., Vertes, A.A., and Yukawa, H. (2004). Metabolic Engineering of *Corynebacterium glutamicum* for fuel ethanol production under oxygen-deprivation conditions. J. Mol. Microbiol. Biotechnol. *8*, 243-254.

Ishige, T., Krause, M., Bott, M., Wendisch, V.F., and Sahm, H. (2003). The phosphate starvation stimulon of *Corynebacterium glutamicum* determined by DNA microarray analyses. J. Bacteriol. *185*, 4519-4529.

Jakob, K., Satorhelyi, P., Lange, C., Wendisch, V.F., Silakowski, B., Scherer, S., and Neuhaus, K. (2007). Gene expression analysis of *Corynebacterium glutamicum* subjected to long-term lactic acid adaptation. J. Bacteriol. *189*, 5582-5590.

Kabus, A., Niebisch, A., and Bott, M. (2007). Role of cytochrome *bd* oxidase from *Corynebacterium glutamicum* in growth and lysine production. Appl Environ. Microbiol. *73*, 861-868.

Kalinowski, J., Bathe, B., Bartels, D., Bischoff, N., Bott, M., Burkovski, A., Dusch, N., Eggeling, L., Eikmanns, B. J., Gaigalat, L., et al. (2003). The complete *Corynebacterium glutamicum* ATCC 13032 genome sequence and its impact on the production of L-aspartate-derived amino acids and vitamins. J. Biotechnol. *104*, 5-25.

Kataoka, M., Hashimoto, K. I., Yoshida, M., Nakamatsu, T., Horinouchi, S., and Kawasaki, H. (2006). Gene expression of *Corynebacterium glutamicum* in response to the conditions inducing glutamate overproduction. Lett. Appl Microbiol. *42*, 471-476.

Khodursky, A.B., Bernstein, J.A., Peter, B.J., Rhodius, V., Wendisch, V.F., and Zimmer, D.P. (2003). *Escherichia coli* spotted double-strand DNA microarrays: RNA extraction, labeling, hybridization, quality control, and data management. Methods Mol. Biol. *224*, 61-78.

Kocan, M., Schaffer, S., Ishige, T., Sorger-Herrmann, U., Wendisch, V.F., and Bott, M. (2006). Two-component systems of *Corynebacterium glutamicum*: deletion analysis and involvement of the PhoS-PhoR system in the phosphate starvation response. J. Bacteriol. *188*, 724-732.

Koch, D.J., Rückert, C., Albersmeier, A., Hüser, A.T., Tauch, A., Pühler, A., and Kalinowski, J. (2005). The transcriptional regulator SsuR activates expression of the *Corynebacterium glutamicum* sulphonate utilization genes in the absence of sulphate. Mol. Microbiol. *58*, 480-494.

Krings, E., Krumbach, K., Bathe, B., Kelle, R., Wendisch, V.F., Sahm, H., and Eggeling, L. (2006). Characterization of myo-inositol utilization by *Corynebacterium glutamicum*: the stimulon, identification of transporters, and influence on L-lysine formation. J. Bacteriol. *188*, 8054-8061.

Krömer, J.O., Sorgenfrei, O., Klopprogge, K., Heinzle, E., and Wittmann, C. (2004). In-depth profiling of lysine-producing *Corynebacterium glutamicum* by combined analysis of the transcriptome, metabolome, and fluxome. J. Bacteriol. *186*, 1769-1784.

Krug, A., Wendisch, V.F., and Bott, M. (2005). Identification of AcnR, a TetR-type repressor of the aconitase gene *acn* in *Corynebacterium glutamicum*. J. Biol. Chem. *280*, 585-595.

Lange, C., Rittmann, D., Wendisch, V.F., Bott, M., and Sahm, H. (2003). Global expression profiling and physiological characterization of *Corynebacterium glutamicum* grown in the presence of L-valine. Appl Environ. Microbiol. *69*, 2521-2532.

Larisch, C., Nakunst, D., Hüser, A.T., Tauch, A., and Kalinowski, J. (2007). The alternative sigma factor SigB of *Corynebacterium glutamicum* modulates global gene expression during transition from exponential growth to stationary phase. BMC Genomics *8*, 4.

Magasanik, B. (2000). Global regulation of gene expression. Proc. Natl. Acad. Sci. USA. *97*, 14044-14045.

Marx, A., Hans, S., Möckel, B., Bathe, B., and de Graaf, A.A. (2003). Metabolic phenotype of phosphoglucose isomerase mutants of *Corynebacterium glutamicum*. J. Biotechnol. *104*, 185-197.

Mitsuhashi, S., Hayashi, M., Ohnishi, J., and Ikeda, M. (2006). Disruption of malate: quinone oxidoreductase increases L-lysine production by *Corynebacterium glutamicum*. Biosci Biotechnol. Biochem. *70*, 2803-2806.

Möker, N., Brocker, M., Schaffer, S., Krämer, R., Morbach, S., and Bott, M. (2004). Deletion of the genes encoding the MtrA-MtrB two-component system of *Corynebacterium glutamicum* has a strong influence on cell morphology, antibiotics susceptibility and expression of genes involved in osmoprotection. Mol. Microbiol. *54*, 420-438.

Nakamura, J., Hirano, S., Ito, H., and Wachi, M. (2007). Mutations of the *Corynebacterium glutamicum* NCgl1221 gene, encoding a mechanosensitive channel homolog, induce L-glutamic acid production. Appl Environ. Microbiol. *73*, 4491-4498.

Nakunst, D., Larisch, C., Hüser, A.T., Tauch, A., Pühler, A., and Kalinowski, J. (2007). The extracytoplasmic function-type sigma factor SigM of *Corynebacterium glutamicum* ATCC 13032 is involved in transcription of disulfide stress-related genes. J. Bacteriol. *189*, 4696-4707.

Netzer, R., Krause, M., Rittmann, D., Peters-Wendisch, P.G., Eggeling, L., Wendisch, V.F., and Sahm, H. (2004a). Roles of pyruvate kinase and malic enzyme in *Corynebacterium glutamicum* for growth on carbon sources requiring gluconeogenesis. Arch. Microbiol. *182*, 354 -363.

Netzer, R., Peters-Wendisch, P., Eggeling, L., and Sahm, H. (2004b). Cometabolism of a nongrowth substrate: L-serine utilization by *Corynebacterium glutamicum*. Appl Environ. Microbiol. *70*, 7148-7155.

Niebisch, A., Kabus, A., Schultz, C., Weil, B., and Bott, M. (2006). Corynebacterial protein kinase G controls 2-oxoglutarate dehydrogenase activity via the phosphorylation status of the OdhI protein. J. Biol. Chem. *281*, 12300-12307.

Nottebrock, D., Meyer, U., Krämer, R., and Morbach, S. (2003). Molecular and biochemical characterization of mechanosensitive channels in *Corynebacterium glutamicum*. FEMS Microbiol. Lett. *218*, 305-309.

Ohnishi, J., Katahira, R., Mitsuhashi, S., Kakita, S., and Ikeda, M. (2005). A novel *gnd* mutation leading to increased L-lysine production in *Corynebacterium glutamicum*. FEMS Microbiol. Lett. *242*, 265-274.

Ohnishi, J., Mitsuhashi, S., Hayashi, M., Ando, S., Yokoi, H., Ochiai, K., and Ikeda, M. (2002). A novel methodology employing *Corynebacterium glutamicum* genome information to generate a new L-lysine-producing mutant. Appl Microbiol. Biotechnol. *58*, 217-223.

Okino, S., Inui, M., and Yukawa, H. (2005). Production of organic acids by *Corynebacterium glutamicum* under oxygen deprivation. Appl Microbiol. Biotechnol. *68*, 475-480.

Pallerla, S.R., Knebel, S., Polen, T., Klauth, P., Hollender, J., Wendisch, V.F., and Schoberth, S.M. (2005). Formation of volutin granules in *Corynebacterium glutamicum*. FEMS Microbiol. Lett. *243*, 133-140.

Patek, M. (2007). Branched-Chain Amino Acids. *In* Amino Acid Biosynthesis – Pathways, Regulation and Metabolic Engineering (Wendisch VF, ed), Springer, Heidelberg, Germany.

Peters-Wendisch, P.G., Kreutzer, C., Kalinowski, J., Patek, M., Sahm, H., and Eikmanns, B.J. (1998). Pyruvate carboxylase from *Corynebacterium glutamicum*: characterization, expression and inactivation of the *pyc* gene. Microbiology *144*, 915-927.

Peters-Wendisch, P.G., Schiel, B., Wendisch, V.F., Katsoulidis, E., Möckel, B., Sahm, H., and Eikmanns, B.J. (2001). Pyruvate carboxylase is a major bottleneck for glutamate and lysine production by *Corynebacterium glutamicum*. J. Mol. Microbiol. Biotechnol. *3*, 295-300.

Polen, T., and Wendisch, V.F. (2004). Genomewide expression analysis in amino acid-producing bacteria using DNA microarrays. Appl Biochem. Biotechnol. *118*, 215-232.

Polen, T., Schluesener, D., Poetsch, A., Bott, M., and Wendisch, V.F. (2007). Characterization of citrate utilization in *Corynebacterium glutamicum* by transcriptome and proteome analysis. FEMS Microbiol. Lett. *273*, 109-119.

Radmacher, E., Vaitsikova, A., Burger, U., Krumbach, K., Sahm, H., and Eggeling, L. (2002). Linking central metabolism with increased pathway flux: L-valine accumulation by *Corynebacterium glutamicum*. Appl Environ. Microbiol. *68*, 2246-2250.

Radmacher, E., Stansen, K.C., Besra, G.S., Alderwick, L.J., Maughan, W.N., Hollweg, G., Sahm, H., Wendisch, V.F., and Eggeling, L. (2005). Ethambutol, a cell wall inhibitor of *Mycobacterium tuberculosis*, elicits L-glutamate efflux of *Corynebacterium glutamicum*. Microbiology *151*, 1359-1368.

Rey, D.A., Nentwich, S.S., Koch, D.J., Rückert, C., Pühler, A., Tauch, A., and Kalinowski, J. (2005). The McbR repressor modulated by the effector substance S-adenosylhomocysteine controls directly the transcription of a regulon involved in sulphur metabolism of *Corynebacterium glutamicum* ATCC 13032. Mol. Microbiol. *56*, 871-887.

Rhodius, V., Van Dyk, T.K., Gross, C., and LaRossa, R.A. (2002). Impact of genomic technologies on studies of bacterial gene expression. Annu. Rev. Microbiol. *56*, 599-624.

Riedel, C., Rittmann, D., Dangel, P., Möckel, B., Petersen, S., Sahm, H., and Eikmanns, B.J. (2001). Characterization of the phosphoenolpyruvate carboxykinase gene from *Corynebacterium glutamicum* and significance of the enzyme for growth and amino acid production. J. Mol. Microbiol. Biotechnol. *3*, 573-583.

Rittmann, D., Schaffer, S., Wendisch, V.F., and Sahm, H. (2003). Fructose-1,6-bisphosphatase from *Corynebacterium glutamicum*: expression and deletion of the *fbp* gene and biochemical characterization of the enzyme. Arch. Microbiol. *180*, 285-292.

Rittmann, D., Sorger-Herrmann, U., and Wendisch, V.F. (2005). Phosphate starvation-inducible gene *ushA* encodes a 5′ nucleotidase required for growth of *Corynebacterium glutamicum* on media with nucleotides as the phosphorus source. Appl Environ. Microbiol. *71*, 4339-4344.

Ruklisha, M., Paegle, L., and Denina, I. (2007). L-Valine biosynthesis during batch and fed-batch cultivations of *Corynebacterium glutamicum*: Relationship between changes in bacterial growth rate and intracellular metabolism. Process Biochemistry *42*, 634.

Sahm, H., and Eggeling, L. (1999). D-Pantothenate synthesis in *Corynebacterium glutamicum* and use of *panBC* and genes encoding L-valine synthesis for D-pantothenate overproduction. Appl Environ. Microbiol. *65*, 1973-1979.

Schaaf, S., and Bott, M. (2007). Target genes and DNA-binding sites of the response regulator PhoR from *Corynebacterium glutamicum*. J. Bacteriol. *189*, 5002-5011.

Schrumpf, B., Eggeling, L., and Sahm, H. (1992). Isolation and prominent characteristics of an L-lysine hyperproducing strain of *Corynebacterium glutamicum*. Appl Microbiol. Biotechnol. *37*, 566-571.

Schultz, C., Niebisch, A., Gebel, L., and Bott, M. (2007). Glutamate production by *Corynebacterium glutamicum*: dependence on the oxoglutarate dehydrogenase inhibitor protein OdhI and protein kinase PknG. Appl Microbiol. Biotechnol, *in press*.

Seletzky, J. M., Noack, U., Fricke, J., Hahn, S., and Büchs, J. (2006). Metabolic activity of *Corynebacterium glutamicum* grown on L-lactic acid under stress. Appl Microbiol. Biotechnol. *72*, 1297-1307.

Shimizu, H., and Hirasawa, T. (2007). Production of Glutamate and Glutamate-Related Amino Acids: Molecular Mechanism Analysis and Metabolic Engineering. *In* Amino Acid Biosynthesis – Pathways, Regulation and Metabolic Engineering (Wendisch VF, ed), Springer, Heidelberg, Germany, DOI 10.1007/7171_2006_1064.

Siewe, R.M., Weil, B., Burkovski, A., Eikmanns, B.J., Eikmanns, M., and Krämer, R. (1996). Functional and genetic characterization of the (methyl)ammonium uptake carrier of *Corynebacterium glutamicum*. J. Biol. Chem. *271*, 5398-5403.

Silberbach, M., Hüser, A., Kalinowski, J., Pühler, A., Walter, B., Krämer, R., and Burkovski, A. (2005a). DNA microarray analysis of the nitrogen starvation response of *Corynebacterium glutamicum*. J. Biotechnol. *119*, 357-367.

Silberbach, M., Schäfer, M., Hüser, A.T., Kalinowski, J., Pühler, A., Krämer, R., and Burkovski, A. (2005b). Adaptation of *Corynebacterium glutamicum* to ammonium limitation: a global analysis using transcriptome and proteome techniques. Appl Environ. Microbiol. *71*, 2391-2402.

Simic, P., Willuhn, J., Sahm, H., and Eggeling, L. (2002). Identification of *glyA* (encoding serine hydroxymethyltransferase) and its use together with the exporter ThrE to increase L-threonine accumulation by *Corynebacterium glutamicum.* Appl Environ. Microbiol. *68*, 3321-3327.

Sindelar, G., and Wendisch, V.F. (2007). Improving lysine production by *Corynebacterium glutamicum* through DNA microarray-based identification of novel target genes. Appl Microbiol. Biotechnol, *76,* 677-689

Stansen, C., Uy, D., Delaunay, S., Eggeling, L., Goergen, J.L., and Wendisch, V.F. (2005). Characterization of a *Corynebacterium glutamicum* lactate utilization operon induced during temperature-triggered glutamate production. Appl Environ. Microbiol. *71*, 5920-5928.

Tauch, A., Hermann, T., Burkovski, A., Krämer, R., Pühler, A., and Kalinowski, J. (1998). Isoleucine uptake in *Corynebacterium glutamicum* ATCC 13032 is directed by the *brnQ* gene product. Arch. Microbiol. *169*, 303-312.

Tauch, A., Wehmeier, L., Götker, S., Pühler, A., and Kalinowski, J. (2001). Relaxed *rrn* expression and amino acid requirement of a *Corynebacterium glutamicum rel* mutant defective in (p)ppGpp metabolism. FEMS Microbiol. Lett. *201*, 53-58.

Thierbach, G., Kalinowski, J., Bachmann, B., and Pühler, A. (1990). Cloning of a DNA fragment from *Corynebacterium glutamicum* conferring aminoethyl cysteine resistance and feedback resistance to aspartokinase. Appl Microbiol. Biotechnol. *32*, 443-448.

Tricot, C., Stalon, V., and Legrain, C. (1991). Isolation and characterization of *Pseudomonas putida* mutants affected in arginine, ornithine and citrulline catabolism: function of the arginine oxidase and arginine succinyltransferase pathways. J. Gen. Microbiol. *137*, 2911-2918.

Uy, D., Delaunay, S., Germain, P., Engasser, J.M., and Goergen, J.L. (2003). Instability of glutamate production by *Corynebacterium glutamicum* 2262 in continuous culture using the temperature-triggered process. J. Biotechnol. *104*, 173-184.

Wehmeier, L., Schäfer, A., Burkovski, A., Krämer, R., Mechold, U., Malke, H., Pühler, A., and Kalinowski, J. (1998). The role of the *Corynebacterium glutamicum rel* gene in (p)ppGpp metabolism. Microbiology *144*, 1853-1862.

Wendisch, V.F. (2003). Genome-wide expression analysis in *Corynebacterium glutamicum* using DNA microarrays. J. Biotechnol. *104*, 273-285.

Wendisch, V.F. (2006). Genetic regulation of *Corynebacterium glutamicum* metabolism. J. Microbiol. Biotechnol. *16*, 999.

Wennerhold, J., and Bott, M. (2006). The DtxR regulon of *Corynebacterium glutamicum.* J. Bacteriol. *188*, 2907-2918.

Wennerhold, J., Krug, A., and Bott, M. (2005). The AraC-type regulator RipA represses aconitase and other iron proteins from *Corynebacterium* under iron limitation and is itself repressed by DtxR. J. Biol. Chem. *280*, 40500-40508.

Wittmann, C., and Becker, J. (2007). The L-Lysine Story: From Metabolic Pathways to Industrial Production. *In* Amino Acid Biosynthesis – Pathways, Regulation and Metabolic Engineering (Wendisch VF, ed), Springer, Heidelberg, Germany.

Yukawa, H., Omumasaba, C.A., Nonaka, H., Kos, P., Okai, N., Suzuki, N., Suda, M., Tsuge, Y., Watanabe, J., Ikeda, Y., et al. (2007). Comparative analysis of the *Corynebacterium glutamicum* group and complete genome sequence of strain R. Microbiology *153*, 1042-1058.

Zimmer, D.P., Soupene, E., Lee, H.L., Wendisch, V.F., Khodursky, A.B., Peter, B.J., Bender, R.A., and Kustu, S. (2000). Nitrogen regulatory protein C-controlled genes of *Escherichia coli*: scavenging as a defense against nitrogen limitation. Proc. Natl. Acad. Sci. USA. *97*, 14674-14679.

From: Corynebacteria: Genomics and Molecular Biology. Edited by: Andreas Burkovski.

Proteomics of *Corynebacterium glutamicum* and Other Corynebacteria

4

Jörn Kalinowski, Dirk Wolters and Ansgar Poetsch

Abstract

Corynebacteria were subject to several proteomics studies in the past, either aiming to describe comprehensive cytosolic, extracellular and membrane proteomes or comparing proteomes of different species, wildtype and mutant strains, respectively. In these surveys, a variety of techniques were applied, one- and two-dimensional gel electrophoresis as well as liquid chromatography (LC) separations. Identification of proteins was done mostly by protease digestion and mass-spectrometry, employing MALDI or ESI mass spectrometers and subsequent comparison of acquired mass spectra with databases. In the next years, automated separation and online identification by LC-MALDI or LC-ESI will progressively replace the gel-based methods, ultimately leading to real quantitative data.

While mostly hydrophilic proteins can be analyzed comprehensively with routine techniques, separation and identification of membrane proteins is more complex and was investigated in detail in a number of recent studies. This chapter will first review the initial work on hydrophilic proteins and then focus on membrane proteomics as the novel and fast proceeding field of research.

1. Introduction

The term 'proteome' is defined as the complement of all proteins expressed in a cell, a tissue or an organism, at a certain time point and under defined conditions. Knowledge on the expressed proteins, their locations and their abundances in a specific situation delivers important information on how the genomic blueprint is translated into cellular functions. Although proteome analysis techniques (proteomics) such as two-dimensional gel electrophoresis (2-DE) were developed long before (O'Farrell, 1975; Klose, 1975), they became really important in the context of whole genome DNA sequences.

2. Features of the *C. glutamicum* proteome

From the *Corynebacterium glutamicum* genome sequences determined so far, around 3000 protein-coding regions have been predicted (Kalinowski et al., 2003). These proteins can be grouped into three fractions according to bioinformatic analyses. The secreted protein fraction is characterized by predicted secretion signal sequences, the membrane fraction by the presence of transmembrane helices and those proteins that have neither of these probably belong to the cytosolic fraction. With around 2000 proteins the cytosolic fraction is the largest. The membrane fraction comprises 650 proteins and the secreted fraction 311 proteins (Kalinowski et al., 2003).

Beside this, also physical parameters of proteins that are used to separate them in 2-DE, such as molecular mass and isoelectric point, can be calculated by *in silico*. *C. glutamicum* proteins cover a wide range of masses, ranging from 30 amino acids (3.7 kDa) for the leader peptide of the tryptophan operon (TrpL) to 2996 amino acids (316.7 kDa) for the fatty acid synthase IA (Fas-IA). The predicted isoelectric points (the pH at which the proteins have no net charge) range from 3.48 (Cg1238, conserved protein of unknown function) to 13.12 (RpmH, 50S ribosomal protein L34).

3. Comparison of technologies for proteomics

In general proteome analysis is a two step procedure. In the first step the proteins are separated by gel-based or liquid chromatography-based techniques. The gel-based systems use one-dimensional (1-D) or 2-D SDS-polyacrylamide gel electrophoresis (PAGE) and separate proteins according to their masses or charges. Whereas the sieving effect of the SDS-polyacrylamide gel serves to separate proteins according to molecular mass, separation with respect to charge is achieved by pH gradients immobilized in polyacrylamide gels. In an electrophoresis using these pH gradients, the first dimension, proteins move to a point where they have no net charge (isoelectric point). The pH gradient gels in which the proteome is focussed were put on top of the SDS-PAGE, and were separated by another electrophoresis in the second dimension. For imaging after electrophoresis, the protein gels were stained. Stains were either Coomassie Brilliant Blue, silver salts or fluorescent dyes. A very interesting new technique that avoids staining is the direct imaging of proteins due to their fluoresence in the UV range of the spectrum after excitation by laser (Rögener et al., 2003). Figure 4.1 shows a typical 2-D SDS-PAGE image of cytosolic *C. glutamicum* proteins separated in a pH range from 4-7 and stained with Coomassie blue.

In a second step, the proteins are identified. This is generally achieved by a protease fingerprint, e.g. with trypsin. The sequence-specific protease cleaves a protein into a defined number of peptides that can be predicted from its amino acid sequence. Since complete genomes also deliver all possible protein sequences, determination of the exact masses of these proteolytic peptides by mass spectrometry (e.g. MALDI-TOF mass spectrometry) and comparison with the theoretical peptide patterns in databases allows to identify proteins in short time.

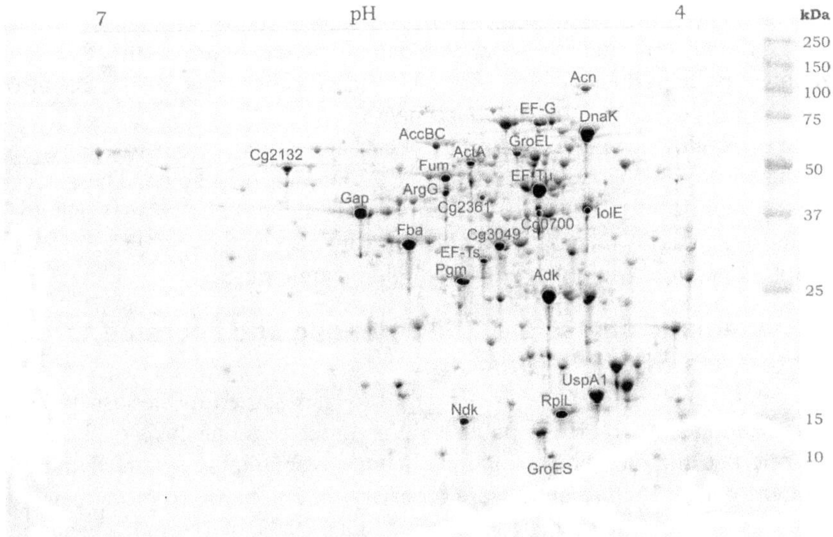

Figure 4.1. Image of a 2-D gel loaded with the cytosolic proteome of *C. glutamicum*. Separation in the first dimension was achieved by an immobilised pH gradient ranging from pH 4-7, separation in the second dimension was achieved by SDS-PAGE. Staining was with Coomassie Brilliant blue. On the right side, the molecular masses of the protein standard are given in kilo Daltons (kDa). Protein spots analyzed by tryptic fingerprinting are marked by white ellipses and abbreviated protein names are given. Abbreviations: AccBC, acetyl-CoA carboxylase subunit; Acn, aconitase; ActA, acyl-CoA transferase; Adk, adenylate kinase; ArgG, argininosuccinate synthase; Cg0700, putative inositol-monophosphate dehydrogenase, Cg2361, putative cell division protein; Cg3049, putative ferredoxin/ferredoxin-NADP reductase; DnaK, chaperone; EF-G, EF-Ts, EF-Tu, Translation elongation factors; Fba, fructose-bisphosphate aldolase; Fum, fumarase; Gap, glyceraldehyde-3-phosphate dehydrogenase; GroES, chaperonin; IolE, putative sugar-phosphate isomerase; Ndk, nucleoside diphosphate kinase; Pgm, phosphoglycerate mutase; RplL, 50S ribosomal protein L7/L12; UspA1, universal stress protein.

Alternatively, protease cleavage can be applied to complex protein samples and separation of peptides is achieved by liquid-chromatography. This allows to automatize the process further and makes it possible to directly couple separation and analysis by mass spectrometry. Different procedures have been applied to *C. glutamicum* proteomes, MudPIT (Washburn et al., 2001) and LC-MALDI (Brand et al., 2005).

Gel-based approaches are well-established and their limitations are well-known. A typical 2-D gel is capable of separating proteins with masses between 10 and 100 kDa. The immobilized pH gradients theoretically allow to separate all proteins well, but have practical limitations with respect to the more basic proteins. In addition, it is almost impossible to analyze more hydrophobic (membrane) proteins with this procedure.

Staining is another critical point. Silver stains generally interfere with mass spectrometry but are more sensitive than Coomassie blue stains. Fluorescent stains combine high sensitivity and compatibility with mass spectrometry, but are considerably more expensive. The technique of direct visualization of proteins due to their intrinsic UV fluorescence is promising, but also not commercially available at the moment.

For high-throughput proteome analysis, liquid-chromatography in conjunction with mass spectrometry is the method of choice. These techniques avoid some limitations caused by very large, basic or hydrophobic proteins, simply by analyzing peptides and are also better suited for quantification of peptides (proteins) within a high dynamic range.

4. Systematic analysis of the cytosolic and secreted proteomes of *C. glutamicum*

Systematic analysis of *C. glutamicum* proteomes started in the last decade, when Hermann et al. (1998) published the first 2-D gel analysis of the cytoplasmic and the membrane fractions. After silver-staining several hundred protein spots in each fraction were observed. Additionally, 10 proteins were identified by N-terminal microsequencing, but the lack of a complete genome sequence prohibited identification of proteins in a larger scale at that time. In two consecutive studies (Hermann et al., 2001; Schaffer et al., 2001) proteins were analyzed by trypsin digestion and mass spectrometry. With the help of the first genomic sequences that became available at that time, more than 150 different proteins were identified among the soluble proteins and 13 different in the membrane fraction. It was first noted that several proteins gave rise to more than one spot and also differences in predicted and observed mass and isoelectric points were recorded (Schaffer et al., 2001) suggesting possible post-translational modifications. The most recent study on systematic identification of soluble proteins in *C. glutamicum* and in its close relative *Corynebacterium efficiens* (Hansmeier et al., 2006b) listed 265 different proteins, 211 from the cytosol and 54 secreted ones. Among the 2-D gel based studies this is currently the highest number, but did not come close to the estimated number of around 1500 expressed genes in *C. glutamicum* (J. Kalinowski, unpublished results). A full compilation of *C. glutamicum* proteins is only possible by liquid-chromatography-based techniques that yielded many more identifications (LC-MALDI: 445 proteins, Brand et al., 2005; MudPIT: >1000, Fischer et al., 2006).

5. Protein modification in *C. glutamicum*

Organisms display a high number of different post-translational modifications in proteins during their maturation. These modifications can either arise by proteolytic processing or by covalent modification. Left aside the formation of disulfide bonds which allow correct folding and modifications by cofactors (vitamins) that are essential for function of specific enzymes, covalent modifications can be acetylations, glycosylations, hydroxylations, methylations, nucleotidylations, phosphorylations, and ADP-ribosylations, to name the most abundant. In *C. glutamicum* especially two different modifications of proteins have been analyzed by proteomics methods, phosphorylation and glycosylation.

5.1 Phosphorylation

In a pioneering work, Bendt et al. (2001) analyzed global protein phosphorylation in *C. glutamicum* in two different ways: by immunoblotting of 2-D gels of the cytoplasmic proteome with phosphoserine-specific antibodies and by radiolabelling of proteins with phosphoric acid (^{33}P). With antibodies, 90 spots were detected, whereas the radioactive method yielded 60 signals. This indicates that protein phosphorylation is an abundant phenomenon in *C. glutamicum*. Altogether 31 proteins were detected with both methods, most of them represent enzymes of the central carbon metabolism, chaperones and other proteins that are abundant. It is expected that this is only the 'tip of the iceberg' since phosphorylated proteins of low abundance or from other compartments of the cell escaped this analyses.

In a recent study (Krüger et al., 2006), protein phosphorylation stoichiometry in *C. glutamicum* was assessed by two analytical strategies, element mass spectrometry and simultaneous monitoring of ^{31}P isotopes. It was calculated that the cytoplasmatic proteome contains approximately 0.01 mol of phosphorous per mol of protein, indicating that only a smaller fraction of *C. glutamicum* proteins is indeed phosphorylated.

At the moment, it is not possible to resolve this apparent disagreement, since the phosphorylation of none of the proteins identified in the study by Bendt et al. (2001) has yet been proven in an individual case. However, phosphorylation has indeed been shown in case of the *C. glutamicum* OdhI protein that is part of the 2-oxoglutarate dehydrogenase complex (Niebisch et al., 2006). The unphosphorylated OdhI protein strongly inhibits this tricarboxylic acid cycle enzyme and leads to a lack in glutamine utilization. This mechanism is a regulatory principle that was unknown before and more surprises might await their disclosure on this field of research.

5.2 Glycosylation

For a long time, protein glycosylation has been considered to be a specific feature of eukaryotes. However, in the last years more and more examples of bacterial protein glycosylation systems have been discovered. The first glycosylated protein found in *C. glutamicum* was the secreted Rpf2 protein (Hartmann et al., 2004). Rpf2 is similar to the essential resuscitation-promoting factor (Rpf) of *Micrococcus luteus*. The Rpf2 protein was purified from concentrated cell supernatants and modified Rpf2 variants with reduced mobility when compared with the calculated size of Rpf2 in SDS-gels were found. A Western blot-based enzyme immunoassay demonstrated glycosylation of these Rpf2 variants leading to higher molecular masses. Galactose and mannose were identified as two components of the oligosaccharide portion of the Rpf2 glycoprotein by capillary gas chromatography coupled to mass spectrometry.

A later study by Mahne et al. (2006) disvovered the mechanism by which *C. glutamicum* is able to glycosylate secreted proteins as O-glycosylation occurring at serine or threonine residues. First they showed by 2-D gel electrophoresis and immunoassays of the secretory subproteome of *C. glutami-*

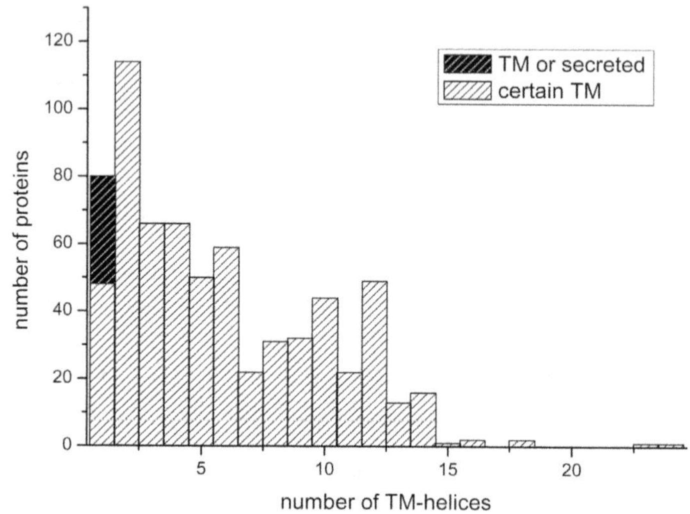

Figure 4.2. Number of proteins vs. number of TM helices predicted with TMHMM from the *C. glutamicum* genome. Proteins that may either be secreted or membrane proteins are represented as black shaded bars.

cum that at least three secreted proteins are glycosylated. Beside Rpf2 these are a putative lipoprotein LppS and a protein of hitherto unknown function (Cg1849). In the same study, the *C. glutamicum* gene *cg1014* was found to encode a protein similar to eukaryotic protein-O-mannosyltransferases (PMTs). Deletion of the gene *cg1014* (*pmt*) caused a complete loss of glycosylation of secreted proteins including Rpf2. Because the *pmt* gene as well as *rpf* genes are present in the genomes of all actinobacteria, this work provided new opportunities to elucidate the molecular mechanisms of bacterial protein glycosylation and of Rpf activity also in pathogenic growth and infection. In this respect, the fast-growing and non-pathogenic *C. glutamicum* can be regarded as a model system for bacteria of the *Mycobacterium tuberculosis* complex.

6. Membrane proteomics with *C. glutamicum*

6.1 Features of the *C. glutamicum* membrane proteome

According to the TMHMM v2.0 predictor (Krogh et al., 2001), *C. glutamicum* contains about 650 genes coding for integral membrane proteins (IMPs), which is about 21% of the total genome. Figure 4.2 displays the number of proteins plotted against the number of predicted transmembrane (TM) helices. While the results from state-of-the-art transmembrane helices prediction software like TMHMM can be overall considered as very accurate (Chen et al., 2002), it is not always possible to discriminate signal peptides from TM helices. On the basis of this fact, proteins with only one predicted TM helix that could not be unequivocally allocated to the secreted or TM proteins are represented as black bars. It can be seen that for 2 TM helices the number of proteins is highest and that the majority of membrane proteins contain less than 10 TM helices. However, a noticeable number of membrane proteins possess 12 TM helices – these proteins are often parts of ABC-transporters. Two proteins, the predicted subunits of a Na^+/H^+-antiporter complex possess the impressive number of 23 or 24 TM helices, respectively. Grouping of membrane proteins into functional categories was done during annotation of the first genome sequence (Ikeda and Nakagawa, 2003; Kalinowski et al., 2003). The largest fraction is that of hypothetical proteins; together with proteins of unknown function, these are slightly more than 50% of all predicted membrane proteins. The next biggest classes are transport and binding proteins (~18%), unclassified (~7%), and cell envelope proteins (~6%). Energy metabolism, protein fate, and cellular processes each comprise 3-5% of all membrane proteins, while the other categories contain only 2% or less. Although for some membrane proteins the functions could be experimentally clarified, it still must be said that for the majority of membrane proteins the function is not known or only deduced from sequence analysis. In comparison to cytosolic proteins, the number of proteins without functional evidence is higher, most likely due to the experimental difficulties

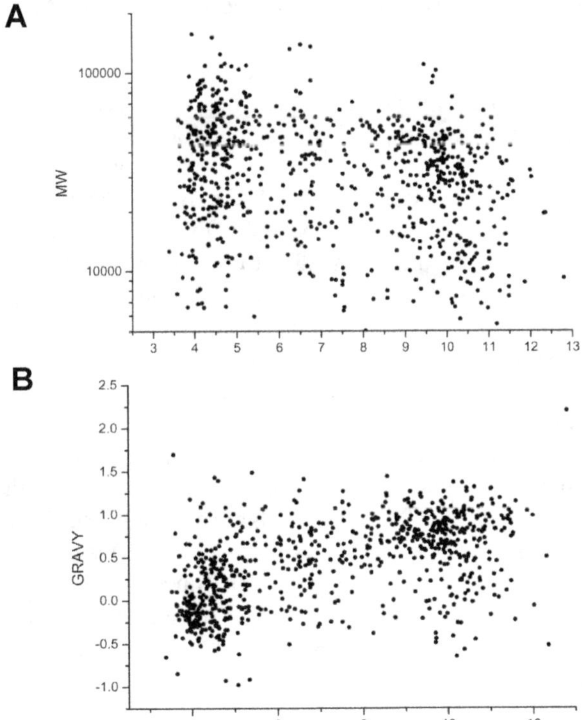

Figure 4.3. The pI, MW, and GRAVY were calculated *in silico* for the predicted *C. glutamicum* membrane proteins and visualized as virtual 2-D gel maps. a) virtual 2-D gel for C. *glutamicum* membrane proteins. b) GRAVY-socre vs. pI for the *C. glutamicum* membrane proteins.

when dealing with membrane proteins. Based on the TMHMM predictions for integral membrane proteins, the complete *C. glutamicum* membrane proteome can be visualized in a virtual 2-D gel (Figure 4.3). The numbers of proteins in the acidic and basic pH-range are rather equal, while two clusters with centers at pH 4 and 10 exist. Thus, in comparison to the soluble proteome (Hermann et al., 2001), the basic pH-region is more heavily populated. The most acidic membrane protein is Cg2054 (pI 3.57), the most basic is Cg3416 (pI 12.8); for both the function is unknown. Concerning protein molecular weight, the two extremes are Cg1210 (2.7 kDa), which is mostly a single TM helix, and Cg2032 (151 kDa); whereas their function is still unclear, too. A common protein attribute used in membrane proteomics is the GRAVY-score (Grand Average HydropathY; Kyte and Doolittle, 1982). While the mean GRAVY-score is 0.43 for *C. glutamicum* membrane proteins, it can be seen from Figure 4.3 that acidic membrane proteins tend to be more hydrophilic than basic ones. A correlation between protein size and GRAVY can be concluded by an average GRAVY of 0.53 calculated for proteins < 20 kDa, while for proteins > 70 kDa the average GRAVY is 0.15 (Fischer and Poetsch, 2006).

6.2 Requirements and obstacles for bacterial membrane proteomics

The ultimate goal in membrane proteomics is to separate, identify, and quantify all integral membrane proteins present under a certain condition. As laid out in the previous paragraph, the challenge for *C. glutamicum* in particular is to fulfill this task for a membrane proteome that covers a pI range from 3.6 to 12.8, a MW range from 2.7 to 151 kDa and that contains some highly hydrophobic proteins. Since the space occupied by the cytosol is much larger than the cell envelope, the numbers of cytosolic proteins by far outweigh integral membrane proteins; we estimate that IMPs constitute less than 5% of the total cellular protein amount. Prefractionation techniques like membrane isolation and washing have prevailed to enrich the membrane fraction and/or the integral membrane proteins. In actinomycetes, the plasma membrane is tightly linked to the cell wall via layers of peptidoglycans, arabinogalactans and mycolic acids (Eggeling and Sahm, 2001); although it has been shown that it is possible to separate the *C. glutamicum* cell wall from the plasma membrane (Lichtinger et al., 1998), the yields are quite low and *C. glutamicum* membrane fractions usually analyzed with proteomics consist of the plasma membrane, periplasmic, and cell wall proteins. Since the fundamental work of Hoischen and Krämer (1990), it is well recognized that the lipid composition of *C. glutamicum* cell membranes is variable, which has consequences for membrane proteome sample preparation. If proteomics technologies are applied that require the solubilization of membrane proteins for further separation (e.g. 2-D electrophoresis or IEC/SDS-PAGE), the solubilization efficiency of certain membrane proteins can be adversely affected by a change in the lipid environment. Therefore, a thorough screening of different detergents and solubilization conditions is mandatory to achieve optimal recovery of all membrane proteins in this step. The amino acid composition and distribution in membrane proteins, especially of the trypsin cleavage sites lysine and arginine, hampers protein identification in proteomics; these two amino acids are less abundant in integral membrane proteins and almost absent in transmembrane helices. Although it is possible to analyze and identify intact membrane proteins (Whitelegge et al., 2006), in current state-of-the-art high-throughput analysis, the proteins are digested with proteases and their corresponding peptide fragments are used for identification. Most often trypsin is the protease of choice for protein digestion, since it is highly specific (Olsen et al., 2004) and since tryptic peptides usually produce the best MS and MS/MS spectra both in MALDI and ESI ionization. However, we have carried out a theoretical study that unfortunately showed other, detrimental effects of tryptic digestion on the analysis of bacterial integral membrane proteins (Fischer and Poetsch, 2006). *In silico* digestions of the *C. glutamicum* membrane proteome revealed that the mean sequence coverage in the mass range of 600-4000 Da (which is a reasonable assumption for the mass range covered by common MALDI or ESI instruments) is only about 55% for IMPs, whereas it is about 70% for soluble proteins when simulating trypsin proteolysis. Upon closer inspection, it was found that a significant fraction of the TM regions escape detection and membrane protein identification is mainly based on the proteolytic fragments of hydrophilic protein domains, such as TM-helices connecting loops. Another consequence of the lower number of suitable peptides per membrane protein is that small membrane proteins are more difficult to detect and to identify than soluble proteins of the same size. If high membrane protein sequence coverage or the identification of small membrane proteins is required, alternative protein cleavage conditions are advised. In our hands, the sequential digestion with trypsin and cyanogen bromide (Schlüsener et al., 2005) or a combination of trypsin and chymotrypsin (Fischer et al., 2006) gave the best results. The former digestion is more specific, but more time consuming and safety precautions must be taken when working with cyanogen bromide, while the latter can be performed faster and easier, but successful protein identification is often more difficult to achieve (Fischer and Poetsch, 2006).

6.3 Comparison of technologies for membrane proteomics

Sample preparation requires for membrane proteins not only the disruption of covalent and noncovalent protein complexes as for soluble proteins, but also protein extraction from the lipid bilayer and saturation of hydrophobic protein domains with suitable detergents. Although harsh, ionic detergents (e.g. SDS) achieve the best IMP solubilization, they are incompatible with 2-DE at high concentrations and must be diluted prior to IEF (Fanous et al., 2007). We have undertaken a solubilization screening for different nonionic and zwitterionic detergents with a *C. glutamicum* L-lysine producing strain (Schlüsener et al., 2005): With the zwitterionic detergent ASB-14 88% of the proteins in the membrane fraction were solubilized, with SB3-10 81% were solubilized, and with the nonionic detergent N-dodecyl β-D-maltoside 76% were solubilized. With the zwitterionic detergent CHAPS, only 52% of the proteins were retrieved in the supernatant, while with the reference detergent SDS 90% were retrieved. CHAPS is often the detergent of choice for 2-D PAGE, because it delivers high quality gels with only a minimum of band streaking and has been applied to *C. glutamicum* membranes before (Fanous et al., 2007; Schaffer et al., 2001). Nevertheless, new detergents have been sought for improved solubility and detection of membrane proteins on 2-D gels. The group of Rabilloud has synthesized a variety of 2-D compatible detergents (Chevallet et al., 1998) and demonstrated an improved detection of some membrane proteins (Tastet et al., 2003). These new surfactants have not been tested with *C. glutamicum* membranes yet, but may yield results improved over ASB-14 for strains or growth conditions differing from our experiments with the L-lysine producer. It should be mentioned that there have been attempts to predict detergent efficiency (Tastet et al., 2003) and membrane protein solubility (Santoni et al., 2000a), yet so far the best approach is still the elaborate testing of various surfactants for every membrane sample. Furthermore, it must be said that in general the application of 2-D PAGE for membrane proteins has not been very successful (for an overview see Santoni et al., 2000b). It is evident that the IEF-separation is problematic for membrane proteins, since it leads to aggregation and precipitation of the proteins during focusing, which later precludes their transfer into the SDS gel (see Klein et al., 2005, for experimental evidence and a thorough analysis of this problem).

Detection of membrane proteins on 2-D gels appears to be limited to proteins with a low GRAVY or a low number of TM domains; Santoni et al. (2000b) suggest the number of TM helices / kDa as a score to predict the amenability of an IMP for 2-D PAGE, proteins with a score above 0.25 are undetectable on 2-D gels. Fortunately, the situation is much less severe for β-barrel membrane proteins, which makes 2-D PAGE an attractive alternative for cell wall proteome analysis, but such a targeted study using this technique has only been published for the related *Mycobacterium tuberculosis* (Ragas et al., 2007). A 2-D gel electrophoresis technique which avoids IEF-separation and its problems is the 16 BAC/SDS-PAGE (2-D BAC; Hartinger et al., 1996). It has been demonstrated that this kind of electrophoresis is well-suited for hydrophobic membrane proteins, among them even proteins with 12 TMHs and a GRAVY-score up to 0.742 have been identified (Zahedi et al., 2005). In this procedure the different protein binding properties of the cationic detergent 16 BAC (benzyldimethyl-n-hexadecylammoniumchloride) in the first separation dimension and the anionic detergent SDS in the second dimension are harnessed to obtain substantial protein mobility differences, resulting in protein scattering along a diagonal in the SDS-gels. 2-D BAC was used for a quantitative comparison of a *C. glutamicum* L-lysine producer and the wild-type (Schlüsener et al., 2007). The obtained 2-D BAC gels were highly reproducible and could be used for the relative protein quantification. However, one major drawback was the limited separation capacity of the system; in the diagonal of the 2-D BAC gels proteins are concentrated in a rather small area (60 cm^2) - a serious obstacle for correct quantification and protein identification due to the overlapping of different protein spots. A powerful technology for the separation and analysis of membrane protein complexes is blue native electrophoresis (BN-PAGE; Schägger and Jagow, 1991), which has been used extensively in studies for both eukaryotic and prokaryotic membranes (see Krause, 2006, for an overview). Corncerning *C. glutamicum*, BN-PAGE has been only employed for analysis of a pre-purified complex of the succinate:menaquinone oxidoreductase (Kurokawa and Sakamoto, 2005), but it is also applicable for the separation of solubilized protein complexes from crude membrane preparations (A. Poetsch, unpublished results).

A technique which provides the separation of intact membrane proteins is the consecutive combination of anion exchange chromatography and SDS-PAGE (AIEC/SDS-PAGE; Schlüsener et al., 2005). The technique was developed to meet the following requirements: intact IMPs, including proteins with high GRAVY and high number of TM helices should be mapped similar to 2-D PAGE or 16 BAC PAGE, but circumvent their limitations, it should offer a fast screening and comparison of membrane proteomes, and it should be compatible with high throughput protein identification technologies like MALDI-TOF. In AIEC/SDS-PAGE the two separation dimensions are orthogonal, i.e. proteins are separated according to their size in SDS-PAGE and according to their charge during AIEC, the latter in analogy to IEF. The resolution power of AIEC/SDS-PAGE is superior to 16 BAC but inferior to 2-D PAGE. On the routinely used 15 x 20 cm gels, about 356 protein bands were visible; while on the 2 D BAC gels of size 15 x 15 cm only about 50 spots could be discriminated (Schlüsener et al., 2007). The chromatography is compatible with nonionic and zwitterionic detergents and moderate amounts of lipids and salts. IMPs with a GRAVY-score up to 0.8 and up to 13 TM helices were retrieved after AIEC/SDS-PAGE from a solubilized *C. glutamicum* membrane fraction (Schlüsener et al., 2005). The technology is suitable for a quantitative comparison of different samples by for instance densitometric analysis of the protein bands from stained gels (Schlüsener et al., 2007), though the reproducibility of the chromatography is not as good as IEF.

Around the millennium, several gel-free proteomics technologies, such as MudPIT, have been developed. In MudPIT (Multidimensional Protein Identification Technology) a complex protein mixture is denatured and digested with proteases. The two separation dimensions usually included in one MudPIT capillary column consist of a strong cation exchanger (SCX) and reverse phase (RP) material (see Figure 4.4). Consecutive salt steps applying increasing salt concentrations followed by reversed phase gradients are alternated, until peptides are completely depleted from the SCX resin. The clear superiority of MudPIT over 2-D electrophoresis in identifying membrane proteins was demonstrated already in the first publication (Washburn et al., 2001), where over 100 membrane proteins from a yeast sample were identified. Since trypsin alone is not an ideal protease for membrane protein digestion, in an analysis of the *C. glutamicum* membrane fraction, two different cleavage

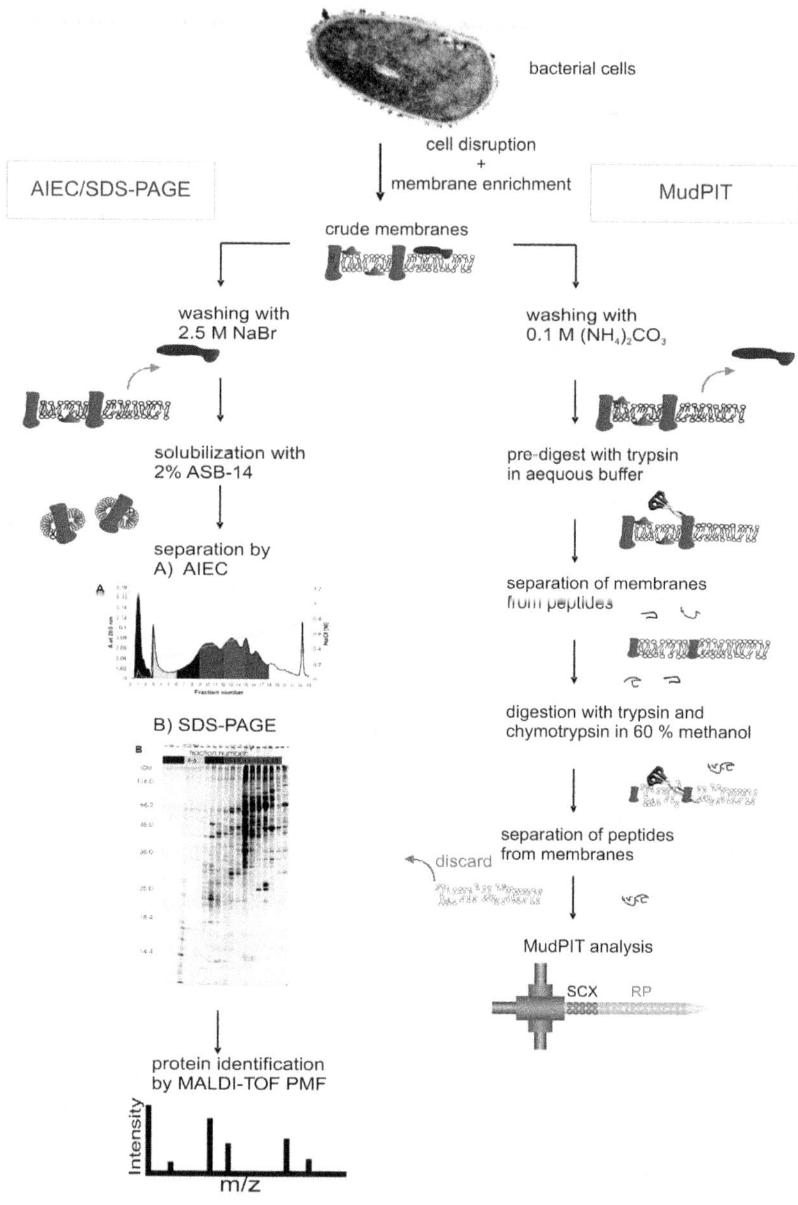

Figure 4.4. Overview of membrane sample preparation and analysis with the AIEC/SDS-PAGE and MudPIT technology

strategies were employed (Fischer et al., 2006). Membranes were either treated with cyanogen bromide in 90% formic acid and thereupon digested with trypsin, or the membranes were pre-digested with trypsin in aequous buffer and then digested again with trypsin and chymotrypsin simultaneously in 60% methanol. Besides a distinct number of cytosolic or membrane-associated proteins, the combined data analysis from both digests yielded 326 integral membrane proteins (~50 % of all predicted) covering membrane proteins both with small and large numbers of transmembrane helices. Also, membrane proteins with a high GRAVY-score were identified, and basic and acidic membrane proteins were evenly represented. Furthermore, when the combined trypsin/chymotrypsin digestion was used, a substantial amount of peptides corresponding to TM-helices were identified. The observed protein set was unbiased in respect to GRAVY, number of TM-helices and pI, though small membrane proteins were apparently underrepresented, which is an inherent and general problem in shotgun proteomics approaches.

6.4 Membrane proteomics with *C. glutamicum*

In the initial efforts to separate and analyze the *C. glutamicum* membrane proteome, 2-D PAGE was applied. Upon membrane solubilization with the detergent CHAPS, 700 protein spots could be discriminated on a silver stained gel of the membrane fraction, but only a few proteins such as peripheral subunits of the ATP synthase complex, were identified with N-terminal microsequencing (Hermann et al., 1998). Later, also a protocol was described for the extraction of *C. glutamicum* cell surface proteins, which uses a short incubation of intact cells with the harsh detergent lauroyl sarcosinate (Hermann et al., 2000). Significantly more proteins were identified after a *C. glutamicum* genome sequencing project was launched and protein detection was achieved by more powerful mass spectrometry (Hermann et al., 2001), though not much progress was made in respect to membrane proteins. The most comprehensive 2-D gel reference map for the membrane fraction was presented for *C. glutamicum* cells grown in glucose minimal medium: from 660 spots 22 protein species were identified by MALDI-TOF PMF (Schaffer et al., 2001). However no integral membrane protein could be detected, the spots only corresponded to hydrophilic, peripheral subunits of membrane proteins. Examples are the subunits A and B of the succinate dehydrogenase, the subunit α of the ATP synthase, the maltose binding protein AmyE, and ATP-binding components of the glutamate and glutamine transporters. In a comparative study investigating the stress response to the herbicide 2,4-dichlorophenoxy acetic acid, the complete proteome was resolved on 2-D gels covering the pH ranges between 3.4 and 5.4, and between 4.9 and 6.9 (Fanous et al., 2007). Three proteins related to the cell envelope increased their abundance during growth with the herbicide, suggesting that they play a pivotal role in the cells, adaptation to the toxic compound. Two proteins, the glutamine-dependent amidotransferase (Cg2410) and a glycosyltransferase (Cg1672) are involved in cell envelope synthesis. The amidotransferase catalyzes the synthesis of UDP-N-acetylglucosamine, while the glycosyltransferase may catalyze the polymerization or crosslinking of the peptidoglycan. The third protein was the ATPase component (Cg2695) of an uncharacterized ABC transport system. The protein contains domains found in cobalt transporters and multidrug exporters. In light of these findings, one can speculate that the bacterium defends against the herbicide by remodeling the cell wall to restrict its penetration and by inducing the expression of an ABC transport system that eliminates the compound from the cytosol.

The very limited coverage of membrane proteins with 2-D PAGE was the reason to develop the AIEC/SDS-PAGE technique for *C. glutamicum*. A mastergel displaying 356 discrete gel bands was established for cells from an L-lysine producing strain harvested in the late stationary phase (Schlüsener et al., 2005). The excised gel slices were treated with trypsin and cyanogen bromide in sequence for an improved identification of membrane proteins by mass spectrometry. The majority of proteins could be successfully identified with MALDI-TOF PMF; in case protein identification failed (mainly due to a protein mixture in the respective gel band), the more powerful LC-ESI MS/MS was used to separate the peptide mixture and to obtain partial peptide sequences by fragmentation in an ion trap. Altogether 170 different proteins were identified, among them 50 integral membrane proteins. 32 membrane proteins were identified by MALDI-TOF PMF, while for 18 proteins LC-ESI MS/MS was required. The different functional classes were represented in the membrane proteome, most membrane proteins were grouped into the transport and binding proteins class (10 proteins) and the hypothetical proteins (22 proteins). The L-lysine exporter was not detected, although its presence was confirmed in a later study (Fischer et al., 2006, see below). Since the true power of the AIEC/SDS-PAGE technology lies in a fast visualization of the membrane proteome composition, it was applied for various comparative studies with *C. glutamicum*.

The membrane proteome of the wild-type was compared with an L-lysine-producing strain obtained by genome-based strain reconstruction (Schlüsener et al., 2007). The producer harbors point mutations in 4 genes directly involved in L-lysine overproduction (*lysC, zwf, hom, pyc*). The compared cells were cultivated in a fermenter and harvested in the stationary phase. Interestingly, changes in the membrane proteome were found though the coding genes were not targeted by the metabolic engineering. The succinate dehydrogenase complex showed the most drastic change, its abundance increased by a factor of 4 in the producer. Furthermore, the maltose binding protein (AmyE, Cg2705) and a subunit of the cytochrome bc-aa_3 supercomplex increased their level twice, while the periplasmic component of an ABC-type transporter with unknown function (Cg0924) slightly decreased. These results suggest high consumption of carbon sources in the L-lysine-producing cells and an increased activity of the oxidative phosphorylation in the stationary phase. Although transcripts for membrane proteins were not analyzed, it is worth mentioning that in a transcriptome analysis of a genetically very similar strain no regulation of central metabolic enzymes was observed during L-lysine production for cells harvested at the mid-exponential phase (Hayashi et al., 2006).

AIEC/SDS-PAGE and DNA microarrays were used in combination to characterize citrate utilization in *C. glutamicum* (Polen et al., 2007). Cells were grown on citrate or glucose as carbon source and harvested during exponential growth. Bacteria grown on citrate displayed elevated levels of the succinate dehydrogenase and the ATP synthase complexes. The most prominent increase emerged for the citrate transporter genes *tctCBA* (*cg3125-*

cg3127) and *citM* (*cg0088*). For the tricarboxylate binding protein, TctC, a fivefold increase was detected, while attempts to identify CitM and to confirm its regulation on the proteome level failed. A threefold and fourfold decreased expression on citrate was determined for *ptsS* and *ptsF*, respectively; although the PtsF protein was earlier identified on the AIEC/SDS-PAGE mastergel (Schlüsener et al., 2005), here no significant alteration in abundance was detected. One conserved integral membrane protein with unknown function (Cg0896) increased twofold on citrate, while the corresponding transcript level only slightly changed. The higher level of TCA-cycle proteins and the ATP synthase complex may be explained by the higher energy and carbon demand for gluconeogenesis. The down-regulation of sugar uptake systems indicates a catabolite repression effect during growth on citrate (Polen et al., 2007). Interestingly, in this study the different carbon source affected the expression of transport proteins more drastically than enzymes of the TCA cycle or glycolysis/gluconeogenesis and one may speculate that regulation mainly on the transcript level is a general feature for transport proteins in *C. glutamicum* and other bacteria.

Another study focused on the protease FtsH. After its deletion no growth defects or other phenotypes could be observed. Therefore proteomics was applied to discover direct or indirect proteolysis targets (Lüdke et al., 2007). Since FtsH is a membrane embedded protease, it was tried to identify proteolytic processing by comparing AIEC/SDS-PAGE gels of the deletion mutant and the wild type. FtsH was identified in a gel band only in the wild type, proving that under the analyzed growth conditions the protein was present in substantial amounts. However, no strong indication for proteolysis of an integral membrane protein was obtained except that the higher abundance of the succinate dehydrogenase complex, especially the subunit A, in the deletion strain suggests that this protein complex is at least indirectly affected by FtsH.

It was possible to obtain the most comprehensive coverage of a bacterial membrane proteome so far by using MudPIT with a new protein digestion procedure (Fischer et al., 2006). In total about 50% of all predicted IMPs were identified, and for IMPs above 60 kDa even 80% were identified. 22% of all identified IMPs were classified as transport and binding proteins and 7% were related to energy metabolism, especially for the protein complexes from the respiratory chain almost all constitutive subunits were identified. Notably, no subunit of the cytochrome *bd* menaquinol oxidase was detected, consistent with the absence of its biogenesis protein Cyd, and previous reports showing an expression of this enzyme only under microaerobic or copper-deficient conditions (Bott and Niebisch, 2003). An analysis of this data to determine the main protein inventory of the *C. glutamicum* membrane is presented in the next paragraph.

6.5 The main protein inventory of the *C. glutamicum* membrane

The high coverage of the membrane proteome achieved with MudPIT motivated us to attempt in this chapter the first presentation of the most abundant integral membrane proteins detected in a *C. glutamicum* L-lysine

Table 4.1. Relative abundance for the 50 most abundant integral membrane proteins identified in an L-lysine producing strain (Fischer et al., 2006). The relative abundance was estimated by calculating the ratio between the detected peptides per protein (spectral counting) in a MudPIT experiment and the protein mass.

Relative abundance	MW [kDa]	ID	description
134	72571	Cg1537	(*ptsG*) glucose-specific enzyme II BC component of PTS (EC 2.7.1.69)
130	4680	Cg3008	(*porA*) porin
69	70502	Cg2120	(*ptsF*) sugar specific PTS system, fructose/mannitol-specific transport protein (EC 2.7.1.69)
44	39632	Cg2409	(*ctaC*) cytochrome c oxidase chain II (EC 1.9.3.1)
43	34786	Cg3138	membrane protease subunit, stomatin/prohibitin homolog
33	11532	Cg0561	(*secE*) SecE subunit of protein translocation complex
18	14535	Cg1001	(*mscL*) large conductance mechanosensitive channel
16	59809	Cg2403	(*qcrB*) cytochrome b, membrane protein (EC 1.10.2.2)
15	45184	Cg2404	(*qcrA1*) rieske iron-sulfur protein
15	65102	Cg2780	(*ctaD*) probable cytochrome c oxidase polypeptide subunit (EC 1.9.3.1)
15	31153	Cg2405	(*qcrC*) cytochrome c1
15	10342	Cg1408	putative membrane protein
13	65759	Cg0417	(*capD*) probable DTDP-glucose 4,6-dehydratase transmembrane protein (EC 4.2.1.46)
12	69148	Cg2925	(*ptsS*) enzyme II sucrose protein (EC 2.7.1.69)
11	33764	Cg2703	sugar permease
11	27836	Cg2977	putative membrane protein
11	28373	Cg0445	(*sdhCD*) succinate dehydrogenase CD
10	21095	Cg1364	(*atpF*) ATP synthase b chain (EC 3.6.1.34)
10	10155	Cg2735	putative membrane protein
8	47509	Cg0451	putative membrane protein
8	17380	Cg1312	putative membrane protein
8	13056	Cg0952	putative integral membrane protein
6	47903	Cg0647	(*secY*) preprotein translocase subunit SecY
6	41650	Cg3067	putative membrane protein
6	16824	Cg2428	putative membrane protein
6	36193	Cg3429	preprotein translocase subunit YidC

6	36409	Cg1362	(*atpB*) ATP synthase F0 subunit 6 (EC 3.6.1.34)
5	38530	Cg0780	membrane protein ribonuclease BN-like family
5	41293	Cg2157	(*terC*) tellurium resistance membrane protein
5	58476	Cg0812	(*dtsR1*) acetyl/propionyl-CoA carboxylase beta chain (EC 6.4.1.2,6.4.1.3)
5	21569	Cg0359	putative membrane protein
5	29763	Cg2412	putative membrane protein
5	52734	Cg3365	(*rmpC*) putative ribitol transport membrane protein
5	8123	Cg1363	(*atpE*) ATP synthase c CHAIN (EC 3.6.1.34)
5	28443	Cg2424	putative membrane protein
5	48185	Cg1345	(*narK*) putative nitrate/nitrite transporter
5	10854	Cg2362	putative membrane protein
4	18381	Cg0514	hypothetical membrane protein
4	17380	Cg1312	putative membrane protein
4	56647	Cg1314	(*putP*) proline transport system
4	15820	Cg2607	putative membrane protein
4	29422	Cg0047	conserved hypothetical protein
4	8098	Cg1786	(*secG*) preprotein translocase SecG subunit
3	69082	Cg0569	cation-transporting ATPase (EC 3.6.1.-)
3	17740	Cg3357	(*trpP*) tryptophan-specific permease, 5-methyltryptophan resistance
3	33778	Cg1238	putative membrane protein
3	36944	Cg2183	ABC-type peptide transport system, permease component
3	15545	Cg2408	putative membrane protein
3	29098	Cg1341	(*narI*) respiratory nitrate reductase 2 gamma chain (EC 1.7.99.4)

producer (Fischer et al., 2006). A rough quantification was achieved by calculating the number of peptides subjected to MS/MS scans per protein and by dividing this number by the protein mass for normalization. Although this "spectral count" approach is a good approximation for the protein quantity in a sample (Liu et al., 2004), the protein abundance rankings in Table 4.1 are not without uncertainty, especially for small proteins with only a few number of detected peptides. According to our analysis, the by far most abundant integral membrane protein classes are the sugar transporters, PtsG, PtsF, PtsS, a sugar permease (Cg2703), and a putative ribitol transporter (RmpC). We also detected a high amount of the porin PorA, which appears to be the major component of the copurified cell wall. Other prominent membrane proteins are the subunits of the oxidative phosphorylation complexes: the bc_1-aa_3

supercomplex, the succinate dehydrogenase, the ATP synthase, and the nitrate reductase. The components of the protein translocating pore, SecE, SecY, SecG, YidC were identified, which agrees well with the observed high protein secretion activity of *C. glutamicum*. Of the mechanosensitive channels, only MscL was highly abundant, which suggests that under the examined growth conditions this channel and not MscS is mainly responsible for osmoprotection. Membrane proteins conferring metal cation transport were present, the tellurium resistance protein TerC, and a cation transporting ATPase. Other rather strongly expressed transporters are a putative nitrate/nitrite transporter, the proline transport system, and a tryptophan specific permease. High amounts of a membrane protease subunit (Cg3138) were identified both with MudPIT and AIEC/SDS-PAGE, yet its physiological importance remains to be clarified. Interestingly, the lysine exporter LysE was not among the 50 most abundant proteins, though it was identified in this work. It could be that a large number of transporters was not required at this late growth stage, yet it has been reported that overexpression of *lysE* improves amino acid production (Vrljic et al., 1996). For a substantial fraction of the highly abundant membrane proteins the purpose is unknown. Elucidation of their function should be beneficial both for our understanding of bacterial physiology and for strain engineering.

7. Proteomics of related bacteria

In this paragraph, selected proteome studies for closely related bacteria will be outlined, if the used technology and/or the obtained results are of particular interest for *C. glutamicum*.

7.1 *Corynebacterium efficiens*

The two species *C. glutamicum* and *C. efficiens* possess remarkable similarities and differences. Both organisms belong to the non-pathogenic soil bacteria, secrete glutamic acid and their genomes show a remarkably similar gene content and a highly syntenic gene order. Despite of this, *C. efficiens* grows at higher temperatures and the guanine plus cytosine content of the *C. efficiens* genome is about 10% higher. It was therefore of high interest to study their proteomes in order to compare them and find out what is conserved and what can be interpreted as a special adaptation in the two bacteria. Hansmeier et al. (2006b) established comprehensive reference maps of the secreted, the cell surface and the cytosolic proteomes of *C. efficiens* strain YS-314 and identified 89, 49 and 164 different proteins in the three proteome fractions, summing up to 177 different proteins. In the same work, a total of 245 different proteins were identified in *C. glutamicum* ATCC 13032 that was cultivated under the same conditions, 54 in the secreted fraction, 36 in the cell surface fraction and 211 in the cytosolic fraction. The study used 2-D gels and identification by tryptic fingerprints and MALDI mass spectrometry. Beside a large number of orthologous proteins that were identified in both species, it was found out that the abundances of chaperone proteins which are especially important at stress situations was consistently higher in *C. efficiens*, probably reflecting an adaptation of this species towards higher temperature.

7.2 *Corynebacterium diphtheriae*

Since also the genome sequence of the founder of the genus *Corynebacterium* and the etiological agent of the acute, communicable disease diphtheria was available, Hansmeier and colleagues (2006a) also established a reference map of the type strain, *Corynebacterium diphtheriae* C7 (toxin-negative). Since in this case interaction with human surfaces was of exceptional interest, only soluble proteins from the cell surface and the secreted fractions were separated by 2-D gel electrophoresis and proteins were identified by subsequent tryptic fingerprinting and MALDI mass spectrometry followed by comparisons to peptide databases derived from the genome sequence. The analyses identified 107 extracellular protein spots and 53 protein spots from the cell surface preparation, summing up to 85 different proteins. Several of the identified proteins are encoded in contigous regions of the genome, so-called pathogenicity islands, and are potential virulence factors. In addition, several iron-transporting proteins were identified, with a hitherto unknown one as the most abundant in the cell surface proteome. This finding might provide a new aspect into pathogenicity of this organism, since iron plays an essential role in virulence and also regulates expression of the diphtheriae toxin.

7.3 *Corynebacterium jeikeium*

Based on its recently determined genome sequence, comprehensive 2-D proteome maps of the skin pathogen *Corynebacterium jeikeium* have been made available recently (Hansmeier et al., 2007). As in *C. diphtheriae*, cell surface proteins were detergent extracted from intact cells prior to 2-D electrophoresis. Apart from probably *C. jeikeium*-specific putative virulence factors, 64 proteins were detected in this subproteome that share homologies to *C. glutamicum* proteins. While 30 of these proteins, such as the solute-binding components of glutamate (GluB), methionine (MetQ), enterochelin (FeuS) ABC transporters, and trehalose corynomycolyl transferases (CmtA and CmtC) were detected also on 2-D gels for *C. glutamicum*, *C. efficiens*, or *C. diphtheriae*, the other 34 proteins were not found maybe due to differences in the respective growth conditions.

7.4 *Mycobacteria*

As stated earlier in this chapter, an enrichment of the membrane fraction is indispensable for an efficient and comprehensive analysis of the *C. glutamicum* membrane proteome. It was shown for *Mycobacterium bovis* membranes that the classical Triton X-114 phase partitioning can be used to further purify the plasma membrane fraction upon sodium carbonate washing (Mattow et al., 2007). The extracted membrane proteins were separated with 1D SDS-PAGE and 2-D electrophoresis in parallel and analyzed with LC-ESI-MS/MS and MALDI-MS/MS. In total, 125 proteins were identified, of which 54 harbored TM helices. It can be concluded from these results that the enrichment of IMPs was very efficient with this approach, but the total number of identified IMPs is about equal (Schlüsener et al., 2005) or less (Fischer et al., 2006) than previously achieved for *C. glutamicum*.

The cytosolic and membrane proteomes of two *Mycobacterium avium* subsp. *paratuberculosis* strains were subjected to offline MudPIT (Radosevich et al., 2007). If the membrane fraction data is analyzed with the earlier described "spectral counting" approach, the following membrane proteins with homologs in *C. glutamicum* were among the 100 most abundant proteins in the two samples: subunits of the ATP synthase and the succinate dehydrogenase complexes, a Fe-S oxidoreductase (gi|41398762), the Rieske Fe-S protein, the protease FtsH, the subunit SecA of the Sec-translocon, and hypothetical proteins (gi|41395855, gi|41398016, gi|41398596); for three hypothetical membrane proteins no homolog was found in *C. glutamicum*. These results show that the membrane composition is in part similar to *C. glutamicum* (e.g. oxidative phosphorylation), but a striking difference is the absence of any carbon source transporters from the most abundant proteins in *M. avium* membranes. The membrane proteome of *M. tuberculosis* has been examined with LC-ESI MS/MS, too (Xiong et al., 2005), in total 100 integral membrane proteins and 249 other proteins were unambiguously identified. In this report, no quantitative analysis was undertaken, yet according to the qualitative data various proteins of the oxidative phosphorylation were the best represented protein class.

Acknowledgements

We are grateful to the members of the BMBF project 0312843 "Neue Methoden zur Proteomanalyse" for numerous contributions and to the Bundesministerium für Bildung und Forschung (BMBF) for granting the project.

References

Bendt, A.K., Burkovski, A., Schaffer, S., Bott, M., Farwick M., and Hermann, T. (2003). Towards a phosphoproteome map of *Corynebacterium glutamicum*. Proteomics 3, 1637-1646.

Bott, M., and Niebisch, A. (2003). The respiratory chain of *Corynebacterium glutamicum*. J. Biotechnol. *104*, 129-153.

Brand, S., Hahner, S., and Ketterlinus, R. (2005). Protein profiling and identification in complex biological samples using LC-MALDI. Drugplus Int. *3*, 6-8.

Chen, C.P., Kernytsky, A., and Rost, B. (2002). Transmembrane helix predictions revisited. Protein Sci. *11*, 2774-2791.

Chevallet, M., Santoni, V., Poinas, A., Rouquie, D., Fuchs, A., Kieffer, S., Rossignol, M., Lunardi, J., Garin, J., and Rabilloud, T. (1998). New zwitterionic detergents improve the analysis of membrane proteins by two-dimensional electrophoresis. Electrophoresis *19*, 1901-1909.

Eggeling, L., and Sahm, H. (2001). The cell wall barrier of *Corynebacterium glutamicum* and amino acid efflux. J. Biosci. Bioeng. *92*, 201-213.

Fanous, A., Weiland, F., Lück, C., Görg, A., Friess, A., and Parlar, H. (2007). A proteome analysis of *Corynebacterium glutamicum* after exposure to the herbicide 2,4-dichlorophenoxy acetic acid (2,4-D). Chemosphere *69*, 25-31.

Fischer, F., and Poetsch, A. (2006). Protein cleavage strategies for an improved analysis of the membrane proteome. Proteome Sci. *4*, 2.

Fischer, F., Wolters, D., Rögner, M., and Poetsch, A. (2006). Toward the complete membrane proteome: high coverage of integral membrane proteins through transmembrane peptide detection. Mol. Cell. Proteomics *5*, 444-453.

Hansmeier, N., Chao, T.C., Kalinowski, J., Pühler, A., and Tauch, A. (2006a). Mapping and comprehensive analysis of the extracellular and cell surface proteome of the human pathogen *Corynebacterium diphtheriae*. Proteomics *6*, 2465-2476.

Hansmeier N., Chao T.C., Pühler, A., Tauch, A., and Kalinowski, J. (2006b). The cytosolic, cell surface and extracellular proteomes of the biotechnologically important soil bacterium *Corynebacterium efficiens* YS-314 in comparison to those of *Corynebacterium glutamicum* ATCC 13032. Proteomics *6*, 233-250.

Hansmeier, N., Chao, T.C., Daschkey, S., Müsken, M., Kalinowski, J., Pühler, A., and Tauch, A. (2007). A comprehensive proteome map of the lipid-requiring nosocomial pathogen *Corynebacterium jeikeium* K411. Proteomics *7*, 1076-1096.

Hartinger, J., Stenius, K., Hogemann, D., and Jahn, R. (1996). 16-BAC/SDS-PAGE: a two-dimensional gel electrophoresis system suitable for the separation of integral membrane proteins. Anal. Biochem. *240*, 126-133.

Hartmann, M., Barsch, A., Niehaus, K., Pühler, A., Tauch, A., and Kalinowski J. (2004). The glycosylated cell surface protein Rpf2, containing a resuscitation-promoting factor motif, is involved in intercellular communication of *Corynebacterium glutamicum*. Arch. Microbiol. *182*, 299-312.

Hayashi, M., Ohnishi, J., Mitsuhashi, S., Yonetani, Y., Hashimoto, S., and Ikeda, M. (2006). Transcriptome analysis reveals global expression changes in an industrial L-Lysine producer of *Corynebacterium glutamicum*. Biosci. Biotechnol. Biochem. *70*, 546-550.

Hermann, T., Wersch, G., Uhlemann, E.M., Schmid, R., and Burkovski, A. (1998). Mapping and identification of *Corynebacterium glutamicum* proteins by two-dimensional gel electrophoresis and microsequencing. Electrophoresis *19*, 3217-3221.

Hermann, T., Finkemeier, M., Pfefferle, W., Wersch, G., Krämer, R., and Burkovski, A. (2000). Two-dimensional electrophoretic analysis of *Corynebacterium glutamicum* membrane fraction and surface proteins. Electrophoresis *21*, 654-659.

Hermann, T., Pfefferle, W., Baumann, C., Busker, E., Schaffer, S., Bott, M., Sahm, H., Dusch, N., Kalinowski, J., Pühler, A., *et al.* (2001). Proteome analysis of *Corynebacterium glutamicum*. Electrophoresis *22*, 1712-1723.

Hoischen, C., and Krämer, R. (1990). Membrane alteration is necessary but not sufficient for effective glutamate secretion in *Corynebacterium glutamicum*. J. Bacteriol. *172*, 3409-3416.

Ikeda, M., and Nakagawa, S. (2003). The *Corynebacterium glutamicum* genome: features and impacts on biotechnological processes. Appl. Microbiol. Biotechnol. *62*, 99-109.

Kalinowski, J., Bathe, B., Bartels, D., Bischoff, N., Bott, M., Burkovski, A., Dusch, N., Eggeling, L., Eikmanns, B.J., Gaigalat, L., Goesmann, A., Hartmann, M., Huthmacher, K., Krämer, R., Linke, B., McHardy, A.C., Meyer, F., Möckel, B., Pfefferle, W., Pühler, A., Rey, D.A., Rückert, C., Rupp, O., Sahm, H., Wendisch, V.F., Wiegräbe, I., and Tauch, A. (2003). The complete *Corynebacterium glutamicum* ATCC 13032 genome sequence and its impact on the production of L-aspartate-derived amino acids and vitamins. J. Biotechnol. *104*, 5-25.

Klein, C., Garcia-Rizo, C., Bisle, B., Scheffer, B., Zischka, H., Pfeiffer, F., Siedler, F., and Oesterhelt, D. (2005). The membrane proteome of *Halobacterium salinarum*. Proteomics *5*, 180-197.

Klose, J. (1975). Protein mapping by combined isoelectric focusing and electrophoresis of mouse tissues. A novel approach to testing for induced point mutations in mammals. Humangenetik *26*, 231-243.

Krause, F. (2006). Detection and analysis of protein-protein interactions in organellar and prokaryotic proteomes by native gel electrophoresis: (Membrane) protein complexes and supercomplexes. Electrophoresis *27*, 2759-2781.

Krogh, A., Larsson, B., von Heijne, G., and Sonnhammer, E.L. (2001). Predicting transmembrane protein topology with a hidden Markov model: application to complete genomes. J. Mol. Biol. *305*, 567-580.

Krüger, R., Kübler, D., Pallisse, R., Burkovski, A., and Lehmann, W.D. (2006). Protein and proteome phosphorylation stoichiometry analysis by element mass spectrometry. Anal. Chem. *78*, 1987-1994.

Kurokawa, T., and Sakamoto, J. (2005). Purification and characterization of succinate: menaquinone oxidoreductase from *Corynebacterium glutamicum*. Arch. Microbiol. *183*, 317-324.

Kyte, J., and Doolittle, R.F. (1982). A simple method for displaying the hydropathic character of a protein. J. Mol. Biol. *157*, 105-132.

Lichtinger, T., Burkovski, A., Niederweis, M., Krämer, R., and Benz, R. (1998). Biochemical and biophysical characterization of the cell wall porin of *Corynebacterium glutamicum*: the channel is formed by a low molecular mass polypeptide. Biochemistry *37*, 15024-15032.

Liu, H., Sadygov, R.G., and Yates, J.R., 3rd (2004). A model for random sampling and estimation of relative protein abundance in shotgun proteomics. Anal. Chem. *76*, 4193-4201.

Lüdke, A., Krämer, R., Burkovski, A., Schlüsener, D., and Poetsch, A. (2007). A proteomic study of *Corynebacterium glutamicum* AAA+ protease FtsH. BMC Microbiol. *7*, 6.

Mahne M., Tauch A., Pühler A., and Kalinowski J. (2006). The *Corynebacterium glutamicum* gene *pmt* encoding a glycosyltransferase related to eukaryotic protein-O-mannosyltransferases is essential for glycosylation of the resuscitation promoting factor (Rpf2) and other secreted proteins. FEMS Microbiol. Lett. *259*, 226-233.

Mattow, J., Siejak, F., Hagens, K., Schmidt, F., Koehler, C., Treumann, A., Schaible, U.E., and Kaufmann, S.H. (2007). An improved strategy for selective and efficient enrichment of integral plasma membrane proteins of mycobacteria. Proteomics *7*, 1687-1701.

Niebisch A., Kabus A., Schultz C., Weil B., and Bott M. (2006). Corynebacterial protein kinase G controls 2-oxoglutarate dehydrogenase activity via the phosphorylation status of the OdhI protein. J. Biol. Chem. *281*, 12300-12307.

O'Farrell, P.H. (1975). High resolution two-dimensional electrophoresis of proteins. J. Biol. Chem. *250*, 4007-4021.

Olsen, J.V., Ong, S.E., and Mann, M. (2004). Trypsin cleaves exclusively C-terminal to arginine and lysine residues. Mol. Cell. Proteomics *3*, 608-614.

Polen, T., Schlüsener, D., Poetsch, A., Bott, M., and Wendisch, V.F. (2007). Characterization of citrate utilization in *Corynebacterium glutamicum* by transcriptome and proteome analysis. FEMS Microbiol. Lett. *273*, 109-119.

Radosevich, T.J., Reinhardt, T.A., Lippolis, J.D., Bannantine, J.P., and Stabel, J.R. (2007). Proteome and differential expression analysis of membrane and cytosolic proteins from *Mycobacterium avium* subsp. *paratuberculosis* strains K-10 and 187. J. Bacteriol. *189*, 1109-1117.

Ragas, A., Roussel, L., Puzo, G., and Riviere, M. (2007). The *Mycobacterium tuberculosis* cell-surface glycoprotein apa as a potential adhesin to colonize target cells via the innate immune system pulmonary C-type lectin surfactant protein A. J. Biol. Chem. *282*, 5133-5142.

Rögener, J., Lutter, P., Reinhardt, R., Bluggel, M., Meyer, H.E., and Anselmetti, D. (2003). Ultrasensitive detection of unstained proteins in acrylamide gels by native UV fluorescence. Anal. Chem. *75*, 157-159.

Santoni, V., Kieffer, S., Desclaux, D., Masson, F., and Rabilloud, T. (2000a). Membrane proteomics: use of additive main effects with multiplicative interaction model to classify plasma membrane proteins according to their solubility and electrophoretic properties. Electrophoresis *21*, 3329-3344.

Santoni, V., Molloy, M., and Rabilloud, T. (2000b). Membrane proteins and proteomics: un amour impossible? Electrophoresis *21*, 1054-1070.

Schaffer, S., Weil, B., Nguyen, V.D., Dongmann, G., Gunther, K., Nickolaus, M., Hermann, T., and Bott, M. (2001). A high-resolution reference map for cytoplasmic and membrane-associated proteins of *Corynebacterium glutamicum*. Electrophoresis *22*, 4404-4422.

Schägger, H., and von Jagow, G. (1991). Blue native electrophoresis for isolation of membrane protein complexes in enzymatically active form. Anal. Biochem. *199*, 223-231.

Schlüsener, D., Fischer, F., Kruip, J., Rögner, M., and Poetsch, A. (2005). Mapping the membrane proteome of *Corynebacterium glutamicum*. Proteomics *5*, 1317-1330.

Schlüsener, D., Rögner, M., and Poetsch, A. (2007). Evaluation of two proteomics technologies used to screen the membrane proteomes of wild-type *Corynebacterium glutamicum* and an L-lysine-producing strain. Anal. Bioanal.Chem. *389.4*, 1055-1064.

Tastet, C., Charmont, S., Chevallet, M., Luche, S., and Rabilloud, T. (2003). Structure-efficiency relationships of zwitterionic detergents as protein solubilizers in two-dimensional electrophoresis. Proteomics *3*, 111-121.

Vrljic, M., Sahm, H., and Eggeling, L. (1996). A new type of transporter with a new type of cellular function: L-lysine export from *Corynebacterium glutamicum*. Mol. Microbiol. *22*, 815-826.

Washburn, M. P., Wolters, D., and Yates, J. R., 3rd (2001). Large-scale analysis of the yeast proteome by multidimensional protein identification technology. Nat. Biotechnol. *19*, 242-247.

Whitelegge, J., Halgand, F., Souda, P., and Zabrouskov, V. (2006). Top-down mass spectrometry of integral membrane proteins. Expert Rev. Proteomics *3*, 585-596.

Xiong, Y., Chalmers, M. J., Gao, F.P., Cross, T.A., and Marshall, A.G. (2005). Identification of *Mycobacterium tuberculosis* H37Rv integral membrane proteins by one-dimensional gel electrophoresis and liquid chromatography electrospray ionization tandem mass spectrometry. J. Proteome Res. *4*, 855-861.

Zahedi, R.P., Meisinger, C., and Sickmann, A. (2005). Two-dimensional benzyldimethyl-n-hexadecylammonium chloride/SDS-PAGE for membrane proteomics. Proteomics *5*, 3581-3588.

From: Corynebacteria: Genomics and Molecular Biology. Edited by: Andreas Burkovski.

Metabolic Network Analysis and Design in *Corynebacterium glutamicum* 5

Christoph Wittmann and Elmar Heinzle

Abstract

Methods of metabolic flux analysis and design of metabolic networks for the overproduction of metabolites are reviewed and discussed with respect to their importance for *Corynebacterium glutamicum*. These methods are relevant for an in-depth understanding of the *in vivo* function of metabolism and its directed engineering. The methods discussed here are based on solid material balances and well established genomic and biochemical knowledge about metabolic reactions. Monitoring of metabolic fluxes yields very reliable and quantitative genome wide data reflecting the genomic background and environmental influences on cells studied. Particularly flux split ratios between glycolysis and pentose phosphate pathway and activities around the pyruvate node are elucidated that are most relevant for the supply of precursors and NADPH. The paper outlines also methods for design of metabolic pathways optimal for the overproduction of metabolites.

1. Introduction

The discovery of the glutamate producing soil bacterium *Corynebacterium glutamicum* (Kinoshita et al., 1957) has initiated a deep investigation of its biochemistry and genetics from early on. Throughout the past decades extensive studies on enzymes and their encoding genes in central carbon metabolism and production pathways, preferentially towards amino acids, have provided valuable information for strain and process optimization and contributed to our current understanding of this micro-organism. Within short time, all essential amino acids but methionine could be produced by different mutants of *C. glutamicum* (Nakayama et al., 1978). Today we know that the metabolism of *C. glutamicum* it is a complex network of several hundred interconnected reactions and pathways. Metabolic engineers face this complexity as a severe problem in identifying promising genetic targets for the improvement and optimization of production strains. Stimulated by the enormous economical potential of *C. glutamicum* for industrial amino acid production, novel approaches have emerged in recent years which allow a detailed quantitative analysis of this complex metabolic network and display key technologies in investigating the metabolism of *C. glutamicum*, visualizing its response to genetic or environmental perturbation and guiding rational strain engineering (Kelle et al., 2005; Wittmann and de Graaf, 2005; Wendisch et al., 2006a; Wendisch et al., 2006b; Takors et al., 2007; Wittmann and Becker, 2007). Among those, the probably most powerful and straightforward method is metabolic flux analysis, which allows the quantification of *in vivo* pathway fluxes in the central carbon and nitrogen metabolism (Kelleher, 2001; Wiechert, 2001; Wittmann, 2007). Application of metabolic flux analysis to *C. glutamicum* has recently unravelled fascinating characteristics of the *C. glutamicum* metabolism which otherwise could not have been observed (Wittmann and de Graaf, 2005). Additionally, different *in silico* approaches such as elementary flux mode or extreme pathway analysis have been developed (Schuster et al., 2007). These provide valuable information on the properties of metabolic networks such as theoretical optimum yields, robustness or potential effects of genetic modifications (Krömer et al., 2005). The present chapter provides an overview on the state-of-art of metabolic network analysis and design of *C. glutamicum* explaining the different concepts as well as experimental and computational tools available today. The potential of metabolic network analysis for the analysis and optimization of *C. glutamicum* is highlighted by various examples from the literature.

2. Metabolic network of *C. glutamicum*

2.1 Topology of central carbon metabolism

The central catabolic network of *C. glutamicum* comprises the pathways of glycolysis, pentose phosphate pathway (PPP), tricarboxylic acid (TCA) cycle and glyoxylate cycle (Figure 5.1). The Entner-Doudoroff pathway has not been detected (Vallino and Stephanopoulos, 1993). Moreover *C. glutamicum* exhibits an extensive set of enzymes for the inter-conversion of C_4 metabolites of the TCA cycle and C_3 metabolites of glycolysis. This involves pyruvate carboxylase (Peters-Wendisch et al., 1998), phosphoenol pyruvate carboxylase (Peters-Wendisch et al., 1996; Peters-Wendisch et al., 1998) and malic enzyme (Dominguez et al., 1998) as carboxylating enzymes. Moreover, PEP carboxykinase (Jetten and Sinskey, 1993), oxaloacetate decarboxylase (Jetten and Sinskey, 1995) and probably phosphoenol pyruvate synthetase

Figure 5.1 (opposite). Central metabolic network of *C. glutamicum* involving glucose PTS system (1), glucose 6-phosphate dehydrogenase (2), 6-phosphogluconate dehydrogenase (3), phosphoribose isomerase (4), ribulose 5-phosphate epimerase (5), transketolase (6), transaldolase (7), transketolase (8), glucose 6-phosphate isomerase (9), phosphofructokinase (10), fructose 1,6-bisphosphatase (11), fructose 1,6-bisphosphate aldolase (12), triose phosphate isomerase (13), glyceraldehyde 3-phosphate dehydrogenase (14), phosphoglycerate kinase (15), phospoglycerate mutase (16), enolase (17), pyruvate kinase (18), pyruvate dehydrogenase (19), oxaloacetate decarboxylase (20), pyruvate carboxylase (21), phosphoenol pyruvate carboxykinase (22), phosphoenol pyruvate carboxylase (23), malic enzyme (24), citrate synthase

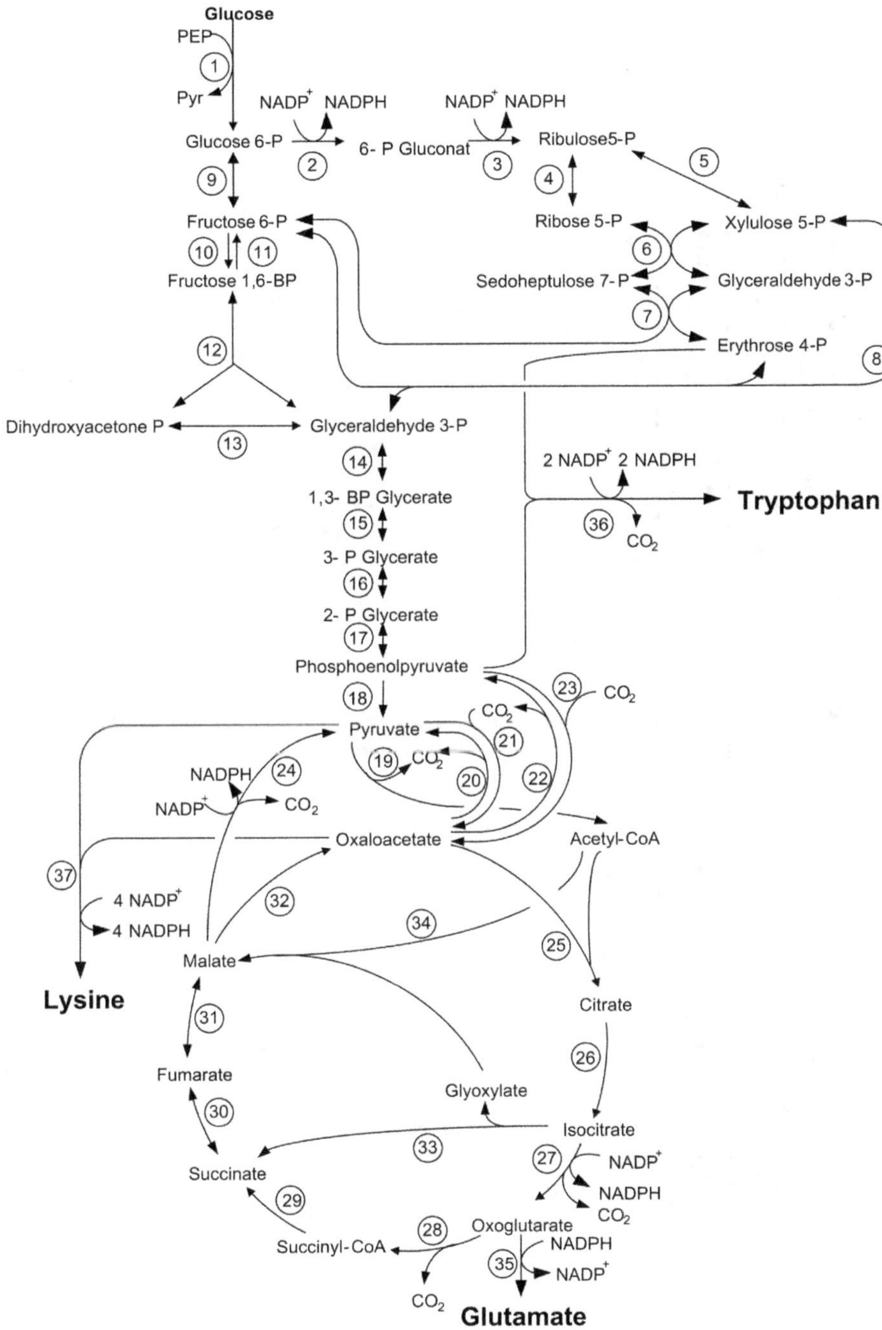

(25), aconitase (26), isocitrate dehydrogenase (27), oxoglutarate dehydrogenase (28), succinate thiokinase (29), succinate dehydrogenase (30), fumarase (31), malate dehydrogenase (32), isocitrate lyase (33), malate synthase (34), reactions consuming intermediary metabolites for anabolic purposes (35-37).

Table 5.1. Precursor demand for biomass formation in *C. glutamicum*. Amino acid composition was determined in continuous culture (Marx et al., 2003). Nucleotide contents are calculated from the GC content of *C. glutamicum* of 56 %. The peptidoglycan content was estimated via the fraction of diaminopimelate (Dap). The precursor demand for peptidoglycan considers the sugar backbone, whereas the demand for the peptide side chain is attributed to the protein fraction. Due to their large pool sizes the intracellular glutamine, Glnintra, and glutamate, Gluintra, are listed separately. Other constituents are those for *E. coli* and are taken from the literature (Neidhardt et al., 1990; Yang et al., 2005). All data are given in μmol (g dry cell mass)$^{-1}$.

Precursor	Demand (μmol/g)	G6P	F6P	R5P	E4P	GAP	PGA	PEP	PYR	AcCoA	OAA	AKG	CO_2	NADPH
Ala	606								1					1
Arg	189											1	1	4
Asx	399										1			1
Cys	87						1							5
Glu	360											1		1
Gln	147											1		1
Glu_{intra}	250											1		1
Gln_{intra}	49											1		1
Gly	361						1							1
His	71			1										1
Ile	202								1		1		-1	5
Leu	440								2	1			-2	2
Lys	202								1		1		-1	4
Met	146										1			8
Phe	133				1			2					-1	2
Pro	170											1		3
Ser	225						1							1
Thr	275										1			3
Trp	54			1	1			1					-1	2
Tyr	81				1			2					-1	2
Val	284								2				-1	2
Dap	146								1		1			4

Protein	0	0	125	268	0	673	482	2604	440	1370	1165	-1647	10548
RNA			630			368				262		368	427
DNA			100			50				50		50	200
Lipids					129	129			2116				3612
LPS	51	16	24			24	24		329				470
Peptidoglycan		292					146		292				146
Glycogen	154												
C_1-units						49							49
Polyamines											59		180
Total	205	308	879	268	129	1293	652	2604	3177	1682	1224	-1229	15623

(Jetten et al., 1994), catalyzing the reverse decarboxylation reactions have been found in *C. glutamicum.*

2.2 Cellular demand for anabolism

Various anabolic reactions drain carbon via precursor compounds from the intermediary metabolism. These anabolic reactions have to be taken into account in all studies on metabolic fluxes, where growing cells are investigated. The precursors for the formation of biomass in *C. glutamicum* are glucose 6-phosphate, fructose 6-phosphate, ribose 5-phosphate, erythrose 4-phosphate, glyceraldehyde 3-phosphate, 3-phosphoglycerate, pyruvate, phosphoenol pyruvate, acetyl-CoA, 2-oxoglutarate, and oxaloacetate. Today extensive knowledge is available on the biochemical pathways leading from the precursors to the biomass constituents. Together with a thorough analysis of the cellular composition of *C. glutamicum*, we exactly know the demand for precursors and reduction equivalents for growth (Table 5.1). Overall, 16.4 mmol NADPH is needed for the synthesis of 1 g of biomass. It is obvious that this high anabolic demand competes with the overproduction of NADPH-demanding products such as lysine or tryptophan. The same holds for the supply of precursor compounds, e.g. phosphoenol pyruvate (for aromatic amino acids) or oxaloacetate (for lysine), which are needed for growth as well as for production.

2.3 Industrially relevant production pathways

C. glutamicum is the most important micro-organism today for industrial amino acid production. The two largest products are glutamate and lysine with annual production volumes of 1,500,000 t/a (Kimura, 2003) and 850,000 t/a (Wittmann and Becker, 2007), respectively. Additionally, tryptophan has recently received increasing interest as additional supplement for animal feed. Its market size with about 15.000 t/a is still small, but steadily increasing (Ikeda, 2006). The metabolic pathways for biosynthesis of these industrially important products have been studied extensively throughout the past decades. They are closely linked to the central carbon metabolic routes via the supply of precursors, cofactors or energy (Figure 5.1). Glutamate is synthesized by glutamate dehydrogenase from the TCA cycle intermediate 2-oxoglutarate requiring ammonium and NADPH. The preceding reaction of isocitrate dehydrogenase produces NADPH, so coupling of both reactions results in $NADP^+$ regeneration during glutamate production. Glutamate overproduction depends on efficient replenishment of the TCA cycle and flux direction towards the product glutamate at the 2-oxoglutarate branch point (Kimura, 2003). Interestingly *C. glutamicum* has a split pathway for the biosynthesis of lysine (Schrumpf et al., 1991; Sonntag et al., 1993). Production is linked to the central metabolism via the demand for the precursors pyruvate and oxaloacetate and the high requirement of 4 NADPH per molecule of lysine. Due to this, efficient NADPH supply especially in the PPP (Wittmann and Heinzle, 2002) as well as efficient anaplerotic reactions (Peters-Wendisch et al., 2001) are crucial for lysine overproduction. In addition to lysine biosynthesis, the pathway is also required to supply the pathway interme-

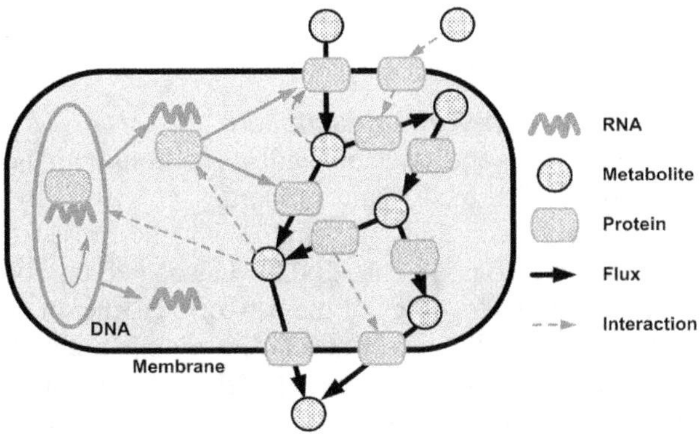

Figure 5.2. Interacting levels of a metabolic network comprising the major components DNA, RNA, proteins and metabolites as well as their complex interactions.

diate diaminopimelate as an important cell wall constituent. Tryptophan is synthesized from erythrose 4-phosphate, phosphoenol pyruvate and ribose 5-phosphate further consuming 2 NADPH. The PPP supplying both cofactors and building blocks is of central importance (Ihnen and Demain, 1969; Ikeda and Katsumata, 1999).

3. Tools and concepts for metabolic network analysis

3.1 Modelling concepts

In this context goals of modelling metabolic networks are deepening understanding, quantification of otherwise inaccessible variables, e.g. *in vivo* metabolic rates or fluxes, or rational design of organisms optimal with respect to reaction network structure and activity. The type of model and depth of modelling is depending on the goal attempted. A major characteristic of any metabolic network model is its topology as described above for *C. glutamicum*. A comprehensive model describing a whole cell would include processes on all levels indicated in Figure 5.2. Modern bioinformatic tools allow a direct retrieval of required information from internet databanks that are based on complete sequencing of *C. glutamicum* and annotation of the identified open reading frames to gene function. Annotation is partly based on direct biochemical knowledge of *C. glutamicum* genes and partly on homology with related genes of other organisms stored in databanks. At present the knowledge is still incomplete and surprising new reactions and correspondingly new metabolites may be identified particularly in mutants with redirected metabolic fluxes (e.g. (Krömer et al., 2006a). The presently most comprehensive data bank for many organisms and also for *C. glutamicum* is KEGG (Kyoto Encyclopedia of Genes and Genomes (Tomii and Kanehisa, 1998; Yamagata et al., 2001)). The collection of regulatory network information has been initiated by the Bielefeld group (CoryneRegNet (Baumbach et al., 2007)). Since knowledge about regulatory networks in *C. glutamicum* is still limited, corresponding databanks of *Escherichia coli* (Ecocyc (Keseler et al., 2005); RegulonDB (Salgado and Collado-Vides, 2004; Salgado et al.,

2006)) are useful. The data and knowledge stored in these data bases provides substantial support for modelling.

3.1.1 Metabolic reaction network description

Elements of metabolic networks are biochemical reactions with their specific substrates and products. Thermodynamic properties, i.e. Gibb's free enthalpy, determine the equilibrium constant and therefore whether these reactions are reversible or irreversible (Beard et al., 2002; Mancuso et al., 2004). These properties together with the topology of the network, i.e. the set of reactions, are the basis of any quantitative description of metabolic networks both for metabolic flux analysis and for a flux based design of optimal networks. A metabolic network is characterized by internal reactions between internal metabolites and transport or exchange fluxes providing connection with the environment. For prokaryotic organisms the boundary of the metabolic network is defined by the cytoplasm membrane. Reactions are defined by the reactants and their stoichiometric coefficients, ν_i.

$$\nu_A A + \nu_B B \rightleftarrows \nu_C C + \nu_D D \tag{1}$$

For quantitative description of the activity of such a network we set up material balances for all metabolites. At constant volume the balance for each metabolite is

$$\frac{dc_{M,k}}{dt} = \sum_{i=1}^{n} \nu_{i,k} r_i \tag{2}$$

Where $c_{M,k}$ is concentration of metabolite M_k (mol L^{-1}) and r_i are the n reaction rates connecting this metabolite with other metabolites. $\nu_{i,k}$ is the stoichiometric coefficient of metabolite M_k of reaction i, having positive values for products and negative values for substrates. A full dynamic description of the whole network would require definition of all reaction rates as function of concentrations of influencing metabolites. At steady state, however, equation 2 simplifies to

$$0 = \mathbf{Sr} \tag{3}$$

The stoichiometric matrix S has m columns representing all reactions and n rows representing all balanced or internal metabolites. This linear system of equations is applied for purposes of flux analysis as well as design determining e.g. extreme pathways or elementary modes.

3.1.2 Metabolic flux analysis using metabolite balancing

Metabolic fluxes represent an essential characteristic of the physiological state of a cell or a whole culture. In-depth analysis of metabolic fluxes requires a combination of metabolite balancing and isotopomer balancing. The basis of metabolite balancing is Equation 3. Starting from this equation one can separate reactions into measurable and non-measurable ones

$$0 = \mathbf{S}_m \mathbf{r}_m + \mathbf{S}_{nm} \mathbf{r}_{nm} \quad (4)$$

Index *m* characterizes measurable and *nm* non-measurable reaction rates. If the matrix $\mathbf{S}_{nm}$ is square and full rank, the non-measurable rates $\mathbf{r}_{nm}$ can be calculated as follows

$$\mathbf{r}_{nm} = \mathbf{S}_{nm}^{-1}(-\mathbf{S}_m \mathbf{r}_m) \quad (5)$$

If the system is over determined a least square regression can be made

$$\mathbf{r}_{nm} = \left(\mathbf{S}_{nm}^T \mathbf{S}_{nm}\right)^{-1} \mathbf{S}_{nm}^T (-\mathbf{S}_m \mathbf{r}_m) = \mathbf{S}_{nm}^{\#}(-\mathbf{S}_m \mathbf{r}_m) \quad (6)$$

T indicates the transpose #, the so-called pseudo-inverse. All these calculations can be carried out easily using available software, e.g. MATLAB or MATHCAD.

Typical measured rates are substrate uptake and metabolite secretion rates which are determined from extra cellular metabolome analysis. Additional measurable rates are anabolic reactions that are determined from biomass formation rate knowing the composition of biomass, e.g. proteins, DNA, RNA, lipids and carbohydrates. Biosynthetic reactions yielding monomers can usually be lumped together and the eventual requirement of precursors, e.g. pyruvate, oxaloacetate and acetyl-CoA, is represented by yield coefficients derived from data supplied in Table 5.1. Most systems are, however, still underdetermined. Co-metabolite, e.g. NADH and NADPH balancing extends the applicability of the method but there are still some uncertainties concerning the exact fate of these metabolites. Metabolite balancing can be applied for any type of cultivation, e.g. also for industrially relevant fed-batch cultures provided the concentration of intracellular metabolites do not change significantly (Chassagnole et al., 2003). However, this kind of analysis alone does not allow a more detailed determination of network fluxes comprising circular pathways, reversible reactions and parallel pathways, e.g. the flux splitting between glycolysis and pentose phosphate pathway (PPP) and anaplerotic reactions around pyruvate.

3.1.3 Metabolic flux analysis using isotopically labelled substrates

Metabolite balancing can be extended by the application of isotopically labelled substrates and balancing of isotopomers. Isotopomers are chemically identical species but with different isotope composition. In most cases labelling with ^{13}C provides richest information (Wiechert, 2001; Wittmann and Heinzle, 2001c; Yang et al., 2005). The transformation of individual carbon atoms in biochemical reactions is well known and well defined. A systematic formulation of atom transfer was provided by Zupke and Stephanopoulos (1994) introducing the concept of Atom Mapping Matrices (AMM). Using this concept one can calculate the abundance of each product atom knowing the atom abundances in the substrates converted. Unfortunately, with this method one can only calculate NMR spectra and total ^{13}C abundance but not mass spectra. This concept was significantly expanded by Schmidt et al. (1997)

who introduced isotopomer vectors describing the abundance of isotopomers and isotopomer mapping matrices (IMM). From isotopomer abundance one can directly calculate mass and NMR spectra. Some numerical problems of this methodology were solved by the introduction of cumomers and cumomer mapping matrices (Wiechert et al., 1999) which are derived from isotopomers and isotopomer matrices by transformation and more recently by the so-called elementary metabolite units (Antoniewicz et al., 2007). For the purpose of flux determination, mass isotopomer distribution analysis can be made of intracellular and extra cellular metabolites as well as of monomers contained in biological polymers, e.g. amino acids in proteins. Mass isotopomers are chemically identical molecules only differing in their molecular mass. The existence of stable isotopes other than ^{13}C requires correction (Wittmann and Heinzle, 1999; van Winden et al., 2002) These methods allow the determination of a series of important physiological parameters, e.g. flux split ratio glycolysis – pentose phosphate pathway (de Graaf et al., 2001), flux split at the pyruvate node, anaplerotic pathway activities, activity of parallel lysine synthesis pathways (Wittmann and Heinzle, 2001a) or reversibilities of reactions of the upper glycolysis (Kiefer et al., 2004; Wittmann et al., 2004a). Using modern cultivation and analytical methods such analysis can even be carried out at a microtiter plate scale (Fischer et al., 2004; Sauer, 2004; Wittmann et al., 2004b). Another extension of metabolic flux analysis is via dynamic analysis of ^{13}C labelling in central metabolites after application of a substrate shift (Noh et al., 2007). This requires parameter estimation using a dynamic isotopomer model.

3.1.4. Dynamic models based on material balances

Overall phenomenological models are interesting to describe dynamic growth phenomena but do not provide any detailed description of network activities (Gayen and Venkatesh, 2007). Dynamic metabolic network models require the understanding of all interactions influencing reaction rates and allow a most comprehensive and directed optimization of a pathway or of a whole cell. Dynamic models as specified in Equation 2 represent a set of usually non-linear differential equations and appropriate definition of reaction rates, r_i

$$r_i = f(c_j, pH, T) \qquad (7)$$

Reaction rates are influenced by concentration of substrates, products and effectors. For most pathways of *C. glutamicum* kinetic information is, however, not available. Lacking kinetic information leads to the application of power law (Heinzle et al., 2007) or lin-log kinetics (Visser and Heijnen, 2003; Magnus et al., 2006). Detailed metabolic models, as have been developed for the lysine biosynthetic pathways (Yang et al., 1999; Brik Ternbach et al., 2005; Magnus et al., 2006), allow a direct calculation of optimal combinations of network activities, e.g. enzyme concentrations. More comprehensive models would also allow the description of regulatory and signalling phenomena. Presently, the information is, however, rather limited and far from allowing setting up such a comprehensive model comprising various types of regulatory phenomena. Dynamic models are also needed for dynamic labelling experiments as has been shown for *E. coli* (Noh et al., 2007).

3.1.5 Extreme pathways and elementary modes

Equation 3 combined with information about reversibilities of reactions as provided by their equilibrium constant is the basis for the determination of extreme pathways (Schilling et al., 1999) and elementary modes (Schuster et al., 1999) that provide an excellent basis for metabolic network design. For each metabolic network there exists a unique set of extreme pathways and elementary modes. Extreme pathways represent a subset of elementary modes (Papin et al., 2004). Each elementary mode represents functional unit composed of a minimal set of enzymes or reactions.

3.2 Experimental tools

3.2.1 Cultivation and sampling

Appropriate experimental techniques are essential for any quantitative physiological study. Metabolic flux analysis is most precisely done at metabolite steady-state conditions which are best guaranteed in continuous cultures where all parameters are at steady state. However, most commonly metabolic flux analysis is carried out in shake flasks. If flux analysis using expensive labelled substrates is carried out, it is useful to use small scale bioreactors to reduce costs (Kumar et al., 2003; Wittmann et al., 2004b). In shaken bioreactors dissolved oxygen supply plays a critical role (Duetz et al., 2000; John et al., 2003; Wittmann et al., 2003). To elucidate metabolic fluxes in different phases of production processes and gain more insight into industrially relevant conditions, different flux approaches have been developed and applied. An interesting concept describes a novel Sensor Reactor approach, in which a 300-L production reactor and a 1-L Sensor Reactor are run in a parallel master/slave modus, thus generating a series of ^{13}C-based metabolic flux maps that document large-scale fermentation courses in detail (Drysch et al., 2003; El Massaoudi et al., 2003; Drysch et al., 2004). Another promising approach is based on respirometric flux analysis, i.e. combining parallel tracer studies on differently labelled substrates with labelling analysis of carbon dioxide formed by mass spectrometry involving a novel set of parallel miniaturized bioreactor with on-line respirometric MS analysis at the mL scale (Yang et al., 2006b; Yang et al., 2006a). This method has been applied to study fluxes of lysine producing *C. glutamicum*, yielding excellent agreement to flux data from conventional flux analysis. The developed method is applicable to study growing or non-growing cells, primary and secondary metabolism and immobilized cells and is also interesting to be coupled to large scale bioprocesses, providing precise flux estimates within a few minutes. For determination of flux distribution of large numbers of mutants or for the application of very expensive labelled substrates microtiter plate reactors can be applied

. *In-vivo* kinetic information in large scale reactors is obtained in stimulus-response measurements as carried out for *C. glutamicum* to study the valine and leucine biosynthesis pathways (Magnus et al., 2006). Stimuli are most informative for kinetics close to the stimulus entry point but are, however, attenuated further down metabolic paths.

Sampling and reaction quenching is most easy when metabolic fluxes are measured by metabolite balancing and when ^{13}C labelling in the monomers of biopolymers are analyzed, since changes are slow enough to get authentic samples. The determination of concentrations of intracellular metabolites with short time constants, e.g. metabolites of glycolysis, requires not only fast quenching of intracellular reactions but also of transport fluxes. This seems to be very difficult because of the cold shock phenomenon. A possible solution is whole culture broth quenching at high or low temperature. This is, however, complicating the following analysis and amplifies errors because of the necessary difference calculation between broth and intracellular content. Leakage because of the cold-shock phenomenon is however less influencing the intracellular labelling state of metabolites (Noh et al., 2007). Intracellular metabolites with medium time constants, e.g. amino acids can be quantified reliably using a fast filtration method (Wittmann et al., 2004c).

3.2.2. Isotopomer analysis for ^{13}C labelling experiments

^{13}C labelling flux analysis methods combine metabolite balancing with isotopomer or cumomer modelling and ^{13}C labelling analysis using either ^{13}C NMR, GC-MS, LC-MS (Matsuda et al., 2003; Noh et al., 2007), MALDI-TOF-MS (Wittmann and Heinzle, 2001b; Wittmann and Heinzle, 2001a), capillary electrophoresis MS (Toya et al., 2007), membrane inlet MS (Yang et al., 2006b) or a combination of different techniques (Arauzo-Bravo and Shimizu, 2001; Shirai et al., 2006).

4. Pathway Fluxes in *C. glutamicum*

The major findings on metabolic fluxes and their regulation in *C. glutamicum* result from the fully integrated approaches combining metabolite balancing and ^{13}C labelling and thus are quite recent. The first flux studies in this organism, however, have been initiated soon after its discovery (Shiio et al., 1960) and contributed to a large extent to the later developments. Some of these pioneering works have unravelled important characteristics of the metabolic behaviour of *C. glutamicum*. An outstanding contribution at that time was the elucidation of fluxes through central metabolic pathways, i.e. citric acid cycle, anaplerosis and glyoxylate shunt in *Microbacterium ammoniaphilum*, a subspecies of *C. glutamicum*, by combining stable isotope labelling with ^{13}C NMR analysis and mathematical equations for flux calculation (Walker et al., 1982). Subsequent labelling studies investigated fluxes through the dual lysine pathway of *C. glutamicum* (Ishino et al., 1984), selected flux parameters in a histidine producing strain (Ishino et al., 1986), in glucose-grown *Corynebacterium melassecola* (Rollin et al., 1995) and in the ammonium assimilatory pathway (Tesch et al., 1999).

In 1993, Vallino and Stephanopoulos (Vallino and Stephanopoulos, 1993) applied metabolite balancing to study fluxes in different phases of a lysine production process of *C. glutamicum*. In their important contribution, they obtained a detailed insight into flux changes, e.g. through PPP, glycolysis, TCA cycle or anaplerosis, when cells switched from growth to production. Subsequently, metabolite balancing was applied to study the flexibility of the metabolic network of *C. glutamicum* at the glucose 6-phosphate (Vallino and Stephanopoulos, 1994) and the pyruvate node (Vallino and Stephanopoulos, 1993) as well as the response of *C. glutamicum* to osmotic stress (Varela et al., 2003), to different oxygen levels (Dominguez et al., 1993) or to growth on different nutrients such as lactate, glucose, fructose or substrate mixtures (Cocaign-Bousquet and Lindley, 1995; Cocaign-Bousquet et al., 1996; Pons et al., 1996). Similarly, selected flux parameters in a pantothenate overproducing strain were estimated using this approach (Chassagnole et al., 2001).

Most of the above studies could only resolve parts of the metabolic network of *C. glutamicum*. Additionally they are based on assumptions, e.g. on the absence of certain pathways or closed balances over ATP, NADH or NADPH, which might not be valid. As stated above, metabolite balancing cannot be applied to distinguish between parallel pathways with the same overall stoichiometry or to resolve exchange fluxes through reversible or bidirectional reactions. Thusly, important information cannot be obtained using this approach. As example, the alternative enzymes catalyzing anaplerotic carboxylation, i.e. pyruvate carboxylase and phosphoenol pyruvate carboxylase, or the reverse decarboxylation, i.e. phosphoenol pyruvate carboxykinase, malic enzyme or oxaloacetate decarboxylase, cannot be distinguished. Also the fluxes through the dual branches of the lysine biosynthetic pathway cannot be resolved, but only be determined as a lumped flux. Similarly, also most of the labelling studies could only resolve a few selected flux parameters. Despite these limitations, inherently incorporated in the two approaches, the above studies have provided very useful information on the metabolism of *C. glutamicum* and significantly stimulated the recent developments of the comprehensive state-of-art approaches for metabolic flux analysis of *C. glutamicum* today available. The most important contributions on metabolic fluxes in *C. glutamicum* are presented in detail in the following chapter.

4.1 Metabolic fluxes on different nutrients

Most of the flux studies of *C. glutamicum* have been carried out with glucose as substrate, so that most of our current knowledge on the physiology of this organism is based on metabolism of this substrate (Wittmann and de Graaf, 2005). *C. glutamicum* can, however, utilize quite a number of additional carbon sources. Among those are glucose, fructose and sucrose which are used for the industrial production of amino acids in the form of starch, raw sugar and molasses and organic acids such as acetate or lactate which might be major by-products of industrial production and reutilized during the fermentation. Knowledge on the metabolic pathways employed for growth and production on these alternative substrates appears quite crucial for opti-

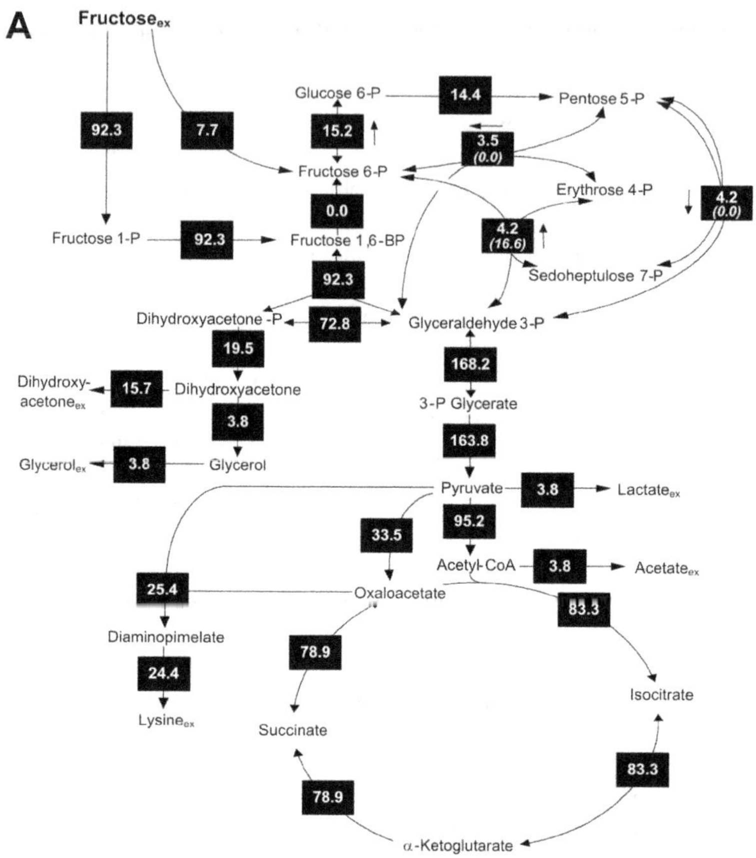

Figure 5.3. *In vivo* carbon flux distribution in the central metabolism of *C. glutamicum* ATCC 21526 during lysine production on glucose (A) and on fructose (B, Opposite) estimated from the best fit to the experimental results using a comprehensive approach of combined metabolite balancing and isotopomer modelling for ^{13}C tracer experiments with labelling measurement by GC/MS (Kiefer et al., 2004). All fluxes are expressed as a molar percentage of the mean specific uptake rate for glucose (1.77 mmol g^{-1} h^{-1}) and fructose (1.93 mmol g^{-1} h^{-1}), respectively. Anabolic fluxes are not shown.

mization of production strains. Selected flux studies have addressed this important topic of fluxes on different nutrients and contributed significantly to our current understanding of the physiology of *C. glutamicum.*

Comparative studies on glucose and fructose previously revealed a significantly reduced yield for lysine and for biomass on fructose as compared to glucose (Kiefer et al., 2002). Additionally, elevated levels of dihydroxyacetone, glycerol and lactate were observed. These are typical overflow metabolites of *C. glutamicum* and point at a limitation of glycolysis for fructose-grown cells. Depending on the carbon source dramatic differences resulted for the distribution of flux through the central metabolism in a comparative flux study of a lysine producing strain (Figure 5.3 A and B) (Kiefer et al.,

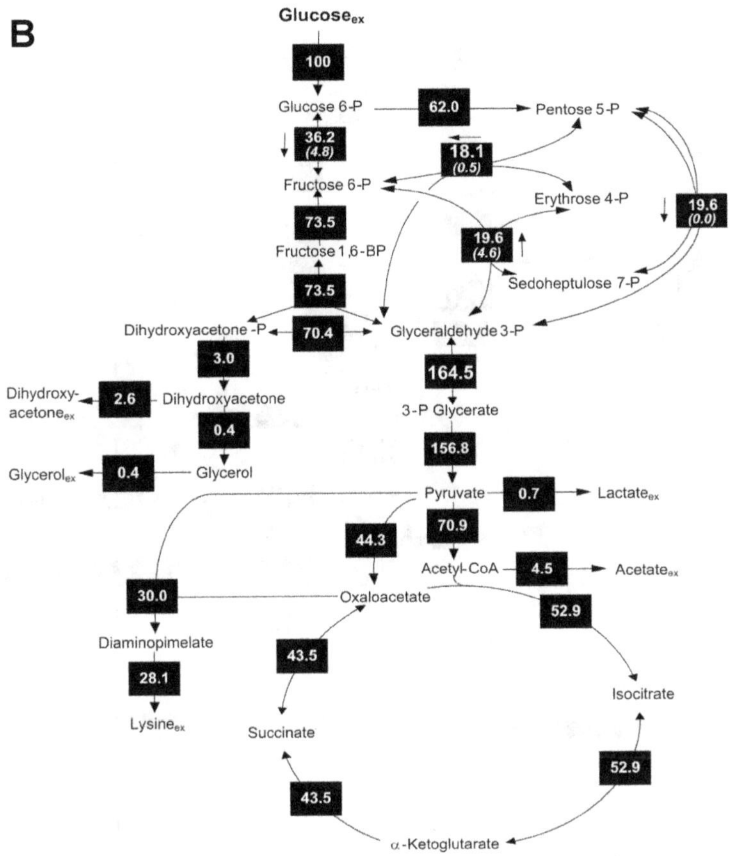

2004). On fructose, the flux into the pentose phosphate pathway was only 14.4 % and thus much lower as compared to glucose (62 %). This is due to the fact that fructose is almost exclusively taken up via the PTS^{Frc} (92 %) and enters metabolism at the level of fructose 1,6-bisphosphate. Caused by the inactivity of fructose 1,6-bisphosphatase (FBPase) the carbon flux is exclusively directed towards the lower glycolytic chain, probably triggering the enhanced by-product accumulation. Through the absence of an active FBPase no gluconeogenic back flux from fructose 1,6-bisphosphate into the PPP is possible. According to this the overall NADPH supply on fructose (112 %) was much lower then on glucose (177 %). Since lysine production has a high demand for NADPH, this seems to be the major reason for the lower product yield on fructose. Balancing for NADPH showed an apparent undersupply by the PPP and the TCA cycle, indicating the additional contribution of decarboxylating malic enzyme to supply of NADPH. Similar findings were also derived from a flux study of fructose-grown *C. melassecola* ATCC 17965 (Dominguez et al., 1998). Since malic enzyme consumes the lysine precursor oxaloacetate, this could be an additional reason for the lower lysine production. The data not only revealed fascinating characteristics of the metabolism

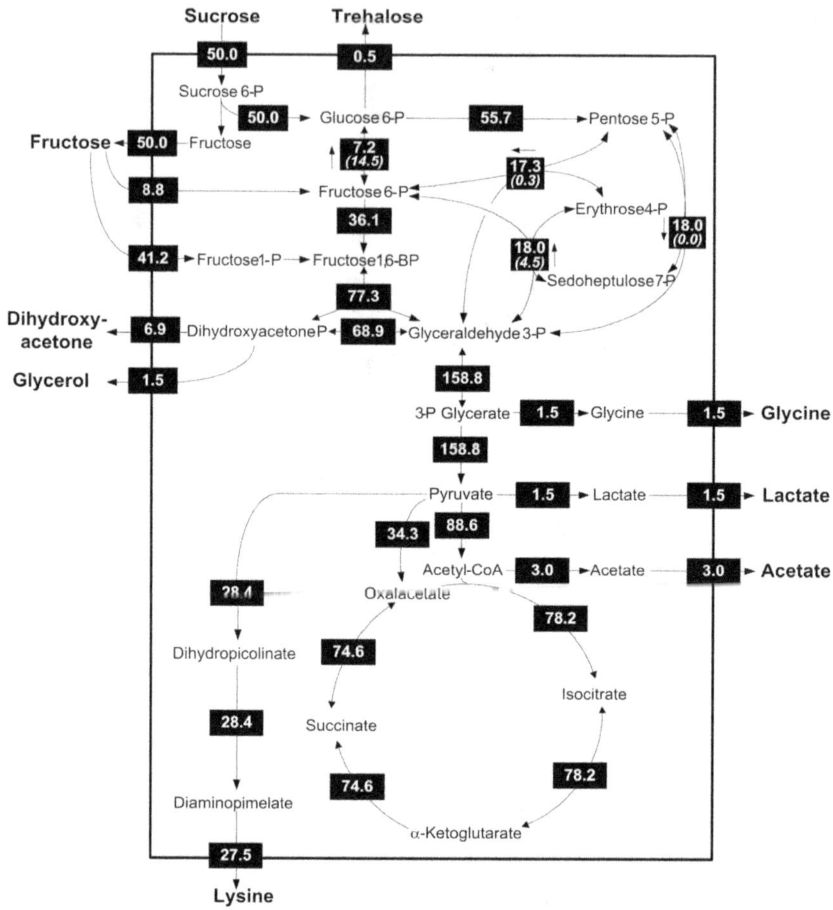

Figure 5.4. *In vivo* carbon flux distribution in the central metabolism of *C. glutamicum* ATCC 21526 during lysine production on sucrose. The fluxes were estimated from the best fit to the experimental results using a comprehensive approach of combined metabolite balancing and isotopomer modelling for three independent ^{13}C tracer experiments on [1-$^{13}C^{Fru}$] sucrose, [1-$^{13}C^{Glc}$] sucrose, and [$^{13}C_6^{Fru}$] sucrose, respectively, and labelling measurement of by GC-MS (Wittmann et al., 2004a). All fluxes are expressed as a molar percentage of the mean specific sucrose uptake rate (0.8 mmol g^{-1} h^{-1}), which is set to 50 %. This facilitates the comparison with other flux maps, which are usually derived from hexose substrates. Anabolic fluxes are not shown.

of *C. glutamicum*, but also suggested different targets for genetic optimization towards increased lysine production as described below.

Metabolic flux analysis of lysine producing *C. glutamicum* was also carried out on sucrose (Wittmann et al., 2004a). Due to the high complexity of uptake and utilization of this disaccharide with three potential entry points into the metabolism, parallel tracer studies on [1-^{13}C fructose]-sucrose, [1-^{13}C glucose]-sucrose, and [$^{13}C_6$ fructose]-sucrose were combined. This allowed the first insight into fluxes of molasses based production processes containing sucrose as major carbon source (Figure 5.4). *C. glutamicum* utilized 27.4 % of the substrate for lysine production. The strain further showed a high PPP flux. Interestingly, the glucose part of the substrate was completely directed into the PPP. Additionally, a small fraction from the fructose part contributed to the PPP flux involving the glucose 6-phosphate isomerase operating in the gluconeogenic direction. As also observed on fructose, increased flux into the undesired by-products dihydroxyacetone and glycerol was observed. The relatively high TCA cycle flux, resulted in a significant NADPH supply by isocitrate dehydrogenase, but due to the elevated loss of carbon as CO_2, displays an interesting target.

Interestingly *C. glutamicum* reveals non-diauxic growth with simultaneous metabolization of both acetate and glucose. Completely different flux situations are present on either of these carbon sources (Wendisch et al., 2000). While on glucose the flux pattern agreed with previous findings with a high relative flux through the PPP and an inactive glyoxylate cycle, a different picture was observed for acetate-grown cells with an active glyoxylate shunt and a strong gluconeogenic net flux through PEP carboxykinase to supply anabolic precursors of upper glycolysis and of the PPP. Drastic differences resulted also for pyruvate dehydrogenase, which was highly active on glucose (109%), weakly active (12%) during growth on acetate/glucose and inactive during growth on acetate. In addition to the carbon sources studied so far, *C. glutamicum* can utilize quite a number of additional compounds. Interesting substrates to be studied in the future comprise for example novel industrially relevant substrates such as glycerol or xylose, the utilization of which has been recently genetically implemented into *C. glutamicum* (Kawaguchi et al., 2006).

4.2 Metabolic fluxes during growth and production

Knowledge about the response of *C. glutamicum* to different metabolic burden caused e.g. by over production of single products is important to understand its metabolism and the underlying regulatory mechanisms (Wittmann and de Graaf, 2005). Previous studies revealed that the central metabolism of *C. glutamicum* exhibits an enormous flexibility in response to changing demands for growth or production. As example it responds to the increased demand for lysine production by appropriate rearrangement of fluxes throughout the central metabolism (Kim et al., 2006). In comparison to the wild type, lysine producing *C. glutamicum* ATCC 13032 *lysC*^{fbr} has an increased flux through PPP and through anaplerosis, leading to enhanced supply of NADPH and oxaloacetate. One should notice that the flux changes observed are due to only one single nucleotide difference in the entire genome of the two strains, changing the kinetic behaviour of only one single enzyme. The cellular response observed involves the total metabolic network as a functional unit of interconnected pathways and reactions. Also in chemostat *C. glutamicum* exhibits a remarkable flexibility (Marx et al., 1997). As an example the relative flux into the PPP increased drastically from 25% for glutamate production and 66% for lysine production with the glycolytic flux accordingly decreasing. Interesting differences between different lysine producing strains were also observed in batch culture probably linked to different demand of precursors for growth and lysine production (Sonntag et al., 1995; Wittmann and Heinzle, 2001a). While lysine flux was similar, anaplerotic net flux and PPP flux were significantly higher for *C. glutamicum* ATCC 21253 as compared to strain MH 20-22B, whereas TCA cycle flux was markedly decreased. This is due to the higher biomass yield and by-product formation of *C. glutamicum* ATCC 21253. A comparative study of different lysine producing mutants from a genealogy of classical strain development showed that a number of key fluxes are closely correlated to production. This involves an increased flux through PPP and anaplerosis with increasing flux through the production pathway (Wittmann and Heinzle, 2002). Overall, the flexibility of the flux partitioning between PPP and glycolysis is a general characteristic of this organism, probably serving to maintain a constant intracellular NADPH concentration.

The Sensor Reactor system described above was used to elucidate metabolic fluxes in different phases of production processes and gain more insight into industrially relevant conditions (El Massaoudi et al., 2003). This approach allowed to analyze different phases of lysine production by *C. glutamicum* in a pilot scale process and show that the C_4-decarboxylation flux at the anaplerotic node significantly decreased (70%) in parallel with threefold increased lysine formation during the investigated subsequent phases of exponential growth (Drysch et al., 2003). Subsequently the same concept was extended to leucine limited fed-batch production of lysine by *C. glutamicum*. In agreement to the batch process, analysis of anaplerotic reactions revealed that selective lysine production was accompanied by a significant reduction of decarboxylating reactions, whereas carboxylating fluxes remained constant (Drysch et al., 2003). Respirometric flux analysis with parallel reactors fed with differently labelled substrates and analysis of carbon dioxide isotopomers allowed estimation of full flux maps of *C. glutamicum* with excellent agreement to flux data from conventional flux analysis (Yang et al., 2006b; Yang et al., 2006a). Besides the carbon balance also the redox balance using mass spectrometric oxygen uptake rate measurement were nearly closed. The developed method is applicable to study growing or non-growing cells, primary and secondary metabolism and immobilized cells and is also interesting to be coupled to large scale bioprocesses, providing precise flux estimates within a few minutes. In another approach labelling analysis of free intracellular acids, which continuously update their labelling pattern to the actual fluxes was applied to investigate a lysine production process (Wittmann et al., 2004c). It could be shown that the phase shift from growth to lysine production was accompanied by a decrease in glucose uptake flux, the redirection of flux from the tricarboxylic acid (TCA) cycle towards anaplerotic carboxylation and lysine biosynthesis. While production of lysine has been most intensively studied, selected flux studies have been also carried out for other industrially relevant products such as glutamate. Here, changes in the anaplerotic supply and at the oxoglutarate branch point could be related to differences in the production behaviour. Recently, flux analysis has been also applied to resolve the different pathways

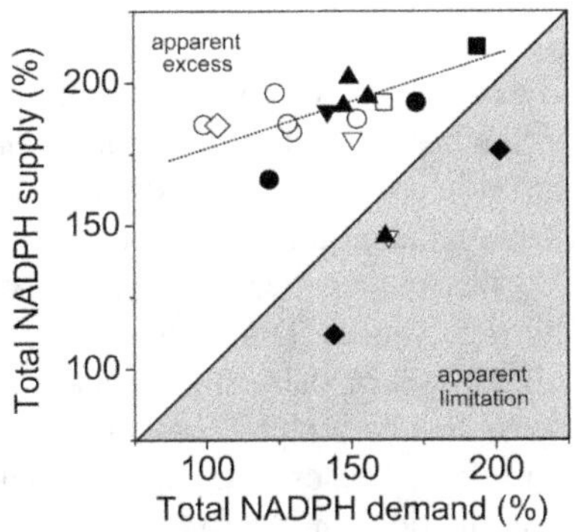

Figure 5.5. Overall flux through NADPH supply and demand of different *C. glutamicum* strains under various cultivation conditions as revealed by ^{13}C flux analysis. NADPH supply considers G6P dehydrogenase, 6PG dehydrogenase, and isocitrate dehydrogenase. Lysine production (4 NADPH/lysine) and anabolism are considered as NADPH-consuming reactions. Hereby, the NADPH demand for biomass formation, 16.4 mmol (g cell dry mass)$^{-1}$, was based on a detailed analysis of cellular composition (Wittmann and de Graaf, 2005). Studies involved flux analysis of (i) non-producing *C. glutamicum* ATCC 13032 in batch culture on glucose (*open diamond*) (Sonntag et al., 1995), (ii) non-producing *C. glutamicum* ATCC 13032 and lysine producing *C. glutamicum* ATCC 13287, ATCC 21253, ATCC 21526, and ATCC 21543 in batch culture on glucose (*open circle*) (Wittmann and Heinzle, 2002), (iii) lysine producing *C. glutamicum* ATCC 21253 during maximum production phase in batch culture on glucose (*open square*) (Wittmann and Heinzle, 2001a), (iv) non-producing *C. glutamicum* LE4 and lysine-producing *C. glutamicum* MH 20–22B in continuous culture on glucose (*closed circles*) (Marx et al., 1997), (v) lysine-producing *C. glutamicum* MH 20–22B in continuous culture on glucose (*closed square*) (Marx et al., 1996), lysine-producing *C. glutamicum* ATCC 21526 in batch culture on glucose, fructose (*closed diamond*) (Kiefer et al., 2004) and on sucrose (*closed triangle*) (Wittmann et al., 2004a), and lysine-producing *C. glutamicum lysC*fbr and *C. glutamicum lysC*^{fbr} P_{EFTU}fbp during batch culture on glucose (open triangle) (Becker et al. 2005).

within the methionine biosynthesis of *C. glutamicum* (Krömer et al., 2006b) (Krömer et al., 2006b). Concerning pantothenate production by *C. glutamicum*, metabolic flux analysis during the fermentation demonstrated that the P-ilvEM3 mutation successfully reoriented the carbon flux towards pantothenate biosynthesis (Hüser et al., 2005).

4.3 Redox fluxes – supply of NADPH

NADPH is needed as cofactor for the biosynthesis of different amino acids including the industrially relevant products lysine, tryptophan or methionine. Due to this the NAPDH metabolism of *C. glutamicum* has received huge interest, especially for the production of lysine, where the supply of this cofactor has turned out to be a key characteristic of efficient producer strains. In lysine producing *C. glutamicum* glucose 6-phosphate dehydrogenase, 6-phosphogluconate dehydrogenase and isocitrate dehydrogenase are NADPH generating reactions whereas NADPH consuming reactions comprise lysine production and growth with a stoichiometric demand of 16.4 mmol NADPH (g biomass)$^{-1}$ (Wittmann and de Graaf, 2005).

Metabolic flux analysis has provided a quantitative insight into the NADPH metabolism of lysine producing *C. glutamicum* by establishing NADPH balances under various physiological conditions.

It has turned out, as shown in Figure 5.5, that the NADPH metabolism of *C. glutamicum* is highly flexible e.g. depending on metabolic growth state (Marx et al., 1997), the nutrient source (Dominguez et al., 1998; Wendisch et al., 2000; Wittmann et al., 2004a) or the genetic background. In most cases an apparent NADPH excess results which seems to increase with diminishing NADPH oxidation. In selected cases, such as maximal lysine production during batch culture or production on fructose allowing only a small PPP, however, an apparent limitation for NADPH is observed (Wittmann et al., 2004a). The obvious imbalance between NADPH supply and demand points at so far unassigned metabolic reactions either consuming or supplying NADPH. Possible candidates for this are NADPH oxidase (Matsushita et al., 2001) or a metabolic cycle at the pyruvate node involving malic enzyme (de Graaf, 2000; Petersen et al., 2000). Concerning the supply of NADPH a close correlation of lysine production with the flux through the PPP could be observed in various studies. Comparison of different lysine producing mutants (Figure 5.6) showed that increasing lysine flux is linked to increasing PPP flux (Wittmann and Heinzle, 2002). In contrast, the contribution of isocitrate dehydrogenase decreases with increasing lysine production. In summary, it turns out that (i) lysine production is likely limited by NADPH availability and that (ii) the PPP is the major pathway for supply of NADPH. In light of this, different approaches have successfully redirected the flux through the PPP in order to increase lysine producing by *C. glutamicum*.

The importance of the PPP has also been shown for tryptophan production by *C. glutamicum*, whereby this pathway does not only supply NADPH, but also erythrose 4-phosphate as product precursor. In case of glutamate production, NADPH consuming glutamate dehydrogenase is preceded by NADPH forming isocitrate dehydrogenase in the biosynthetic pathway balancing supply and demand during synthesis of this product.

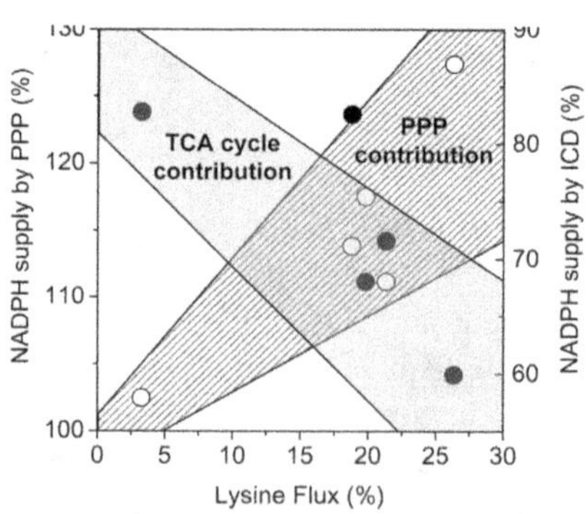

Figure 5.6. NADH supply by pentose phosphate pathway (PPP, open circles) and isocitrate dehydrogenase (ICD, closed circles) and lysine flux (molar % of glucose uptake) (Wittmann and Heinzle, 2002).

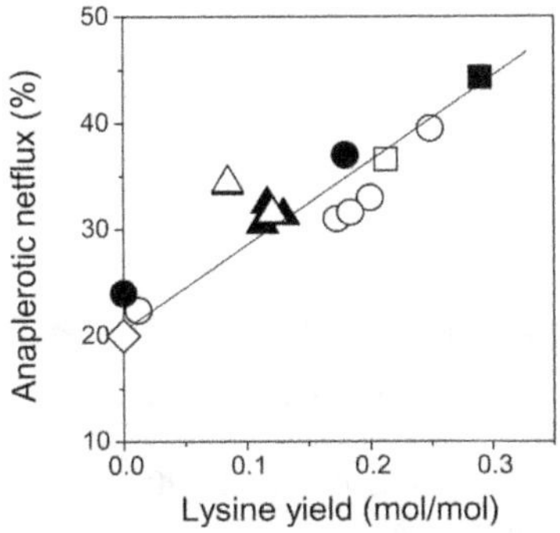

Figure 5.7. Correlation between lysine yield and anaplerotic net flux in different *C. glutamicum* strains under various cultivation conditions as revealed by ^{13}C flux analysis. Studies involved flux analysis of (i) non-producing *C. glutamicum* ATCC 13032 in batch culture on glucose (*open diamond*) (Sonntag et al., 1995), (ii) non-producing *C. glutamicum* ATCC 13032 and lysine-producing *C. glutamicum* ATCC 13287, ATCC 21253, ATCC 21526, and ATCC 21543 in batch culture on glucose (*open circle*) (Wittmann and Heinzle, 2002), (iii) lysine-producing *C. glutamicum* ATCC 21253 during maximum production phase in batch culture on glucose (*open square*) (Wittmann and Heinzle, 2001a), (iv) non-producing *C. glutamicum* LE4 and lysine-producing *C. glutamicum* MH 20–22B in continuous culture on glucose (*closed circles*) (Marx et al., 1997), (v) lysine-producing *C. glutamicum* MH 20–22B in continuous culture on glucose (*closed square*) (Marx et al., 1996), lysine-producing *C. glutamicum* ATCC 21526 in batch culture on glucose, fructose (*closed diamond*) (Kiefer et al., 2004) and on sucrose (*closed triangle*) (Wittmann et al., 2004a), and lysine-producing *C. glutamicum lysC*^{fbr} and *C. glutamicum lysC*^{fbr} P_{EFTU}fbp during batch culture on glucose (*open triangle*) (Becker et al., 2005).

4.4 Anaplerotic fluxes – supply of building blocks

As described above, *C. glutamicum* possesses a rich set of enzymes interconnecting C_3 pools of glycolysis and C_4 pools of the TCA cycle. These reactions are of specific importance, because different industrially relevant products such as lysine, threonine, isoleucine or methionine require oxaloacetate as precursor. Metabolic flux analysis revealed an extensive *in vivo* activity of the C_4 decarboxylating reactions even during pure growth on glucose (Wittmann and de Graaf, 2005). This indicates a highly flexible metabolic reaction cycle around the pyruvate node, which is also present in various other microorganisms (Sauer and Eikmanns, 2005). This cycle could be involved in wasting excess ATP (Sauer et al., 1997; Wittmann and Heinzle, 2001a) or equilibrating the intracellular pool sizes of metabolites around the pyruvate node (Sauer and Eikmanns, 2005). Additionally a contribution to the NADPH metabolism has been hypothesized (Cocaign-Bousquet and Lindley, 1995; de Graaf et al., 2000). Different flux studies revealed that the fluxes around the pyruvate node are indeed directly related to the production of e.g. lysine. Clearly, improved lysine formation is based on an enhanced anaplerotic net flux providing the product precursor oxaloacetate (Figure 5.7). In contrast decarboxylating flux is decreasing with increased production as observed in a comparative study

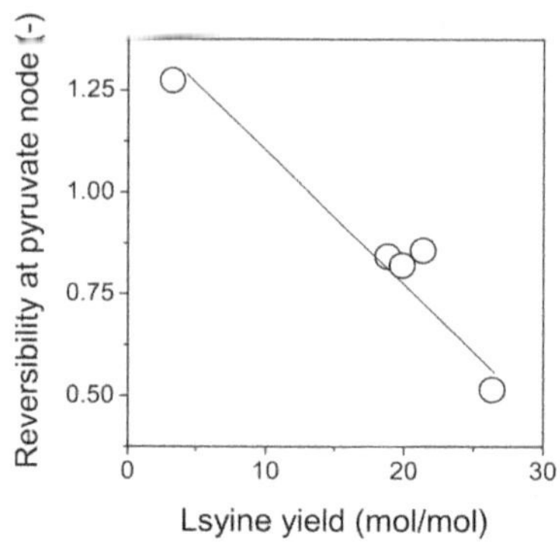

Figure 5.8. Reversibility an the pyruvate node as function of lysine yield (Wittmann and Heinzle, 2002).

of different lysine producing mutants (Figure 5.8). These flux data suggest that the enzymes involved display important targets for metabolic engineering, due to the fact that they either deliver or withdraw the lysine precursor oxaloacetate. Indeed, reduction of decarboxylating flux by deletion of PEP carboxykinase leads up to 20% increased lysine accumulation (Petersen et al., 2001; Riedel et al., 2001). Similarly also engineering of pyruvate carboxylase through introduction of a point mutation is beneficial for lysine production (Ohnishi et al., 2002). Whereas most flux studies consider the different carboxylating and decarboxylating enzymes as two lumped reactions, selected studies with extended labelling analysis and specific experimental design could resolve the fluxes around the pyruvate node in more detail. As example, the wild type of *C. glutamicum* growing in chemostat, revealed pyruvate carboxylase as major enzyme of anaplerotic carboxylation (Figure 5.9). PEP carboxylase, originally assumed to be the sole anaplerotic enzyme in *C. glutamicum*, was almost inactive. Excess oxaloacetate was recycled back by PEP carboxykinase, whereas oxaloacetate decarboxylase exhibited only minor activity. In a recent study, metabolic fluxes around the pyruvate node were also investigated for a glutamate producing strain of *C. glutamicum* during pure growth as well as during production (Shirai et al., 2007). It was found, that phosphoenol pyruvate carboxylase (PEPcx) was the sole active anaplerotic enzyme during the growth phase. In the glutamate overpro-

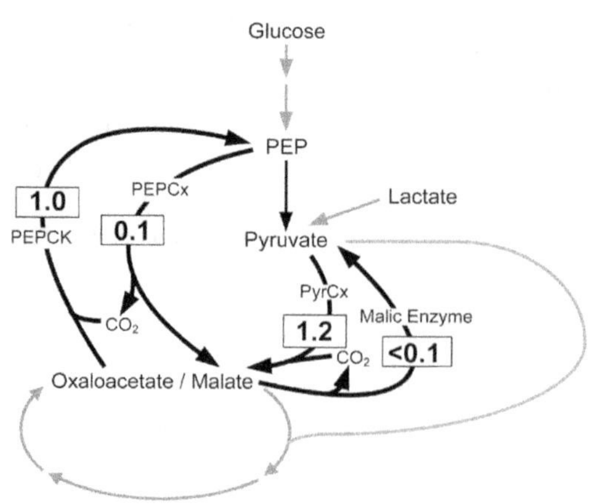

Figure 5.9: Metabolic flux distribution in the anaplerosis of *C. glutamicum* ATCC 13032 involving PEP carboxylase, pyruvate carboxylase, PEP carboxykinase, malic enzyme, and oxaloacetate decarboxylase (Petersen et al., 2001). All values are given in mmol (g cell dry mass)$^{-1}$ h^{-1}.

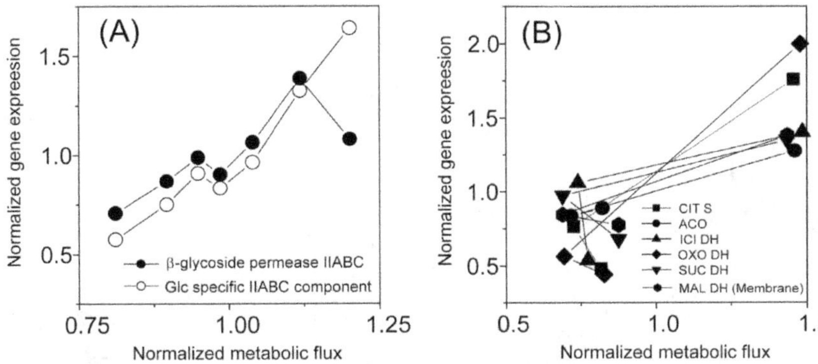

Figure 5.10. Coordination of gene expression and metabolic fluxes in glycolysis (A) and lysine biosynthesis (B) during batch cultivation of lysine-producing *C. glutamicum* ATCC 13287. For substrate uptake, glucose uptake flux and the expression of PTS components were considered (Krömer et al., 2004).

duction phase, the reaction catalyzed by pyruvate carboxylase was activated probably in response to the increased demand for replenishment of thc TCA cycle.

4.5 Flux analysis integrated into systems biology approaches

Metabolic flux analysis can be regarded as the first systems oriented approach to *C. glutamicum* physiology since it considers comprehensive metabolic network models of the biosynthetic and central metabolic pathways. The recently available genome sequence of *C. glutamicum* and post-genomics methods such as transcriptomics, proteomics and metabolomics now allow a characterization of metabolic and regulatory properties of this bacterium on a truly global level (Wendisch et al., 2006b). The first study integrating metabolic flux analysis into systems-oriented analysis of *C. glutamicum* investigated a lysine production process (Krömer et al., 2004). Through combination of pathway fluxes (fluxome) with gene expression levels (transcriptome) as well as intra and extra cellular metabolite levels (metabolome), a very detailed, global picture of lysine production could be obtained. This includes a quantitative insight into possible correlation between the *in vivo* activity of an enzyme, i.e. the metabolic flux, and the expression level of the corresponding gene. As example, proteins of the phosphotransferase system for glucose uptake and of the TCA cycle showed a close correlation between flux and gene expression, indicating that amplification of the corresponding genes probably leads to an increased flux (Figure 5.10). In many cases including lysine biosynthetic pathway, however, no clear correlation was observed underlining the additional contribution of e.g. regulation on the metabolite level. Interestingly, the strongest changes in gene expression during the cultivation were observed for enzymes at the entry points into different pathways, e.g. glycolysis, PPP or TCA cycle. Probably, these connecting nodes are specifically regulated on the transcriptional level. In summary, such novel systems-oriented approaches provide a deep global

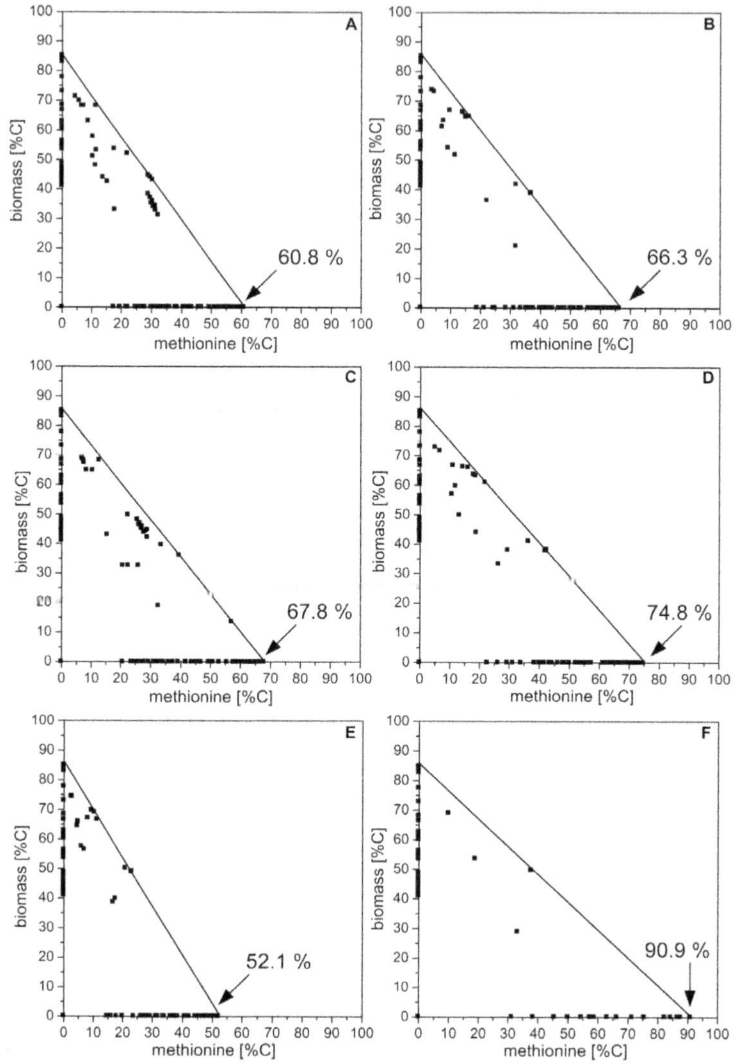

Figure 5.11. Metabolic pathway analysis of *C. glutamicum* for methionine production with different carbon and sulfur sources: carbon yield for biomass and methionine for the obtained elementary modes of C. glutamicum utilizing thiosulfate (A), thiosulfate and formate (B), sulfide (C), sulfide and formate (D), formate (E), and methanethiol (F). The number given indicates the maximal theoretical carbon yield for methionine for each scenario. The strait line connects the modes with maximal biomass and maximal methionine yields (Krömer et al., 2006b).

understanding of the metabolic network and thus seem very promising for the alteration of complex regulatory networks e.g. towards better salt, temperature or pH tolerance of production strains (Kelle et al., 2005).

5. Pathway design

Flux based design of metabolic networks for the optimal production of metabolites can vary from purely empirical methods based on expert experience and knowledge to methods based on complex mathematical models. In the first method hypotheses for improved production are created and tested after implementation. Ideally, such tests comprise metabolic flux analysis to provide a detailed picture of any genetic modification carried, e.g. gene deletion, gene over expression or gene mutation. Additional knowledge created is directly used for further improvement. Most improvements in corynebacteria in the past followed this route mostly even without detailed flux analysis (Ikeda, 2003). In general, one starts with amplifying biosynthetic and transport genes, if known. Feedback regulation is often limiting pathway fluxes. Therefore, this is eliminated in a next step, e.g. feedback inhibition of the aspartokinase by lysine and threonine. High producing strains require also modifications of the precursor supply. This requires modification of central pathways that is much more difficult to predict. The other extreme, optimization of a metabolic network entirely based on a comprehensive mathematical model is still not possible though first attempts in this direction have been reported (Shimizu et al., 1999).

New genome wide stoichiometry based modelling of metabolic pathways is now possible and has been applied to study possible routes for the production of methionine (Krömer et al., 2006b). In this respect elementary mode analysis is very useful to first calculate potential yields and the routes leading to maximal yields (Figure 5.11). This analysis showed that the supply of reduced sulphur sources is beneficial for high carbon yield. This is mainly caused by the large amount of NADPH is needed for reduction to sulphide oxidation state. Ideally, methanethiol should be used resulting in highest possible carbon yield. This provides a sound bases for further optimization or allows judgement of the general economical feasibility of a project.

6. Outlook

Presently there is a tendency to favour *E. coli* as major metabolite producing organism because of its fast growth rate, extensive physiological knowledge and excellent genetic engineering tools. With the further development of physiological knowledge and improved tools *C. glutamicum* seems to become more attractive again in the near future because of its inherent capacity of providing large amount of biosynthetically important NADPH and other precursors. Corynebacteria species may be also interesting for the production of nucleotides. Besides biotechnological production corynebacteria are also interesting for medical reasons. It will therefore be important to increase biochemical knowledge which is significantly supported by the analysis of metabolic flux distribution of organisms of interest under various physiological conditions.

References

Abbouni, B., Elhariry, H.M., and Auling, G. (2004). Overproduction of NAD^+ and 5'-inosine monophosphate in the presence of 10 microM Mn^{2+} by a mutant of *Corynebacterium ammoniagenes* with thermosensitive nucleotide reduction (nrd(ts)) after temperature shift. Arch. Microbiol. *182*, 119-125.

Antoniewicz, M.R., Kelleher, J.K., and Stephanopoulos, G. (2007). Elementary metabolite units (EMU): a novel framework for modeling isotopic distributions. Metab. Eng. *9*, 68-86.

Arauzo-Bravo, M.J., and Shimizu, K. (2001). Estimation of bidirectional metabolic fluxes from MS and NMR data using positional representations. Genome Inform. *12*, 63-72.

Baumbach, J., Wittkop, T., Rademacher, K., Rahmann, S., Brinkrolf, K., and Tauch, A. (2007). CoryneRegNet 3.0--an interactive systems biology platform for the analysis of gene regulatory networks in corynebacteria and *Escherichia coli*. J. Biotechnol. *129*, 279-289.

Beard, D.A., Babson, E., Curtis, E., and Qian, H. (2004). Thermodynamic constraints for biochemical networks. J. Theor. Biol. *228*, 327-333.

Beard, D.A., Liang, S.D., and Qian, H. (2002). Energy balance for analysis of complex metabolic networks. Biophys. J. *83*, 79-86.

Becker, J., Klopprogge, C., Herold, A., Zelder, O., Bolten, C.J., and Wittmann, C. (2007). Metabolic flux engineering of L-lysine production in *Corynebacterium glutamicum*-over expression and modification of G6P dehydrogenase. J. Biotechnol. *132*, 99-109

Becker, J., Klopprogge, C., Zelder, O., Heinzle, E., and Wittmann, C. (2005). Amplified expression of fructose 1,6-bisphosphatase in *Corynebacterium glutamicum* increases in vivo flux through the pentose phosphate pathway and lysine production on different carbon sources. Appl. Environ. Microbiol. *71*, 8587-8596.

Bolten, C.J., Kiefer, P., Letisse, F., Portais, J.C., and Wittmann, C. (2007). Sampling for metabolome analysis of microorganisms. Anal. Chem. *79*, 3843-3849.

Brik Ternbach, M., Bollman, C., Wandrey, C., and Takors, R. (2005). Application of model discriminating experimental design for modeling and development of a fermentative fed-batch L-valine production process. Biotechnol. Bioeng. *91*, 356-368.

Chassagnole, C., Diano, A., Letisse, F., and Lindley, N.D. (2003). Metabolic network analysis during fed-batch cultivation of *Corynebacterium glutamicum* for pantothenic acid production: first quantitative data and analysis of by-product formation. J. Biotechnol. *104*, 261-272.

Chassagnole, C., Rais, B., Quentin, E., Fell, D.A., and Mazat, J.P. (2001). An integrated study of threonine-pathway enzyme kinetics in *Escherichia coli*. Biochem. J. *356*, 415-423.

Chen, R., and Yang, H. (2000). A highly specific monomeric isocitrate dehydrogenase from *Corynebacterium glutamicum*. Arch. Biochem. Biophys. *383*, 238-245.

Cocaign-Bousquet, M., Guyonvarch, A., and Lindley, N.D. (1996). Growth rate-dependent modulation of carbon flux through central metabolism and the kinetic consequences for glucose-limited chemostat cultures of *Corynebacterium glutamicum*. Appl. Environ. Microbiol. *62*, 429-436.

Cocaign-Bousquet, M., and Lindley, N.D. (1995). Pyruvate overflow and carbon flux within the central metabolic pathways of *Corynebacterium glutamicum* during growth on lactate. Enzyme Microb. Technol. *17*, 260-267.

Dauner, M., and Sauer, U. (2000). GC-MS analysis of amino acids rapidly provides rich information for isotopomer balancing. Biotechnol. Prog. *16*, 642-649.

de Graaf, A.A. (2000). Metabolic flux analysis of *Corynebacterium glutamicum*, In Bioreaction Engineering, K. Schügerl, and K.H. Bellgard, eds. (Berlin, Heidelberg: Springer Verlag), pp. 506-555.

de Graaf, A.A., Eggeling, L., and Sahm, H. (2001). Metabolic engineering for L-lysine production by *Corynebacterium glutamicum*. Adv. Biochem. Eng. Biotechnol. *73*, 9-29.

de Graaf, A.A., Mahle, M., Mollney, M., Wiechert, W., Stahmann, P., and Sahm, H. (2000). Determination of full ^{13}C isotopomer distributions for metabolic flux analysis using heteronuclear spin echo difference NMR spectroscopy. J. Biotechnol. *77*, 25-35.

Dominguez, H., Nezondet, C., Lindley, N.D., and Cocaign, M. (1993). Modified carbon flux during oxygen limited growth of *Corynebacterium glutamicum* and the consequences for amino acid overproduction. Biotechnol. Lett. *15*, 449-454.

Dominguez, H., Rollin, C., Guyonvarch, A., Guerquin-Kern, J.L., Cocaign-Bousquet, M., and Lindley, N.D. (1998). Carbon-flux distribution in the central metabolic pathways of *Corynebacterium glutamicum* during growth on fructose. Eur. J. Biochem. *254*, 96-102.

Drysch, A., El Massaoudi, M., Mack, C., Takors, R., de Graaf, A.A., and Sahm, H. (2003). Production process monitoring by serial mapping of microbial carbon flux distributions using a novel Sensor Reactor approach: II--(13)C-labeling-based metabolic flux analysis and L-lysine production. Metab. Eng. *5*, 96-107.

Drysch, A., El Massaoudi, M., Wiechert, W., de Graaf, A.A., and Takors, R. (2004). Serial flux mapping of *Corynebacterium glutamicum* during fed-batch L-lysine production using the sensor reactor approach. Biotechnol. Bioeng. *85*, 497-505.

Duetz, W.A., Ruedi, L., Hermann, R., O'Connor, K., Büchs, J., and Witholt, B. (2000). Methods for intense aeration, growth, storage, and replication of bacterial strains in microtiter plates. Appl. Environ. Microbiol. *66*, 2641-2646.

Eikmanns, B.J., Rittmann, D., and Sahm, H. (1995). Cloning, sequence analysis, expression, and inactivation of the *Corynebacterium glutamicum* icd gene encoding isocitrate dehydrogenase and biochemical characterization of the enzyme. J. Bacteriol. *177*, 774-782.

El Massaoudi, M., Spelthahn, J., Drysch, A., de Graaf, A., and Takors, R. (2003). Production process monitoring by serial mapping of microbial carbon flux distributions using a novel sensor reactor approach: I--Sensor reactor system. Metab. Eng. *5*, 86-95.

Fischer, E., Zamboni, N., and Sauer, U. (2004). High-throughput metabolic flux analysis based on gas chromatography-mass spectrometry derived 13C constraints. Anal. Biochem. *325*, 308-316.

Fujio, T., Teshiba, S., and Maruyama, A. (1997). Construction of a plasmid carrying both CTP synthetase and a fused gene formed from cholinephosphate cytidylyltransferase and choline kinase genes and its application to industrial CDP-choline production: enzymatic production of CDP-choline from orotic acid (Part II). Biosci. Biotechnol. Biochem. *61*, 960-964.

Gayen, K., and Venkatesh, K.V. (2007). A phenomenological model to represent the kinetics of growth by *Corynebacterium glutamicum* for lysine production. J. Ind. Microbiol. Biotechnol. *34*, 363-372.

Goto, S., Bono, H., Ogata, H., Fujibuchi, W., Nishioka, T., Sato, K., and Kanehisa, M. (1997). Organizing and computing metabolic pathway data in terms of binary relations. Pac. Symp. Biocomput., 175-186.

Heinzle, E., Matsuda, F., Miyagawa, H., Wakasa, K., and Nishioka, T. (2007). Estimation of metabolic fluxes, expression levels and metabolite dynamics of a secondary metabolic pathway in potato using label pulse-feeding experiments combined with kinetic network modelling and simulation. Plant J. *50*, 176-187.

Hüser, A.T., Chassagnole, C., Lindley, N.D., Merkamm, M., Guyonvarch, A., Elisakova, V., Patek, M., Kalinowski, J., Brune, I., Puhler, A., and Tauch, A. (2005). Rational Design of a *Corynebacterium glutamicum* Pantothenate Production Strain and Its Characterization by Metabolic Flux Analysis and Genome-Wide Transcriptional Profiling. Appl. Environ. Microbiol. *71*, 3255-3268.

Ihnen, E.D., and Demain, A.L. (1969). Glucose-6-phosphate dehydrogenase and its deficiency in mutants of *Corynebacterium glutamicum*. J. Bacteriol. *98*, 1151-1158.

Ikeda, M. (2003). Amino acid production processes. Adv. Biochem. Eng. Biotechnol. *79*, 1-35.

Ikeda, M. (2006). Towards bacterial strains overproducing L-tryptophan and other aromatics by metabolic engineering. Appl. Microbiol. Biotechnol. *69*, 615-626.

Ikeda, M., and Katsumata, R. (1999). Hyperproduction of tryptophan by *Corynebacterium glutamicum* with the modified pentose phosphate pathway. Appl. Environ. Microbiol. *65*, 2497-2502.

Ikeda, M., Okamoto, K., and Katsumata, R. (1999). Cloning of the transketolase gene and the effect of its dosage on aromatic amino acid production in *Corynebacterium glutamicum*. Appl. Microbiol. Biotechnol. *51*, 201-206.

Ishino, S., Kuga, T., Yamaguchi, K., Shirahata, K., and Araki, K. (1986). 13C NMR studies of histidine fermentation with a *Corynebacterium glutamicum* mutant. Agric. Biol. Chem. *50*, 307-310.

Ishino, S., Yamaguchi, K., Shirahata, K., and Araki, K. (1984). Involvement of meso-a,e-diaminopimelate D-dehydrogenase in lysine biosynthesis in *Corynebacterium glutamicum*. Agric. Biol. Chem. *48*, 2557-2560.

Jetten, M., and Sinskey, A.J. (1993). Characterization of phosphoenolpyruvate carboxykinase from *Corynebacterium glutamicum*. FEMS Microbiol. Lett. *111*, 183-188.

Jetten, M.S., Pitoc, G.A., Follettie, M.T., and Sinskey, A.J. (1994). Regulation of phospho(enol)-pyruvate- and oxaloacetate-converting enzymes in *Corynebacterium glutamicum*. Appl. Microbiol. Biotechnol. *41*, 47-52.

Jetten, M.S., and Sinskey, A.J. (1995). Purification and properties of oxaloacetate decarboxylase from *Corynebacterium glutamicum*. Antonie Van Leeuwenhoek *67*, 221-227.

John, G.T., Klimant, I., Wittmann, C., and Heinzle, E. (2003). Integrated optical sensing of dissolved oxygen in microtiter plates: a novel tool for microbial cultivation. Biotechnol. Bioeng. *81*, 829-836.

Kanehisa, M., Goto, S., Hattori, M., Aoki-Kinoshita, K.F., Itoh, M., Kawashima, S., Katayama, T., Araki, M., and Hirakawa, M. (2006). From genomics to chemical genomics: new developments in KEGG. Nucleic Acids Res. *34*, D354-357.

Kawaguchi, H., Vertes, A.A., Okino, S., Inui, M., and Yukawa, H. (2006). Engineering of a xylose metabolic pathway in *Corynebacterium glutamicum*. Appl. Environ. Microbiol. *72*, 3418-3428.

Kelle, R., Hermann, T., and Bathe, B. (2005). L-Lysine Production, In Handbook of Cornyebacterium glutamicum, L. Eggeling, and M. Bott, eds. (Boca Raton: CRC Press), pp. 465-488.

Kelleher, J.K. (2001). Flux estimation using isotopic tracers: common ground for metabolic physiology and metabolic engineering. Metab. Eng. *3*, 100-110.

Keseler, I.M., Collado-Vides, J., Gama-Castro, S., Ingraham, J., Paley, S., Paulsen, I.T., Peralta-Gil, M., and Karp, P.D. (2005). EcoCyc: a comprehensive database resource for *Escherichia coli*. Nucleic Acids Res. *33*, Database Issue, D334-337.

Kiefer, P., Heinzle, E., and Wittmann, C. (2002). Influence of glucose, fructose and sucrose as carbon sources on kinetics and stoichiometry of lysine production by *Corynebacterium glutamicum*. J. Ind. Microbiol. Biotechnol. *28*, 338-343.

Kiefer, P., Heinzle, E., Zelder, O., and Wittmann, C. (2004). Comparative metabolic flux analysis of lysine-producing *Corynebacterium glutamicum* cultured on glucose or fructose. Appl. Environ. Microbiol. *70*, 229-239.

Kim, H.M., Heinzle, E., and Wittmann, C. (2006). Deregulation of aspartokinase by single nucleotide exchange leads to global flux rearrangement in the central metabolism of *Corynebacterium glutamicum*. J. Microbiol. Biotechnol. *16*,1174-1179.

Kimura, E. (2003). Metabolic Engineering of Glutamate Production. Adv. Biochem. Eng. Biotechnol. *79*, 37-57.

Kinoshita, S., Shigezo, U., and Shimono, M. (1957). Studies on the amino acid fermentation: Part I: Production of L-glutamic acid by various microorganisms. J. Gen. Appl. Microbiol. 3, 193-205.

Klapa, M.I., Aon, J.C., and Stephanopoulos, G. (2003a). Ion-trap mass spectrometry used in combination with gas chromatography for high-resolution metabolic flux determination. Biotechniques *34*, 832-836, 838, 840 passim.

Klapa, M.I., Aon, J.C., and Stephanopoulos, G. (2003b). Systematic quantification of complex metabolic flux networks using stable isotopes and mass spectrometry. Eur. J. Biochem. *270*, 3525-3542.

Koizumi, S., Yonetani, Y., Maruyama, A., and Teshiba, S. (2000). Production of riboflavin by metabolically engineered *Corynebacterium ammoniagenes*. Appl. Microbiol. Biotechnol. *53*, 674-679.

Krömer, J.O., Heinzle, E., Schröder, H., and Wittmann, C. (2006a). Accumulation of homolanthionine and activation of a novel pathway for isoleucine biosynthesis in *Corynebacterium glutamicum* McbR deletion strains. J. Bacteriol. *188*, 609-618.

Krömer, J.O., Sorgenfrei, O., Klopprogge, K., Heinzle, E., and Wittmann, C. (2004). In-depth profiling of lysine-producing *Corynebacterium glutamicum* by combined analysis of the transcriptome, metabolome, and fluxome. J. Bacteriol. *186*, 1769-1784.

Krömer, J.O., Wittmann, C., Schröder, H., and Heinzle, E. (2006b). Metabolic pathway analysis for rational design of L-methionine production by *Escherichia coli* and *Corynebacterium glutamicum*. Metab. Eng. *8*, 353-369.

Kumar, S., Wittmann, C., and Heinzle, E. (2004). Minibioreactors. Biotechnol. Lett. *26*, 1-10.

Magnus, J.B., Hollwedel, D., Oldiges, M., and Takors, R. (2006). Monitoring and modeling of the reaction dynamics in the valine/leucine synthesis pathway in *Corynebacterium glutamicum*. Biotechnol. Prog. *22*, 1071-1083.

Marx, A., de Graaf, A., Wiechert, W., Eggeling, L., and Sahm, H. (1996). Determination of the fluxes in the central metabolism of *Corynebacterium glutamicum* by Nuclear Magnetic Resonance Spectroscopy combined with metabolite balancing. Biotechnol. Bioeng. *49*, 111-129.

Marx, A., Eikmanns, B.J., Sahm, H., de Graaf, A.A., and Eggeling, L. (1999). Response of the central metabolism in *Corynebacterium glutamicum* to the use of an NADH-dependent glutamate dehydrogenase. Metab. Eng. *1*, 35-48.

Marx, A., Hans, S., Möckel, B., Bathe, B., and de Graaf, A.A. (2003). Metabolic phenotype of phosphoglucose isomerase mutants of *Corynebacterium glutamicum*. J. Biotechnol. *104*, 185-197.

Marx, A., Striegel, K., de Graaf, A., Sahm, H., and Eggeling, L. (1997). Response of the central metabolism of *Corynebacterium glutamicum* to different flux burdens. Biotechnol. Bioeng. *56*, 168-180.

Matsuda, F., Morino, K., Miyashita, M., and Miyagawa, H. (2003). Metabolic flux analysis of the phenylpropanoid pathway in wound-healing potato tuber tissue using stable isotope-labeled tracer and LC-MS spectroscopy. Plant Cell Physiol. *44*, 510-517.

Matsushita, K., Otofuji, A., Iwahashi, M., Toyama, H., and Adachi, O. (2001). NADH dehydrogenase of *Corynebacterium glutamicum*. Purification of an NADH dehydrogenase II homolog able to oxidize NADPH. FEMS Microbiol. Lett. *204*, 271-276.

Mori, H., Iida, A., Fujio, T., and Teshiba, S. (1997). A novel process of inosine 5'-monophosphate production using overexpressed guanosine/inosine kinase. Appl. Microbiol. Biotechnol. *48*, 693-698.

Nakayama, K., Araki, K., and Kase, H. (1978). Microbial production of essential amino acid with *Corynebacterium glutamicum* mutants. Adv. Exp. Med. Biol. *105*, 649-661.

Neidhardt, F.C., Ingraham, J., and Schaechter, M. (1990). Physiology of the Bacterial Cell: A Molecular Approach (Sunderland, MA: Sinauer Associates).

Nöh, K., Gronke, K., Luo, B., Takors, R., Oldiges, M., and Wiechert, W. (2007). Metabolic flux analysis at ultra short time scale: isotopically non-stationary 13C labeling experiments. J. Biotechnol. *129*, 249-267.

Ohnishi, J., Katahira, R., Mitsuhashi, S., Kakita, S., and Ikeda, M. (2005). A novel gnd mutation leading to increased L-lysine production in *Corynebacterium glutamicum*. FEMS Microbiol. Lett. *242*, 265-274.

Ohnishi, J., Mitsuhashi, S., Hayashi, M., Ando, S., Yokoi, H., Ochiai, K., and Ikeda, M. (2002). A novel methodology employing *Corynebacterium glutamicum* genome information to generate a new L-lysine-producing mutant. Appl. Microbiol. Biotechnol. *58*, 217-223.

Papin, J.A., Stelling, J., Price, N.D., Klamt, S., Schuster, S., and Palsson, B.O. (2004). Comparison of network-based pathway analysis methods. Trends Biotechnol. *22*, 400-405.

Park, S.M., Shaw-Reid, C., Sinskey, A.J., and Stephanopoulos, G. (1997). Elucidation of anaplerotic pathways in *Corynebacterium glutamicum* via ^{13}C-NMR spectroscopy and GC-MS. Appl. Microbiol. Biotechnol. *47*, 430-440.

Peters-Wendisch, P.G., Kreutzer, C., Kalinowski, J., Patek, M., Sahm, H., and Eikmanns, B.J. (1998). Pyruvate carboxylase from *Corynebacterium glutamicum*: characterization, expression and inactivation of the pyc gene. Microbiology *144*, 915-927.

Peters-Wendisch, P.G., Schiel, B., Wendisch, V.F., Katsoulidis, E., Möckel, B., Sahm, H., and Eikmanns, B.J. (2001). Pyruvate carboxylase is a major bottleneck for glutamate and lysine production by *Corynebacterium glutamicum*. J. Mol. Microbiol. Biotechnol. *3*, 295-300.

Peters-Wendisch, P.G., Wendisch, V.F., de Graaf, A.A., Eikmanns, B.J., and Sahm, H. (1996). C_3-carboxylation as an anaplerotic reaction in phosphoenolpyruvate carboxylase-deficient *Corynebacterium glutamicum*. Arch. Microbiol. *165*, 387-396.

Petersen, S., de Graaf, A.A., Eggeling, L., Möllney, M., Wiechert, W., and Sahm, H. (2000). In vivo quantification of parallel and bidirectional fluxes in the anaplerosis of *Corynebacterium glutamicum*. J. Biol. Chem. *275*, 35932-35941.

Petersen, S., Mack, C., de Graaf, A.A., Riedel, C., Eikmanns, B.J., and Sahm, H. (2001). Metabolic consequences of altered phosphoenolpyruvate carboxykinase activity in *Corynebacterium glutamicum* reveal anaplerotic regulation mechanisms in vivo. Metab. Eng. *3*, 344-361.

Pons, A., Dussap, C.G., Pequignot, C., and Gros, J.B. (1996). Metabolic Flux Distribution in Corynebacterium melassecola Atcc 17965 for Various Carbon Sources. Biotechnol. Bioeng. *51*, 177-189.

Riedel, C., Rittmann, D., Dangel, P., Mockel, B., Petersen, S., Sahm, H., and Eikmanns, B.J. (2001). Characterization of the phosphoenolpyruvate carboxykinase gene from *Corynebacterium glutamicum* and significance of the enzyme for growth and amino acid production. J. Mol. Microbiol. Biotechnol. *3*, 573-583.

Rollin, C., Morgant, V., Guyonvarch, A., and Guerquin-Kern, J.L. (1995). ^{13}C-NMR studies of Corynebacterium melassecola metabolic pathways. Eur. J. Biochem. *227*, 488-493.

Salgado, H., Gama-Castro, S., Peralta-Gil, M., Diaz-Peredo, E., Sanchez-Solano, F., Santos-Zavaleta, A., Martinez-Flores, I., Jimenez-Jacinto, V., Bonavides-Martinez, C., Segura-Salazar, J., et al. (2006). RegulonDB (version 5.0): *Escherichia coli* K-12 transcriptional regulatory network, operon organization, and growth conditions. Nucleic Acids Res. *34*, D394-397.

Sauer, U. (2004). High-throughput phenomics: experimental methods for mapping fluxomes. Curr. Opin. Biotechnol. *15*, 58-63.

Sauer, U., and Eikmanns, B.J. (2005). The PEP-pyruvate-oxaloacetate node as the switch point for carbon flux distribution in bacteria. FEMS Microbiol. Rev. *29*, 765-794.

Sauer, U., Hatzimanikatis, V., Bailey, J.E., Hochuli, M., Szyperski, T., and Wuthrich, K. (1997). Metabolic fluxes in riboflavin-producing Bacillus subtilis. Nat. Biotechnol. *15*, 448-452.

Schaub, J., Schiesling, C., Reuss, M., and Dauner, M. (2006). Integrated Sampling Procedure for Metabolome Analysis. Biotechnol. Prog. *22*, 1434-42.

Schilling, C.H., Schuster, S., Palsson, B.O., and Heinrich, R. (1999). Metabolic pathway analysis: basic concepts and scientific applications in the post-genomic era. Biotechnol. Prog. *15*, 296-303.

Schmidt, K., Carlsen, M., Nielsen, J., and Villadsen, J. (1997). Modeling isotopomer distributions in biochemical networks using isotopomer mapping matrices. Biotechnol. Bioeng. *55*, 831-840.

Schrumpf, B., Schwarzer, A., Kalinowski, J., Pühler, A., Eggeling, L., and Sahm, H. (1991). A functionally split pathway for lysine synthesis in Corynebacterium glutamicium. J. Bacteriol. *173*, 4510-4516.

Schuster, S., Dandekar, T., and Fell, D.A. (1999). Detection of elementary flux modes in biochemical networks: a promising tool for pathway analysis and metabolic engineering. Trends Biotechnol. *17*, 53-60.

Schuster, S., von Kamp, A., and Pachkov, M. (2007). Understanding the roadmap of metabolism by pathway analysis. Methods Mol. Biol. *358*, 199-226.

Shiio, I., Otsuka, S.I., and Tsunoda, T. (1960). Glutamic acid formation from glucose by bacteria. J. Biochem. *47*, 414-421.

Shimizu, H., Takiguchi, N., Tanaka, H., and Shioya, S. (1999). A maximum production strategy of lysine based on a simplified model derived from a metabolic reaction network. Metab. Eng. *1*, 299-308.

Shirai, T., Fujimura, K., Furusawa, C., Nagahisa, K., Shioya, S., and Shimizu, H. (2007). Study on roles of anaplerotic pathways in glutamate overproduction of *Corynebacterium glutamicum* by metabolic flux analysis. Microb. Cell Fact. *6*, 19.

Shirai, T., Matsuzaki, K., Kuzumoto, M., Nagahisa, K., Furusawa, C., Shioya, S., and Shimizu, H. (2006). Precise metabolic flux analysis of coryneform bacteria by gas chromatography-mass spectrometry and verification by nuclear magnetic resonance. J. Biosci. Bioeng. *102*, 413-424.

Shirai, T., Nakato, A., Izutani, N., Nagahisa, K., Shioya, S., Kimura, E., Kawarabayasi, Y., Yamagishi, A., Gojobori, T., and Shimizu, H. (2005). Comparative study of flux redistribution of metabolic pathway in glutamate production by two coryne-form bacteria. Metab. Eng. *7*, 59-69.

Soga, T. (2007). Capillary electrophoresis-mass spectrometry for metabolomics. Methods Mol. Biol. *358*, 129-137.

Sonntag, K., Eggeling, L., De Graaf, A.A., and Sahm, H. (1993). Flux partitioning in the split pathway of lysine synthesis in *Corynebacterium glutamicum*. Quantification by ^{13}C- and ^{1}H-NMR spectroscopy. Eur. J. Biochem. *213*, 1325-1331.

Sonntag, K., Schwinde, J., de Graaf, A., Marx, A., Eikmanns, B., Wiechert, W., and Sahm, H. (1995). 13C NMR studies of the fluxes in the central metabolism of *Corynebacterium glutamicum* during growth and overproduction of amino acids in batch cultures. Appl. Microbiol. Biotechnol. *44*, 489-495.

Szyperski, T., Glaser, R.W., Hochuli, M., Fiaux, J., Sauer, U., Bailey, J.E., and Wuthrich, K. (1999). Bioreaction network topology and metabolic flux ratio analysis by biosynthetic fractional ^{13}C labeling and two-dimensional NMR spectroscopy. Metab. Eng. *1*, 189-197.

Takors, R., Bathe, B., Rieping, M., Hans, S., Kelle, R., and Huthmacher, K. (2007). Systems biology for industrial strains and fermentation processes--example: amino acids. J. Biotechnol. *129*, 181-190.

Tesch, M., de Graaf, A.A., and Sahm, H. (1999). In vivo fluxes in the ammonium-assimilatory pathways in *Corynebacterium glutamicum* studied by ^{15}N nuclear magnetic resonance. Appl. Environ. Microbiol. *65*, 1099-1109.

Toya, Y., Ishii, N., Hirasawa, T., Naba, M., Hirai, K., Sugawara, K., Igarashi, S., Shimizu, K., Tomita, M., and Soga, T. (2007). Direct measurement of isotopomer of intracellular metabolites using capillary electrophoresis time-of-flight mass spectrometry for efficient metabolic flux analysis. J. Chromatogr. A *1159*, 134-141.

Vallino, J.J., and Stephanopoulos, G. (1993). Metabolic flux distributions in *Corynebacterium glutamicum* during growth and lysine overproduction. Biotechnol. Bioeng. *41*, 633-646.

Vallino, J.J., and Stephanopoulos, G. (1994). Carbon flux distributions at the glucose-6-phosphate branch point in *Corynebacterium glutamicum* during lysine over-production. Biotechnol. Prog. *10*, 327-334.

van Winden, W.A., Wittmann, C., Heinzle, E., and Heijnen, J.J. (2002). Correcting mass isotopomer distributions for naturally occurring isotopes. Biotechnol. Bioeng. *80*, 477-479.

Varela, C., Agosin, E., Baez, M., Klapa, M., and Stephanopoulos, G. (2003). Metabolic flux redistribution in *Corynebacterium glutamicum* in response to osmotic stress. Appl. Microbiol. Biotechnol. *60*, 547-555.

Velagapudi, V.R., Wittmann, C., Lengauer, T., Talwar, P., and Heinzle, E. (2006). Metabolic Screening of Saccharomyces cerevisiae Single Knockout Strains Reveals Unexpected Mobilization of Metabolic Potential. Process Biochem. *41*, 2170-2179.

Velagapudi, V.R., Wittmann, C., Schneider, H., and Heinzle, E. (2007). Metabolic Flux Screening of Saccharomyces cerevisiae Single Knockout Strains on Glucose and Galactose Supports Elucidation of Gene Function. J. Biotechnol. *132*, 395-404.

Visser, D., and Heijnen, J.J. (2003). Dynamic simulation and metabolic redesign of a branched pathway using linlog kinetics. Metab. Eng. *5*, 164-176.

Waagepetersen, H.S., Qu, H., Sonnewald, U., Shimamoto, K., and Schousboe, A. (2005). Role of glutamine and neuronal glutamate uptake in glutamate homeostasis and synthesis during vesicular release in cultured glutamatergic neurons. Neurochem. Int. *47*, 92-102.

Walker, T.E., Han, C.H., Kollman, V.H., London, R.E., and Matwiyoff, N.A. (1982). 13C nuclear magnetic resonance studies of the biosynthesis by Microbacterium ammoniaphilum of L-glutamate selectively enriched with carbon-13. J. Biol. Chem. *257*, 1189-1195.

Wendisch, V.F., Bott, M., and Eikmanns, B.J. (2006a). Metabolic engineering of *Escherichia coli* and *Corynebacterium glutamicum* for biotechnological production of organic acids and amino acids. Curr. Opin. Microbiol. *9*, 268-274.

Wendisch, V.F., Bott, M., Kalinowski, J., Oldiges, M., and Wiechert, W. (2006b). Emerging *Corynebacterium glutamicum* systems biology. J. Biotechnol. *124*, 74-92.

Wendisch, V.F., de Graaf, A.A., Sahm, H., and Eikmanns, B.J. (2000). Quantitative determination of metabolic fluxes during coutilization of two carbon sources: comparative analyses with *Corynebacterium glutamicum* during growth on acetate and/or glucose. J. Bacteriol. *182*, 3088-3096.

Wiechert, W. (2001). ^{13}C metabolic flux analysis. Metab. Eng. *3*, 195-206.

Wiechert, W., Mollney, M., Isermann, N., Wurzel, M., and de Graaf, A.A. (1999). Bidirectional reaction steps in metabolic networks: III. Explicit solution and analysis of isotopomer labeling systems. Biotechnol. Bioeng. *66*, 69-85.

Wittmann, C. (2007). Fluxome analysis using GC-MS. Microb. Cell Fact. *6*, 6.

Wittmann, C., and Becker, J. (2007). The L-lysine story: From metabolic pathways to industrial production, In: Microbiology Monographs (Berlin/Heidelberg: Springer). *5*, 39-70.

Wittmann, C., and de Graaf, A. (2005). Metabolic flux analysis in *Corynebacterium glutamicum*, In Handbook of *Corynebacterium glutamicum*, L. Eggeling, and M. Bott, eds. (Boca Raton: CRC Press), pp. 277-304.

Wittmann, C., Hans, M., and Heinzle, E. (2002). In vivo analysis of intracellular amino acid labelings by GC/MS. Anal. Biochem. *307*, 379-382.

Wittmann, C., and Heinzle, E. (1999). Mass spectrometry for metabolic flux analysis. Biotechnol. Bioeng. *62*, 739-750.

Wittmann, C., and Heinzle, E. (2001a). Application of MALDI-TOF MS to lysine-producing *Corynebacterium glutamicum*: a novel approach for metabolic flux analysis. Eur. J. Biochem. *268*, 2441-2455.

Wittmann, C., and Heinzle, E. (2001b). MALDI-TOF MS for quantification of substrates and products in cultivations of *Corynebacterium glutamicum*. Biotechnol. Bioeng. *72*, 642-647.

Wittmann, C., and Heinzle, E. (2001c). Modeling and experimental design for metabolic flux analysis of lysine-producing Corynebacteria by mass spectrometry. Metab. Eng. *3*, 173-191.

Wittmann, C., and Heinzle, E. (2002). Genealogy profiling through strain improvement by using metabolic network analysis: metabolic flux genealogy of several generations of lysine-producing corynebacteria. Appl. Environ. Microbiol. *68*, 5843-5859.

Wittmann, C., Kiefer, P., and Zelder, O. (2004a). Metabolic fluxes in *Corynebacterium glutamicum* during lysine production with sucrose as carbon source. Appl. Environ. Microbiol. *70*, 7277-7287.

Wittmann, C., Kim, H.M., and Heinzle, E. (2004b). Metabolic network analysis of lysine producing *Corynebacterium glutamicum* at a miniaturized scale. Biotechnol. Bioeng. *87*, 1-6.

Wittmann, C., Kim, H.M., John, G., and Heinzle, E. (2003). Characterization and application of an optical sensor for quantification of dissolved O_2 in shake-flasks. Biotechnol. Lett. *25*, 377-380.

Wittmann, C., Krömer, J.O., Kiefer, P., Binz, T., and Heinzle, E. (2004c). Impact of the cold shock phenomenon on quantification of intracellular metabolites in bacteria. Anal. Biochem. *327*, 135-139.

Yang, C., Hua, Q., and Shimizu, K. (1999). Development of a kinetic model for L-lysine biosynthesis in *Corynebacterium glutamicum* and its application to metabolic control analysis. J. Biosci. Bioeng. *88*, 393-403.

Yang, T.H., Heinzle, E., and Wittmann, C. (2005). Theoretical aspects of ^{13}C metabolic flux analysis with sole quantification of carbon dioxide labeling. Comput. Biol. Chem. *29*, 121-133.

Yang, T.H., Wittmann, C., and Heinzle, E. (2006a). Respirometric ^{13}C flux analysis--Part II: in vivo flux estimation of lysine-producing *Corynebacterium glutamicum*. Metab. Eng. *8*, 432-446.

Yang, T.H., Wittmann, C., and Heinzle, E. (2006b). Respirometric ^{13}C flux analysis, Part I: design, construction and validation of a novel multiple reactor system using on-line membrane inlet mass spectrometry. Metab. Eng. *8*, 417-431.

Zupke, C., and Stephanopoulos, G. (1994). Modeling of Isotope Distributions and Intracellular Fluxes in Metabolic Networks Using Atom Mapping Matrices. Biotechnol. Prog. 10, 489-498.

From: Corynebacteria: Genomics and Molecular Biology. Edited by: Andreas Burkovski.

Plasmids and Promoters in Corynebacteria and Their Applications

6

Jan Nešvera and Miroslav Pátek

Abstract

The discovery of small plasmids in corynebacteria initiated genetic analysis and molecular breeding of industrially important *Corynebacterium* strains. The constructed plasmid vectors for corynebacteria include cloning vectors, expression vectors and promoter-probe vectors. They were particularly used for the isolation of various genes and their overexpression and inactivation, which allowed their metabolic role to be deduced. Using promoter-probe vectors, many promoters of corynebacteria were localized and characterized. A systematic promoter sequence analysis and mutational studies resulted in the definition of promoter consensus sequences. In addition to the promoters of housekeeping genes, promoters recognized by alternative sigma factors of RNA polymerase have recently been characterized. Plasmid vectors for corynebacteria are important tools, not only for the detailed analysis of gene functions and their regulation, but also for the improvement of industrial strains. The gene manipulations within the chromosome using integrative plasmid vectors and graded expression from altered promoters, both of which result in stable recombinant strains, are considered a promising approach to strain breeding.

1. Introduction

Genetic manipulations in the most intensively studied corynebacteria, the amino acid producers, were initiated in 1984 with the isolation of small plasmids from strains of *Corynebacterium glutamicum* (Miwa et al., 1984; Ozaki et al., 1984; Santamaría et al., 1984). These plasmids served as a basis for the construction of various cloning vectors carrying suitable selectable markers, mostly antibiotic resistance determinants. Efficient genetic analyses and manipulations in corynebacteria were made possible by the development and optimization of DNA transfer methods (transformation, electrotransformation and conjugation). The isolation of genes, investigation of their ex-

Table 6.1 Native plasmids of corynebacteria with determined complete nucleotide sequences

Plasmid family	Plasmid	Size (bp)	Marker(s)[a]	Host strain (former designation)	GenBank Accession Number	Reference
A) RC-replicating plasmids						
pIJ101/pJV1	pBL1[b]	4,447	-	*C. glutamicum* ATCC 13869 (*B. lactofermentum*)	AF092037	Santamaría et al., 1984
	pCC1	4,109	-	*C. callunae* ATCC 15991	AJ308231	Sandoval et al., 1984; Venkova-Canova et al., 2004
	pAG3	4,603	-	*C. glutamicum* 22220 (*C. melassecola*)	AY172684	Takeda et al., 1990; Tauch et al., 2003b
	pCG2	6,758	-	*C. glutamicum* ATCC 31832	AY172685	Ozaki et al., 1984; Tauch et al., 2003b
	pXZ608	5,949	-	*C. glutamicum* 227	AF479770	Lei et al., 2002
pNG2	pCG1[c]	3,069	-	*C. glutamicum* ATCC 31808	AB027714	Ozaki et al., 1984
	pGA1	4,823	-	*C. glutamicum* LP-6	X90817	Sonnen et al., 1991; Nešvera et al., 1997
	pYM2	4,369	-	*C. efficiens*	AB084384	
	pGA2	19,218	-	*C. glutamicum* LP-6	AY172687	Sonnen et al., 1991; Tauch et al., 2003b
	pAG1	19,751	Tc^r	*C. glutamicum* 22243 (*C. melassecola*)	AF121000	Takeda et al., 1990; Tauch et al., 2000b
	pTET3	27,856	Sm^r/Sp^r, Su^r, Tc^r	*C. glutamicum* LP-6	AJ420072	Tauch et al., 2002a
	pCG4	29,371	Sm^r/Sp^r, Su^r	*C. glutamicum* ATCC 31830	AF164956	Katsumata et al., 1984; Tauch et al., 2003b
	pCE2	23,743	-	*C. efficiens* YS-314	AP005225	Nishio et al., 2003
	pCE3	48,762	-	*C. efficiens* YS-314	AP005226	Nishio et al., 2003
	pNG2	15,100	Em^r	*C. diphtheriae* S601	AF492560	Schiller et al. 1980; Tauch et al., 2003a
	pCJ84	7,580	-	*C. jeikeium* CJ84	AY048596	Tauch et al., 2004
	pK43	7,617	-	*C. jeikeium* K43	AF364477	Tauch et al., 2004
	pK64	10,904	-	*C. jeikeium* K64	AY079086	Tauch et al., 2004

	pA505	11,946	-	*C. jeikeium*	AY26399	Pitcher et al., 1990; Tauch et al., 2004
	pB85776	14,875	-	*C. jeikeium* B85776	AF486522	Tauch et al., 2004
	pA501	19,934	Cm^r	*C. jeikeium*	AY266269	Pitcher et al., 1990; Tauch et al., 2004
	pLEW279a	34,606	Sm^r/Sp^r, Su^r, Tc^r	*Corynebacterium* sp. L2-79-05	DQ390458	Williams et al., 2006
Unclassified	pCR1	1,488	-	*C. renale*	X99132	Srivastava et al., 2006
B) Theta-type plasmids						
ColE2-P9	pXZ1014	2,444	-	*C. glutamicum* 1014	X72691	
	pXZ10145	4,885	Cm^r	*C. glutamicum* 1014	U85507	Zheng et al., 1987
IncW	pLEW279b	29,854	-	*Corynebacterium* sp. L2-79-05	DQ390457	Williams et al., 2006
	pCRYA4[d]	~ 48,000	-	*C. glutamicum* LP-6	AY172686[d]	Tauch et al., 2003b
Unclassified	pCR2	3,195	-	*C. renale*	EF488047	Walia et al., 2007
	pCGR1	49,120	-	*C. glutamicum* R	AP009045	Yukawa et al., 2007
C) Plasmids coding for replication initiation proteins of two different types						
pNG2, IncW	pKW4	14,323	-	*C. jeikeium*	AF401314	Kerry-Williams and Noble 1986; Tauch et al., 2004
	pTP10	51,409	Cm^r, Em^r, Km^r, Tc^r	*C. striatum* M82B	AF139896	Tauch et al. 2000a

[a]Antibiotic resistance markers: Cm^r, chloramphenicol; Em^r, erythromycin; Km^r kanamycin; Sm^r, streptomycin; Sp^r, spectinomycin; Su^r, sulfamethoxaazole; Tc^r, tetracycline.
[b]Complete nucleotide sequence of highly similar plasmids pAM330 (4,448 bp; GenBank D00038; Miwa et al., 1984; Yamaguchi et al., 1986) and pGX1901 (4,457 bp; GenBank X03987; Filpula et al., 1986; Smith et al., 1986) from the same host strain were also determined.
[c]Complete nucleotide sequence of highly similar plasmid pSR1 (3,054 bp; GenBank Z22927; Yoshihama et al., 1985; Archer and Sinskey, 1993) from *C. glutamicum* ATCC 19223 was also determined.
[d]Only the 1,856 bp replication region of pCRY4 was sequenced.

pression and studies of their regulatory DNA sequences led to the mapping of promoters as the main transcriptional signals. The analysis of particular promoters (Vašicová et al., 1999) and attempts to generalize their structure (Pátek et al., 1996; Pátek et al., 2003) contributed to understanding the mechanisms of transcription initiation in corynebacteria. Knowledge of the precise positions and consensus sequence of the corynebacterial promoters facilitated purposeful modulation of gene expression and the application of rational metabolic design, particularly in the amino acid producers (Krämer, 1996; de Graaf et al., 2001). Advances in the genetic analysis of *C. glutamicum*, whose complete genome sequence was determined by two groups (Ikeda and Nakagawa, 2003; Kalinowski et al., 2003), made this bacterium a suitable Gram-positive model organism for basic studies and facilitated the efficient improvement of industrial strains.

2. Native plasmids of corynebacteria

The first native (indigenous) plasmids of corynebacteria were detected in the human pathogen *Corynebacterium diphtheriae* (Schiller et al., 1980). To develop cloning vectors for gene manipulations in the amino acid-producing corynebacteria, native plasmids of these bacteria, classified into species *C. glutamicum*, *Corynebacterium callunae* and *Corynebacterium efficiens*, were searched for. In 1984, the small cryptic plasmids pBL1 and pCC1 were found in *C. glutamicum* ATCC-13869 (Santamaría et al., 1984) and in *C. callunae* ATCC 15991 (Sandoval et al., 1984), respectively. Since then, 25 native plasmids of various *C. glutamicum* strains (Tauch et al., 2003b; Yukawa et al., 2007), 2 plasmids of *C. callunae* ATCC 15991 and 3 plasmids of *C. efficiens* have been isolated, ranging in size from 2.4 to 95 kb (Tauch, 2005). Native plasmids were also detected in the human pathogens *Corynebacterium striatum* (formely misclassified as *Corynebacterium xerosis*) (Kono et al., 1983; Tauch et al., 2000a) and *Corynebacterium jeikeium* (Kerry-Williams and Noble, 1986; Tauch et al., 2004) and in a naphthalene-degrading strain of *Corynebacterium renale* (Nath and Deb, 1995). It has recently been found that bacterial strains isolated from poultry litter and classified as members of the genus *Corynebacterium* are the major reservoir of antibiotic resistance integrons (Nandi et al., 2004). Although only two plasmids from one of these strains (*Corynebacterium* sp. L2-79-05) have been characterized so far (Williams et al., 2006), it is likely that many more yet undescribed plasmids are present in these natural isolates of corynebacteria.

The complete nucleotide sequences of 30 native plasmids from various strains of the genus *Corynebacterium* have been determined so far. Analysis of these nucleotide sequences revealed genes coding for replication initiator proteins, which made it possible to suggest the mode of replication of these plasmids. Circular bacterial plasmids mainly use two modes of DNA replication, namely rolling circle (RC) mode and theta-type mode (del Solar et al., 1998). As shown in Table 6.1, most of the plasmids from corynebacteria use RC mode for their replication. Only 6 plasmids out of 31 listed in Table 6.1 are assumed to replicate in theta-type mode. This group consists of small plasmids pXZ10142, pXZ10145 and pCR2 (2.4 – 4.9 kb) and large plasmids pLEW279b, pCGR1 and pCRY4 (30 – 49 kb). Two plasmids, pTP10 from *C. striatum* and pKW4 from *C. jeikeium*, were found to code for two different replication initiator proteins, one typical for RC mode, the other typical for theta-type mode (Tauch et al., 2000a; Tauch et al., 2004). The actual mechanism of their replication has not yet been analyzed, so it is not known if both of the replication proteins are functional and if the plasmids can switch their mode of replication.

The degree of amino acid sequence similarity of replication initiator proteins has been used as a basic criterion for the classification of plasmids (especially those replicating by the RC mechanism) into distinct families. As shown in Table 6.1, five small RC plasmids of corynebacteria (pBL1, pCC1, pAG3, pCG2 and pXZ608) have been classified into the pIJ101/pJV1 family of RC-replicating plasmids defined by Khan (1997). Other authors classify the plasmids of this family into the cluster III of group III of RC plasmids (i.e. into a subfamily of the pC194 family) in the *Database of Plasmid Replicons* (http://www.essex.ac.uk/bs/staff/osborn/DPR_home.htm). All other RC plasmids of corynebacteria (with the exception of the smallest corynebacterial plasmid pCR1 from *C. renale*) form a new group of RC replicons, named the pNG2 family (Tauch et al., 2003a). The pNG2 family consists of 17 plasmids of corynebacteria shown in Table 6.1 and of two plasmids of the related species *Brevibacterium stationis* (plasmid pBY503) and *Arcanobacterium pyogenes* (plasmid pAP2). It should be mentioned that Tauch and his coworkers (Tauch et al., 2003b) defined four distinct families of *C. glutamicum* plasmids: the pBL1 and pCG1 families of RC-replicating plasmids and the pXZ10142 and pCRY4 families of theta-type replicating plasmids. In this review, we classify plasmids of corynebacteria into the general plasmid families defined regardless of their hosts. We have therefore included the *C. glutamicum* plasmids of the pBL1, pCG1, pXZ10142 and pCRY4 families (Tauch et al., 2003b) into the pIJ101/pJV1, pNG2, ColE2-P9 and IncW families, respectively.

As shown in Table 6.1, most of the sequenced plasmids of corynebacteria are cryptic. Antibiotic resistance determinants were found predominantly on larger plasmids. Among the small plasmids, only the plasmid pXZ10145 replicating by theta-type mode carries a chloramphenicol resistance determinant on transposon Tn*45* (Shen et al., 1993). The construction of cloning vectors for corynebacteria, based on native small cryptic plasmids, therefore required the insertion of antibiotic resistance determinants coming from other plasmids. Most of the constructed plasmid vectors contain replicons of the small cryptic plasmids pBL1, pCG1 and pGA1 from *C. glutamicum* or the minimal replicon of the plasmid pNG2 from *C. diphtheriae*. The basic characteristics of the native plasmids used for vector construction are described in more detail below.

3. Plasmids of corynebacteria used for vector construction

3.1 Plasmid pBL1

The small cryptic plasmid pBL1 was isolated from *C. glutamicum* ATCC 13869 (originally designated *Brevibacterium lactofermentum*) (Santamaría et al., 1984). Highly similar (or even identical) plasmids, designated pAM330 (Miwa et al., 1984), pBL25 (Kato et al., 1989), pBL100 (Shaw and Hartley, 1988), pGX1901 (Smith et al., 1986), pWS101 (Yoshihama et al., 1985) and pX18 (Yeh et al., 1986), were independently isolated from various strains of amino acid-producing corynebacteria. The copy number of these plasmids in *C. glutamicum* cells was estimated to be between 10 and 30 copies per chromosome (Miwa et al., 1984; Santamaría et al., 1984). The complete nucleotide sequences of pBL1 (GenBank AF092037), pAM330 (Yamaguchi et al., 1986) and pGX1901 (Filpula et al., 1986) were determined, revealing only negligible differences.

Analysis of the nucleotide sequence of pBL1 revealed five open reading frames (ORFs) on its DNA (Figure 6.1A). It was found that a 1.8-kb DNA fragment sufficient for autonomous replication of pBL1 contains two ORFs (Fernandez-Gonzalez et al., 1994). The observed accumulation of single-strand DNA intermediates of pBL1 in *C. glutamicum* cells indicates that pBL1 replicates by RC mode (Fernandez-Gonzalez et al., 1994). The deduced amino acid sequence of the gene product of the larger ORF present on the pBL1 minimal replicon shows a high degree of similarity with the replication initiator proteins (Rep proteins) encoded by RC-replicating plasmids classified into the pIJ101/pJV1 family, and the gene was therefore designated *rep*. The Rep protein of pBL1 contains amino acid sequences of motifs II (HuHuuu; u = hydrophobic residues) and III (uxxYuxK), universally conserved in Rep proteins of RC plasmids from all basic families (Ilyina and Koonin, 1992), as well as of the C-terminal motif V (wxeyexaxxgrraxxwsrglr) typical for RC plasmids from the pIJ101/pJV1 family (Billington et al., 1998). The putative double-strand origin (*dso*) of RC replication was localized, according to the sequence similarity with the *dso* of plasmid pJV1, just upstream of the pBL1 *rep* gene (Fernandez-Gonzalez et al., 1994).

It was found that the constructed *C. glutamicum-Escherichia coli* shuttle vectors, containing the whole pBL1 sequence, inhibited growth and caused filamentation of *E. coli* host cells. Deletion of the 1.2-kb HindIII-AccI DNA fragment, located outside the pBL1 minimal replicon and carrying the *orf3* gene, abolished these defects. The *orf3* gene product of pBL1 is therefore responsible for the growth defects of *E. coli* host cells (Goyal et al., 1996). Therefore, pBL1 DNA lacking the *orf3* gene should be used for the construction of optimized *C. glutamicum-E. coli* shuttle vectors.

Plasmid constructs containing the pBL1 replicon were found to replicate in *C. callunae* (Sandoval et al., 1985), *C. diphtheriae* (Tauch et al., 2002c), *Arthrobacter* sp. B3728 (Shaw and Hartley, 1988), *Brevibacterium methylicum* (Nešvera et al., 1994), *Rhodococcus* sp. R312 (Duran, 1998) and *Bifidobacterium animalis* (Argnani et al., 1996).

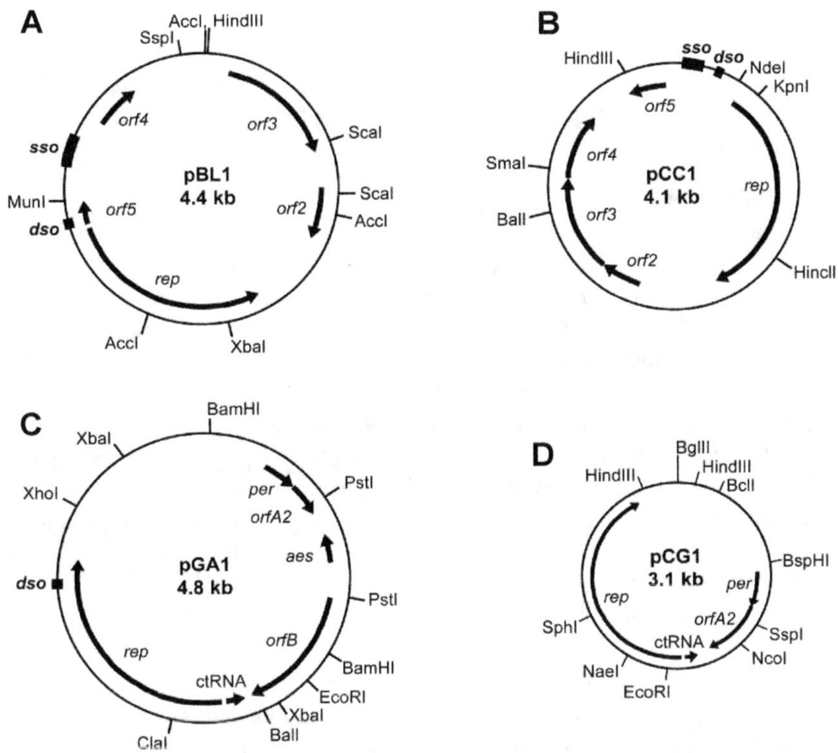

Figure 6.1. Maps of small cryptic plasmids from corynebacteria. (A) pBL1 from *C. glutamicum* (formerly *B. lactofermentum*) ATCC 13869, (B) pCC1 from *C. callunae* ATCC 15991, (C) pGA1 from *C. glutamicum* LP-6, (D) pCG1 from *C. glutamicum* ATCC 31808. Coding regions are marked by arrows indicating the direction of transcription. *dso*, double-strand origin of RC replication; *sso*, single-strand origin of RC replication; *rep*, initiator of RC replication; *per*, positive effector of plasmid replication; *aes*, accessory element of stable maintenance; ctRNA, countertranscribed regulatory RNA.

3.2 Plasmid pCC1

The small cryptic plasmid pCC1 was isolated from *C. callunae* NRRL B-2244 (= ATCC 15991) (Sandoval et al., 1984). The analysis of its complete nucleotide sequence (GenBank AJ308231) revealed five *orf*s longer than 200 bp (Figure 6.1B). The largest *orf* was found to be sufficient for the replication and stable maintenance of pCC1 (Venkova-Canova et al., 2004). The deduced amino acid sequence of its gene product showed significant similarity to that of Rep proteins of RC plasmids classified into the pIJ101/pJV1 family and the gene was therefore designated *rep*. According to these data, pCC1 replicates by the RC mechanism. The Rep protein of pCC1 showed the highest degree of similarity to the Rep proteins of plasmids pAG3 and pBL1 from *C. glutamicum* (34% and 32% identity, respectively). The putative double-strand origin (*dso*) of RC replication was localized, according to the sequence similarity with the *dso* of plasmid pJV1, just upstream of the pCC1 *rep* gene.

Despite the fact that related plasmids are mostly incompatible, replicons of pCC1 and pBL1 coding for similar Rep proteins were found to be compatible (Venkova-Canova et al., 2004). Vectors based on plasmids pBL1 and pCC1 can therefore serve as useful tools for the development of biplasmid systems in corynebacteria.

3.3 Plasmid pGA1

The small cryptic plasmid pGA1 was isolated from *C. glutamicum* LP-6 (Sonnen et al., 1991). It was found to be present in about 35 copies per chromosome in *C. glutamicum* cells. Analysis of the complete nucleotide sequence of pGA1 revealed five ORFs on its DNA (Figure 6.1C). Its minimal replicon (1.7 kb) contains the *rep* gene coding for the initiator of replication by the RC mode (Nešvera et al., 1997) and the double-strand origin of RC replication, located atypically within the distal part of the *rep* gene (Abrhámová et al., 2002). The deduced amino acid sequence of the pGA1 *rep* gene product shows a high degree of similarity with the Rep proteins of many other plasmids of corynebacteria. Of the several conserved motifs typical for the Rep proteins of RC plasmids from basic families (Ilyina and Koonin, 1992), only motif III (uxxYuxK) was identified in the Rep proteins of these plasmids from corynebacteria, which were thus classified into a new group of RC plasmids, named the pNG2 family (Tauch et al., 2003a). The original uxxYuxK motif is only present in the Rep proteins of small plasmids pGA1, pCG1 and pYM2, while the larger plasmids of the pNG2 family possess the uxxYuxxxK motif. The pNG2 family can thus be divided into two distinct subfamilies.

Small (ca. 89 nt in length) countertranscribed RNA (ctRNA), encoded by the region upstream of the pGA1 *rep* gene, was found to be responsible for the negative control of plasmid copy number. In addition to the ctRNA, the sequence within the *rep* leader region itself seems to act as a second regulatory element of a novel type, negatively influencing expression of the pGA1 *rep* gene (Venkova-Canova et al., 2003).

The plasmid pGA1 contains two genes, *per* and *aes*, whose products positively influence stable plasmid maintenance. Derivatives of pGA1 devoid of the *per* gene exhibited a significant decrease in their copy number in *C. glutamicum* cells and displayed high segregational instability. Introduction of the *per* gene *in trans* into the cells harboring these deletion plasmids markedly increased their copy number and stability. The *per* gene product therefore positively influences pGA1 plasmid copy number (Nešvera et al., 1997). The small *aes* gene, when present *in trans*, was shown to increase the segregational stability of pGA1 derivatives carrying the main stability determinant *per*. The *aes* gene product therefore acts as an accessory element involved in the stable maintenance of pGA1 (Venkova et al., 2001).

Plasmid constructs containing the pGA1 replicon were found to replicate in *C. callunae* (Sonnen et al., 1991), *C. diphtheriae* (Tauch et al., 2002c), *Rhodococcus erythropolis* (Veselý et al., 2003) and *Clavibacter michiganensis* ssp. *michiganensis* (Tauch, 2005).

3.4 Plasmid pCG1

The small cryptic plasmid pCG1 was isolated from *C. glutamicum* ATCC 31808 (Ozaki et al., 1984). Plasmids with identical restriction maps, designated pHM1519 (Miwa et al., 1984), pCG100 (Shaw and Hartley, 1988; Trautwetter and Blanco, 1991) and pSR1 (Yoshihama et al., 1985) were independently isolated from other *C. glutamicum* strains. The complete nucleotide sequence of plasmids pCG1 (GenBank AB027714) and pSR1 (Archer and Sinskey, 1993), were determined showing only minor differences. Analysis of the pCG1 sequence revealed three ORFs, designated according to their homology with the genes of the plasmid pGA1 (Figure 6.1D), while only two ORFs were detected on the pSR1 DNA (Archer and Sinskey, 1993). It was found that the plasmid pCG100 is present in *C. glutamicum* cells in about 30 copies per chromosome. Its 1.9-kb NcoI-BglII fragment is sufficient for autonomous plasmid replication (Trautwetter and Blanco, 1991). This fragment contains the *rep* gene coding for the protein highly similar to the Rep proteins of RC plasmids from corynebacteria classified into the pNG2 family. Since the uxxYuxK motif typical for small plasmids of the pNG2 family is present in the pCG1 Rep protein, pCG1 can be classified, together with pGA1 and pYM2, into a subfamily of small plasmids. Analysis of the pCG1 DNA sequence revealed its additional high similarity with the DNA sequence of the plasmid pGA1 also outside the *rep* gene. The pCG1 homolog of the pGA1 *per* gene, coding for the positive regulator of plasmid copy number, was found to complement the deletion of *per* in pGA1 (Nešvera et al., 1997). The leader sequence of the pCG1 *rep* gene is highly similar (92% identity) to that of pGA1 coding for regulatory ctRNA (Venkova-Canova et al., 2003). Plasmid pCG1 therefore very probably also codes for a ctRNA that negatively controls plasmid copy number.

Plasmid constructs containing the pCG1 replicon were found to replicate in *C. callunae*, *Corynebacterium ammoniagenes*, *Corynebacterium pilosum*, *Arthrobacter ilicis*, *Brevibacterium stationis*, *Rhodococcus fascians* (Schäfer et al., 1990), *C. diphtheriae* (Tauch et al., 2002c), *Arthrobacter* sp. B3728 (Shaw and Hartley, 1988), *Brevibacterium methylicum* (Nešvera et al., 1994), *R. erythropolis* (Veselý et al., 2003), *Rhodococcus* sp. R312 (Duran, 1998) and *Bifidobacterium animalis* (Argnani et al., 1996). The broad host range and very small size of pCG1 make this plasmid an extraordinarily suitable basis for the construction of vectors used in various corynebacteria and related species.

3.5 Plasmid pNG2

The erythromycin resistance plasmid pNG2 from the human pathogen *C. diphtheriae* S601 is the first plasmid isolated from corynebacteria (Schiller et al., 1980). It replicates in *C. glutamicum*, *Corynebacterium ulcerans* (Serwold-Davis et al., 1987), *Corynebacterium pseudotuberculosis* (Songer et al., 1991), *Arcanobacterium pyogenes* (Jost et al., 1997), in various *Mycobacterium* species (Radford and Hodgson, 1991) and even in *Escherichia coli* (Serwold-Davis et al., 1987). It was found that merely a 1.8-kb fragment of pNG2 is sufficient for its replication (Radford and Hodgson,

Table 6.2. Plasmid cloning vectors for corynebacteria

Vector	Size (kb)	*C. glutamicum* replicon	Selection[a] markers	Cloning sites	Further characteristics	Reference
A) Autonomously replicating *C. glutamicum*/*E. coli* shuttle vectors						
pEK0	6.1	pBL1	Km^r	EcoRI, SacI, KpnI, SmaI, BamHI, SalI	-	Eikmanns et al., 1991a
pEC5	7.2	pBL1	Cm^r	EcoRI, SacI, KpnI, SmaI, SalI, HindIII	-	Eikmanns et al., 1991a
pEBM2	8.0	pBL1	Km^r, *lacZα*	pUC18 MCS	mobilizable	Tauch et al., 1998
pMJ-A[b]	5.2	pBL1	Km^r, *lacZα*	pUC18 MCS	-	Jakoby et al., 1999
pULMJ55	8.6	pBL1	Hm^r	BglII	positive selection of recombinants (Cm^s , Km^s)	Cadenas et al., 1996
pCRB1[c]	4.1	pBL1	Cm^r, *lacZα*	pUC18 MCS	-	Nakata et al., 2004
pECKA	5.4	pBL1	Km^r, *lacZα*	pUC19 MCS	-	Elišáková et al., 2005
pJC1	6.1	pCG1	Km^r	BamHI, XbaI, SalI, PstI	-	Cremer et al., 1990
pSRK21	5.8	pCG1	Km^r, *lacZα*	pUC19 MCS	-	Veselý et al., 2003
pEC-K18*mob2*[d]	5.7	pGA1	Km^r, *lacZα*	pUC18 MCS	mobilizable	Tauch et al., 2002c
pSCCD1	5.2	pCC1	Sm^r/Sp^r	SalI, XbaI, BamHI, SmaI, SacI	compatible with pBL1 and pCG1 replicons	Venkova-Canova et al., 2004
pWK0	6.0	pNG2	Km^r	BamHI, XbaI, SalI, PstI, SphI, EcoRI	low copy number	Reinscheid et al., 1994
pBHK18[e]	3.3	pNG2	Km^r, *lacZα*	pUC18 MCS	low copy number	Kirchner and Tauch, 2003
pEP2	3.1	pNG2	Km^r	EcoRI, BamHI, SalI, PstI	low copy number	Radford and Hodgson, 1991

B) Autonomously replicating *C. glutamicum* vectors for self-cloning						
pAJ228	7.6	pBL1	Tp^r	ClaI, HpaI, XhoI, XmaI	-	Takagi et al., 1986
pCG11	6.9	pCG1	Sm^r/Sp^r	PstI, StuI, NheI, SpeI, BclI, ScaI	-	Ikeda and Katsumata, 1998
pSELF2000X	5.8	pGA1	*alr*	EcoRI, XhoI, BamHI	growth of Δ*alr* mutant without D-alanine	Tauch et al., 2000b
C) Vectors for integration into the *C. glutamicum* chromosome						
pSKFT2	7.0	pBL1	Km^r, *lacZα*	HindIII, PstI, BamHI, SmaI, EcoRI	*ts* replicon, integration into chromosome at 34 °C	Nakamura et al., 2006
pEM1	3.3	-	Km^r	pUC19 MCS	mobilizable	Schrumpf et al., 1991
pEM1dppc	3.9	-	Km^r	pUC19 MCS	mobilizable, site-specific integration	Vašicová et al., 1998
pA3253	4.7	-	Km^r, *lacZα*	pBluescript II SK MCS	Φ16 site-specific integration function	Moreau et al., 1999
pKX15	5.0	-	Km^r	pUC4-KIXX MCS	site-specific integration into *rrnD*	Amador et al., 2000
pK18*mob*[f]	3.8	-	Km^r, *lacZα*	pUC18 MCS	mobilizable	Schäfer et al., 1994
pK18*mobsacB*[f]	5.7	-	Km^r, *lacZα*	pUC18 MCS	mobilizable, *sacB* gene for selection of double cross-over events	Schäfer et al., 1994
pK18*mob2*[g]	3.0	-	Km^r, *lacZα*	pUC18 MCS	mobilizable	Kirchner and Tauch, 2003

[a]Antibiotic resistance markers: Cm^r, chloramphenicol; Gm^r, gentamycin; Hm^r, hygromycin; Km^r, kanamycin; Sm^r, streptomycin; Sp^r, spectinomycin; Tc^r, tetracycline; Tp^r, trimethoprim.
[b]The pMJ-A derivative pMJ1, containing Cm^r determinant, has been constructed as well (Jakoby et al., 1999).
[c] The pCRB1 derivatives pCRB2, pCRB3 and pCRB4 containing Km^r, Gm^r and Sp^r determinants, respectively, have been constructed as well (Nakata et al., 2004).
[d]The pEC-K18mob2 derivatives pEC-C18mob2, pEC-T18mob2 and pEC-S18mob2, containing Cm^r, Tc^r and Sm^r determinants, respectively, have been constructed as well (Kirchner and Tauch, 2003).
[e]The pBHK18 derivatives pBHC18, pBHT18 containing Cm^r and Tc^r determinants, respectively, and the mobilizable derivatives pBHK18*mob2*, pBHC18*mob2* and pBHT18*mob2* have been constructed as well (Kirchner and Tauch, 2003).
[f] The pK18*mob* and pK18*mobsacB* derivatives pK19*mob* and pK19*mobsacB*, containing the pUC19 MCS, have been constructed as well (Schäfer et al., 1994)
[g] The pK18*mob2* derivatives pC18*mob2* and pT18*mob2* containing Cm^r and Tc^r determinants, respectively, have been constructed as well (Kirchner and Tauch, 2003).

1991). This pNG2 minireplicon, designated pEP2, contains the *rep* gene coding for the initiator of replication by the RC mode (Zhang et al., 1994). Analysis of the complete nucleotide sequence of pNG2 revealed 17 ORFs. The deduced amino acid sequence of the pNG2 Rep protein was found to be highly similar to that of the Rep proteins of many plasmids from corynebacteria and pNG2 was defined as the type plasmid of a new family of RC-replicating plasmids (Tauch et al., 2003a).

4. Plasmid vectors for corynebacteria

Since 1984, when small plasmids were discovered in amino-acid producing corynebacteria (Miwa et al., 1984; Ozaki et al., 1984; Santamaría et al., 1984), various vectors have been constructed for gene analysis and manipulations in corynebacteria. As the small native plasmids of corynebacteria are mostly cryptic (Table 6.1), the development of vectors required the insertion of selectable markers, predominantly antibiotic resistance determinants, from other sources. The heterologous kanamycin (*aph* gene coding for aminoglycoside phosphotransferase) and chloramphenicol (*cat* gene coding for chloramphenicol acetyltransferase) resistance determinants as well as streptomycin/spectinomycin and tetracycline resistance determinants coming from large plasmids pCG4 and pAG1 of *C. glutamicum* are the most frequent antibiotic resistance markers used for the construction of vectors for corynebacteria. Plasmid vectors carrying determinants of resistance to clinically utilized antibiotics, however, may spread these resistance genes by horizontal transfer to other bacteria when recombinant strains are applied at industrial scale. To avoid potential medical risk caused by these plasmid vectors, alternative selectable markers have been searched for. It was found that vectors containing the *alr* gene coding for alanine racemase permit growth of Δ*alr* *C. glutamicum* strains in the absence of D-alanine (Tauch et al., 2002b). The *alr* gene can thus be applied as an alternative selection marker in industrial *C. glutamicum* strains.

Most of the autonomously replicating vectors for corynebacteria are based on small cryptic plasmids pBL1, pCG1 and pGA1 from *C. glutamicum* or on the minimal replicon of the plasmid pNG2 from *C. diphtheriae*. Plasmid vectors suitable for use in *C. glutamicum* have recently been compiled in the *Handbook of Corynebacterium glutamicum* (Eggeling and Reyes, 2005). We present here examples of cloning vectors, expression vectors and promoter-probe vectors, which have been used for analysis and gene manipulations in corynebacteria.

4.1 Cloning vectors

Plasmid cloning vectors for corynebacteria include autonomously replicating *C. glutamicum-E. coli* shuttle vectors and *C. glutamicum* vectors for self-cloning and vectors for integration into the *C. glutamicum* chromosome (Table 6.2).

C. glutamicum-E. coli shuttle vectors have mostly been constructed by joining replicons of native plasmids from corynebacteria to common *E. coli* plasmid vectors often containing the multiple cloning sites (MCS) within the *lacZα* gene fragment, which allows the direct selection of recombinant plasmids (blue/white colonies) in *E. coli*. The complete 3-kb plasmid pCG1 (or almost identical plasmids pHM1519 or pSR1) or minimal replicons of plasmids pBL1 (or almost identical plasmid pAM330), pCC1 and pNG2 have been used for these constructions. The vectors based on the plasmid pGA1 require the presence of the *per* gene, coding for a positive effector of replication (Nešvera et al., 1997), in addition to the minimal replicon to ensure the stable maintenance of the constructed vectors in cells of corynebacteria (Tauch et al., 2002c). Most of the constructed *C. glutamicum-E. coli* shuttle vectors are present in the cells of corynebacteria in 10 – 50 copies per chromosome. Since a high dosage of some cloned genes could be detrimental to the growth of the host cells, low copy number vectors based on a minireplicon of plasmid pNG2 from *C. diphtheriae* were also constructed (Reinscheid et al., 1994; Kirchner and Tauch, 2003). All these plasmid vectors can be transferred to the cells of corynebacteria by transformation (currently performed by electrotransformation). To permit the transfer of *C. glutamicum-E. coli* shuttle vectors to the cells of corynebacteria by conjugation, mobilizable vectors were constructed containing the origin of conjugative transfer (*oriT*) and the *mob* region of large broad-host-range conjugative plasmid RP4 (Schäfer et al., 1994). The *E. coli* S17-1 strain with an RP4 derivative integrated into the chromosome, which provides the transfer functions necessary for mobilization (Simon et al., 1983), is commonly used as a donor in the conjugative transfer of mobilizable vectors into corynebacteria. Since the frequency of DNA transfer by conjugation was found to be much higher than that by electrotransformation, conjugation is used preferentially when rare events of integration of cloned genes into the chromosome of corynebacteria are required.

Safety measures in industrial applications of genetically modified organisms may require the absence of heterologous DNA in the constructed recombinant strains. Self-cloning systems containing only *C. glutamicum* DNA have therefore been developed. For instance, the plasmid vector pCG11 (Ikeda and Katsumata, 1998), consisting of the pCG1 replicon and the streptomycin/spectinomycin resistance determinant from the *C. glutamicum* plasmid pCG4, has been used for the amplification of genes involved in tryptophan biosynthesis with the aim of improving the *C. glutamicum* tryptophan producer (Ikeda and Katsumata, 1999). The other self-cloning system, suitable for gene manipulations in industrial strains of *C. glutamicum*, is devoid of any antibiotic resistance gene. It consists of the cloning vector pSELF2000X, carrying the *alr* gene coding for alanine racemase, and the Δ*alr C. glutamicum* host strain. There is strong selection for plasmid-harboring Δ*alr* mutant strains, since plasmidless cells cannot grow in the absence of external D-alanine (i.e. in common complex media) (Tauch et al., 2002b).

Various vectors were constructed to promote the integration of cloned fragments into the chromosome. These vectors are the basic tools used for disruptions and replacements of genes in the chromosome of corynebacteria. They are based, with a single exception, on common *E. coli* plasmid vectors

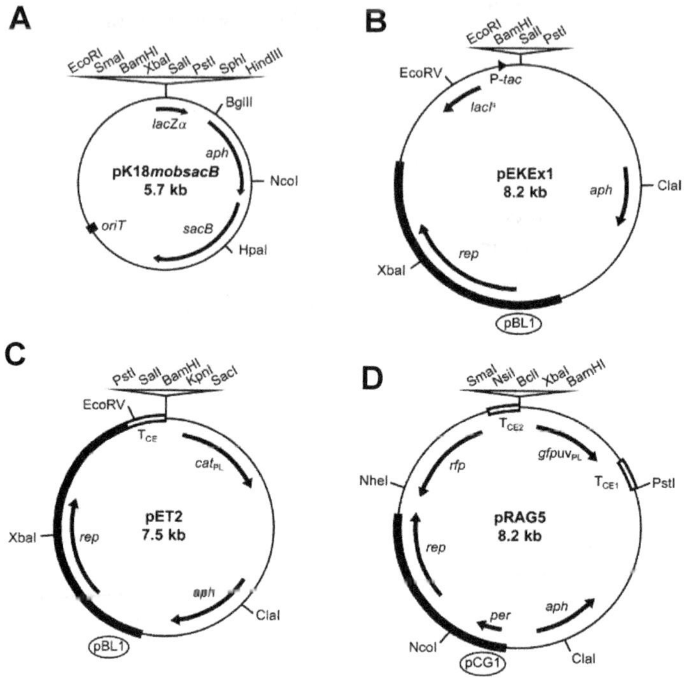

Figure 6.2. Maps of plasmid vectors for corynebacteria. (A) Mobilizable *E. coli* cloning vector pK18*mobsacB* for integration into chromosome of corynebacteria, (B) *C. glutamicum*/*E. coli* shuttle expression vector pEKEx1 (C) *C. glutamicum*/*E. coli* shuttle promoter-probe vector pET2, (D) *C. glutamicum*/*E. coli* shuttle promoter-probe vector pRAG5 with *rfp* reference gene. Regions coming from plasmids of corynebacteria (pBL1 and pCG1, repectively) are shown as thick black lines. The genes are shown inside the maps as arrows. *aph*, kanamycin resistance determinant; *lacZ*α, ≥-complementing β-galactosidase; *sacB*, levansucrase from *B. subtilis*; *oriT*, origin of conjugative transfer; *lacI*^{q}, lactose repressor; P-*tac*, promoter inducible by IPTG; *rep*, initiator of RC replication; *per*, positive effector of plasmid replication; *cat*$_{\mathrm{PL}}$, promoterless reporter gene coding for chloramphenicol acetyltransferase; *gfp*uv$_{\mathrm{PL}}$, promoterless reporter gene coding for green fluorescent protein; *rfp*, reference gene coding for red fluorescent protein; $\mathrm{T_{CE}}$, $\mathrm{T_{CE1}}$ and $\mathrm{T_{CE2}}$, tandems of transcriptional terminators.

that are non-replicating in corynebacteria. The recently constructed vector pSFKT2, exhibiting temperature-sensitive replication due to its mutation within the *rep* gene of the pBL1 replicon, is the only exception. This vector can replicate in *C. glutamicum*, though only at 25 °C, while at 34 °C it can only be selected if its integration into the chromosome was successful, via recombination of the cloned fragment with its chromosomal homolog (Nakamura et al., 2006). Most of the integrative vectors for corynebacteria contain *oriT* and the *mob* region of conjugative plasmid RP4 and can thus be transferred from *E. coli* S17-1 (Simon et al., 1983) to corynebacteria by conjugation. The higher frequency of conjugation, compared to that of electrotransformation, increases the probability of the selection of rare events of integration. The mobilizable vector pK18*mobsacB* (Figure 6.2A), carrying the conditionally lethal *sacB* selection marker from *Bacillus subtilis* (Jäger et al., 1992), was found to be very useful for introducing defined mutations or deletions into the *C. glutamicum* chromosome (Schäfer et al., 1994). Since the levansucrase encoded by *sacB* is lethal for *C. glutamicum* in the presence of sucrose, only those clones in which the rare double cross-over event (involving the excision of plasmid sequences from the chromosome) occurred, can be selected on sucrose-containing media.

In addition to the vectors integrated into the chromosomal regions homologous to those cloned in the plasmid, vectors promoting integration into specific sites on the chromosome were constructed. These vectors contain chromosomal DNA sequences, whose disruptions on the chromosome via homologous recombination have no effect on the viability of the recombinant strains. For instance, the vector pEM1dppc integrates into the non-coding sequence downstream of the *ppc* gene coding for phosphoenolpyruvate carboxylase (Vašicová et al., 1998), while the vector pKX15 integrates into one of several copies of *C. glutamicum* genes coding for 16S rRNA (Amador et al., 2000). In addition, integrative plasmid vectors have been constructed, carrying regions of the DNAs of various corynephages involved in their site-specific integration. The site-specific integration plasmid vector pA3253, containing the minimal DNA region required for integration of the temperate corynephage Φ16, is an example of one of these vectors (Moreau et al., 1999).

4.2 Expression vectors

Plasmid vectors that ensure the controlled expression of the cloned genes from strong regulated promoters in corynebacteria are shown in Table 6.3. They include *C. glutamicum-E. coli* shuttle vectors as well as vectors for integration into the *C. glutamicum* chromosome. The strong *E. coli* promoters P-*lac*, P-*tac* and P-*trc* as well as the *lac* operator-repressor system were found to function in *C. glutamicum* and were used in most of the constructed vectors. As an example, the *C. glutamicum-E. coli* shuttle vector pEKEx1 is shown in Figure 6.2B. It carries the *lacI*^{q} gene coding for the *lac* repressor and the strong regulated P-*tac* promoter induced by isopropyl-β-D-thiogalactopyranoside (IPTG) (Eikmanns et al., 1991a). Regulated P-*tac* is also present in the vector pBKGEXm2 containing the glutathion-*S*-transferase fusion casette. Using this expression vector, the products of the cloned genes can be isolated as fusion proteins by affinity chromatography (Srivastava and Deb, 2002). The P_RP_L promoter-operator-*c*I repressor system of bacteriophage λ was also found to function in *C. glutamicum*. The expression vector pEC901 containing the P_RP_L promoters and the *c*I857 gene coding for the temperature-sensitive repressor was constructed and the high level of expression of the genes cloned under the P_RP_L promoters was observed after temperature pulses (40 °C) (Tsuchiya and Morinaga, 1988).

4.3 Promoter-probe vectors

A number of promoter-probe vectors for testing the activity of promoters and their regulation in corynebacteria have been constructed (Table 6.4).

Table 6.3. Plasmid expression vectors for corynebacteria

Vector	Size (kb)	*C. glutamicum* replicon	Selection marker	Cloning sites	Promoter, regulatory gene	Induction conditions	Reference
A) Autonomously replicating *C. glutamicum/E. coli* shuttle vectors							
pEKEx1	8.2	pBL1	Km^r	EcoRI, BamHI, SalI, PstI	P-*tac*, *lacI*^{q}	IPTG	Eikmanns et al., 1991a
pXMJ19	6.6	pBL1	Cm^r	pUC19 MCS	P-*tac*, *lacI*^{q}	IPTG	Jakoby et al., 1999
pCRA1	5.3	pBL1	Cm^r	EcoRI, SacI, KpnI, XhoI, PstI SmaI, BamHI, XbaI, SalI	P-*lac*		Nakata et al., 2004
pBKGEXm2[a]	7.3	pBL1	Km^r	BamHI, EcoRI, SmaI, SalI, XhoI, NotI	P-*tac*, *lacI*^{q}	IPTG	Srivastava and Deb, 2002
pZ8-1	7.0	pCG1	Km^r	EcoRI, BamHI, SalI, PstI	P-*tac*		Dusch et al., 1999
pVWEx1	8.9	pCG1	Km^r	BamHI, PstI, XbaI, SalI,	P-*tac*, *lacI*^{q}	IPTG	Peters-Wendisch et al., 2001
pEC901	8.5	pCG1	Km^r	BamHI	P_L/P_R (λ), cI857	40 °C	Tsuchiya and Morinaga, 1988
pEC-XK99E[b]	7.0	pGA1	Km^r	pTrc99A MCS	P-*trc*, *lacI*^{q}	IPTG	Kirchner and Tauch, 2003
pAPE12	4.6	pNG2	Km^r	EcoRI, SalI, BamHI	P-*trc*, *lacI*^{q}	IPTG	Guillouet et al., 1999
B) Vector for integration into the *C. glutamicum* chromosome							
pXK99E[c]	4.4	-	Km^r	pTrc99A MCS	P-*trc*, *lacI*^{q}	IPTG	Kirchner and Tauch, 2003

[a]Expression of glutathione-S-transferase fusion proteins

[b]The pEC-XK99E derivatives pEC-XC99E and pEC-XT99A containing Cm^r and Tc^r determinants, respectively, have been constructed as well (Kirchner and Tauch, 2003).

[c]The pXK99E derivatives pXC99E and pXT99A containing Cm^r and Tc^r determinants, respectively, have been constructed as well (Kirchner and Tauch, 2003).

Various promoterless reporter genes coding for easily detected and quantified products were used for the construction of these vectors. Promoters on DNA fragments inserted upstream of the reporter genes, and thus forming transcriptional fusions, drive transcription of the reporter genes. Construction of an efficient promoter-probe vector also requires the insertion of a transcriptional terminator(s) upstream of the cloning sites to prevent readthrough from the vector to the reporter gene. The following reporter genes coding for various enzymes were shown to be active in corynebacteria: *lacZ* coding for *E. coli* β-galactosidase (Eikmanns et al., 1991a), *cat* coding for chloramphenicol acetyltransferase from Tn*9* (Eikmanns et al., 1991a; Vašicová et al., 1998), *aph* coding for aminoglycoside phosphotransferase from Tn*5*, (Barák et al., 1990; Cadenas et al., 1991; Zupancic et al., 1995), *uidA* coding for *E. coli* β-glucuronidase (Bardonnet and Blanco, 1991) and *amy* coding for α-amylase from *Streptomyces griseus* (Cadenas et al., 1996). As an example of autonomously replicating multicopy *C. glutamicum-E. coli* shuttle promoter-probe vectors, the frequently used vector pET2, carrying the promoterless *cat* reporter gene (Vašicová et al., 1998), is shown in Figure 6.2C. To ensure the transcriptional fusion of the tested promoter to a reporter gene in a single copy in the cell, the integrative promoter-probe vector pRIM2 was constructed (carrying the *cat* reporter gene). It contains a replicon of an *E. coli* vector that is non-replicating in corynebacteria and a non-coding sequence downstream of the *C. glutamicum ppc* gene coding for phosphoenolpyruvate carboxylase, which allows integration of the vector into the *C. glutamicum* chromosome via homologous recombination without any detrimental effect on cell growth (Vašicová et al., 1998).

Novel reporter genes have recently been used in promoter-probe vectors for corynebacteria. The insertion of the promoter-carrying fragment upstream of a promoterless tyrosinase operon (*melC1melC2*) from *Streptomyces glaucescens* resulted in the production of easily detected melanine in corynebacteria (Adham et al., 2003). The use of the *gfp* reporter gene, coding for green fluorescent protein from the jellyfish *Aequorea victoria*, allows promoter activity to be tested by estimating fluorescence intensities in living cells. Promoter-probe vectors carrying the reporter genes coding for various forms of GFP protein have recently been used for testing promoter activity in corynebacteria in vivo (Letek et al., 2006; Knoppová et al., 2007). The vector pRAG5 was also constructed (Figure 6.2D), carrying the *rfp* reference gene (coding for red fluorescence protein from coral *Discosoma* sp.) under the strong constitutive *C. glutamicum* promoter, in addition to the *gfp*uv reporter gene (Knoppová et al., 2007). The vector pRAG5 is suitable for normalized measurements of promoter activities during the growth of bacterial batch cultures, since estimation of the GFP/RFP fluorescence ratio in strains carrying the plasmid pRAG5 with the tested promoters upstream of *gfp*uv avoids the influence of plasmid copy number variations on the promoter activity assay.

In addition to promoter-probe vectors for corynebacteria, the terminator-probe vector pUT2 was also constructed for screening termination signals in

Table 6.4. Promoter-probe vectors and a terminator-probe vector for corynebacteria

Vector	Size (kb)	*C. glutamicum* replicon	Selection marker	Cloning sites	Reporter	Further characteristics	Reference
A) Autonomously replicating vectors							
pEKpllacz	9.4	pBL1	Km^r	SalI, BamHI, SmaI, KpnI	*lacZ*	-	Eikmanns et al., 1991a
pEKplCm	7.4	pBL1	Km^r	BamHI, SalI	*cat*	-	Eikmanns et al., 1991a
pET2	7.5	pBL1	Km^r	PstI, SalI, BamHI, KpnI, SacI	*cat*	-	Vašicová et al., 1998
pSK1Cat	6.3	pCG1	Km^r	SmaI, BamHI, SalI, PstI, HindIII	*cat*	-	Park et al., 2004
pULMJ95	6.9	pBL1	Km^r	BamHI, KpnI, EcoRI	*amy*	-	Cadenas et al., 1996
pULMJ88	5.9	pBL1	Hm^r	BglII	*aph* (Km^r)	-	Cadenas et al., 1991
pJUP05	10.0	pBL1	Cm^r	BamHI, KpnI, BglII,	*aph* (Km^r)	-	Barák et al., 1990
pPROBE17	8.0	pBL1	Cm^r	EcoRV	*aph* (Km^r)	-	Zupancic et al., 1995
pUT3	7.4	pBL1	Km^r	pUC18 MCS	*uidA*	-	Bardonnet and Blanco, 1991
pUT2	7.4	pBL1	Km^r	pUC18 MCS	*uidA*	terminator-probe vector	Bardonnet and Blanco, 1991
pEMel-1	13.0	pCG1	Km^r	EcoRI-NdeI	*melC1-melC2*	replacement of EcoRI-NdeI fragment (P-kan) necessary	Adham et al., 2003
pEGFP	12.0	pCG1	Km^r	EcoRI-NdeI	*egfp2*	replacement of EcoRI-NdeI fragment (P-kan) necessary	Letek et al., 2006

pPRE11	7.6	pCG1	Km^r	KpnI, BamHI, XbaI	rs*gfp*	-	Knoppová et al., 2007
pEPR1	7.3	pCG1	Km^r	SmaI, NsiI, BclI, XbaI, BamHI	*gfp*uv	-	Knoppová et al., 2007
pRAG5	8.2	pCG1	Km^r	SmaI, NsiI, BclI, XbaI, BamHI	*gfp*uv	*rfp* reference gene under constitutive P45 promoter	Knoppová et al., 2007
B) Vector for integration into the *C. glutamicum* chromosome							
pRIM2	5.3	/	Km^r	HindIII, SphI, PstI, XbaI, BamHI, SmaI, KpnI	*cat*	-	Vašicová et al., 1998

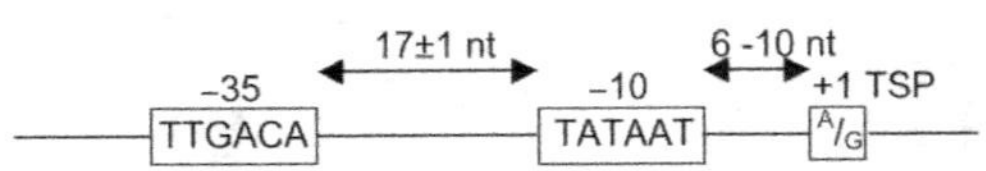

Figure 6.3. *Escherichia coli* consensus pattern of the housekeeping promoter sequences. The main elements and parameters of the core promoter are shown: –35 hexamer, –10 hexamer, transcriptional start point (TSP at +1) and spacing between these elements in nucleotides (nt).

corynebacteria. The presence of a transcriptional terminator on the cloned fragment can be detected by the loss of activity of the *uidA* reporter gene coding for β-glucuronidase (Bardonnet and Blanco, 1991).

5. Promoters of corynebacteria

Promoters are signal DNA sequences which determine the specific sites for the binding of DNA-dependent RNA polymerase and for the initiation of gene transcription. The strength of each promoter (i.e., frequency of transcription initiation that occurs at the promoter) is basically defined by its degree of similarity to the ideal promoter described by the consensus sequence. The transcriptional activity of every promoter may be further downregulated or upregulated by diverse specific or global factors that react to changes in environmental conditions and the physiological state of the cell. The frequency of transcription initiation may therefore range over about four orders of magnitude. The activity of promoters is thus an important factor, which allows the organism to precisely tune gene expression to the level required by the particular conditions.

The basic structure of a bacterial promoter, called the core promoter, recognized by the primary *E. coli* sigma factor σ^{70} (coded by *rpoD*), consists of two conserved hexamers centered approximately at positions −35 and −10 relative to the transcriptional start point and of the non-specific spacer of 17±1 nucleotides (nt) between them (Figure 6.3). Conserved elements within the promoters in other bacteria and those recognized by other sigma factors are, due to their similarities and for simplicity's sake, mostly designated as the −35 and −10 regions (elements), although the centers of these elements may slightly differ. The activity of the core promoters is purely regulated by the presence of the respective active sigma factor. The core promoter and the flanking sequences (upstream and downstream regions), reaching as far as several hundreds of bases away, may contain further sequence signals (e.g. sequences binding regulatory proteins), which can mediate the activation or repression of the promoter activity. This review is focused on the description of *C. glutamicum* core promoters and sigma factors as the essential regulatory proteins necessary for promoter function, whereas the modulation of promoter activity by other transcriptional factors is described in other parts of the book. At the end of this chapter, examples of practically important promoters are given.

5.1 Sigma factors of RNA polymerase

The bacterial RNA polymerase (RNAP) core enzyme, which has non-specific affinity to DNA, consists of five subunits (α_2, β, β´ and ω). The RNAP holoenzyme additionally contains the σ subunit (sigma factor), which enables the enzyme to specifically recognize promoters. Different sigma factors within the bacterial cell direct RNAP to different classes of promoters. Sigma factors thus function as positive global regulators that ensure the coordinated expression of physiologically or developmentally related sets of genes (regulons).

Promoters and transcriptional factors including sigma factors are, among corynebacteria, best described in *C. glutamicum*. According to an analysis of their genome sequences, *C. glutamicum* ATCC 13032, as well as *C. glutamicum* strain R, code for 7 sigma factors: primary SigA (σ^A) and alternative SigB, SigC, SigD, SigE, SigH and SigM, named according to their homologues in the related actinomycete *Mycobacterium tuberculosis*. SigA is the principle sigma factor, which is responsible for the transcription of the majority of genes essential for exponential growth in a minimal medium under optimum conditions (housekeeping genes) (Halgasova et al., 2001). SigB of *C. glutamicum* is a non-essential primary-like sigma factor, which plays a role in the transcription of genes involved in response to some environmental stresses (Halgasova et al., 2001) and the genes relevant for the transition period (between the exponential and stationary growth phases) and the stationary phase (Larisch et al., 2007). The other five sigma factors are involved in the regulation of extracytoplasmic functions and are therefore called extracytoplasmic-function (ECF) factors (Helmann, 2002). These sigma factors are not essential; however, they play an important role in adaptation to various stress conditions.

SigH was shown to control the promoters of genes involved in heat-shock response, oxidative stress and the stationary growth phase (Engels et al., 2004). SigM recognizes the promoters of the genes which function in thiol oxidative stress and heat stress (Nakunst et al., 2007). SigE is most probably involved in the response to heat shock (Muffler et al., 2002) and acid shock (Jakob et al., 2007). The functions of SigC and SigD have not yet been studied. *C. efficiens* also contains 7 sigma factors homologous to those in *C. glutamicum*. There are 9 annotated sigma factors in *C. diphtheriae*, primary σ^A, primary-like σ^B and 7 ECF sigma factors. The same number of sigma factors has been found by analysis of the *C. jeikeium* genome sequence. Knowledge of the promoters of housekeeping genes as well as the promoters recognized by alternative sigma factors in these pathogenic corynebacteria is still scarce.

5.2 Promoters of housekeeping genes

The positions of more than 50 promoters of housekeeping genes (that are most probably recognized by σ^A) have been localized by determining their respective transcriptional start points (TSPs) within the *C. glutamicum* genome. Using computer-aided analysis of the 70-nt promoter sequences (10 nt downstream and 60 nt upstream of the TSPs), a promoter consensus sequence for *C. glutamicum* was derived (Pátek, 2005; Pátek et al., 1996; Pátek et al., 2003). The core hexamer TANAAT of the −10 region had already been derived from 33 sequences in an early study of *C. glutamicum* promoters (Pátek et al., 1996). In contrast, no clear −35 consensus sequence could be derived even from the analysis of more than 50 *C. glutamicum* promoters (Pátek, 2005; Pátek et al., 2003). There are *C. glutamicum* promoters in which the sequences in the −35 region closely resemble the *E. coli* consensus (e.g., TTGACG in P-*aceA*, TTGATA in P-*glnA* and ATGACA in P-*brnE*), but this canonical −35 hexamer is not apparent in most of the others. The statistical consensus ttG$^c/_g$ca could be defined for the −35 region, but the nucleotides at most positions (in lower case) are only weakly conserved. The same feature

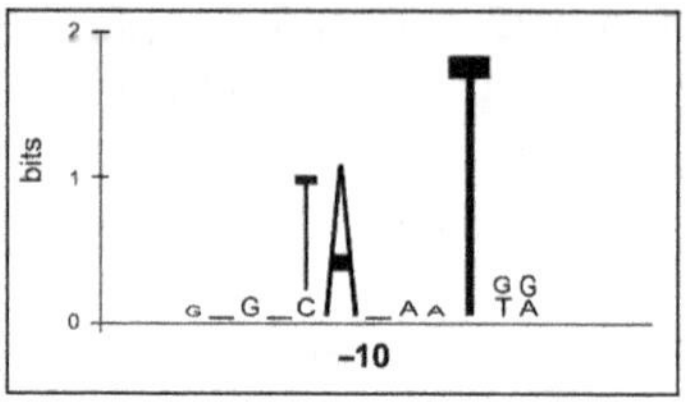

Figure 6.4. Logo representing the consensus sequence of the extended −10 region of *C. glutamicum* promoters of house keeping genes. The y-axis corresponds to the bits of information. The logo was generated by submitting all sequences shown in Figure 6.5 and sequences analyzed previously (Pátek et al., 2003) to the WebLogo program (Crooks et al., 2004).

	Sequence (−10 … TSP)
P-*aceA*	AACCCAAACCGATTGACGCACCAATGCCCGATGG**AGCAAT**GTGTGAACC**A**
P-*aceB*	ACGTGATGTGCATCGGTTTGCGTGGTGGCGTGGTTCA**CACATT**GCTCCA**T**
P-*amt*	AAACAGAAATCTATAGAACGATAGGTAAAAACTGGAC**TAGGTT**TATCTA**T**
P-*argS*	AAGTATGGGTCGTATTCTGTGCGACGGGTGTACCTCGGC**TAGAAT**TTCT**C**
P-*askP1*	AAACACTCCTCTGGCTAGGTAGACACAGTTTATAAAGG**TAGAGT**TGAGC**G**GG**T**
P-*askP2*	GAATACGCTCGTGCATTCAATGTGCCACTTCGCGTACGC**TCGTCT**TATA**G**
P-*brnE*	TGGGAAAGGCTGCTAAATGACAACTGATTTCTCCTCTAT**TCTCCT**TGTT**G**
P-*brnF*	TGCGCAAACTGGCAACAAAACTACCCGGCAATTGTGT**GATGAT**TGTAGT**G**
P-*dapA*	GCGGGGTTGTTTAACCCCCAAATGAGGGAAGAAGG**TAACCT**TGAACTCT**A**
P-*dapB1*	AGGTATGGATATCAGCACCTTCTGAACGGGTACGTC**TAGACT**GGTGGGC**G**
P-*dapB2*	GTTTGAAAAACTCTTCGCCCCACGAAAATGAAGGAG**CATAAT**GGGAATC**A**
P-*fda*	AGGAAATATCACACGACAAAAGTTGAGTGATGCAGG**CATAAT**TGGCTAT**A**
P-*gap*	TGATTTTGCATCTGCTGCGAAATCTTTGTTTCCCCGC**TAAAGT**TGAGGA**C**
P-*gdh*	TTTTAATTCTTTGTGGTCATATCTGTGCGACACTGC**CATAAT**TGAAGGT**G**
P-*glnA*	TTTCAGAAAAGTTTTGATAGATCGACAGGTAATGCAT**TATACT**GACAAC**G**
P-*gltA*	CTCACTTCGGATATGGCTAAACCGCATTTATCGGTA**TAGCGT**GTTAACC**G**
P-*gltB*	ACCCTTTTGTTACTTGCGTTGCAGGTAGTGTCCCTGA**TTTTCT**TATTAT**C**
P-*hom*	ATTTCGGGGCTTTAAAGCAAAAATGAACAGCTTGGTC**TATAGT**GGCTAG**G**
P-*ilvA*	CCTATGCCAAAGTAGGTGCAATTCTAGGAGAAGAT**TACACT**AGTCAACC**A**
P-*ilvB*	TTTCGTGCGTTGTGGAATTGGAAACTCGATGTGTG**TAGCAT**GACACACC**A**
P-*ilvC*	GTGACTAGTCAAACACCGTCTAATTACATGTGTGTGG**TAGAAC**AATAAT**G**
P-*leuA*	CTCCCCCGAGTTGCTACCCACACCACAAAGTTGTTG**TATGCT**TCACCAC**A**
P-*lpdA*	CGCACCCGGATGCACGACAATGACCCACTAAACACG**TATCCT**TGAATGC**G**
P-*lrp*	GTAGTTTTGTTGCCAGTTTGCGCACCTCAACTAGGC**TATTGT**GCAATAT**A**
P-*lysE*	TAAGAACCAATCATTTTACTTAAGTACTTCCATAGGT**CACGAT**GGTGAT**C**
P-out (P-45)	TGGTCAGGGATTTTTTCCCGAGGGCACTAATTTTGC**TAAAGT**AAGTGAC**G**
P-*pgk*	CAATTGAATACCGGTGCCAGCGCCACACAATGTGTGG**CAATCT**GGGACA**G**
P-*pta1*	CTTTGCAAACCGGGCTGTACGCAAGGCGGACGAACGC**TAAACT**ATGTAA**G**
P-*pyc*	AAACAAAAACCGATGTTTGATTGGGGGAATCGGGGGT**TACGAT**ACTAGG**A**
P-*secE*	CCTAATAAATATTGCGAGGGTTCGCGGGATTAATG**TACTCT**CGAAGGTT**G**
P-*sigA*	AACAGGCCCCTTTGTGACATCGGCGCAGTTGTTCAAC**TATAAT**GGAACG**C**
P-*thrB*	AAGCCTGTTGTTAAGGCAATCAACAGTGTGATCCGCC**TCGAAA**GGGACT**A**
P-*thrC*	ATATTTGAGACGGTGTGGGGGAGTATTGTGTCACCCCT**TGGGAT**AGGGT**T**
P-*thrE*	CTTAGCGTATGTGTACATCACAATGGAATTCGGGGC**TAGAGT**ATCTGGT**G**
P-*orfMP*	GACAGGTGCTACTTCGCGAGCAACTCTTTAGTCAAC**TACCCT**GAATCAA**G**
P-*ptsG*	TTTGTCTGTATTCAGCTTGTTTTGGGTGGGTTTCCGGCT**TATCAT**GATG**G**
P-*mepA*	GTAGCACGAAAACTTGTACAGGGGTGGGGTTACCCCC**TAAGGT**GGTC**ACA**
P-*ppmA*	AATTAGAAATCCCCTTTTAATGTGATATCACTTTCTGT**TAAACT**GATAT**C**
P-*betP*	ACCCATTTTTCGGTTTGCTTTCCAGGTTTCCCCAAGT**AAAGGT**GAGTT**TT**
P-*proP*	ACTTGAAAACGATCTGACTTTTCACCCCGAACCTTAC**TAAGGT**CGAT**TCA**
P-*cspB*	TCTCCAATTCGCCTGGCAAAAATCTCAATTGTCGCT**TACAGT**TTTTCTC**A**
P-*cg0771*	GACGTGAGACATTTCCTCCAATACCTCTCCAGTGATA**TTCTGT**CGGGCA**G**
P-*cg0922*	AGTTTTCCCCTATTTACACACACCTGACCGAAGCTG**TAAGGT**TTGCCTA**A**
P-*cg0527*	ATTCACTCAATCAATTCACTTAGGCTTGCCATACCTA**TATATT**GAGTGA**A**
P-*cg2782*	GCGCTAACCTATGCTGACAGCCTTGCGGAAGTTGTG**TACGTT**AGGGGCC**A**
P-*cg3327*	GGCTAAGGGCCGGTGACAATTTATCAAGCAGTGC**TATAAT**AATAGGGGC**A**
P-*cg0445*	AGCGCCCCGTGACTGGTTAATGCCCCGATCTGGGACG**TACAGT**AACAAC**G**

Figure 6.5. Promoters of housekeeping genes from *C. glutamicum*. Experimentally determined TSPs are in bold and underlined, putative −10 hexamers are shown in bold. A consensus sequence within the -35 regions is not apparent. References: P-*ptsG* (Engels and Wendisch, 2007), P-*mepA*, P-*ppmA*, P-*betP* and P-*proP* (Möker et al., 2004), P-*cspB* (Hansmeier et al., 2006), cg0771, cg0922, cg0527, cg2782, cg3327 and cg0445 (Brune et al., 2006). For references to the other promoter sequences, see Pátek et al. (2003).

was found in mycobacterial promoters (Bashyam and Tyagi, 1998). Although the binding affinity of RNAP to the promoters is variable depending on the level of conformity of the particular promoter sequence to the consensus, there is some lower limit of informational content which ensures specific recognition of the promoter by RNAP. In *E. coli*, the sequence conservation of σ^{70} promoters is 6.5±0.1 bits in standard promoters and 6.7±0.2 bits in promoters lacking the –35 element (Shultzaberger et al., 2007). In *C. glutamicum*, the low conservation of the –35 region and therefore the low informational content of the upstream recognition sequence is most probably compensated for by an extended –10 region, similarly to some *E. coli* promoters. Statistical analysis and mutational analysis (Vašicová et al., 1999) of *C. glutamicum* promoters resulted in the proposal of an extended –10 region which fulfils the requirements of a specific recognition site for RNAP. The sequence $TGTG^{C}/_{G}TATAATGG$ thus defines a hypothetical strong *C. glutamicum* promoter (Pátek, 2005). A logo representing the position-specific weight of nucleotides in the extended –10 promoter region of housekeeping promoters is shown in Figure 6.4. The significance of the individual positions within the –10 region, depicted by the size of the most frequent nt, was mostly confirmed by mutagenesis (Vašicová et al., 1999) Pátek et al., unpublished). The selected well-defined promoter sequences of *C. glutamicum* housekeeping genes are shown in Figure 6.5. It is not possible to prove *in vivo* that these promoters are recognized exclusively by σ^A, because this sigma factor is essential and knock-out mutants could not be constructed. The dependence of these promoters on σ^A could be indicated by their function in the exponential growth phase and documented by overexpression of the *sigA* gene. Conclusive evidence could be obtained in the future from an *in vitro* transcription system, like in *M. tuberculosis* (Beaucher et al., 2002).

Seven promoters were also precisely localized on the *C. glutamicum* plasmid pGA1 (Figure 6.6). According to the sequences of their putative –10 elements and their activity during the exponential growth phase, these promoters are most probably σ^A-dependent. The –35 motif of plasmid promoters is much more similar to the consensus sequence found in both Gram-positive (*B. subtilis*) and Gram-negative (*E. coli*) bacteria than that of the chromosomal *C. glutamicum* promoters. This feature of plasmid promoters might assist the expression of plasmid genes that have been spread by horizontal transfer into various hosts.

	–35		–10		TSP
P-*aes1*	TAATCCATC**TTGCCT**	TTCCGCTGCAGATAAAT	**TACGCT**	GAAAAA	**C**
P-*aes2*	TTTTGTTCC**TTGCCG**	ATGCACTTTCTGCGAGC	**TACTGT**	GAAGAA	**G**
P-*ctRNA*	GCGCTAGCT**GTGACT**	GTGTCCTGCGGATCGGC	**TAGAGT**	CATGTCTT	**G**
P-*I*	AAAAGTTTT**TTGAGC**	AATCGCAGAAAAAGTTT	**CACAAT**	AAACGA	**A**
P-*orfB*	AACGCATAC**TTCTCA**	ATACCTAACGCATACCG	**TATTAT**	TAATTCTT	**A**
P-*per*	AATTGGGGC**TAGAAT**	TTTTAACGAACGTTCGT	**TATAAT**	GGTGTC	**A**
P-*rep*	TAAGGTAGT**TAAGCG**	TTCATTTACGAAGAAAA	**CACGAT**	AAGCTGC	**A**

Figure 6.6. Promoters found on the *C. glutamicum* plasmid pGA1. Experimentally determined TSPs are in bold and underlined, the putative –35 and –10 hexamers are in bold (Nešvera et al., 1997; Venkova et al., 2001; Venkova-Canova et al., 2003).

		-10		TSP
cg0096	CAATGCGATGATCGTCGGAAACTACCTGAC	**TACGCT**	CGGCCGCCCAA	**T**
cg1083 cgR10	TAAGCCTGCAGCCGACGGGATTAAGGCAGC	**TAACAT**	TGAGACA	**C**
cg1417	GACGTCGAAAAGCAATGAATTTAATGCTTT	**TAACCT**	GGATTT	**T**
cg2418 ilvE	CTGACTAGTGTATCTGTCAGGTAGCAGGTG	**TACCTT**	AAAATC	**C**
cg3141 hmp	CATCATATTAAGGCCAAATTGCTTGGATCC	**TGGGAT**	TTATTTA	**A**
cg3330	ACGTGAAAGGCACCTAAAGCGCATTAACGG	**TAAAGT**	GCGAGAGG	**T**

Figure 6.7. Sequences of SigB-controlled promoters. TSPs are in bold and underlined, putative –10 hexamers are shown in bold. A consensus sequence within the –35 region is not apparent (Larisch et al., 2007).

5.3 SigB-dependent promoters

The *C. glutamicum sigB* gene is mainly expressed during the transition of cultures from the exponential growth phase to the stationary phase and in the early stationary phase, whereas expression of *sigA* simultaneously decreases (Larisch et al., 2007).

Using the microarray technique in the global gene expression analysis of the *sigB*-deficient and *sigB*-proficient strains, genes controlled by σ^B were revealed (Larisch et al., 2007). These genes, more highly expressed in the transition phase (only in the presence of σ^B), fall into several categories, e.g., genes involved in amino acid metabolism and transport, stress defense mechanisms, membrane processes and regulatory processes.

Six σ^B-dependent promoters exhibiting strongly elevated transcriptional activity in the transition phase were localized by determination of the respective TSPs. It is apparent from Figure 6.7 that the sequences of these promoters contain –10 elements which conform to the consensus of σ^A-dependent promoters. Interestingly, five of these six –10 hexamer sequences have an identical counterpart within the defined promoters of house keeping genes (P-*aes1*, P-75, P-*dapA*, P-*thrC* and P-*gap*) (Pátek et al., 2003). In addition, P-*ilvE* was found to be very active during the exponential phase and is therefore also probably σ^A-dependent (Pátek et al., unpublished).

In the SigB-deficient strain, the genes whose expression increased during the transition phase in the SigB-proficient strain displayed transcription profiles very similar to those of housekeeping genes, according to real-time RT-PCR (Larisch et al., 2007). This could be explained by the presence of another (σ^A-dependent) promoter in front of the genes, or more probably, by an overlapping recognition specificity of σ^A and σ^B. The expression of *sigB* is also induced by various extracytoplasmic stress factors (Halgasova et al., 2001). The growth and survival of a *sigB* knock-out mutant of *C. glutamicum* CCM 251 (formerly named *Brevibacterium flavum*) showed that σ^B plays an important role in responses to acid, salt, ethanol, cold and heat stresses (Halgasova et al., 2002). The *sigB* gene itself is transcribed from a promoter that is probably recognized by an ECF sigma factor (see below).

SigB is therefore most probably a back-up sigma factor, which drives transcription from promoters identical or very similar to σ^A-dependent promoters under some stress conditions or during the transition phase, in which various stresses affect cell metabolism and growth. Promoter structures and regulatory mechanisms which lead to different recognition of the promoters by RNAP with σ^A and σ^B remain to be elucidated.

P-clpP1P2 GTTTCA**TGGAAA**TACGCGGGTAGTCTGGTGAC**ATTGAA**CCAA**A**
P-clpC AAAGTC**TGGAAG**TTTT---GCCCAATAAGGGC**GTTAAA**GTGGG**T**GAA
P-clgR TAAACT**GGGAAC**AAAT--TTTAGGGAAAGGGA**GTTGAA**CCTAA**C**

Figure 6.8. Promoter sequences of the *clp* genes of *C. glutamicum*. TSPs pertinent to the housekeeping promoters and the respective putative –10 hexamers are shaded, the TSPs pertinent to σ^H-dependent promoters are in bold and underlined, the respective putative –35 and –10 elements are in bold. Transcription from the upstream *clpC* TSP (T) is driven by both the housekeeping promoter and the σ^H-dependent promoter. The dashes indicate gaps introduced to align these elements (Engels et al., 2004).

5.4 Promoters dependent on extracytoplasmic-function (ECF) sigma factors

ECF sigma factors are involved in the regulation of the transcriptional responses of cells to external signals perceived by extracytoplasmic sensory domains of trans-membrane proteins. This includes cell envelope functions and responses to environmental stress factors (Helmann, 2002). The transcription initiation of each gene coding for an ECF sigma factor is either autoregulated or regulated by one or more other sigma factors. The σ factors thus form a transcriptional cascade or even a regulatory net. Most ECF sigma factors in bacteria are cotranscribed with one or more negative regulators. These include anti-sigma factors that bind the cognate sigma factors and inhibit their function. Changing environmental conditions produce a signal which leads to a release of the σ factor, which can then bind to RNAP and stimulate transcription from the respective promoters. Bacteria code for various number of ECF sigma factors: 3 in *E. coli*, 7 in *B. subtilis*, 10 in *M. tuberculosis* (Rodrigue et al., 2006) and an astonishing 51 in *Streptomyces coelicolor*.

Only a few promoters have been unequivocally proved to be controlled by a particular ECF σ factor in corynebacteria. Three σ^H-dependent promoters were localized in a study of the genes involved in the proteolytic activity of the ATP-dependent protease Clp, which degrades non-functional proteins within the *C. glutamicum* cells (Engels et al., 2004). The respective transcriptional start points of the genes *clpP1P2* (proteolytic Clp protease subunits), *clpC* (Clp protease ATPase subunit) and *clgR* (transcriptional regulator of *clpP1P2* and *clpC*) were detected after severe heat stress in the wild-type strain but not in the σ^H knock-out strain. In addition to the stress-induced promoter, a housekeeping promoter (σ^A-dependent) controlled by ClgR or by HspR (repressor of chaperone gene expression) drives the transcription of each of these genes (Figure 6.8). Transcription of the genes involved in the formation of the ATP-dependent protease Clp is thus governed by a complex regulatory network (Engels et al., 2004). Sequences of the promoters P-*clpP1P2*, P-*clpC* and P-*clgR* contain the –35 and –10 elements GGAA and GTT with a spacing of 18 to 21 nt (Figure 6.8), which is typical for promoters recognized by ECF σ factors σ^H and σ^E in *M. tuberculosis* (Rodrigue et al., 2006) and by σ^R in *S. coelicolor* (Paget et al., 2001). The *sigH* gene itself was found to be most probably autoregulated in *C. glutamicum* (Kim et al., 2005), although motifs typical for an ECF promoter are not present upstream of the *sigH* gene.

P-*trxB* CAACTGA**TGGAAG**T-TTTTCAAAGTGTCTGAC**GTTGAA**AACGGT**G**
P-*trxB1* CTTGGCC**GGGAAT**A-ACTACAGTCCGCTGAAA**GTTGGT**CTATATAA**G**
P-*trxC* TGAATGT**CGGATT**CCCCAGGAGTCCCGTCATT**GTTAAT**TTAGGA**G**
P-*sufR* TGGACAC**GGGAAT**GGAATTAGGGAACACTTGT**GTTGTC**TAAAGGT**G**

Figure 6.9. Sequences of the σ^M-dependent promoters. TSPs pertinent to σ^M-dependent promoters are in bold and underlined and the respective putative –35 and –10 elements are in bold. The dashes indicate gaps introduced to align these elements. TSP probably pertinent to the additional σ^A-dependent promoter of *sufR*, and the respective putative –35 and –10 hexamers are shaded (Nakunst et al., 2007).

Four promoters apparently recognized by σ^M have recently been uncovered in *C. glutamicum* (Nakunst et al., 2007). The promoters were localized by determining the respective TSPs using the RACE technique (rapid amplification of cDNA ends) upstream of the genes, which were differentially expressed in wild-type and Δ*sigM* strains according to microarray transcriptome analysis. The respective genes *trxB* (thioredoxin reductase), *trxB1* (thioredoxin), *trxC* (thioredoxin) and *sufR* (transcriptional regulator of the *suf* operon) are involved in disulfide stress response. The sequences of the promoters P-*trxB*, P-*trxB1*, P-*trxC* and P-*sufR* also contain the –35 and –10 elements GGAA and GTT with a 19-20 nt spacing (Figure 6.9) which were found in σ^H-dependent *C. glutamicum* promoters (Figure 6.8). Therefore the possibility cannot be excluded that the recognition specificities of σ^H and σ^M overlap, as was described for some σ factors in *B. subtilis* and *M. tuberculosis* (Dainese et al., 2006; Qiu and Helmann, 2001).

As in σ^H-dependent *clp* promoters, a TSP active during exponential growth without stress conditions, probably initiating transcription from the σ^A-dependent promoter, was detected within the promoter region of *sufR* (Figure 6.9).

In total, the transcription of 23 genes mainly involved in the response to heat and oxidative stress was found to be σ^M-dependent. Transcription of the *sigM* gene itself increased significantly after heat, cold and disulfide (oxidative) stress but not in the Δ*sigH* strain. The *sigM* gene is therefore probably transcribed from the σ^H-dependent promoter, which is also suggested by the determination of its TSP (Nakunst et al., 2007) (Figure 6.10).

Two overlapping promoters of different type were also found by TSP determination to drive transcription of the heat-shock gene *dnaK* in *C. glutamicum* (Barreiro et al., 2004). The putative –10 hexamer of the weaker promoter conforms to σ^A-dependent promoters, whereas the –35 and

-35 -35 -10 -10
P-*sigM* CGGGTCAGG**TGGAAT**CCGTGGTGATTCTGGTCGATGAG**GTTCGT**TCCTC**A**

Figure 6.10. Sequence of the *sigM* promoter. TSP is in bold and underlined; two alternative pairs of –35 and –10 elements are in bold or underlined, respectively. Neither pair seems very similar to the σ^H-dependent promoter (Figure 6.8): in the bold pair, the spacing is too large; in the underlined pair, the –35 element does not match the pattern and the distance between the –10 hexamer and TSP is too short. Since the recognition patterns of the σ^H- and σ^M-dependent promoters are similar, *sigM* expression might also be autoregulated (Nakunst et al., 2007).

P-*dnaK* TCTAGT**TGGAAC**AACTTTGTGGCATTTAC**CGTTGC**TATAT**A**TGTAAGCTTTGAGT

Figure 6.11. Sequence of the *dnaK* promoter. TSP pertinent to the supposed σ^A-dependent promoter and the respective putative –35 and –10 hexamers are shaded, the TSP pertinent to an ECF σ-dependent promoter (σ^H or σ^M) is in bold and underlined and the respective putative –35 and –10 elements are in bold (Barreiro et al., 2004).

–10 elements of the stronger promoter, which is highly induced by heat shock, display the GGAA - N_{19} - GTT sequence pattern found in both σ^H- and σ^M-dependent promoters (Figure 6.11). The response of σ^A-dependent P-*dnaK* to heat shock is probably induced by HspR-mediated derepression, as in P-*clgR*, whereas the response of the other promoter is positively regulated by a stress sigma factor (σ^H or σ^M). A highly similar sequence pattern was also identified within the *dnaK* promoter regions of *C. efficiens, C. diphtheriae* and *C. jeikeium* (Barreiro et al., 2004).

Another promoter which is induced under stress conditions in *C. glutamicum* is P-*sigB* (Figure 6.12). The –35 and –10 elements of P-*sigB* are also similar to those found in promoters recognized by *C. glutamicum* σ^H and σ^M. These elements are also highly similar to those found in P-*sigB* in other corynebacteria and in the downstream *sigB* promoter in *M. tuberculosis*. Two promoters drive transcription of *sigB* in *M. tuberculosis*; σ^F recognizes the upstream promoter, whereas as many as three σ factors (σ^E, σ^H and σ^L) support initiation from the downstream promoter (shown in Figure 6.12) at an identical TSP (Dainese et al., 2006). It is very probable, therefore, that the *sigB* gene is also controlled by at least a single stress σ factor (σ^H or σ^M) in *C. glutamicum*.

Promoters recognized by σ^C, σ^D and σ^E have not yet been described in corynebacteria. The σ^C factor could not be inactivated in *C. glutamicum* strain R (Yukawa et al., 2007) and might therefore be essential like σ^A, however its function in corynebacteria is not known. The σ^D factor is involved in the expression of ribosome-associated genes in the stationary phase in *M. tuberculosis* (Raman et al., 2001). The σ^E factor is induced after exposure of the cells to heat shock, detergents and oxidative stress in *M. tuberculosis* (Manganelli et al., 1999). The consensus sequence of σ^E-dependent promoters in *M. tuberculosis*, $GG^A/_G{}^A/_CC - N_{(15\text{-}19)} - {}^G/_CGTTG$ (Rodrigue et al., 2006), is quite similar to the sequences of the promoters recognized by σ^H and σ^M in corynebacteria. Transcription profiling using a DNA array showed that the *sigE* transcript was formed in response to heat shock in *C. glutamicum* (Muffler

	-35	-10	TSP	
C. glutamicum	CGCTT**GGGAAC**TTTTTGTGGAAGCAGTC**CGTTGA**ACCTCTT**G**			(1)
C. efficiens	TTAAT**GGGAAC**TTTATAAATGGTCGTCC**CGTTAA**ATGTTATG			
C. diphtheriae	GGTAA**TGGAAC**TAGATTTATCGATTCGC**CGTTGT**TCCATG**T**G			(2)
C. jeikeium	GAGTC**GGGAAC**AATTCAACGAAGATGTC**CGTTAA**CCCCTTAG			
M. tuberculosis	TGGGT**GGGAAC**TCAAAGTCCGGCTTTGT**CGTTAA**ACCCCA**T**G			(3)

Figure 6.12. Promoters of the *sigB* gene. Experimentally determined TSPs are in bold and underlined, putative -35 and -10 elements are in bold. References: (1) Halgasova et al., 2001; Oguiza et al., 1996; (2) Oram et al., 2006; (3) Dainese et al., 2006.

et al., 2002). Transcription fusion of *C. glutamicum* P-*sigE* with a reporter gene showed that promoter activity was induced after heat shock and during the stationary growth phase (Pátek et al., unpublished). These results suggest that σ^E together with σ^H, σ^M and σ^B form a regulatory cascade or network, enabling the cells to respond coordinately to temperature, oxidative and growth-phase-induced stresses.

The promoter sequences of four genes involved in osmotic stress response seem to be σ^A-dependent, and their expression is regulated by a two-component MtrA-MtrB system in *C. glutamicum* (Möker et al., 2004). We may conclude that the stress genes in *C. glutamicum* are expressed either from σ^A-dependent promoters regulated by repression or from promoters recognized by ECF σ factors or from a combination of both. Transcription profiling using a DNA array showed that at least some stress σ factors are also expressed to a certain basal level during optimal exponential growth and that the changes of physiological state of the cells due to different growth phases or stresses alter the concentration of individual σ factors rather than strictly control their presence (Engels et al., 2004; Larisch et al., 2007; Nakunst et al., 2007). Moreover, some σ regulons overlap and individual σ factors probably recognize the same promoters but with different efficiency in these cases. The promoter activity therefore seems to be regulated in a complex way in which environmental and physiological signals, concentrations of σ and other transcriptional factors and overlapping recognition specificity and efficiency of the σ factors play major roles in fine tuning the current level of gene expression to adapt the cells to particular conditions.

5.5 Overlapping transcriptional and translational start sites

The translation of bacterial mRNA is initiated by the binding of the Shine-Dalgarno (SD) sequence located at the 5′ end of the mRNA to the complementary 3′ end of 16S rRNA in most cases. The SD sequences are similar to the GGAGG motif located 5 to 13 nt from the initiation codon in most bacteria. The promoters of several genes in *C. glutamicum* were found very close to the translational start sites and the respective TSPs were localized at the first nt of the AUG or GUG translational start codon. Such an arrangement was found in the genes *brnF*, *lrp*, *lpdA*, *ilvA*, *leuA* (leader peptide) (summarized in Pátek et al., 2003), *betP* (Möker et al., 2004) and in three genes (*cg0527*, *cg2782* and *cg3327*) controlled by the DtxR regulator (Brune et al., 2006). Transcripts of *ilvB* (leader peptide) (Keilhauer et al., 1993) and *lysE* (Bellmann et al., 2001) start just 1 nt upstream of the respective initiation codon. Such unusual arrangements appear quite often in *C. glutamicum*. Overlap of the TSP and translation initiation codon was also observed in other actinomycetes (Zahrt and Deretic, 2000) and in archaea. In the rare cases of such leaderless transcripts in *E. coli*, the 5′ terminal AUG start codon was found sufficient to support gene expression and the presence of a downstream box, which might substitute for the function of the SD motif, was not proved (Moll et al., 2002). There are no data available on the possible regulatory effects of leaderless transcripts in corynebacteria.

6. Applications

6.1 Plasmids and promoters in the genetic analysis of corynebacteria

Plasmid vectors are essential tools used for genetic analysis and manipulations in bacteria. The *E. coli* plasmid and phage-derived vectors (e.g., cosmids) were applied in the initial experiments aimed at determining the function of the cloned genes of corynebacteria by heterologous complementation of *E. coli* mutants. The genes involved e.g. in the biosynthesis of lysine and threonine were isolated by this technique (Marquez et al., 1985; Mateos et al., 1987). The optimization of DNA transfer into *C. glutamicum* cells by electrotransformation facilitated the use of *C. glutamicum-E. coli* shuttle cloning vectors for testing the function of cloned genes by homologous complementation of defined *C. glutamicum* auxotrophic mutants (Follettie et al., 1988; Kalinowski et al., 1990).

Promoter-probe vectors, sometimes called promoter-trap or promoter-assay vectors, can be used for the screening (promoter trapping) and analysis (promoter assay) of promoters and their regulatory functions.

The promoter-probe vector pEKplCm (Eikmanns et al., 1991a) was used for the first wider screening of *C. glutamicum* promoters (Pátek et al., 1996). Ten chromosomal promoters and three *C. glutamicum* phage promoters were mapped by determining the respective TSPs by primer extension analysis. Their strength (activity), ranging within two orders of magnitude in the exponential growth phase, was determined by measuring the activity of the chloramphenicol acetyltransferase (*cat*) reporter enzyme. All isolated promoter fragments were also active in *E. coli*. The newly determined promoters and all other known promoter sequences (33 sequences in total) were aligned at the TSPs and analyzed using an unbiased computer protocol. The resulting promoter consensus sequence (TANaaT for –10 and ttGcca for –35) was similar to the general sequence pattern of the bacterial promoters (Figure 6.3) of house keeping genes, although the level of conservation (particularly in the –35 region) was lower than in *E. coli*. The screening of promoters was also carried out in the *C. glutamicum* strains formerly designated *B. lactofermentum* (Cadenas et al., 1991) and *B. flavum* (Zupancic et al., 1995) using the vectors pULMJ88 and pPROBE17, respectively.

The promoter-probe vector pET2 (Vašicová et al., 1998) was found useful in the detailed mutational and deletion analysis of the promoter of the *dapA* gene coding for dihydrodipicolinate synthase. This analysis largely confirmed the derived statistical consensus sequence, showing that the sequence of the –35 hexamer had a very limited effect on promoter activity. The vital positions were T_1 and T_6 in the –10 hexamer TAACCT and the TGN motif immediately upstream of the –10 hexamer contributed strongly to P-*dapA* activity (Vašicová et al., 1999).

These results were confirmed by the promoter mutations inserted into the *C. glutamicum* chromosome by means of the integrative promoter-probe vector pRIM2 (Vašicová et al., 1998). The deletions of sequences upstream and downstream of the core *dapA* promoter had a weak effect on the transcriptional activity and P-*dapA* functioned essentially constitutively over all growth phases. Alterations or deletions within a stretch of six T's upstream of the −35 region, which may induce a DNA bend or serve as an UP element, markedly affected the strength of P-*dapA* (Vašicová et al., 1999).

The effects of the RamA regulator on the activities of enzymes encoded by the genes *pta* (phosphotransacetylase), *ack* (acetate kinase), *aceA* (isocitrate lyase) and *aceB* (malate synthase), essential for the growth of *C. glutamicum* on acetate, were also studied using transcriptional fusions of the respective promoters with the *cat* reporter gene in pET2 (Cramer et al., 2006). Up to a 50-fold increase in the activity of CAT was found in the strains with promoters P-*pta-ack*, P-*aceA* and P-*aceB* grown on acetate (as compared to growth on glucose), whereas no or very low promoter activity was detected in the *ramA* knock-out strain. These results proved that RamA activates the transcription of the *pta-ack* operon and the *aceA* and *aceB* genes.

The promoter-probe vector pET2 was also used to analyze the transcriptional regulation of the *pst* and *ugp* operons under phosphate starvation conditions (Ishige et al., 2003), the heat shock proteins *dnaK* and *groES-groEL* (Barreiro et al., 2004) and the pH-regulated membrane ATP synthase-coding operon (Barriuso-Iglesias et al., 2006).

To test the regulation of the genes *gntP* (gluconate permease) and *gntK* (gluconate kinase) involved in gluconate utilization, their promoters were cloned in the promoter-probe vectors pEMel-1 and pEGFP carrying the *melC* operon (responsible for the production of melanin) and *egfp2* (green fluorescent protein) gene as reporters, respectively (Letek et al., 2006). The promoters were repressed in the presence of sucrose and various monosaccharides in a concentration-dependent manner. A CRP-type GlxR protein was found to bind to the promoter regions of both genes in the presence of cAMP, thus mediating the carbon catabolite regulation.

The properties of plasmids and promoters useful for genetic analysis as well as for biotechnological applications are conveniently combined in expression vectors. Regulated as well as constitutive promoters are often used for the overexpression of genes to study gene functions. *E. coli* regulated promoters P-*lac*, P-*tac* and P-*trc* were applied in *C. glutamicum* due to the well described mechanisms of their induction and due to the availability of various expression vectors. According to some results, these promoters display certain disadvantages when used in corynebacteria. P-*lac*, P-*tac* and P-*trc* are not very strong in *C. glutamicum* and show leaky expression in some *C. glutamicum* strains, including the type strain ATCC 13032. Moreover, most *C. glutamicum* strains exhibit a low permeability for IPTG, the common inducer (Pátek et al., 2003). In spite of these drawbacks, P-*trc* was successfully used for the overexpression of *sodA* (superoxide dismutase) from *Corynebacterium melassecola* (Merkamm and Guyonvarch, 2001) and *malE* (malic enzyme) from *C. glutamicum* (Gourdon et al., 2000). The IPTG-induced overexpression of the *pyc* gene (pyruvate carboxylase) from P-*tac* in the vector pVWEx1 increased glutamate, lysine and threonine production by the recombinant strain (Peters-Wendisch et al., 2001).

Attempts to use homologous promoters in corynebacteria were also done. Promoters of the genes *gntP* and *gntK* were used for regulated overexpression of the *divIVA* gene (translationally fused to *gfp*) from the plasmids pEPDG and pEKDG in *C. glutamicum* (Letek et al., 2006). The presence of gluconate in minimal medium resulted in an altered morphology; the cells were larger, rounded, and tended to grow in a polar fashion. However, a reliable *C. glutamicum* strong regulated promoter which could be both fully repressed and easily induced at will is clearly missing in the armoury for gene manipulation in corynebacteria.

6.2 Plasmids and promoters in the biotechnology of corynebacteria

Plasmid vectors for corynebacteria are important tools, not only for the detailed analysis of gene function, but also for the improvement of producing strains. Cloning homologous or heterologous genes involved in biosynthesis pathways into multicopy plasmid cloning vectors was found to result in the synthesis of a higher amount of the gene products due to higher gene dosage. *C. glutamicum* strains producing threonine (Eikmanns et al., 1991b), tryptophan (Matsui et al., 1988), tyrosine (Ito et al., 1990a) and phenylalanine (Ito et al., 1990b) were constructed using this approach. Whole operons (Matsui et al., 1988), individual genes coding for the key enzymes (Ito et al., 1990a) or a combination of genes not linked in the chromosome but involved in a single metabolic pathway (Koffas et al., 2003) were cloned on multicopy plasmid cloning vectors and a substantial increase in the activities of their products was observed. The cloning of two genes involved in one pathway into two different compatible plasmid vectors (based on plasmids pBL1 and pCG1, respectively) has also been applied (Ito et al., 1990b). The cloning of wild-type genes in multicopy plasmid vectors and subsequent oversynthesis of their gene products due to higher gene dosage was found to have no negative effects on the growth of *C. glutamicum* cell cultures and stable maintenance of the recombinant plasmids was observed (Cremer et al., 1991). On the other hand, the cloning of mutant genes coding for feedback-resistant enzymes in multicopy plasmid vectors led to a high instability of the resulting recombinant plasmids. This phenomenon is probably elicited by the deleterious effect of a high cellular concentration of some intermediate(s) or of a final product of the derepressed metabolic pathway (Cremer et al., 1991; Eikmanns et al., 1991b). To avoid such negative consequences of cloning in multicopy plasmids, low-copy-number vectors based on a minireplicon of plasmid pNG2 from *C. diphtheriae* were developed. These vectors were applied in cloning of mutant genes involved in biosynthesis of threonine (Reinscheid et al., 1994). The use of specific mutant host strains and plasmid vectors carrying the genes complementing the mutation is another approach ensuring the stable maintenance of recombinant plasmids. The serine-requiring *C. glutamicum* mutant strain deficient in 3-phosphoglycerate dehydrogenase (PDG) and harboring the recombinant plasmid, which carried the *pdg* gene and the cloned genes involved in the biosynthesis of tryptophan, was used as stable tryptophan producer (Ikeda et al., 1994). The other system includes the Δ*alr* *C. glutamicum* strain requiring D-alanine for growth and the cloning vector pSELF2000X, which carries the *alr* gene coding for alanine racemase. Since plasmidless cells cannot grow in common complex media without their supplementing with D-alanine, there is a strong selective pressure to maintain the plasmids in the cultures of Δ*alr* strains (Tauch et al., 2002b).

To broaden the range of carbon and energy sources and to utilize a cheap substrate, the *xylA* gene (xylose isomerase) from *E. coli* was introduced on a high-copy-number plasmid into *C. glutamicum* cells. Expression of the *xylA* gene from the constitutive P-*trc* resulted in a xylose-consuming strain which produced lactic and succinic acids under oxygen limitation with concomitant arrested cell replication. Such a strain seems promising for the development of ethanol production from lignocellulosic biomass (Kawaguchi et al., 2006). Further metabolic engineering which equipped *C. glutamicum* with the abilities to catabolise additional sugars (cellobiose and galactose) may lead to the development of a cost-effective biomass converter (Barrett et al., 2004; Kotrba et al., 2003).

Using a similar approach, the *amy* gene (α-amylase) from *Streptomyces griseus* was introduced on a plasmid into *C. glutamicum* to construct a starch-utilizing lysine producer (Seibold et al., 2006).

Sophisticated expression systems based on recombinant plasmids, the easy induction of a gene expression with a simple action (e.g., addition of IPTG, increase of temperature), which function quite well in the laboratory, proved clumsy, unstable and non-economic in industrial scale. The presence of plasmids in industrial strains may result in unstable producers, which raise concerns about spreading resistance and other heterologous genes by horizontal gene transfer. Genetic intervention within the bacterial chromosome using integrative vectors thus seems to be a safe and practical alternative. For example, a *C. glutamicum* lysine producer was improved in this way. Introduction of a second copy of the *dapA* gene (dihydrodipicolinate synthase) into the genome resulted in increased enzyme activity and excretion of lysine (de Graaf et al., 2001). At the same time, a decreased metabolic flux towards homoserine diminished the levels of by-produced threonine and isoleucine. Introduction of a *dapA* copy with the mutant promoter P-*dapA* further enhanced lysine production (Pfefferle et al., 2003; Vašicová et al., 1999).

The purposeful mutagenesis of promoters within a chromosome is another strategy in the molecular breeding of producing strains. Such alterations of the promoters of the genes *ilvD* and *ilvE* involved in valine, isoleucine and leucine biosynthesis, was aimed at strengthening their activity (up-mutations) whereas alterations in the promoters of genes *ilvA* and *leuA* were aimed at reducing their transcription (down-mutations). Combinations of particular promoter mutants in the plasmid-free strain resulted in tuning the whole pathway to an overall higher flux of the intermediates and in an increased production of valine (authors' unpublished results).

Thanks to current techniques of fast DNA sequencing, the native plasmids of corynebacteria may provide more hints about acquired heterologous genes, the functions of mobile elements and in the future, the role of horizontal gene transfer in the evolution of genomes. In particular, their role in the development of pathogenic species of corynebacteria awaits elucidation. The plasmid vectors available are intensively utilized as irreplaceable tools in genetic analysis and biotechnological applications and further new special-purpose vectors should be constructed. In promoter studies, it is expected that the sequences recognized by alternative σ factors in RNAP will be uncovered and that the regulatory cascades or network formed by transcription signals will be elucidated.

Acknowledgements

Work in the authors´ laboratory was supported by grants 525/04/0548 and 204/06/0330 from the Scientific Council of the Czech Republic and by Institutional Research Concept No. AV0Z50200510. The authors thank J. Jonák for critical reading of the manuscript.

References

Abrhámová, Z., Pátek, M., and Nešvera, J. (2002). Atypical location of double-strand origin of replication (*nic* site) on the plasmid pGA1 from *Corynebacterium glutamicum*. Folia Microbiol. *47*, 307-310.

Adham, S.A., Rodriguez, S., Ramos, A., Santamaria, R.I., and Gil, J.A. (2003). Improved vectors for transcriptional/translational signal screening in corynebacteria using the *melC* operon from *Streptomyces glaucescens* as reporter. Arch. Microbiol. *180*, 53-59.

Amador, E., Martin, J.F., and Castro, J.M. (2000). A *Brevibacterium lactofermentum* 16S rRNA gene used as target site for homologous recombination. FEMS Microbiol. Lett. *185*, 199-204.

Archer, J.A., and Sinskey, A.J. (1993). The DNA sequence and minimal replicon of the *Corynebacterium glutamicum* plasmid pSR1: evidence of a common ancestry with plasmids from *C. diphtheriae*. J. Gen. Microbiol. *139*, 1753-1759.

Argnani, A., Leer, R.J., van Luijk, N., and Pouwels, P.H. (1996). A convenient and reproducible method to genetically transform bacteria of the genus *Bifidobacterium*. Microbiology *142*, 109-114.

Barák, I., Koptides, M., Jucovič, M., Šišová, M., and Timko, J. (1990). Construction of a promoter-probe shuttle vector for *Escherichia coli* and brevibacteria. Gene *95*, 133-135.

Bardonnet, N., and Blanco, C. (1991). Improved vectors for transcriptional signal screening in corynebacteria. FEMS Microbiol. Lett. *68*, 97-102.

Barreiro, C., Gonzalez-Lavado, E., Pátek, M., and Martín, J.F. (2004). Transcriptional analysis of the *groES-groEL1*, *groEL2*, and *dnaK* genes in *Corynebacterium glutamicum*: characterization of heat shock-induced promoters. J. Bacteriol. *186*, 4813-4817.

Barrett, E., Stanton, C., Zelder, O., Fitzgerald, G., and Ross, R.P. (2004). Heterologous expression of lactose- and galactose-utilizing pathways from lactic acid bacteria in *Corynebacterium glutamicum* for production of lysine in whey. Appl. Environ. Microbiol. *70*, 2861-2866.

Barriuso-Iglesias, M., Barreiro, C., Flechoso, F., and Martín, J.F. (2006). Transcriptional analysis of the F0F1 ATPase operon of *Corynebacterium glutamicum* ATCC 13032 reveals strong induction by alkaline pH. Microbiology *152*, 11-21.

Bashyam, M.D., and Tyagi, A.K. (1998). Identification and analysis of "extended -10" promoters from mycobacteria. J. Bacteriol. *180*, 2568-2573.

Beaucher, J., Rodrigue, S., Jacques, P.E., Smith, I., Brzezinski, R., and Gaudreau, L. (2002). Novel *Mycobacterium tuberculosis* anti-sigma factor antagonists control sigmaF activity by distinct mechanisms. Mol. Microbiol. *45*, 1527-1540.

Bellmann, A., Vrljic, M., Pátek, M., Sahm, H., Krämer, R., and Eggeling, L. (2001). Expression control and specificity of the basic amino acid exporter LysE of *Corynebacterium glutamicum*. Microbiology *147*, 1765-1774.

Billington, S.J., Jost, B.H., and Songer, J.G. (1998). The *Arcanobacterium (Actinomyces) pyogenes* plasmid pAP1 is a member of the pIJ101/pJV1 family of rolling circle replication plasmids. J. Bacteriol. *180*, 3233-3236.

Brune, I., Werner, H., Hüser, A.T., Kalinowski, J., Pühler, A., and Tauch, A. (2006). The DtxR protein acting as dual transcriptional regulator directs a global regulatory network involved in iron metabolism of *Corynebacterium glutamicum*. BMC Genomics *7*, 21.

Cadenas, R.F., Martín, J.F., and Gil, J.A. (1991). Construction and characterization of promoter-probe vectors for Corynebacteria using the kanamycin-resistance reporter gene. Gene *98*, 117-121.

Cadenas, R.F., Fernandez-Gonzalez, C., Martín, J.F., and Gil, J.A. (1996). Construction of new cloning vectors for *Brevibacterium lactofermentum*. FEMS Microbiol. Lett. *137*, 63-68.

Cramer, A., Gerstmeir, R., Schaffer, S., Bott, M., and Eikmanns, B.J. (2006). Identification of RamA, a novel LuxR-type transcriptional regulator of genes involved in acetate metabolism of *Corynebacterium glutamicum*. J. Bacteriol. *188*, 2554-2567.

Cremer, J., Eggeling, L., and Sahm, H. (1990). Cloning the *dapA dapB* cluster of the lysine-secreting bacterium *Corynebacterium glutamicum*. Mol. Gen. Genet. *220*, 478-480.

Cremer, J., Eggeling, L., and Sahm, H. (1991). Control of the lysine biosynthesis sequence in *Corynebacterium glutamicum* as analyzed by overexpression of the individual corresponding genes. Appl. Environ. Microbiol. *57*, 1746-1752.

Crooks, G.E., Hon, G., Chandonia, J.M., and Brenner, S.E. (2004). WebLogo: a sequence logo generator. Genome Res. *14*, 1188-1190.

Dainese, E., Rodrigue, S., Delogu, G., Provvedi, R., Laflamme, L., Brzezinski, R., Fadda, G., Smith, I., Gaudreau, L., Palu, G., and Manganelli, R. (2006). Posttranslational regulation of *Mycobacterium tuberculosis* extracytoplasmic-function sigma factor sigma L and roles in virulence and in global regulation of gene expression. Infect. Immun. *74*, 2457-2461.

de Graaf, A.A., Eggeling, L., and Sahm, H. (2001). Metabolic engineering for L-lysine production by *Corynebacterium glutamicum*. Adv. Biochem. Engn. Biotechnol. *73*, 9-29.

del Solar, G., Giraldo, R., Ruiz-Echevarria, M.J., Espinosa, M., and Diaz-Orejas, R. (1998). Replication and control of circular bacterial plasmids. Microbiol. Mol. Biol. Rev. *62*, 434-464.

Duran, R. (1998). New shuttle vectors for *Rhodococcus* sp. R312 (formerly *Brevibacterium* sp. R312), a nitrile hydratase producing strain. J. Basic Microbiol. *38*, 101-106.

Dusch, N., Pühler, A., and Kalinowski, J. (1999). Expression of the *Corynebacterium glutamicum panD* gene encoding L-aspartate-alpha-decarboxylase leads to pantothenate overproduction in *Escherichia coli*. Appl. Environ. Microbiol. *65*, 1530-1539.

Eggeling, L., and Reyes, O. (2005). Experiments. In Handbook of *Corynebacterium glutamicum*, L. Eggeling, and M. Bott, eds. (Boca Raton, FL, CRC Press), pp. 535-566.

Eikmanns, B.J., Kleinertz, E., Liebl, W., and Sahm, H. (1991a). A family of *Corynebacterium glutamicum/Escherichia coli* shuttle vectors for cloning, controlled gene expression, and promoter probing. Gene *102*, 93-98.

Eikmanns, B.J., Metzger, M., Reinscheid, D., Kircher, M., and Sahm, H. (1991b). Amplification of three threonine biosynthesis genes in *Corynebacterium glutamicum* and its influence on carbon flux in different strains. Appl. Microbiol. Biotechnol. *34*, 617-622.

Elišáková, V., Pátek, M., Holátko, J., Nešvera, J., Leyval, D., Goergen, J.L., and Delaunay, S. (2005). Feedback-resistant acetohydroxy acid synthase increases valine production in *Corynebacterium glutamicum*. Appl. Environ. Microbiol. *71*, 207-213.

Engels, V., and Wendisch, V.F. (2007). The DeoR-type regulator SugR represses expression of *ptsG* in *Corynebacterium glutamicum*. J. Bacteriol. *189*, 2955-2966.

Engels, S., Schweitzer, J.E., Ludwig, C., Bott, M., and Schaffer, S. (2004). *clpC* and *clpP1P2* gene expression in *Corynebacterium glutamicum* is controlled by a regulatory network involving the transcriptional regulators ClgR and HspR as well as the ECF sigma factor sigmaH. Mol. Microbiol. *52*, 285-302.

Fernandez-Gonzalez, C., Cadenas, R.F., Noirot-Gros, M.F., Martín, J.F., and Gil, J.A. (1994). Characterization of a region of plasmid pBL1 of *Brevibacterium lactofermentum* involved in replication via the rolling circle model. J. Bacteriol. *176*, 3154-3161.

Filpula, D., Ally, A.H., and Nagle, J. (1986). Complete nucleotide sequence of a native plasmid of *Brevibacterium lactofermentum*. Nucleic Acids Res. *14*, 5114.

Follettie, M.T., Shin, H.K., and Sinskey, A.J. (1988). Organization and regulation of the *Corynebacterium glutamicum hom-thrB* and *thrC* loci. Mol. Microbiol. *2*, 53-62.

Gourdon, P., Baucher, M.F., Lindley, N.D., and Guyonvarch, A. (2000). Cloning of the malic enzyme gene from *Corynebacterium glutamicum* and role of the enzyme in lactate metabolism. Appl. Environ. Microbiol. *66*, 2981-2987.

Goyal, D., Wachi, M., Kijima, N., Kobayashi, M., Yukawa, H., and Nagai, K. (1996). A cryptic plasmid pBL1 from *Brevibacterium lactofermentum* causes growth inhibition and filamentation in *Escherichia coli*. Plasmid *36*, 62-66.

Guillouet, S., Rodal, A.A., An, G., Lessard, P.A., and Sinskey, A.J. (1999). Expression of the *Escherichia coli* catabolic threonine dehydratase in *Corynebacterium glutamicum* and its effect on isoleucine production. Appl. Environ. Microbiol. *65*, 3100-3107.

Halgasova, N., Bukovska, G., Timko, J., and Kormanec, J. (2001). Cloning and transcriptional characterization of two sigma factor genes, *sigA* and *sigB*, from *Brevibacterium flavum*. Curr. Microbiol. *43*, 249-254.

Halgasova, N., Bukovska, G., Ugorcakova, J., Timko, J., and Kormanec, J. (2002). The *Brevibacterium flavum* sigma factor SigB has a role in the environmental stress response. FEMS Microbiol. Lett. *216*, 77-84.

Hansmeier, N., Albersmeier, A., Tauch, A., Damberg, T., Ros, R., Anselmetti, D., Pühler, A., and Kalinowski, J. (2006). The surface (S)-layer gene *cspB* of *Corynebacterium glutamicum* is transcriptionally activated by a LuxR-type regulator and located on a 6 kb genomic island absent from the type strain ATCC 13032. Microbiology *152*, 923-935.

Helmann, J.D. (2002). The extracytoplasmic function (ECF) sigma factors. Adv. Microb. Physiol. *46*, 47-110.

Ikeda, M., and Katsumata, R. (1998). A novel system with positive selection for the chromosomal integration of replicative plasmid DNA in *Corynebacterium glutamicum*. Microbiology *144*, 1863-1868.

Ikeda, M., and Katsumata, R. (1999). Hyperproduction of tryptophan by *Corynebacterium glutamicum* with the modified pentose phosphate pathway. Appl. Environ. Microbiol. *65*, 2497-2502.

Ikeda, M., and Nakagawa, S. (2003). The *Corynebacterium glutamicum* genome: features and impacts on biotechnological processes. Appl. Microbiol. Biotechnol. *62*, 99-109.

Ikeda, M., Nakanishi, K., Kino, K., and Katsumata, R. (1994). Fermentative production of tryptophan by a stable recombinant strain of *Corynebacterium glutamicum* with a modified serine-biosynthetic pathway. Biosci. Biotechnol. Biochem. *58*, 674-678.

Ilyina, T.V., and Koonin, E.V. (1992). Conserved sequence motifs in the initiator proteins for rolling circle DNA replication encoded by diverse replicons from eubacteria, eucaryotes and archaebacteria. Nucleic Acids Res. *20*, 3279-3285.

Ishige, T., Krause, M., Bott, M., Wendisch, V.F., and Sahm, H. (2003). The phosphate starvation stimulon of *Corynebacterium glutamicum* determined by DNA microarray analyses. J. Bacteriol. *185*, 4519-4529.

Ito, H., Sato, K., Enei, H., and Hirose, Y. (1990a). Improvement in microbial production of L-tyrosine by gene dosage effect of *aroL* gene coding shikimate kinase. Agr. Biol. Chem. *54*, 823-824.

Ito, H., Sato, K., Matsui, K., Sano, K., Enei, H., and Hirose, Y. (1990b). Molecular breeding of a *Brevibacterium lactofermentum* L-phenylalanine producer using a cloned prephenate dehydratase gene. Appl. Microbiol. Biotechnol. *33*, 190-195.

Jäger, W., Schäfer, A., Pühler, A., Labes, G., and Wohlleben, W. (1992). Expression of the *Bacillus subtilis sacB* gene leads to sucrose sensitivity in the Gram-positive bacterium *Corynebacterium glutamicum* but not in *Streptomyces lividans*. J. Bacteriol. *174*, 5462-5465.

Jakob, K., Satorhelyi, P., Lange, C., Wendisch, V.F., Silakowski, B., Scherer, S., and Neuhaus, K. (2007). Gene expression analysis of long-term lactic acid adapted *Corynebacterium glutamicum*. J. Bacteriol. *189*, 5582-5590.

Jakoby, M., Ngouoto-Nkili, C.E., and Burkovski, A. (1999). Construction and application of new *Corynebacterium glutamicum* vectors. Biotechnol. Techniques *13*, 437-441.

Jost, B.H., Billington, S.J., and Songer, J.G. (1997). Electroporation-mediated transformation of *Arcanobacterium* (*Actinomyces*) *pyogenes*. Plasmid *38*, 135-140.

Kalinowski, J., Bachmann, B., Thierbach, G., and Pühler, A. (1990). Aspartokinase genes *lysC* alpha and *lysC* beta overlap and are adjacent to the aspartate beta-semialdehyde dehydrogenase gene *asd* in *Corynebacterium glutamicum*. Mol. Gen. Genet. *224*, 317-324.

Kalinowski, J., Bathe, B., Bartels, D., Bischoff, N., Bott, M., Burkovski, A., Dusch, N., Eggeling, L., Eikmanns, B.J., Gaigalat, L., et al. (2003). The complete *Corynebacterium glutamicum* ATCC 13032 genome sequence and its impact on the production of L-aspartate-derived amino acids and vitamins. J. Biotechnol. *104*, 5-25.

Kato, F., Hara, N., Matsuyama, K., Hattori, K., Ishii, M., and Marata, A. (1989). Isolation of plasmids from *Brevibacterium*. Agr. Biol. Chem. *53*, 879-881.

Katsumata, R., Ozaki, A., Oka, T., and Furuya, A. (1984). Protoplast transformation of glutamate-producing bacteria with plasmid DNA. J. Bacteriol. *159*, 306-311.

Kawaguchi, H., Vertes, A.A., Okino, S., Inui, M., and Yukawa, H. (2006). Engineering of a xylose metabolic pathway in *Corynebacterium glutamicum*. Appl. Environ. Microbiol. *72*, 3418-3428.

Keilhauer, C., Eggeling, L., and Sahm, H. (1993). Isoleucine synthesis in *Corynebacterium glutamicum*: molecular analysis of the *ilvB-ilvN-ilvC* operon. J. Bacteriol. *175*, 5595-5603.

Kerry-Williams, S.M., and Noble, W.C. (1986). Plasmids in group JK coryneform bacteria isolated in a single hospital. J. Hyg. (London) *97*, 255-263.

Khan, S.A. (1997). Rolling-circle replication of bacterial plasmids. Microbiol. Mol. Biol. Rev. *61*, 442-455.

Kim, T.H., Kim, H.J., Park, J.S., Kim, Y., Kim, P., and Lee, H.S. (2005). Functional analysis of *sigH* expression in *Corynebacterium glutamicum*. Biochem. Biophys. Res. Commun. *331*, 1542-1547.

Kirchner, O., and Tauch, A. (2003). Tools for genetic engineering in the amino acid-producing bacterium *Corynebacterium glutamicum*. J. Biotechnol. *104*, 287-299.

Knoppová, M., Phensaijai, M., Veselý, M., Zemanová, M., Nešvera, J., and Pátek, M. (2007). Plasmid vectors for testing in vivo promoter activities in *Corynebacterium glutamicum* and *Rhodococccus erythropolis*. Curr. Microbiol. *55*, 234-239.

Koffas, M.A., Jung, G.Y., and Stephanopoulos, G. (2003). Engineering metabolism and product formation in *Corynebacterium glutamicum* by coordinated gene overexpression. Metab. Engn. *5*, 32-41.

Kono, M., Sasatsu, M., and Aoki, T. (1983). R plasmids in *Corynebacterium xerosis* strains. Antimicrob. Agents Chemother. *23*, 506-508.

Kotrba, P., Inui, M., and Yukawa, H. (2003). A single V317A or V317M substitution in Enzyme II of a newly identified beta-glucoside phosphotransferase and utilization system of *Corynebacterium glutamicum* R extends its specificity towards cellobiose. Microbiology *149*, 1569-1580.

Krämer, R. (1996). Genetic and physiological approaches for the production of amino acids. J. Biotechnol. *45*, 1-21.

Larisch, C., Nakunst, D., Hüser, A.T., Tauch, A., and Kalinowski, J. (2007). The alternative sigma factor SigB of *Corynebacterium glutamicum* modulates global gene expression during transition from exponential growth to stationary phase. BMC Genomics *8*, 4.

Lei, C., Ren, Z., Yang, W., Chen, Y., Chen, D., Liu, M., Yan, W., and Zheng, Z. (2002). Characterization of a novel plasmid pXZ608 from *Corynebacterium glutamicum*. FEMS Microbiol. Lett. *216*, 71-75.

Letek, M., Valbuena, N., Ramos, A., Ordonez, E., Gil, J.A., and Mateos, L.M. (2006). Characterization and use of catabolite-repressed promoters from gluconate genes in *Corynebacterium glutamicum*. J. Bacteriol. *188*, 409-423.

Manganelli, R., Dubnau, E., Tyagi, S., Kramer, F.R., and Smith, I. (1999). Differential expression of 10 sigma factor genes in *Mycobacterium tuberculosis*. Mol. Microbiol. *31*, 715-724.

Marquez, G., Sousa, J.M., and Sanchez, F. (1985). Cloning and expression in *Escherichia coli* of genes involved in the lysine pathway of *Brevibacterium lactofermentum*. J. Bacteriol. *164*, 379-383.

Mateos, L.M., del Real, G., Aguilar, A., and Martín, J.F. (1987). Cloning and expression in *Escherichia coli* of the homoserine kinase (*thrB*) gene from *Brevibacterium lactofermentum*. Mol. Gen. Genet. *206*, 361-367.

Matsui, K., Ishida, M., Tsuchiya, M., and Sano, K. (1988). Construction of tryptophan-producing recombinant strains of *Brevibacterium lactofermentum* using the engineered *trp* operons. Agr. Biol. Chem. *52*, 1863-1865.

Merkamm, M., and Guyonvarch, A. (2001). Cloning of the *sodA* gene from *Corynebacterium melassecola* and role of superoxide dismutase in cellular viability. J. Bacteriol. *183*, 1284-1295.

Miwa, K., Matsui, H., Terabe, M., Nakamori, S., Sano, K., and Momose, H. (1984). Cryptic plasmids in glutamic acid-producing bacteria. Agr. Biol. Chem. *48*, 2901-2903.

Möker, N., Brocker, M., Schaffer, S., Krämer, R., Morbach, S., and Bott, M. (2004). Deletion of the genes encoding the MtrA-MtrB two-component system of *Corynebacterium glutamicum* has a strong influence on cell morphology, antibiotics susceptibility and expression of genes involved in osmoprotection. Mol. Microbiol. *54*, 420-438.

Moll, I., Grill, S., Gualerzi, C.O., and Blasi, U. (2002). Leaderless mRNAs in bacteria: surprises in ribosomal recruitment and translational control. Mol. Microbiol. *43*, 239-246.

Moreau, S., Blanco, C., and Trautwetter, A. (1999). Site-specific integration of corynephage phi16: construction of an integration vector. Microbiology *145*, 539-548.

Muffler, A., Bettermann, S., Haushalter, M., Horlein, A., Neveling, U., Schramm, M., and Sorgenfrei, O. (2002). Genome-wide transcription profiling of *Corynebacterium glutamicum* after heat shock and during growth on acetate and glucose. J. Biotechnol. *98*, 255-268.

Nakamura, J., Kanno, S., Kimura, E., Matsui, K., Nakamatsu, T., and Wachi, M. (2006). Temperature-sensitive cloning vector for *Corynebacterium glutamicum*. Plasmid *56*, 179-186.

Nakata, K., Inui, M., Kos, P.B., Vertes, A.A., and Yukawa, H. (2004). Vectors for genetic engineering of corynebacteria. In Fermentation Biotechnology, B. C. Saha, ed. (Washington, D. C., American Chemical Society), pp. 175-191.

Nakunst, D., Larisch, C., Hüser, A.T., Tauch, A., Pühler, A., and Kalinowski, J. (2007). The extracytoplasmic function-type sigma factor SigM of *Corynebacterium glutamicum* ATCC 13032 is involved in transcription of disulfide stress-related genes. J. Bacteriol. *189*, 4696-4707.

Nandi, S., Maurer, J.J., Hofacre, C., and Summers, A.O. (2004). Gram-positive bacteria are a major reservoir of Class 1 antibiotic resistance integrons in poultry litter. Proc. Natl. Acad. Sci. USA *101*, 7118-7122.

Nath, N., and Deb, J.K. (1995). Partial characterization of small plasmids from *Corynebacterium renale*. Plasmid *34*, 229-233.

Nešvera, J., Hochmannová, J., Pátek, M., Šroglová, A., and Bečvářová, V. (1994). Transfer of the broad-host-range IncQ plasmid RSF1010 and other plasmid vectors to the Gram-positive methylotroph *Brevibacterium methylicum* by electrotransformation. Appl. Microbiol. Biotechnol. *40*, 864-866.

Nešvera, J., Pátek, M., Hochmannová, J., Abrhámová, Z., Bečvářová, V., Jelínková, M., and Vohradský, J. (1997). Plasmid pGA1 from *Corynebacterium glutamicum* codes for a gene product that positively influences plasmid copy number. J. Bacteriol. *179*, 1525-1532.

Nishio, Y., Nakamura, Y., Kawarabayasi, Y., Usuda, Y., Kimura, E., Sugimoto, S., Matsui, K., Yamagishi, A., Kikuchi, H., Ikeo, K., and Gojobori, T. (2003). Comparative complete genome sequence analysis of the amino acid replacements responsible for the thermostability of *Corynebacterium efficiens*. Genome Res. *13*, 1572-1579.

Oguiza, J.A., Marcos, A.T., Malumbres, M., and Martín, J.F. (1996). Multiple sigma factor genes in *Brevibacterium lactofermentum*: characterization of *sigA* and *sigB*. J. Bacteriol. *178*, 550-553.

Oram, D.M., Jacobson, A.D., and Holmes, R.K. (2006). Transcription of the contiguous *sigB*, *dtxR*, and *galE* genes in *Corynebacterium diphtheriae:* evidence for multiple transcripts and regulation by environmental factors. J. Bacteriol. *188*, 2959-2973.

Ozaki, A., Katsumata, R., Oka, T., and Furuya, A. (1984). Functional expression of the genes of *Escherichia coli* in Gram-positive *Corynebacterium glutamicum*. Mol. Gen. Genet. *196*, 175-178.

Paget, M.S., Molle, V., Cohen, G., Aharonowitz, Y., and Buttner, M.J. (2001). Defining the disulphide stress response in *Streptomyces coelicolor* A3(2): identification of the sigmaR regulon. Mol. Microbiol. *42*, 1007-1020.

Park, S.D., Lee, S.N., Park, I.H., Choi, J.S., Jeong, W.K., Kim , Y., and Lee, H.S. (2004). Isolation and characterization of transcriptional elements from *Corynebacterium glutamicum*. J. Microbiol. Biotechnol. *14*, 789-795.

Pátek, M. (2005). Regulation of Gene Expression. In Handbook of *Corynebacterium glutamicum*, L. Eggeling, and M. Bott, eds. (Boca Raton, FL, CRC Press), pp. 81-98.

Pátek, M., Eikmanns, B.J., Pátek, J., and Sahm, H. (1996). Promoters from *Corynebacterium glutamicum*: cloning, molecular analysis and search for a consensus motif. Microbiology *142*, 1297-1309.

Pátek, M., Nešvera, J., Guyonvarch, A., Reyes, O., and Leblon, G. (2003). Promoters of *Corynebacterium glutamicum*. J. Biotechnol. *104*, 311-323.

Peters-Wendisch, P.G., Schiel, B., Wendisch, V.F., Katsoulidis, E., Möckel, B., Sahm, H., and Eikmanns, B.J. (2001). Pyruvate carboxylase is a major bottleneck for glutamate and lysine production by *Corynebacterium glutamicum*. J. Mol. Microbiol. Biotechnol. *3*, 295-300.

Pfefferle, W., Möckel, B., Bathe, B., and Marx, A. (2003). Biotechnological manufacture of lysine. Adv. Biochem. Engn. Biotechnol. *79*, 59-112.

Pitcher, D., Johnson, A., Allerberger, F., Woodford, N., and George, R. (1990). An investigation of nosocomial infection with *Corynebacterium jeikeium* in surgical patients using a ribosomal RNA gene probe. Eur. J. Clin. Microbiol. Infect. Dis. *9*, 643-648.

Qiu, J., and Helmann, J.D. (2001). The -10 region is a key promoter specificity determinant for the *Bacillus subtilis* extracytoplasmic-function sigma factors sigma(X) and sigma(W). J. Bacteriol. *183*, 1921-1927.

Radford, A.J., and Hodgson, A.L. (1991). Construction and characterization of a *Mycobacterium-Escherichia coli* shuttle vector. Plasmid *25*, 149-153.

Raman, S., Song, T., Puyang, X., Bardarov, S., Jacobs, W.R., Jr., and Husson, R.N. (2001). The alternative sigma factor SigH regulates major components of oxidative and heat stress responses in *Mycobacterium tuberculosis*. J. Bacteriol. *183*, 6119-6125.

Reinscheid, D.J., Kronemeyer, W., Eggeling, L., Eikmanns, B.J., and Sahm, H. (1994). Stable expression of *hom*-1-*thrB* in *Corynebacterium glutamicum* and its effect on the carbon flux to threonine and related amino acids. Appl. Environ. Microbiol. *60*, 126-132.

Rodrigue, S., Provvedi, R., Jacques, P.E., Gaudreau, L., and Manganelli, R. (2006). The sigma factors of *Mycobacterium tuberculosis*. FEMS Microbiol. Rev. *30*, 926-941.

Sandoval, H., Aguilar, A., Paniagua, C., and Martín, J.F. (1984). Isolation and physical characterization of plasmid pCC1 from *Corynebacterium callunae* and construction of hybrid derivatives. Appl. Microbiol. Biotechnol. *19*, 409-413.

Sandoval, H., del Real, G., Mateos, L.M., Aguilar, A., and Martín, J.F. (1985). Screening of plasmids in non-pathogenic corynebacteria. FEMS Microbiol. Lett. *27*, 93-98.

Santamaría, R., Gil, J.A., Mesas, J.M., and Martín, J.F. (1984). Characterization of endogenous plasmid and development of cloning vectors and a transformation system in *Brevibacterium lactofementum*. J. Gen. Microbiol. *130*, 2237-2246.

Schäfer, A., Kalinowski, J., Simon, R., Seep-Feldhaus, A.H., and Pühler, A. (1990). High-frequency conjugal plasmid transfer from Gram-negative *Escherichia coli* to various Gram-positive coryneform bacteria. J. Bacteriol. *172*, 1663-1666.

Schäfer, A., Tauch, A., Jäger, W., Kalinowski, J., Thierbach, G., and Pühler, A. (1994). Small mobilizable multi-purpose cloning vectors derived from the *Escherichia coli* plasmids pK18 and pK19: selection of defined deletions in the chromosome of *Corynebacterium glutamicum*. Gene *145*, 69-73.

Schiller, J., Groman, N., and Coyle, M. (1980). Plasmids in *Corynebacterium diphtheriae* and diphtheroids mediating erythromycin resistance. Antimicrob. Agents Chemother. *18*, 814-821.

Schrumpf, B., Schwarzer, A., Kalinowski, J., Pühler, A., Eggeling, L., and Sahm, H. (1991). A functionally split pathway for lysine synthesis in *Corynebacterium glutamicum*. J. Bacteriol. *173*, 4510-4516.

Seibold, G., Auchter, M., Berens, S., Kalinowski, J., and Eikmanns, B.J. (2006). Utilization of soluble starch by a recombinant *Corynebacterium glutamicum* strain: Growth and lysine production. J. Biotechnol. *124*, 381-391.

Serwold-Davis, T.M., Groman, N., and Rabin, M. (1987). Transformation of *Corynebacterium diphtheriae*, *Corynebacterium ulcerans*, *Corynebacterium glutamicum*, and *Escherichia coli* with the *C. diphtheriae* plasmid pNG2. Proc. Natl. Acad. Sci. USA *84*, 4964-4968.

Shaw, P.C., and Hartley, B.S. (1988). A host-vector system for an *Arthrobacter* species. J. Gen. Microbiol. *134*, 903-911.

Shen, T., Jia, P., Na, S., and Men, D. (1993). Determination of nucleotide sequence of *Corynebacterium* plasmid pXZ10145. Chin. J. Biotechnol. *9*, 171-178.

Shultzaberger, R.K., Chen, Z., Lewis, K.A., and Schneider, T.D. (2007). Anatomy of *Escherichia coli* sigma70 promoters. Nucleic Acids Res. *35*, 771-788.

Simon, R., Priefer, U., and Pühler, A. (1983). A broad host range mobilization system for *in vivo* genetic engineering: transposon mutagenesis in Gram negative bacteria. Bio/Technology *1*, 784-791.

Smith, M.D., Flickinger, J.L., Lineberger, D.W., and Schmidt, B. (1986). Protoplast transformation in coryneform bacteria and introduction of an alpha-amylase gene from *Bacillus amyloliquefaciens* into *Brevibacterium lactofermentum*. Appl. Environ. Microbiol. *51*, 634-639.

Songer, J.G., Hilwig, R.W., Leeming, M.N., Iandolo, J.J., and Libby, S.J. (1991). Transformation of *Corynebacterium pseudotuberculosis* by electroporation. Am. J. Vet. Res. *52*, 1258-1261.

Sonnen, H., Thierbach, G., Kautz, S., Kalinowski, J., Schneider, J., Pühler, A., and Kutzner, H.J. (1991). Characterization of pGA1, a new plasmid from *Corynebacterium glutamicum* LP-6. Gene *107*, 69-74.

Srivastava, P., and Deb, J.K. (2002). Construction of fusion vectors of corynebacteria: expression of glutathione-S-transferase fusion protein in *Corynebacterium acetoacidophilum* ATCC 21476. FEMS Microbiol. Lett. *212*, 209-216.

Srivastava, P., Nath, N., and Deb, J.K. (2006). Characterization of broad host range cryptic plasmid pCR1 from *Corynebacterium renale*. Plasmid *56*, 24-34.

Takagi, H., Morinaga, Y., Miwa, K., Nakamori, S., and Sano, K. (1986). Versatile cloning vectors constructed with genes indigenous to a glutamic-acid producer, *Brevibacterium lactofermentum*. Agr. Biol. Chem. *50*, 2597-2603.

Takeda, Y., Fujii, M., Nakajyoh, Y., Nishimura, T., and Isshiki, S. (1990). Isolation of a tetracycline resistance plasmid from a glutamate-producing bacterium,

Corynebacterium melassecola. J. Ferment. Bioeng. *70*, 177-179.

Tauch, A. (2005). Native plasmids of amino acid-producing corynebacteria. In Handbook of *Corynebacterium glutamicum*, L. Eggeling, and M. Bott, eds. (Boca Raton, FL, CRC Press), pp. 57-80.

Tauch, A., Hermann, T., Burkovski, A., Krämer, R., Pühler, A., and Kalinowski, J. (1998). Isoleucine uptake in *Corynebacterium glutamicum* ATCC 13032 is directed by the *brnQ* gene product. Arch. Microbiol. *169*, 303-312.

Tauch, A., Krieft, S., Kalinowski, J., and Pühler, A. (2000a). The 51,409-bp R-plasmid pTP10 from the multiresistant clinical isolate *Corynebacterium striatum* M82B is composed of DNA segments initially identified in soil bacteria and in plant, animal, and human pathogens. Mol. Gen. Genet. *263*, 1-11.

Tauch, A., Pühler, A., Kalinowski, J., and Thierbach, G. (2000b). TetZ, a new tetracycline resistance determinant discovered in Gram-positive bacteria, shows high homology to Gram-negative regulated efflux systems. Plasmid *44*, 285-291.

Tauch, A., Götker, S., Pühler, A., Kalinowski, J., and Thierbach, G. (2002a). The 27.8-kb R-plasmid pTET3 from *Corynebacterium glutamicum* encodes the aminoglycoside adenyltransferase gene cassette *aadA9* and the regulated tetracycline efflux system Tet 33 flanked by active copies of the widespread insertion sequence IS6100. Plasmid *48*, 117-129.

Tauch, A., Götker, S., Pühler, A., Kalinowski, J., and Thierbach, G. (2002b). The alanine racemase gene *alr* is an alternative to antibiotic resistance genes in cloning systems for industrial *Corynebacterium glutamicum* strains. J. Biotechnol. *99*, 79-91.

Tauch, A., Kirchner, O., Löffler, B., Götker, S., Pühler, A., and Kalinowski, J. (2002c). Efficient electrotransformation of *Corynebacterium diphtheriae* with a mini-replicon derived from the *Corynebacterium glutamicum* plasmid pGA1. Curr. Microbiol. *45*, 362-367.

Tauch, A., Bischoff, N., Brune, I., and Kalinowski, J. (2003a). Insights into the genetic organization of the *Corynebacterium diphtheriae* erythromycin resistance plasmid pNG2 deduced from its complete nucleotide sequence. Plasmid *49*, 63-74.

Tauch, A., Pühler, A., Kalinowski, J., and Thierbach, G. (2003b). Plasmids in *Corynebacterium glutamicum* and their molecular classification by comparative genomics. J. Biotechnol. *104*, 27-40.

Tauch, A., Bischoff, N., Pühler, A., and Kalinowski, J. (2004). Comparative genomics identified two conserved DNA modules in a corynebacterial plasmid family present in clinical isolates of the opportunistic human pathogen *Corynebacterium jeikeium*. Plasmid *52*, 102-118.

Trautwetter, A., and Blanco, C. (1991). Structural organization of the *Corynebacterium glutamicum* plasmid pCG100. J. Gen. Microbiol. *137*, 2093-2101.

Tsuchiya, M., and Morinaga, Y. (1988). Genetic control systems of *Escherichia coli* can confer inducible expression of cloned genes in coryneform bacteria. Bio/Technology *6*, 428 - 430.

Vašicová, P., Abrhámová, Z., Nešvera, J., Pátek, M., Sahm, H., and Eikmanns, B. (1998). Integrating and autonomously replicating vectors for analysis of promoters in *Corynebacterium glutamicum*. Biotechnol. Techniques *12*, 743-746.

Vašicová, P., Pátek, M., Nešvera, J., Sahm, H., and Eikmanns, B. (1999). Analysis of the *Corynebacterium glutamicum dapA* promoter. J. Bacteriol. *181*, 6188-6191.

Venkova, T., Pátek, M., and Nešvera, J. (2001). Identification of a novel gene involved in stable maintenance of plasmid pGA1 from *Corynebacterium glutamicum*. Plasmid *46*, 153-162.

Venkova-Canova, T., Pátek, M., and Nešvera, J. (2003). Control of *rep* gene expression in plasmid pGA1 from *Corynebacterium glutamicum*. J. Bacteriol. *185*, 2402-2409.

Venkova-Canova, T., Pátek, M., and Nešvera, J. (2004). Characterization of the cryptic plasmid pCC1 from *Corynebacterium callunae* and its use for vector construction. Plasmid *51*, 54-60.

Veselý, M., Pátek, M., Nešvera, J., Čejková, A., Masák, J., and Jirků, V. (2003). Host-vector system for phenol-degrading *Rhodococcus erythropolis* based on *Corynebacterium* plasmids. Appl. Microbiol. Biotechnol. *61*, 523-527.

Walia, R., Deb, J.K., and Mukherjee, K.J. (2007). Development of expression vectors for *Escherichia coli* based on pCR2 replicon. Microb. Cell Factories *6*, 1114.

Williams, L.E., Detter, C., Barry, K., Lapidus, A., and Summers, A.O. (2006). Facile recovery of individual high-molecular-weight, low-copy-number natural plasmids for genomic sequencing. Appl. Environ. Microbiol. *72*, 4899-4906.

Yamaguchi, R., Terabe, M., Miwa, K., Tsuchiya, M., Takagi, H., Morinaga, Y., Nakamori, S., Sano, K., Momose, H., and Yamazaki, A. (1986). Determination of the complete nucleotide sequence of the *Brevibacterium lactofermentum* plasmid pAM330 and the analysis of its genetic information. Agr. Biol. Chem. *50*, 2771-2778.

Yeh, P., Oreglia, J., Prevots, F., and Sicard, A.M. (1986). A shuttle vector system for *Brevibacterium lactofermentum*. Gene *47*, 301-306.

Yoshihama, M., Higashiro, K., Rao, E.A., Akedo, M., Shanabruch, W.G., Follettie, M.T., Walker, G.C., and Sinskey, A.J. (1985). Cloning vector system for *Corynebacterium glutamicum*. J. Bacteriol. *162*, 591-597.

Yukawa, H., Omumasaba, C.A., Nonaka, H., Kos, P., Okai, N., Suzuki, N., Suda, M., Tsuge, Y., Watanabe, J., Ikeda, Y., et al. (2007). Comparative analysis of the *Corynebacterium glutamicum* group and complete genome sequence of strain R. Microbiology *153*, 1042-1058.

Zahrt, T.C., and Deretic, V. (2000). An essential two-component signal transduction system in *Mycobacterium tuberculosis*. J. Bacteriol. *182*, 3832-3838.

Zhang, Y., Praszkier, J., Hodgson, A., and Pittard, A.J. (1994). Molecular analysis and characterization of a broad-host-range plasmid, pEP2. J. Bacteriol. *176*, 5718-5728.

Zheng, Z.X., Ma, C.P., Yan, W.Y., He, P.F., Mao, Y.X., Sun, W., Lei, Z.Z., Zhu, P., and Wu, J.F. (1987). Restriction map of plasmid pXZ10145 of *Corynebacterium glutamicum* and construction of an integrated plasmid. Chin. J. Biotechnol. *3*, 183-188.

Zupancic, T.J., Kittle, J.D., Baker, B.D., Miller, C.J., Palmer, D.T., Asai, Y., Inui, M., Vertes, A., Kobayashi, M., Kurusu, Y., et al. (1995). Isolation of promoters from *Brevibacterium flavum* strain MJ233C and comparison of their gene expression levels in *B. flavum* and *Escherichia coli*. FEMS Microbiol. Lett. *131*, 121-126.

From: Corynebacteria: Genomics and Molecular Biology. Edited by: Andreas Burkovski.

Regulation of Carbon Metabolism in *Corynebacterium glutamicum*

7

Annette Arndt and Bernhard J. Eikmanns

Abstract

The amino acid producing *Corynebacterium glutamicum* grows aerobically on a variety of carbohydrates, organic acids and alcohols as single or combined sources of carbon and energy. Among the substrates metabolized are glucose, L-lactate, acetate and ethanol which all four can also serve as substrates for amino acid production. Based on biochemical and genetic studies, on quantitative determination of metabolic fluxes and/or on genome-based methods such as DNA microarray analyses, this chapter mainly summarizes the present knowledge on the different steps and the regulation of the fundamental pathways in *C. glutamicum* during growth on glucose, L-lactate, acetate and/or ethanol. It becomes evident that carbon metabolism in this organism is subject to complex regulation at the transcriptional and at the post-transcriptional level. So far, eight transcriptional regulators, i.e., SugR, GntR1, GntR2, AcnR, LldR, RamA, RamB and GlxR and the serine/threonine (Ser/Thr) protein kinase G, phospho-Ser/Thr protein phosphatase and their target protein OdhI have been identified to be involved in the regulation of key enzymes in carbon metabolism, i.e., in the regulation of substrate uptake and activation, glycolysis, tricarboxylic acid (TCA) cycle, glyoxylate cycle and/or gluconeogenesis. Although the molecular mechanisms of transcriptional and post-transcriptional regulation of the enzymes involved in carbon metabolism is not yet completely understood, the recent findings led to a much better understanding of the adaptation of *C. glutamicum* to a given carbon source at the molecular level.

1. Introduction

In nature, microorganisms often encounter situations that change frequently and rapidly with respect to e.g., temperature, pH, oxygen concentration and nutrient availability. To get on well with these different situations, microorganisms have evolved a variety of coordinated and adaptive mechanisms and regulatory circuits by which they adjust their physiology to allow optimal growth. This is especially true with respect to the availability of different carbon and energy sources which have to be converted into metabolites of the central metabolism and then serve as fuel for energy conservation or as precursors for the synthesis of new cell material. The adaptation to the availability of a given substrate often is mediated by substrate-specific induction or derepression of catabolic genes. When a mixture of different carbon and energy sources is present instead of a single substrate, many organisms utilize one carbon source preferentially and consume the additional carbon source(s) only when the preferred one is exhausted. Successive utilization of substrates is often represented by a biphasic growth behaviour and the basis of this so-called diauxic growth (Monod, 1949) is the carbon catabolite repression. This global regulatory system represses (or does not activate) the expression of genes coding for enzymes involved in catabolism of a subordinate substrate in the presence of a catabolite generated from the preferred carbon and energy source (for reviews on catabolite repression in Gram-negative and Gram-positive bacteria and in yeast see Paulsen, 1996; Gancedo, 1998; Saier, 1998; Stülke and Hillen, 2000; Brückner and Titgemeyer, 2002).

The predominantly aerobic, non-pathogenic, Gram-positive soil bacterium *Corynebacterium glutamicum* is able to use a variety of carbohydrates, alcohols and organic acids as single or combined carbon and energy sources for growth and also for amino acid production (reviewed in Kinoshita and Tanaka, 1972; Liebl, 2001; Eggeling and Bott, 2005). However, in contrast to other bacteria such as *Escherichia coli* and *Bacillus subtilis*, *C. glutamicum* does not show diauxic growth and preferential utilization of one carbon source on most substrate mixtures. Instead, the organism co-metabolizes the carbon sources present in the mixtures, e.g., glucose with other sugars and organic acids such as fructose, gluconate, L-lactate, pyruvate, acetate, propionate, vanillate, protocatechuate and serine, and shows monophasic growth on these substrate mixtures (Cocaign et al., 1993; Dominguez et al., 1993; Lee et al., 1998; Wendisch et al., 2000; Claes et al., 2002; Netzer et al., 2004; Merkens et al., 2005; Stansen et al., 2005; Frunzke et al., 2007). Diauxic growth of *C. glutamicum* and sequential utilization of carbon sources so far has been described only for the mixtures of glucose/glutamate, which is due to repression of the glutamate uptake system (*gluABCD* gene cluster) in the presence of glucose (Krämer et al., 1990; Kronemeyer et al., 1995), and for the mixtures of glucose/ethanol and acetate/ethanol (Arndt and Eikmanns, 2007; Arndt et al., 2007; Kotrbova-Kozak et al., 2007; see below). Recently, Letek et al. (2006) reported on glucose-, fructose- and sucrose-mediated repression of *gntP* and *gntK*, which code for gluconate permease and gluconate kinase, respectively. These enzymes are essential for utilization of gluconate, however, there is evidence for a parallel and not a sequential utilization of glucose and gluconate (Frunzke et al., 2007). So far, there is also no experimental evidence for a global carbon catabolite repression system in *C. glutamicum*, such as the CRP-dependent system in *E. coli* (Stülke and Hillen, 1999; Brückner and Titgemeyer, 2002) or the CcpA-dependent carbon control system in *B. subtilis* and other low-GC Gram-positive bacteria (Stülke and

Hillen, 2000; Warner and Lolkema, 2003). In addition, no direct evidence has been obtained for involvement of a phosphotransferase system (PTS) component or of glucose kinase in global carbon regulation in *C. glutamicum* (Parche et al., 2001; Moon et al., 2007). Thus, the adaptation of the *C. glutamicum* carbon metabolism to the presence of various carbon and energy sources is clearly different from that in other well studied bacteria such as *E. coli* or *B. subtilis* (for reviews see above).

Eight transcriptional regulators affecting fundamental pathways of carbon metabolism have recently been described for *C. glutamicum*, i.e., SugR, AcnR, LldR, RamA, RamB and GlxR. SugR is a transcriptional regulator of sugar uptake (Engels and Wendisch, 2007), GntR1 and GntR2 are activators of sugar uptake and repressors of gluconate metabolism (Frunzke et al., 2007), AcnR is a repressor of the aconitase gene and was the first identified regulator of a TCA cycle gene in *C. glutamicum* (Krug et al., 2005). LldR is involved in L-lactate utilization (Georgi and Wendisch, 2007) and RamA, RamB and GlxR represent regulators of acetate and ethanol metabolism (Gerstmeir et al., 2004; Kim et al., 2004; Cramer et al., 2006). Recent studies furthermore revealed that the TCA cycle additionally is subject to post-transcriptional regulation, involving the Ser/Thr protein kinase G (PknG), phospho-Ser/Thr protein phosphatase and their substrate protein OdhI (Niebisch et al., 2006; Schultz et al., 2007). Depending on its phosphorylation state, OdhI inhibits the 2-oxoglutarate dehydrogenase complex and thus, regulates the carbon flux through the TCA cycle at the level of the 2-oxoglutarate branch point (reviewed in Bott, 2007).

This chapter deals with various aspects of the carbon metabolism in *C. glutamicum*. It will summarize the present knowledge on the fundamental pathways in *C. glutamicum* when growing on glucose, L-lactate, acetate and ethanol and then focus on what is known about the transcriptional regulation of respective key enzymes and their regulators. Also discussed are the findings on post-transcriptional regulation of the TCA cycle enzyme 2-oxoglutarate dehydrogenase complex. It becomes evident that striking genetic and regulatory peculiarities become important under certain conditions and the present knowledge reveals the regulation of carbon metabolism in *C. glutamicum* to be a fascinating target of metabolic engineering in order to improve metabolite production with this industrially important organism.

2. Sugar metabolism and its regulation in *C. glutamicum*

2.1 Metabolism of glucose and other carbohydrates

C. glutamicum grows on a variety of different sugars, however, in the biotechnological production of amino acids media mainly based on glucose, fructose or sucrose are used. Accordingly, the growth and amino acid production using these sugars and the respective sugar uptake and conversion pathways of *C. glutamicum* have been intensively studied (for reviews see Liebl, 2001; Yokota and Lindley, 2005; Moon et al., 2007).

For glucose, fructose and sucrose uptake, the organism uses typical phosphotransferase systems (for recent reviews see Parche et al., 2001; Yokota and Lindley, 2005; Moon et al., 2007). Due to continued uptake of glucose in PTS-negative strains, the presence of an additional permease for this sugar uptake has been postulated (Dominguez and Lindley, 1996; Dominguez et al., 1997), however, so far no biochemical evidence has been presented. Studies on sugar transport and genome analyses revealed that *C. glutamicum* possesses the general PTS phosphotransferase I (E1) and HPr proteins, which are shared by at least three EII PTS components specific for glucose (enzyme II^{Glc}), fructose (enzyme II^{Fru}) and sucrose (enzyme II^{Suc}) (reviewed in Moon et al. (2007). A fourth PTS EII has been postulated from genome annotation and homology studies; however, the function of this protein has not yet been clarified.

After sugar uptake and phosphorylation, the further catabolism of the sugar phosphates in *C. glutamicum* proceeds via the Embden-Meyerhof-Parnas pathway (glycolysis), the pentose phosphate pathway and the TCA cycle (Figure 7.1) (for reviews see Eikmanns, 2005; Yokota and Lindley, 2005). The metabolic link between glycolysis and the TCA cycle is represented by the so-called PEP-pyruvate-oxaloacetate node, which in *C.*

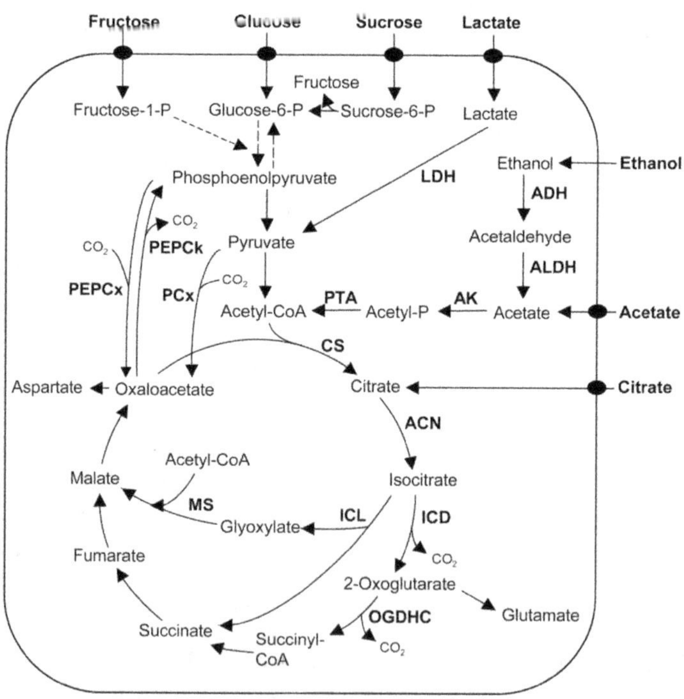

Figure 7.1. Diagram of the central metabolism of *C. glutamicum* during growth on different substrates. Dotted arrows represent pathways consisting of several reactions, uninterrupted arrows represent single reactions. ACN, aconitase; ADH, alcohol dehydrogenase; AK, acetate kinase; ALDH, acetaldehyde dehydrogenase; CS, citrate synthase; ICL, isocitrate lyase; ICD, isocitrate dehydrogenase; LDH, L-lactate dehydrogenase; MS, malate synthase; OGDHC, 2-oxoglutarate dehydrogenase complex; PCx, pyruvate carboxylase; PEPCk, phosphoenolpyruvate carboxykinase; PEPCx, phosphoenolpyruvate carboxylase; PTA, phosphotransacetylase.

glutamicum comprises a surprisingly diverse set of reactions and acts as a highly relevant switch point for carbon flux distribution within the central metabolism (reviewed in Eikmanns, 2005; Sauer and Eikmanns, 2005). At this node, the end products of glycolysis, phosphoenolpyruvate (PEP) and pyruvate either directly serve as precursors for anabolic purposes, enter the TCA cycle via acetyl-CoA (oxidative pyruvate decarboxylation and fueling of the cycle) or enter the cycle via the anaplerotic reactions (oxaloacetate formation by PEP or pyruvate carboxylation). The TCA cycle then serves catabolic and anabolic purposes. On the one hand it is responsible for complete oxidation of acetyl-CoA, it generates ATP and it provides reducing equivalents to the membrane-bound respiratory system (Bott and Niebisch, 2005), on the other hand it provides precursor metabolites for biosynthetic processes such as 2-oxoglutarate and oxaloacetate.

Inspection of the genome sequence of *C. glutamicum* (GenBank accession numbers NC_003450 and NC_006958; Ikeda and Nakagawa, 2003; Kalinowski et al., 2003) revealed that all genes encoding glycolysis, pentose phosphate pathway and TCA cycle are present. Unlike other bacteria, *C. glutamicum* possesses only single (functional) genes for 6-phosphofructokinase and pyruvate kinase, but two genes for glyceraldehyde-3-phosphate dehydrogenase, proposed to have functions in glycolysis and gluconeogenesis, respectively (Omumasaba et al., 2004). A variety of the genes and of the enzymes of glycolysis, pentose phosphate pathway, the PEP-pyruvate-oxaloacetate node and of the TCA cycle have been studied in detail, however, for extensive overviews the reader is referred to recent reviews by Yokota and Lindley (2005) and Eikmanns (2005).

Independent of the proof of the presence of all the enzymes and genes of glycolysis, pentose phosphate pathway, PEP-pyruvate-oxaloacetate node and TCA cycle, the functioning of these pathways during growth on sugars (and other substrates too) was demonstrated by metabolic flux analysis, i.e., by integrating approaches combining ^{13}C-labelling and isotopomer analysis with metabolic balancing (reviewed in Wittmann and de Graaf, 2005). The most important enzymatic controlling point in glycolysis and in the pentose phosphate pathway of *C. glutamicum* is considered to be at the level of pyruvate kinase and glucose-6-phosphate dehydrogenase, respectively (Yokota and Lindley, 2005). Pyruvate kinase is inhibited by ATP and activated by AMP (Ozaki and Shiio, 1969; Jetten et al., 1994), glucose-6-phosphate dehydrogenase is controlled by the concentration ratio of $NADP^+$ to NADPH and thus, the fluxes through glycolysis and the pentose phosphate pathway are assumed to be controlled by the energy status and by the $NADP^+$/NADPH ratio (Jetten et al., 1994; Moritz et al., 2000). As citrate synthase catalyzes the initial reaction of the TCA cycle and is inhibited by ATP, aconitate and isocitrate (Shiio et al., 1977; Eikmanns et al., 1994), this enzyme is assumed to be rate-controlling for the entry of substrates into the cycle. However, other enzymes of the cycle are also subject to inhibitors and activators (reviewed in Eikmanns, 2005) and therefore also may contribute to enzymatic flux control within the cycle. As will be outlined below, the expression of genes for several of the proteins/enzymes in sugar uptake, glycolysis, pentose phosphate pathway, PEP-pyruvate-oxaloacetate node and TCA cycle in *C. glutamicum* change significantly when the cells grow on substrates other than carbohydrates and thus, both enzymatic as well as genetic devices are responsible for flux control in the fundamental pathways in *C. glutamicum.*

2.2 Genetic regulation of sugar metabolism

For growth of *C. glutamicum* on glucose, fructose and different mixtures of glucose and fructose, Dominguez et al. (1997) found that the cells possessed an approximately constant level of total PTS activity and that the consumption rate of each sugar is directly related to the respective PTS transport activity. In response to the sugar present, each individual PTS varied with an about twofold transport capacity. Similarly, when *C. glutamicum* grows on a mixture (1:1; w/v) of glucose and acetate, the glucose uptake rate is reduced about twofold when compared to growth on glucose as sole carbon and energy source (Wendisch et al., 2000). These observations already indicated regulation of the glucose- and of the fructose-specific PTSs in *C. glutamicum.* In fact, Engels and Wendisch (2007) recently identified and characterized the DeoR-type regulator SugR, which represents a repressor of the glucose-specific PTS enzyme II^{Glc} gene *ptsG.* Whereas overexpression of the *sugR* gene in glucose-growing cells of *C. glutamicum* led to decreased glucose utilization and reduced *ptsG* mRNA levels, inactivation of *sugR* had no effect. In contrast to the wild type, however, the *sugR* mutant showed two- to sevenfold higher *ptsG* transcription on gluconeogenic carbon sources, such as pyruvate, acetate or citrate (Engels and Wendisch, 2007). DNA microarray analysis showed that SugR also represses the PTS enzyme II^{Suc} gene (*ptsS*) and the NCgl1859-*fruK-ptsF* operon, encoding a so far uncharacterized DeoR-type transcriptional regulator, fructose-1-phosphate kinase and PTS enzyme II^{Fru}. The authors furthermore showed that SugR binds to the promoter regions of *ptsG, ptsS* and the NCgl1959-*fruK-ptsF* operon and that the binding is prevented by fructose-6-phosphate (Engels and Wendisch, 2007). In agreement with a repressive function on *ptsS* and *ptsF*, overexpression of *sugR* led to a perturbed growth in minimal medium containing sucrose or fructose as sole carbon source (Engels and Wendisch, personal communication). All these results indicate that SugR represses the PTS enzyme II genes during growth on gluconeogenic carbon sources while during growth on substrates providing a high intracellular fructose-6-phosphate concentration, the respective genes are derepressed. Thus, the SugR regulatory system represents a mechanism which allows *C. glutamicum* to adapt sugar uptake to the carbon source available. It should be noted, however, that the expression of *ptsG* in the SugR-deficient mutant was twofold higher during growth on maltose than on glucose and even fourfold higher than during growth on fructose (Engels and Wendisch, 2007). This observation indicates the involvement of a further, so far unidentified regulatory system for tuning *ptsG* expression in response to the sugar available. Here it is noteworthy to mention that Gerstmeir et al. (2004) suggested putative binding sites for RamB in front of *ptsG* and that purified RamA and RamB proteins were shown to bind to the *ptsG* promoter region (Auchter, Schwarz and Eikmanns, unpublished). RamA and RamB are transcriptional regulators of genes involved in acetate and ethanol metabolism of *C. glutamicum* (Gerstmeir et al., 2004; Cramer et al., 2006; see below) and thus, it might well be that both these regulators are also involved in *ptsG* expression control. However, further studies are required to experimentally proof or disproof this hypothesis.

Frunzke et al. (2007) recently described two further, paralogous and functionally equivalent transcriptional regulators, GntR1 and GntR2, which in *C. glutamicum* coordinately regulate glucose uptake and gluconate catabolism. In the absence of gluconate, both these regulators function at the same time as repressors for genes involved in gluconate metabolism (i.e., *gntK, gntP* and *gnd*, encoding gluconate kinase, gluconate permease and 6-phosphogluconate dehydrogenase, respectively) and as activators for *ptsG* and *ptsS*. Thus, the GntR1/GntR2 system is the first example for gluconate-dependent transcriptional control of PTS genes. Gluconate and glucono-δ-lactone were found to be effectors, preventing the binding of GntR1 and GntR2 to the respective operator regions and thus, repression of *ptsG* and derepression of the gluconate genes. In accordance, a mutant lacking both regulators showed a 60% lower glucose uptake rate and growth rate than the wildtyp when glucose was the only carbon and energy source (Frunzke et al., 2007).

Aside from sugar uptake, the fundamental pathways used by *C. glutamicum* for metabolization of sugars are subject to further regulatory systems. By transcriptome and proteome analyses and/or quantitative real-time reverse transcription PCR the expression of genes for enzymes of glycolysis, pentose phosphate pathway and the TCA cycle have been shown to vary with the carbon source (Hayashi et al., 2002; Muffler et al., 2002; Gerstmeir et al., 2003; Han et al., 2007; Polen et al., 2007; Qi et al., 2007). In comparison to growth of *C. glutamicum* on acetate, citrate or aromatic compounds (such as benzoate, gentisate, 4-cresol, phenol or resorcinol), the growth in the presence of glucose gave rise to higher expression of most glycolytic and pentose phosphate pathway genes (e.g., *zwf, tal, tkt, gapA, fba, eno, pyk*) and decreased expression of gluconeogenic (*pck, malE*) and TCA cycle genes (e.g., *gltA, acn, sucCD, sdhCAB, fum*). However, in all these cases the regulators involved are not known and the regulatory mechanism has to be elucidated.

The about 4-fold higher specific activities of aconitase (Table 7.1) and the about fourfold higher mRNA levels of the aconitase gene *acn* in acetate-grown cells of *C. glutamicum* compared with glucose-grown cells prompted Krug et al. (2005) to look for and to analyze the regulatory background for these observations. Downstream of the *acn* gene, the authors found a gene (designated *acnR*) encoding a TetR-type transcriptional regulator, which is cotranscribed with *acn* at a much lower level and which they tested for a regulatory function on *acn*. Deletion of *acnR* led to two- to fivefold higher specific aconitase activities in the cells when grown on citrate, acetate or propionate and compared to wild type cells, whereas overexpression of *acnR*

Table 7.1. Specific activities of enzymes involved in carbon metabolism of *C. glutamicum* cells grown in minimal medium containing glucose (Glc), L-lactate (Lac), acetate (Ac), ethanol (EtOH), Glc plus Ac, Glc plus EtOH or Ac plus EtOH as carbon sources. ADH, alcohol dehydrogenase; ALDH, acetaldehyde dehydrogenase; AK, acetate kinase; PTA, phosphotransacetylase; ICL, isocitrate lyase; MS, malate synthase; CS, citrate synthase; ACN, aconitase; ICD, isocitrate dehydrogenase; LDH, L-lactate dehydrogenase; PEPCk, phosphoenolpyruvate carboxykinase; PCx, pyruvate carboxylase; FBPase, fructose-1,6-bisphosphatase. "-", no data available.

Carbon source	Specific activity [U (mg protein)$^{-1}$] of												
	ADH[a]	ALDH[a]	AK[c]	PTA[c]	ICL[c]	MS[c]	CS[d]	ACN[e]	ICD[f]	LDH[g]	PEPCk[h]	PCx[i]	FBPase
Glc	0.22	0.010	0.21	0.66	0.02	0.04	0.70	0.20	1.07	0.01	0.044	0.016	0.022
Lac	-	-	0.34	1.38	0.08	0.07	0.65	-	1.23	0.13	0.135	0.045	0.041
Ac	0.06	0.030	0.81	2.95	2.07	1.77	0.65	0.32	1.10	0.01	0.072	0.019	-
EtOH	1.85	0.150	4.40	7.65	1.51	1.54	-	-	-	-	-	-	-
Glc/Ac	-	-	0.43	1.87	1.40	1.19	-	-	1.05	-	-	-	-
Glc/EtOH[k]	0.36	0.010	0.28	0.61	0.06	0.17	-	-	-	-	-	-	-
Ac/EtOH[k]	0.09	0.014[b]	1.18	2.19	2.04	2.20	-	-	-	-	-	-	-

[a] Arndt and Eikmanns, 2007; Arndt et al., 2007
[b] Arndt, Auchter and Eikmanns, unpublished
[c] Wendisch et al., 1997; Gerstmeir et al., 2003; Arndt and Eikmanns, unpublished
[d] Eikmanns et al., 1994; Radmacher and Eggeling, 2007
[e] Krug et al., 2005
[f] Eikmanns et al., 1995
[g] Georgi and Wendisch, 2007
[h] Riedel et al., 2001
[i] Peters-Wendisch et al., 1997
[j] Rittmann et al., 2003
[k] Cells were harvested in the first exponential phase of growth

in glucose-grown cells of *C. glutamicum* caused an about twofold lower aconitase activity (Krug et al., 2005). These results indicated a repressor function of AcnR. DNA microarray analyses substantiated this conclusion and additionally revealed that *acn* is the primary target gene of AcnR (Krug et al., 2005). Purified AcnR was shown to bind to the *acn* promoter region and an operator motif was identified in the region 11 to 29 bp upstream of the transcriptional start (Krug et al., 2005). However, although the authors tested a number of metabolites of the central metabolism for their ability to interfere with the binding of AcnR to the *acn* operator, none of the tested compounds had an effect and thus, the effector for AcnR is still unknown (Bott, 2007). Moreover, it is important to mention that the *acnR* mutant of *C. glutamicum* still showed 1.5- to 2.3-fold higher specific aconitase activities on acetate, citrate and propionate compared with glucose, suggesting that regulators other than AcnR are involved in carbon source-dependent regulation of *acn* expression (Krug et al., 2005). Gerstmeir et al. (2004) and Cramer et al. (2006) suggested RamB and RamA binding sites to be present in front of the *acn* gene and thus, RamA and/or RamB are possible candidates as additional regulators of *acn*. However, independent of the still incomplete picture of *acn* expression control, aconitase was suggested to be a major control point of the TCA cycle activity in *C. glutamicum* (Krug et al., 2005), since the levels of the preceeding enzyme, citrate synthase, and that of the isocitrate dehydrogenase in *C. glutamicum* were about identical during growth on glucose and acetate (Table 7.1) (Eikmanns et al., 1994; Eikmanns et al., 1995; Gerstmeir et al., 2003).

3. L-lactate metabolism and its regulation in *C. glutamicum*

C. glutamicum can grow aerobically on L-lactate as sole carbon and energy source (e.g., Cocaign et al., 1993; Wendisch et al., 1997; Gourdon et al., 2000; Seletzky et al., 2006). For L-lactate utilization (see Figure 7.1), the organism requires an uptake system, which so far has not been unequivocally identified, and the quinone-dependent L-lactate dehydrogenase (LldD), which is a peripheral membrane protein and which oxidizes L-lactate to pyruvate (Bott and Niebisch, 2003; Schluesener et al., 2005; Stansen et al., 2005). The formed pyruvate is channelled into the TCA cycle via the reactions of the pyruvate dehydrogenase complex (PDHC) and the anaplerotic pyruvate carboxylase, which both have been shown to be essential for growth of *C. glutamicum* on L-lactate (Peters-Wendisch et al., 1997; 1998; Schreiner et al., 2005). As evidenced by growth analysis of respective mutants, L-lactate utilization by *C. glutamicum* furthermore requires the gluconeogenic enzymes PEP carboxykinase and fructose-1,6-bisphosphatase (FBPase), catalyzing the decarboxylation of oxaloacetate to PEP and the hydrolysis of fructose-1,6-bisphosphate to fructose-6-phosphate and inorganic phosphate (Riedel et al., 2001; Rittmann et al., 2003). Both these enzymes have been purified and characterized and shown to be inhibited by low concentrations of ATP (PEP carboxykinase) or AMP, PEP, phosphate and fructose-1-phosphate (FBPase) (Jetten and Sinskey, 1993; Rittmann et al., 2003). Surprisingly, the protein abundancies and the specific activities of the FBPase did not vary much when *C. glutamicum* was grown on L-lactate or other organic acids (Table 7.1) or different sugars (Rittmann et al., 2003), indicating that the gluconeogenic flux from fructose-1,6-bisphosphate to fructose-6-phosphate is mainly controlled by enzyme inhibition and not by enzyme synthesis, i.e., by genetic regulation. In contrast, the specific PEP carboxykinase activity (Table 7.1) and also the transcript concentration of the PEP carboxykinase gene (*pck*) has been shown to be three- to fivefold higher in L-lactate-grown cells than in glucose-grown cells (Riedel et al., 2001), indicating that the first step of gluconeogenesis is controlled at the transcriptional level. It is not clear whether this regulation is due to *pck* repression in the presence of glucose or to *pck* activation by the presence of L-lactate in the medium. The fact that *pck* expression is also higher in cells grown in LB complex medium without additional carbon source and in minimal medium containing other gluconeogenic substrates (such as acetate or citrate) (Gerstmeir et al., 2003; Polen et al., 2007) suggests a glucose repression mechanism. However, so far a regulatory protein responsible for the observed PEP carboxykinase activity and *pck* expression differences has not been identified.

Above all, utilization of L-lactate in *C. glutamicum* obviously is regulated by expression control of the LldD gene *lldD*. Stansen et al. (2005) recently identified an operon comprising the *lldD* gene and a putative L-lactate permease gene (NCgl2816) and found that the mRNA levels of this operon in *C. glutamicum* were 18-fold higher when the cells were grown on L-lactate instead of pyruvate. Moreover, the authors found that the specific LldD activity was very low in cells grown in minimal medium containing glucose, D-lactate or pyruvate as carbon source and about 10-fold higher in cells grown on L-lactate (Table 7.1) or an L-lactate-glucose mixture. These data already suggested carbon source-dependent transcriptional regulation in the presence of L-lactate in the growth medium. By DNA affinity chromatography, Georgi and Wendisch (2007) recently in fact identified an FadR-type regulator protein which bound to the upstream region of the NCgl2816-*lld*D operon and which the authors designated as LldR. Overexpression of the respective *lldR* gene led to strongly reduced NCgl2816-*lldD* mRNA levels, to strongly reduced LldD activity and to a severe growth defect in medium containing L-lactate as sole carbon and energy source, whereas a *lldR* deletion mutant showed high NCgl2816-*lldD* mRNA levels and high specific LldD activities, independent of the presence or absence of L-lactate in the growth medium (Georgi and Wendisch, 2007). An operator motif (TNGTNNNACNA) typical for FadR-type regulators was identified within the promoter region of the NCgl2816-*lldD* operon and the authors furthermore could show that L-lactate, but not D-lactate, interfered with the binding of LldR to this operator motif (Georgi and Wendisch, 2007). Thus, it can be concluded that in the absence of L-lactate in the growth medium the L-lactate utilization operon NCgl2816-*lldD* is repressed by LldR and this repression is released in the presence of L-lactate.

4. Acetate metabolism and its regulation in *C. glutamicum*

4.1 Metabolism of acetate

The growth of *C. glutamicum* on acetate and on acetate plus glucose has been been thoroughly studied by Wendisch et al. (2000). In case of the experiments with the substrate mixture, acetate and glucose were consumed in parallel (270 nmol acetate and 72 nmol glucose per min and mg protein) and growth rate (0.36 vs. 0.28 per h) as well as biomass yield (0.41 vs. 0.20 g C in dry weight per g C in substrate) were higher, indicating an increased energy metabolism in the presence of glucose. However, the overall carbon consumption rates were very similar (972 vs. 1080 nmol C per min and mg protein) and these observations indicate that the uptake of acetate and/or the pathway from acetate to acetyl-CoA (i.e., acetate activation) are directly or indirectly regulated by the carbon source present in the medium.

Acetate uptake in *C. glutamicum* is catalyzed by a specific secondary carrier according to an acetate/proton symport mechanism and there is no significant contribution by a passive diffusion process (Ebbighausen et al., 1991). The carrier is dependent on the membrane potential and aside from acetate accepts propionate as a substrate for transport. Since the maximal activity (about 70 nmol per mg protein and min) measured in the mid exponential growth phase of *C. glutamicum* cells grown on glucose is much lower than the acetate consumption rate measured during growth on acetate (see above), it has been speculated that the expression of the acetate carrier gene is increased in acetate-grown cells, as are other genes involved in acetate metabolism (see below). However, a gene encoding an acetate/propionate uptake carrier so far has not been identified within the genome of *C. glutamicum* and further studies are necessary to clarify the regulation of acetate (and propionate) uptake at the genetic level.

After uptake, acetate is activated to acetyl-CoA and further metabolized via the TCA cycle and the glyoxylate cycle and, when acetate is the sole carbon and energy source, via gluconeogenesis (Wendisch et al., 2000) (Figure 7.1). Acetate activation in *C. glutamicum* is accomplished by the combined activities of acetate kinase (AK) and phosphotransacetylase (PTA) (Shiio et al., 1969). AK activates acetate to acetyl phosphate which subsequently is converted to acetyl-CoA by PTA. Both enzymes from *C. glutamicum* and their regulation were intensively studied and also mutants defective in either or both of the enzymes were constructed and analyzed (Wendisch et al., 1997; Reinscheid et al., 1999; for a review see Gerstmeir et al., 2003). The key enzymes of the glyoxylate cycle, isocitrate lyase (ICL) and malate synthase (MS), have also been intensively studied in *C. glutamicum* (Shiio et al., 1969; Reinscheid et al., 1994a; 1994b). ICL catalyzes the cleavage of isocitrate to succinate and glyoxylate, and MS condenses glyoxylate with acetyl-CoA to give malate (Figure 7.1). Both enzymes of *C. glutamicum* have been biochemically characterized and it turned out that both are subject to allosteric regulation by several intermediates of the central metabolism (reviewed in Gerstmeir et al., 2003 and in Eikmanns, 2005). The carbon flux distribution between the TCA cycle and the glyoxylate cycle is probably tightly controlled on the enzymatic level by inhibition of ICL and isocitrate dehydrogenase (ICD), both competing for their common substrate isocitrate (see Figure 7.1). The inhibition pattern of ICL and ICD suggests that ICL activity is low when glycolysis is active and that ICD activity is low when the glyoxylate cycle is operative. Additionally, the carbon flux at the isocitrate branch point is probably governed by the intracellular isocitrate concentration since the substrate affinities of ICL and ICD differ by a factor of more than twenty (280 vs. 12 μM; Reinscheid et al., 1994a). However, in contrast to other bacteria (e.g., *E. coli*), the glyoxylate cycle in *C. glutamicum* has been shown to be active on the mixture of acetate and glucose, and in fact, anaplerosis occurred exclusively via the glyoxylate cycle under these conditions (Wendisch et al., 2000). The glyoxylate cycle not only replenishes the TCA cycle for the withdrawal of anabolic precursor metabolites, but additionally provides oxaloacetate for PEP and pyruvate formation. In accordance with these results, a glyoxylate-deficient mutant of *C. glutamicum* showed significantly impaired growth on acetate-glucose mixtures (Wendisch et al., 2000).

In addition to acetate activation and the glyoxylate cycle, *C. glutamicum* requires the operation of gluconeogenesis when growing on acetate as sole carbon source. As already outlined above, the key enzymes of gluconeogenesis, PEP carboxykinase and fructose-1,6-phosphatase from *C. glutamicum* have been intensively studied and the purified enzymes were shown to be affected by intermediates of the central metabolism (Jetten and Sinskey, 1993; Riedel et al., 2001; Rittmann et al., 2003). During growth on acetate, PEP carboxykinase represents the initial enzyme of gluconeogenesis and converts the TCA cycle intermediate oxaloacetate to pyruvate. In principle, two other enzymes, namely malic enzyme and oxaloacetate decarboxylase, may also have gluconeogenic function as they are present in acetate-grown cells of *C. glutamicum* and catalyze the formation of pyruvate from malate or oxaloacetate, respectively (Jetten et al., 1994; Jetten and Sinskey, 1995). However, the inability of a PEP carboxykinase-deficient mutant of *C. glutamicum* to grow on acetate indicates that this enzyme cannot be functionally replaced by malic enzyme and oxaloacetate decarboxylase (Riedel et al., 2001). Fructose-1,6-bisphosphatase circumvents the irreversible phosphofructokinase step of glycolysis and provides fructose-6-phosphate. In addition to these two enzymes, a second glyceraldehyde-3-phosphate dehydrogenase (GAPDH B) has been shown to be present and to support optimal growth on acetate (Omumasaba et. al., 2004). However, whereas GAPDH A was indispensable for growth on glucose, both GAPDH A and B seem to be involved in gluconeogenesis on acetate.

4.2 Genetic regulation of acetate metabolism

The *C. glutamicum* genes for the acetate-activating enzymes AK and PTA (*ack* and *pta*, respectively), those coding for ICL and MS (*aceA* and *aceB*, respectively) as well as those coding for PEP carboxykinase and fructose-1,6-bisphosphatase (*pck* and *fbp*, respectively) have been characterized, and, by gene-directed mutagenesis and analysis of mutants, all six enzymes have been

Table 7.2. Promoter activities determined as specific chloramphenicol acetyltransferase (CAT) activities in cell-free extracts of various *C. glutamicum* strains grown in minimal medium containing glucose (Glc), acetate (Ac), ethanol (EtOH), Glc plus Ac, Glc plus EtOH or Ac plus EtOH. Each strain carries plasmid pET2 with the respective promoter region of *adhA*, the *pta-ack* operon, *aceA* or *aceB*. n.d., not determined.

	Specific CAT activity [U/mg protein] of			
	P_{adhA}[a]	$P_{pta\text{-}ack}$[b]	P_{aceA}[b]	P_{aceB}[b]
Glc	1.57	0.10	0.02	0.04
Ac	0.43	0.32	2.16	1.38
EtOH	6.89	0.58	1.27	1.14
Glc / Ac	n.d.	0.31	3.27	2.38
Glc / EtOH	1.68	0.15	0.05	0.05
Ac / EtOH	0.61	0.48	1.78	1.59

[a] Arndt and Eikmanns, 2007
[b] Gerstmeir et al., 2004; Cramer et al., 2006; Arndt and Eikmanns, unpublished

shown to be essential for the growth of *C. glutamicum* on acetate but not on glucose (Reinscheid et al., 1994a; 1994b; 1999; Riedel et al, 2001; Rittmann et al., 2003). The AK and PTA genes form an operon with *pta* upstream of *ack*, which is expressed from two transcriptional start sites 160 and 47 bp upstream of the translation initiation codon of the *pta* gene (Reinscheid et al., 1999). The *C. glutamicum aceA* and *aceB* genes are clustered on the chromosome, separated by 597 bp and transcribed in divergent directions. Both, *pck* and *fbp* are monocistronic genes and are located on the genome distinct from each other, from the *pta-ack* operon and from the *aceA-aceB* cluster (Kalinowski et al., 2003). With the exception of fructose-1,6-bisphosphatase, the specific activities of all six enzymes (AK, PTA, ICL, MS and PEP carboxykinase) are much higher in the presence of acetate in the growth medium than in its absence (Table 7.1), and, as shown by Northern blot hybridization and transcriptional fusion experiments (Table 7.2), the differences can be explained almost exclusively by coordinated transcriptional regulation of the respective genes (Wendisch et al., 1997; Reinscheid et al., 1999). In addition, DNA microarray analyses of *C. glutamicum* cells grown on acetate, acetate plus glucose and on glucose (see Figure 7.2) and quantitative RT-PCR substantiated this transcriptional regulation of *ack, pta, aceA, aceB* and *pck* (Hayashi et al., 2002; Muffler et al., 2002; Gerstmeir et al., 2003). Using cell extracts of *C. glutamicum* and employing DNA affinity chromatography, two regulatory proteins, specifically binding to the *pta-ack, aceA* and *aceB* promoter regions, were identified and designated as regulators of acetate metabolism A and B, i.e. RamA and RamB (Gerstmeir et al., 2004; Cramer et al., 2006). A further transcriptional regulator, designated as glyoxylate bypass regulator GlxR, was identified in a screen for genes exerting regulatory effects on the activity of the *aceB* promoter (Kim et al., 2004).

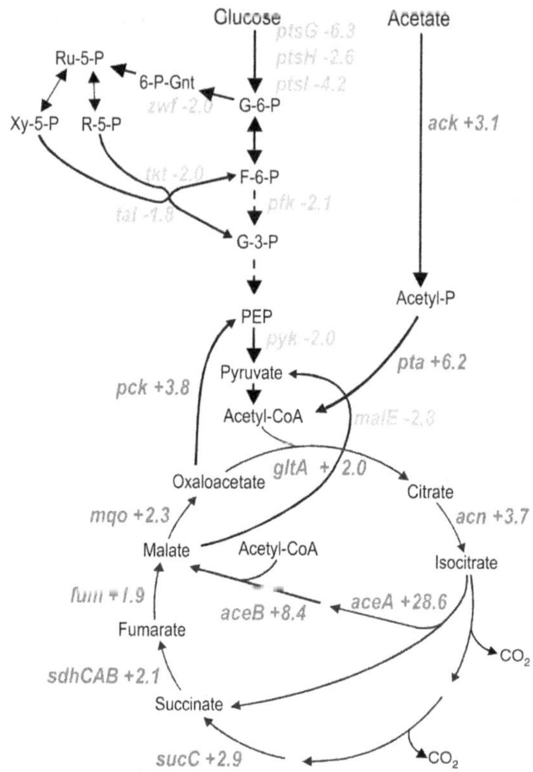

Figure 7.2. Schematic diagram of central metabolism. Given are those genes for enzymes involved in central metabolic pathways which showed different mRNA levels in cells grown on either acetate or glucose. The numbers give the ratios between the mRNA level of acetate-grown cells and that of glucose-grown cells.

RamA is a LuxR-type transcriptional regulator and, as shown by analysis of a *ramA*-deletion mutant, is essential for growth of *C. glutamicum* on acetate as sole carbon and energy source. Inactivation of the *ramA* gene led to very low specific activities of AK and PTA and to abolition of ICL and MS activity, irrespective of the presence of acetate in the growth medium. Comparative transcriptional fusion experiments with the wild type and with the *ramA*-deletion mutant of *C. glutamicum* revealed this deregulation to take place at the level of transcription (Cramer et al., 2006). As evidenced by electrophoretic mobility shift assays (EMSAs) with mutated *pta-ack, aceA* and *aceB* promoter fragments and with respective oligonucleotides, purified RamA binds to single or tandem stretches of A/C/TG$_{4\text{-}6}$T/C or AC$_{4\text{-}5}$A/G/T (Table 7.3) (Cramer et al., 2006). All these data show that RamA represents a transcriptional activator of the genes encoding enzymes involved in acetate metabolism of *C. glutamicum* (Figure 7.3). Since RamA is present in acetate- and (in lower concentrations) in glucose-grown cells, it should be triggered by an effector which is specific for the one or the other growth condition.

However, in spite of testing a variety of metabolic intermediates of acetate and of glucose metabolism for an effect on the binding of RamA to the *aceA/aceB* promoter region (Cramer et al., 2007), the effector for RamA has not been identified so far.

RamB represents a new type of transcriptional regulators with a domain showing some similarity to the so-called HTH-XRE DNA binding domain (Schultz et al., 1998; Gerstmeir et al., 2004). Most of the XRE family proteins are hypothetical proteins and no prediction with respect to a function can be made. However, similar experiments as performed to study the function of RamA showed that RamB is not essential for growth on any substrate tested and represents a transcriptional repressor of the genes encoding AK, PTA, ICL and MS (for a model see Figure 7.3). Whereas the wild type of *C. glutamicum* possessed high specific activities of all four enzymes and high *pta-ack, aceA/aceB* promoter activities when the cells grow in the presence of acetate and low specific enzyme and promoter activities when the cells grow on glucose (Tables 7.1 and 7.2), a defined *ramB*-deletion mutant of *C. glutamicum* showed high specific AK, PTA, ICL and MS activities and high *pta-ack, aceA/aceB* promoter activities irrespective of the substrate (Gerstmeir et al., 2004). Purified RamB protein specifically binds to a highly conserved 13-bp motif (AA/GAACTTTGCAAA; Gerstmeir et al., 2004), however, as in the case with RamA, a metabolite preventing or increasing its binding activity to the *pta-ack, aceA/aceB* promoter regions, i.e., an effector, has not been identified so far.

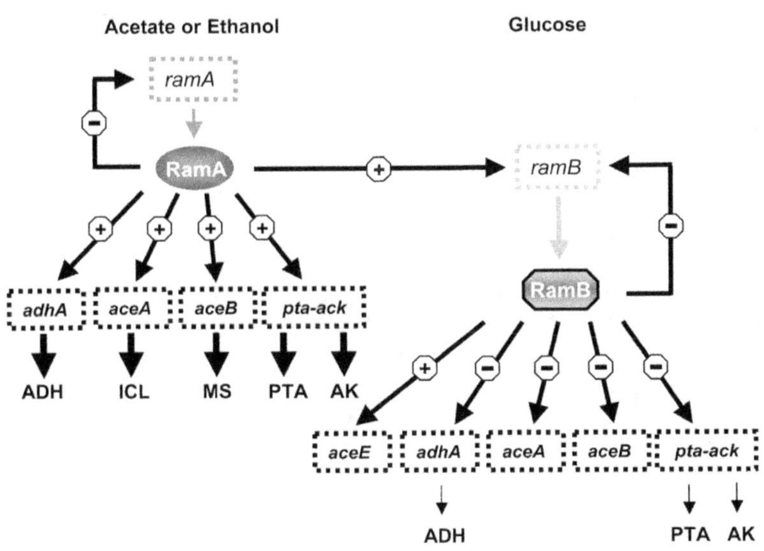

Figure 7.3. Model of the regulatory network involving RamA and RamB in *C. glutamicum* grown in media containing glucose or acetate or ethanol. Activation and repression are indicated by "+" and "-" signs, respectively. The thickness of the arrows to the enzymes give a rough indication of the strength of expression.

Table 7.3 Genes encoding proteins/enzymes of the central metabolism of *C. glutamicum* showing potential and/or functional RamA and RamB binding sequence motifs in their respective promoter regions. cg numbers correspond to Kalinowski et al., 2003.

Accession no.	Gene	Protein	RamA binding motif(s) (Distances (bps) from start codon)	RamB binding motif(s) (Distances (bps) from start codon)
cg0444	*ramB*	Regulator of acetate metabolism B	CGGGGC, (-133); ACCCCG, (-250); ACCCCA, (-395); ACCCCCA, (-424); CGGGGC, (-441); CGGGGC, (-459)	ACAAGTTTGCAAC, (-65)
cg0445	*sdhCAB*	Succinate dehydrogenase	TGGGGGT, (-79); TGGGGT, (-107); CGGGGT, (-252)	
cg0949	*gltA*	Citrate synthase		CAAACTTTGCTAA, (-402); TTTGCAAAGTTTT, (-511)
cg1537	*ptsG*	Phosphotransferase system, glucose specific enzyme II		CAAATTGTGCAAT, (-172)
cg1737	*acn*	Aconitase	AGGGGGT, (-278); TGGGGGT, (-290)	ATTCCTTTGCATC, (-492)
cg2466	*aceE*	Pyruvate dehydrogenase complex, E1 subunit		AAAACTTTGCAAA, (-298)
cg2559	*aceB*	Malate synthase	CGGGGGT, (-253); CGGGGT, (-263); ACCCCT, (-415); CGGGGGT, (-426)	TTTGCAAAGTTCT, (-400)
cg2560	*aceA*	Isocitrate lyase	ACCCCCG, (-180); AGGGGT, (-190); ACCCCG, (-342); ACCCCCG, (-353)	AGAACTTTGCAAA, (-213)

cg2831	*ramA*	Regulator of acetate metabolism A	ACCCCCA, (-103); ACCCCCG, (-169); AGGGGT, (-192); CGGGGGT, (-207); TGGGGT, (-215); CGGGGGT, (-225); TGGGGGT, (-244); AGGGGT, (-297); TGCGGGT, (-317)	
cg3048 *cg3047*	*pta-ack*	Phosphotransacetylase Acetate kinase	TGGGGGGT, (-289); ACCCCA, (-299); TGGGGT, (-334); TGGGGGC, (-346)	AAAACTTTGCAAA, (-144); AAAACTTTGCAAA, (-259)
cg3107	*adhA*	Alcohol dehydrogenase	CGGGGGT, (-190); CGGGGGT, (-245); AGGGGGT, (-297)	TTTTCGAAATTTC, (-264); GCAATTTGCCAAG, (-40)
cg3169	*pck*	Phosphoenolpyruvate carboxykinase	AGGGGGC, (-44); AGGGGGT, (-127); ACCCCCA, (-193); TGGGGT, (-611); TGGGGT, (-684)	TTTTCAAAGTATT, (-582); TTTGCAATGTTTG, (-748)
cg3335	*malE*	Malic enzyme	AGGGGT, (-141); ACCCCT, (-170); ACCCCA, (-225)	TTTGCAAATGTTC, (-133)

DNA affinity chromatography and Western blot experiments with RamA- and RamB-specific antibodies and cell extracts from *C. glutamicum* cells grown in media with different substrates showed higher amounts of RamA protein and lower amounts of RamB protein in acetate-grown cells when compared to glucose-grown cells, raising the question for expression control of the *ramA* and the *ramB* genes itself and for the nature of potential regulators involved (Cramer and Eikmanns, 2007; Cramer et al., 2007). In fact, it could be shown that RamA binds to multiple RamA binding sites within the *ramA* promoter region (Table 7.3), that RamB binds to a typical 13-bp RamB-binding motifs in the *ramB* promoter region (Table 7.3) and that both transcriptional regulators negatively control the expression of their own genes (Cramer and Eikmanns, 2007; Cramer et al., 2007). In addition, the *ramB* expression was found to be subject to carbon source-dependent positive control by RamA (Cramer et al., 2007). From all the results it can be concluded that RamA functions both as an activator and as a repressor and is a master regulator of acetate metabolism in *C. glutamicum.* A model summarizing the present knowledge on the regulatory network involving RamA and RamB in *C. glutamicum* is shown in Figure 7.3. However, since in the RamA-deficient mutant the *ramA* promoter activity was still about twofold higher in acetate-grown cells than in glucose-grown cells (Cramer and Eikmanns, 2007), further studies are necessary to clarify the involvement and the potential role of a further transcription factor controlling *ramA* expression in *C. glutamicum.*

The acetate stimulon as determined by genome-wide and comparative expression analyses with *C. glutamicum* cells grown on acetate, acetate plus glucose and glucose alone revealed, aside from *pta-ack, aceA/aceB* and *pck*, about 50 genes which obviously are under transcriptional control in response to the presence or absence of acetate in the growth medium (Hayashi et al., 2002; Muffler et al., 2002; Gerstmeir et al., 2003). Among these genes there were those for several enzymes involved in central metabolic pathways (glycolysis, pentose phosphate pathway, TCA cycle) (Figure 7.2). Thirteen out of these genes were found to possess typical 13-bp RamB binding motifs in or around the promoter regions, i.e., aside from *ack, pta, aceA, aceB* the genes *acn, sdhCAB, gltA, pck, ptsG, aceE* and *malE,* encoding aconitase, succinate dehydrogenase, citrate synthase, PEP carboxykinase, PTS EII^{Glc}, E1p of the PDHC and malic enzyme, respectively (Table 7.3) (Gerstmeir et al., 2004). Similarly, typical RamA tandem binding motifs were found in or around the promoter regions of the *sdhCAB* operon, *acn* and of *pck* (Table 7.3) (Cramer et al., 2006). However, further studies are required to proof the functionality of all these putative binding sites and to clarify the relevance of RamA and RamB for expression control of all these and the other genes of the acetate stimulon.

Aside from expression control of genes involved in acetate metabolism, both RamA and RamB have additional functions as transcriptional regulators in *C. glutamicum.* As will be outlined below, both regulators are involved in transcriptional regulation of ethanol metabolism of *C. glutamicum.* Moreover,

RamA was shown to activate transcription of the surface-layer protein gene (*cspB*) in *C. glutamicum* ATCC14067 (Hansmeier et al., 2006), and RamB was found to activate the *aceE* gene encoding E1p of the PDHC when the cells grow in glucose- or acetate-containing minimal medium or in complex medium (Blombach et al., 2007). It should also be noted here that genes encoding orthologs of *ramA* and *ramB* have been found in other corynebacteria, e.g., in *Corynebacterium efficiens, Corynebacterium diphtheriae* and *Corynebacterium jeikeium* (Brune et al., 2005). A *ramB*-homologous gene (Rv0465c) and a typical 13-bp RamB binding motif in front of the ICL gene was identified in *Mycobacterium tuberculosis*, a close relative of *C. glutamicum*. However, the nature and a possible regulatory function of the respective gene products has not been investigated in any of these organism.

Aside from RamA and RamB, the GlxR protein has been identified as potential transcriptional regulator of the glyoxylate cycle genes *aceA* and *aceB* (Kim et al., 2004). GlxR belongs to the cAMP receptor protein family of transcriptional regulators and, in the presence of cAMP, binds to the *aceA*/*aceB* intergenic region. *C. glutamicum* cells contain about fourfold higher intracellular cAMP levels when grown in glucose medium than when grown in acetate-medium and thus, Kim et al. (2004) supposed that GlxR represses the glyoxylate bypass genes in the presence of glucose. Surprisingly, overexpression of *glxR* in *C. glutamicum* led to ten- to fifteenfold lower ICL and MS activities in acetate-grown cells and had no effect on the activities in glucose-grown cells. The obvious repression of *aceA* and *aceB* during growth in acetate-medium certainly does not reflect the physiological situation, and is according to Kim et al. (2004) probably due to the multicopy effects of plasmid-bound *glxR*. Thus, the data suggest a cAMP-triggered repressor function of GlxR on *aceA* and *aceB*, however, the binding site(s), the physiological function and the mechanism of expression control by GlxR have to be further experimentally investigated.

Letek et al. (2006) recently reported on cAMP-dependent binding of GlxR to the promoter regions of the gluconate permease and gluconate kinase genes *gntP* and *gntK* and on a weak repression of these genes on *glxR* overexpression. Thus, GlxR seems to be involved in gluconate metabolism of *C. glutamicum*. Additional indications for a broader significance of GlxR as a global regulator for the central metabolism of *C. glutamicum* came from Moon et al. (2007), who describe the successful construction of a *glxR* deletion mutant. This mutant was reported to have severe growth defects and to be affected in the expression of more than one hundred (so far not further specified) genes (Moon et al., 2007).

5. Ethanol metabolism and its regulation in *C. glutamicum*

Although ethanol has been reported to be a co-substrate for the production of L-lysine with a classically obtained mutant of *C. glutamicum* (Pelechova et al., 1980), the growth on and the utilization of ethanol has only recently been investigated in detail with *C. glutamicum* (Arndt et al., 2007). The growth rate of this organism on ethanol-medium is somewhat lower (0.24 vs. 0.28

per h), the biomass yield somewhat higher than on acetate-medium, however, ethanol concentrations higher than 2% (v/v) negatively affected both the growth rate and the yield. In general, ethanol enters the central metabolism via oxidation to acetate and subsequent activation to acetyl-CoA (Figure 7.1). As in the case of the growth with acetate, the glyoxylate cycle plays an essential role as anaplerotic pathway. In accordance with this general scheme, *C. glutamicum* mutants deficient in AK, PTA, ICL and/or MS were unable to grow on ethanol and the specific activities of these enzymes as well as those of alcohol dehydrogenase (ADH) and acetaldehyde dehydrogenase (ALDH) were high in ethanol and low in glucose-grown cells (Table 7.1) (Arndt et al., 2007). Comparative DNA microarray experiments with ethanol- and glucose- or acetate-grown cells of *C. glutamicum* furthermore revealed an increased expression of *pta-ack, aceA*, *aceB* and of a potential ALDH gene (i.e., *cg3096*; Arndt et al., 2007). Although the DNA microarray experiments for ethanol-grown cells indicated higher mRNA levels for 36 genes, they gave no hint to a potential ADH gene. This result probably was due to a hybridization signal which was not high enough to meet the signal-to-noise criterion applied in this study.

The growth of *C. glutamicum* on a mixture of ethanol and glucose is biphasic, with glucose consumption in the first and ethanol consumption in the second growth phase (Arndt et al., 2007; Kotrbova-Kozak et al., 2007). This biphasic growth behaviour is probably due to relatively low ADH, ALDH, AK, PTA, ICL and MS activities in the first and much higher activities of all six enzymes in the second growth phase (Arndt et al., 2007). These results indicated that ethanol catabolism in *C. glutamicum* is subject to carbon source-dependent regulation such as carbon catabolite control. This hypothesis was substantiated by identification and functional analysis of the ADH gene *adhA* (Arndt and Eikmanns, 2007; Kotrbova-Kozak et al., 2007). Inactivation of *adhA* led to the inability of the respective mutant to grow on ethanol or on propanol and to the absence of any ADH activity. In accordance with the inability of the *adhA* mutant to grow on propanol the purified enzyme uses aside from ethanol also n-propanol as a substrate (Kotrbova-Kozak et al., 2007). Transcriptional analysis of *adhA* revealed that it is monocistronic and that it is repressed in the presence of glucose and in the presence of acetate (Table 7.2) (Arndt and Eikmanns, 2007; Kotrbova-Kozak et al., 2007).

Due to their regulatory function in acetate metabolism (see above) and since AK, PTA, ICL and MS are also involved in ethanol utilization, RamA and RamB were likely candidates to be also involved in the expression regulation of *adhA*. In fact, both RamA and RamB were shown to bind to the *adhA* promoter region at typical RamA and RamB binding motifs, respectively (Table 7.3) (Arndt and Eikmanns, 2007). Furthermore, these authors showed that RamA is essential for *adhA* expression and that RamB represses *adhA* expression in cells growing in the presence of glucose or acetate. However, the *adhA* gene in a RamB-deficient mutant was not completely derepressed in the presence of glucose or acetate, indicating either that the intracellular

RamA content is lower and thus *adhA* activation is incomplete under these conditions or that a further transcriptional regulator is involved in catabolite repression of *adhA* (Arndt and Eikmanns, 2007). With respect to the latter hypothesis, it is interesting to note that Arndt and Eikmanns (2007) speculated about the involvement of GlxR and that Kotrbova-Kozak et al. (2007) found an inverted repeat (IR) and a typical cAMP receptor protein binding site (CBS) in the promoter region of *adhA*. While deletion of the putative CBS had no effect on ADH activity, an about fourfold increase in activity was observed upon elimination of the IR, suggesting that the putative CBS is not involved in *adhA* regulation and that the IR is a target of a transcriptional repressor. However, the nature of this additional regulator involved in carbon catabolite repression of *adhA* remains to be identified and functionally characterized.

6. Post-transcriptional regulation of the TCA cycle in *C. glutamicum*

Analysis of the phosphoproteome of *C. glutamicum* revealed about 90 phosphorylated proteins (Bendt et al., 2003). The organism possesses four different eukaryotic-type Ser/Thr protein kinases, designated as PknA, PknB, PknL and PknG. The functional analysis of one of these kinases, PknG, revealed a novel and fascinating regulatory mechanism for control of the 2-oxoglutarate dehydrogenase complex (OGDHC) and thus for control of the TCA cycle activity (Niebisch et al., 2006). A PknG-deficient mutant of *C. glutamicum* was severely impaired in growth on glutamine as sole carbon source. Comparative proteome analysis led to the identification of a small (15 kDa) protein, designated as OdhI, which subsequently was shown to be the substrate of PknG, i.e., to be phosphorylated by the purified protein at Thr14 (Niebisch et al., 2006). Deletion of the *odhI* gene in the *C. glutamicum* wild type background did not affect glutamine utilization, whereas deletion of this gene in the *pknG* mutant abolished the growth defect on glutamine, indicating that unphosphorylated OdhI is responsible for the growth defect. A combination of genetic and biochemical studies then identified the E1o subunit of the OGDHC to be the target of unphosphorylated OdhI (Niebisch et al., 2006). In a concentration-dependent manner, unphosphorylated OdhI inhibits OGDHC and inhibition of this complex can be prevented by PknG-catalyzed phosphorylation of OdhI. In a further study, Schultz et al. (2007) identified a phospho-Ser/Thr protein phosphatase responsible for dephosphorylation of OdhI. According to all the results, the following model for control of OGDHC activity was suggested by Bott (2007): The OGDH inhibitor protein OdhI binds in its unphosporylated state to the EIo subunit (OdhA) of the OGDHC and thereby inhibits the activity of this complex. When OdhI is phosphorylated by the Ser/Thr protein kinase PknG at Thr14, OdhI does not bind to the OGDHC and does not inhibit its activity. Dephosphorylation of OdhI is brought about by the phospho-Ser/Thr protein phosphatase Ppp. The phosphorylation status of OdhI and thus, the activity level of the OGDHC, is determined by the rates of phosphorylation and dephosphorylation. PknG functions a sensor and its phosphorylation activity is thought to be controlled by the absence or presence of specific stimuli, one of which has suggested to be glutamine (Bott, 2007; Schultz et al., 2007). For the phosphatase Ppp it is not yet known whether the protein is constitutively active or whether its activity is also controlled by specific stimuli.

It should be noted that Niebisch et al. (2006) also observed PknG-independent phosphorylation of OdhI, although, to a minor degree. This observation indicates that at least one of the other three Ser/Thr protein kinases present in *C. glutamicum* might be involved in OdhI phosphorylation and possibly in OGDHC regulation. However, this hypothesis has to be experimentally addressed. The key question concerning the OGDHC regulation remains the question for the signals and conditions for phosphorylation and dephosphorylation activities in *C. glutamicum*. It is evident that the proposed mechanism of OGDHC regulation by PknG, Ppp and the phosphorylation status of OdhI is essential for the catabolism of glutamine and glutamate. Of particular importance is the mechanism of OGDHC regulation for the biotechnological production of glutamate, where attenuation of the OGDHC activity was found to be one of the key factors involved (Shirai et al., 2005; Schultz et al., 2007).

7. Concluding remarks and perspectives

The studies summarized in this chapter show that the carbon metabolism in *C. glutamicum* is subject to complex regulation at the transcriptional and at the post-transcriptional level. It is evident that the knowledge on the regulation of the carbon metabolism in *C. glutamicum* gives us the chance to understand the molecular mechanisms of adaptation of this industrially important organism to the extracellular and intracellular nutritional environment. Additionally, this knowledge will allow us in the near future to govern the carbon flux within the central metabolism of *C. glutamicum* more efficiently in desired directions, i.e., in the direction of desired primary metabolites to be produced with this organism. Since global regulatory circuits controlling general functions of the carbon metabolism are likely to be conserved among the members of the *Corynebacterianeae*, the knowledge on the carbon metabolism and its regulation in *C. glutamicum* might be also very helpful for the understanding (and possibly modulation) of the respective pathways and their regulation in related pathogenic bacteria, such as *C. jeikeium*, *C. diphtheriae* or *M. tuberculosis*.

A lot of biochemical and genetic knowledge on the regulation of the carbon metabolism of *C. glutamicum* has been accumulated and it becomes evident that the molecular mechanisms and the regulatory circuits are novel and different when compared to other model organisms such as *E. coli* or *B. subtilis*. However, a variety of questions about the molecular mechanism of repression and/or activation and about the intra- and extracellular signals remain to be answered. In particular, the effectors/metabolites modulating the function or the activities of the transcriptional regulators AcnR, RamA and RamB as well as those metabolites affecting the activities of Ser/Thr protein kinase(s) and phosphatase have to be identified. The roles of RamA and GlxR as potential global regulators of carbon metabolism have to be clarified. Since there are clear indications for additional transcriptional regulators involved in control of sugar uptake, glycolysis and TCA cycle, further studies should aim at the nature of these regulators and their effectors. One of the questions remaining is the exact mechanism of carbon catabolite control observed when *C. glutamicum* grows on mixtures of ethanol and glucose or acetate and glucose. Another one of the most interesting questions is the question for the regulatory interconnection of carbon, nitrogen, phosphorus and sulfur metabolism. As for carbon metabolism, several key regulators have been identified and their function analyzed (for a recent review see Wendisch, 2006). All these findings are the basis to understand the regulation of the *C. glutamicum* metabolism in its full complexity and to elucidate the entire regulatory network of the *C. glutamicum* cell.

Acknowledgements

We thank Andreas Burkovski, Volker Wendisch and Michael Bott for critically reading the manuscript. The support of the BMBF (grants 0313105 "GenoMik" and 0313805G "GenoMik-Plus") and of the business unit "Feed additives" of Evonik Degussa GmbH is gratefully acknowledged.

References

Arndt, A., and Eikmanns, B.J. (2007). The alcohol dehydrogenase gene *adhA* in *Corynebacterium glutamicum* is subject to carbon catabolite repression. J. Bacteriol. *189*, 7408-7416.

Arndt, A., Auchter, M., Ishige, T., Wendisch, V.F., and Eikmanns, B.J. (2007). Ethanol catabolism in *Corynebacterium glutamicum*. J. Mol. Microbiol. Biotechnol., E-pub ahead of print, August 13th, 2007.

Bendt, A.K., Burkovski, A., Schaffer, S., Bott, M., Farwick, M., and Hermann, T. (2003). Towards a phosphoproteome map of *Corynebacterium glutamicum*. Proteomics *3*, 1637-1646.

Blombach, B., Cramer, A., Eikmanns, B.J., and Schreiner, M. (2007). RamB is an activator of the E1p gene in *Corynebacterium glutamicum*. J. Mol. Microbiol. Biotechnol., Epub ahead of print, September 21st, 2007.

Bott, M. (2007). Offering surprises: TCA cycle regulation in *Corynebacterium glutamicum*. Trends Microbiol. *15*, 417-425.

Bott, M., and Niebisch, A. (2003). The respiratory chain of *Corynebacterium glutamicum*. J. Biotechnol. *104*, 129-153.

Bott, M., and Niebisch, A. (2005). Respiratory energy metabolism. In Handbook of *Corynebacterium glutamicum*, L. Eggeling and M. Bott, eds. (Boca Raton, USA: CRC Press.), pp. 305-332.

Brückner, R., and Titgemeyer, F. (2002). Carbon catabolite repression in bacteria: choice of the carbon source and autoregulatory limitation of sugar utilization. FEMS Microbiol. Lett. *209*, 141-148.

Brune, I., Brinkrolf, K., Kalinowski, J., Pühler, A., and Tauch, A. (2005). The individual and common repertoire of DNA-binding transcriptional regulators of *Corynebacterium glutamicum*, *Corynebacterium efficiens*, *Corynebacterium diphtheriae* and *Corynebacterium jeikeium* deduced from the complete genome sequences. BMC Genomics *6*, 86.

Claes, W.A., Pühler, A., and Kalinowski, J. (2002). Identification of two *prpDBC* gene clusters in *Corynebacterium glutamicum* and their involvement in propionate degradation via the 2-methylcitrate cycle. J. Bacteriol. *184*, 2728-2739.

Cocaign, M., Monnet, C., and Lindley, N.D. (1993). Batch kinetics of *Corynebacterium glutamicum* during growth on various carbon substrates: use of substrate mixtures to localise metabolic bottlenecks. Appl. Microbiol. Biotechnol. *40*, 526-530.

Cramer, A., and Eikmanns, B.J. (2007). RamA, the transcriptional regulator of acetate metabolism in *Corynebacterium glutamicum*, is subject to negative autoregulation. J. Mol. Microbiol. Biotechnol. *12*, 51-59.

Cramer, A., Gerstmeir, R., Schaffer, S., Bott, M., and Eikmanns, B.J. (2006). Identification of RamA, a novel LuxR-type transcriptional regulator of genes involved in acetate metabolism of *Corynebacterium glutamicum*. J. Bacteriol. *188*, 2554-2567.

Cramer, A., Auchter, M., Frunzke, J., Bott, M., and Eikmanns, B.J. (2007). RamB, the transcriptional regulator of acetate metabolism in *Corynebacterium glutamicum*, is subject to regulation by RamA and RamB. J. Bacteriol. *189*, 1145-1149.

Dominguez, H., and Lindley, N.D. (1996). Complete sucrose metabolism requires fructose phosphotransferase activity in *Corynebacterium glutamicum* to ensure phosphorylation of liberated fructose. Appl. Environ. Microbiol. *62*, 3878-3880.

Dominguez, H., Cocaign-Bousquet, M., and Lindley, N.D. (1993). Simultaneous consumption of glucose and fructose from sugar mixtures during batch growth of *Corynebacterium glutamicum*. Appl. Microbiol. Biotechnol. *47*, 600-603.

Dominguez, H., Cocaign-Bousquet, M., and Lindley, N.D. (1997).Simultaneous consumption of glucose and fructose from sugar mixtures during batch growth of *Corynebacterium glutamicum*. Appl. Microbiol. Biotechnol. *47*, 600-603.

Ebbighausen, H., Weil, B., and Krämer, R. (1991). Carrier-mediated acetate uptake in *Corynebacterium glutamicum*. Arch. Microbiol. *155*, 505-510.

Eggeling, L., and Bott, M. (2005). Handbook of *Corynebacterium glutamicum* (Boca Raton, USA: CRC Press).

Eikmanns, B.J. (2005). Central metabolism: tricarboxylic acid cycle and anaplerotic reactions. In Handbook of *Corynebacterium glutamicum*, L. Eggeling and M. Bott, eds. (Boca Raton, USA: CRC Press), pp. 241-276.

Eikmanns, B.J., Thum-Schmitz, N., Eggeling, L., Lüdtke, K.-U., and Sahm, H. (1994). Nucleotide sequence, expression and transcriptional analysis of the *Corynebacterium glutamicum gltA* gene encoding citrate synthase. Microbiology *140*, 1817-1828.

Eikmanns, B.J., Rittmann, D., and Sahm, H. (1995). Cloning, sequence analysis, expression, and inactivation of the *Corynebacterium glutamicum icd* gene encoding isocitrate dehydrogenase and biochemical characterization of the enzyme. J. Bacteriol. *177*, 774-782.

Engels, V., and Wendisch, V.F. (2007). The DeoR-type regulator SugR represses expression of *ptsG* in *Corynebacterium glutamicum*. J. Bacteriol. *189*, 2955-2966.

Frunzke, J., Engels, V., Hasenbein, S., Gätgens, C., and Bott, M. (2007). Coordinated regulation of gluconate catabolism and glucose uptake in *Corynebacterium glutamicum* by two functionally equivalent transcriptional regulators, GntR1 and GntR2. Mol. Microbiol. Epub ahead of print, Nov 28th, 2007.

Gancedo, J.M. (1998). Yeast carbon catabolite repression. Microbiol. Mol. Biol. Rev. *62*, 334-361.

Georgi, T., Engels, V., and Wendisch, V.F. (2007). Regulation of L-lactate utilization by the FadR-type regulator LldR of *Corynebacterium glutamicum*. J. Bacteriol., Epub ahead of print, Nov 26th, 2007.

Gerstmeir, R., Wendisch, V.F., Schnicke, S., Ruan, H., Farwick, M., Reinscheid, D., and Eikmanns, B.J. (2003). Acetate metabolism and its regulation in *Corynebacterium glutamicum*. J. Biotechnol. *104*, 99-122.

Gerstmeir, R., Cramer, A., Dangel, P., Schaffer, S., and Eikmanns, B.J. (2004). RamB, a novel transcriptional regulator of genes involved in acetate metabolism of *Corynebacterium glutamicum*. J. Bacteriol. *186*, 2798-2809.

Gourdon, P., Baucher, M.F., Lindley, N.D., and Guyonvarch, A. (2000). Cloning of the malic enzyme gene from *Corynebacterium glutamicum* and role of the enzyme in lactate metabolism. Appl. Environ. Microbiol. *66*, 2981-2987.

Han, S.O., Inui, M., and Yukawa, H. (2007). Expression of *Corynebacterium glutamicum* glycolytic genes varies with carbon source and growth phase. Microbiology *153*, 2190-2202.

Hansmeier, N., Albersmeier, A., Tauch, A., Damberg, T., Ros, R., Anselmetti, D., Pühler, A., and Kalinowski, J. (2006). The surface (S)-layer gene *cspB* of *Corynebacterium glutamicum* is transcriptionally activated by a LuxR-type regulator and located on a 6 kb genomic island absent from the type strain ATCC13032. Microbiology *152*, 923-935.

Hayashi, M., Mizogushi, H., Shiraishi, N., Obayashi, M., Nakagawa, S., Imai, J., Watanabe, S., Ota, T., and Ikeda, M. (2002). Transcriptome analysis of acetate metabolism in *Corynebacterium glutamicum* using a newly developed metabolic array. Biosci. Biotechnol. Biochem. *66*, 1337-1344.

Ikeda, M., and Nakagawa, S. (2003). The *Corynebacterium glutamicum* genome: features and impacts on biotechnological processes. Appl. Microbiol. Biotechnol. *62*, 99-109.

Jetten, M.S.M., and Sinskey, A.J. (1993). Characterization of phosphoenolpyruvate carboxykinase from *Corynebacterium glutamicum*. FEMS Microbiol. Lett. *111*, 183-188.

Jetten, M.S.M., and Sinskey, A.J. (1995). Purification and properties of oxaloacetate decarboxylase from *Corynebacterium glutamicum*. Antonie van Leeuvenhoek *67*, 221-227.

Jetten, M.S.M., Gubler, M.E., Lee, S.H., and Sinskey, A.J. (1994). Structural and functional analysis of pyruvate kinase from *Corynebacterium glutamicum*. Appl. Environ. Microbiol. *60*, 2501-2507.

Kalinowski, J., Bathe, B., Bartels, D., Bischoff, N., Bott, M., Burkovski, A., Dusch, N., Eggeling, L., Eikmanns, B.J., Gaigalat, L., Goesmann, A., Hartmann, M., Huthmacher, K., Krämer, R., Linke, B., McHardy, A.C., Meyer, F., Möckel, B., Pfefferle, W., Pühler, A., Rey, D.A., Rückert, C., Rupp, O., Sahm, H., Wendisch, V.F., Wiegräbe, I., and Tauch, A. (2003). The complete *Corynebacterium glutamicum* ATCC 13032 genome sequence and its impact on the production of L-aspartate-derived amino acids and vitamins. J. Biotechnol. *104*, 5-25.

Kim, H.-J., Kim, T.-H., Kim, Y., and Lee, H.-S. (2004). Identification and characterization of *glxR*, a gene involved in regulation of glyoxylate bypass in *Corynebacterium glutamicum*. J. Bacteriol. *186*, 3453-3460.

Kinoshita, S., and Tanaka, K. (1972). Glutamic acid. In The Microbial Production of Amino Acids, K. Yamada, ed. (New York, USA: John Wiley Press), pp. 263-324.

Kotrbova-Kozak, A., Kotrba, P., Inui, M., Sajdok, J., and Yukawa, H. (2007). Transcriptionally regulated *adhA* gene encodes alcohol dehydrogenase required for ethanol and n-propanol utilization in *Corynebacterium glutamicum*. Appl. Microbiol. Biotechnol. *76*, 1347-1356.

Krämer, R., Lambert, C., Hoischen, C., and Ebbinghausen, H. (1990). Uptake of glutamate in *Corynebacterium glutamicum*. 1. Kinetic properties and regulation by internal pH and potassium. Eur. J. Biochem. *194*, 929-935.

Kronemeyer, W., Peekhaus, N., Krämer, R., Sahm, H., and Eggeling, L. (1995). Structure of the *gluABCD* cluster encoding the glutamate uptake system of *Corynebacterium glutamicum*. J. Bacteriol. *177*, 1152-1158.

Krug, A., Wendisch, V.F., and Bott, M. (2005) Identification of AcnR, a TetR-type repressor of the aconitase gene *acn* in *Corynebacterium glutamicum*. J. Biol. Chem. *280*, 585-595.

Lee, H.W., Pan, J.G., and Lebeault, J.M. (1998). Enhanced L-lysine production in threonine-limited continuous culture of *Corynebacterium glutamicum* by using gluconate as a secondary carbon source with glucose. Appl. Microbiol. Biotechnol. *49*, 9-15.

Letek, M., Valbuena, N., Ramos, A., Ordóñez, E., Gil, J.A., and Mateos, L.M. (2006). Characterization and use of catabolite-repressed promoters from gluconate genes in *Corynebacterium glutamicum*. J. Bacteriol. *188*, 409-423.

Liebl, W. (2001). *Corynebacterium* nonmedical. In The Prokaryotes. An Evolving Electronic Resource for the Microbiolocial Community, M. Dworkin, S. Falkow, E. Rosenberg, K.-H. Schleifer and E. Stackebrandt, eds. (New-York, USA: Springer-Verlag), pp. 796-818.

Merkens, H., Beckers, G., Wirtz, A., and Burkovski A. (2005). Vanillate metabolism in *Corynebacterium glutamicum*. Curr. Microbiol. *51*, 59-65.

Monod, J. (1949). The growth of bacterial cultures. Annu. Rev. Microbiol. *3*, 371-394.

Moon, M.-W., Park, S.-Y., Choi, S.-K., and Lee, J.-K. (2007). The phosphotransferase system of *Corynebacterium glutamicum*: features of sugar transport and carbon regulation. J. Mol. Microbiol. Biotechnol. *12*, 43-50.

Moritz, B., Striegel, K., de Graaf, A.A., and Sahm, H. (2000). Kinetic properties of the glucose-6-phosphate and 6-phosphogluconate dehydrogenases from *Corynebacterium glutamicum* and their application for predicting pentose phosphate pathway flux *in vivo*. Eur. J. Biochem. *267*, 3442-3452.

Muffler, A., Bettermann, S., Haushalter, M., Hörlein, A., Neveling, U., Schramm, M., and Sorgenfrei, O. (2002). Genome-wide transcription profiling of *Corynebacterium glutamicum* after heat shock and during growth on acetate and glucose. J. Biotechnol. *98*, 255-268.

Netzer, R., Krause, M., Rittmann, D., Peters-Wendisch, P.G., Eggeling, L., Wendisch, V.F., and Sahm, H. (2004). Roles of pyruvate kinase and malic enzyme in *Corynebacterium glutamicum* for growth on carbon sources requiring gluconeogenesis. Arch. Microbiol. *182*, 354-363.

Niebisch, A., Kabus, A., Schultz, C., Weil, B., and Bott, M. (2006). Corynebacterial protein kinase G controls 2-oxoglutarate dehydrogenase activity via the phosphorylation status of the OdhI protein. J. Biol. Chem. *281*, 12300-12307.

Omumasaba, C.A., Okai, N., Inui, M., and Yukawa, H. (2004). *Corynebacterium glutamicum* glyceraldehyde-3-phosphate dehydrogenase isoforms with opposite, ATP-dependent regulation. J. Mol. Microbiol. Biotechnol. *8*, 91-103.

Ozaki, H., and Shiio, I. (1969). Regulation of the TCA and glyoxylate cycles in *Brevibacterium flavum*. J. Biochem. *66*, 297-311.

Parche, S., Burkovski A., Sprenger, G.A., Weil, B., Krämer, R., and Titgemeyer, F. (2001). *Corynebacterium glutamicum*: a dissection of the PTS. J. Mol. Microbiol. Biotechnol. *3*, 423-428.

Paulsen, I.T. (1996). Carbon metabolism and its regulation in Streptomyces and other high GC Gram-positive bacteria. Res. Microbiol. *147*, 535-541.

Pelechová, J., Smékal, F., Koura, V., Plachý, J., and Krumphanzl, V. (1980). Biosynthesis of L-lysine in *Corynebacterium glutamicum* on sucrose, ethanol and acetic acid. Folia Microbiol. *25*, 341-346.

Peters-Wendisch, P.G., Wendisch, V.F., Paul, S., Eikmanns, B.J., and Sahm, H. (1997). Pyruvate carboxylase as an anaplerotic enzyme in *Corynebacterium glutamicum*. Microbiology *143*, 1095-1103.

Peters-Wendisch, P.G., Kreutzer, C., Kalinowski, J., Pátek, M., Sahm, H., and Eikmanns, B.J. (1998). Pyruvate carboxylase from *Corynebacterium glutamicum*: characterization, expression and inactivation of the *pyc* gene. Microbiology *144*, 915-927.

Polen, T., Schluesener, D., Poetsch, A., Bott, M., and Wendisch, V.F. (2007). Characterization of citrate utilization in *Corynebacterium glutamicum* by transcriptome and proteome analysis. FEMS Microbiol. Lett. *273*, 109-119.

Qi, S.-W., Chaudhry, M.T., Zhang, Y., Meng, B., Huang, Y., Zhao, K.-X., Poetsch, A., Jiang, C.-Y., Liu, S., and Liu, S.-J. (2007). Comparative proteomes of *Corynebacterium glutamicum* grown on aromatic compounds revealed novel proteins involved in aromatic degradation and a clear link between aromatic catabolism and gluconeogenesis via fructose-1,6-bisphosphatase. Proteomics *7*, 3775-3787.

Radmacher, E., and Eggeling, L. (2007). The three tricarboxylate synthase activities of *Corynebacterium glutamicum* and increase of L-lysine synthesis. Appl. Microbiol. Biotechnol. *76*, 587-595.

Reinscheid, D.J., Eikmanns, B.J., and Sahm, H. (1994a). Characterization of the isocitrate lyase gene from *Corynebacterium glutamicum* and biochemical analysis of the enzyme. J. Bacteriol. *176*, 3474-3483.

Reinscheid, D.J., Eikmanns, B.J. and Sahm, H. (1994b). Malate synthase from *Corynebacterium glutamicum*: sequence analysis of the gene and biochemical characterization of the enzyme. Microbiology *140*, 3099-3108.

Reinscheid, D.J., Schnicke, S., Rittmann, D., Zahnow, U., Sahm, H., and Eikmanns, B.J. (1999). Cloning, sequence analysis, expression and inactivation of the *Corynebacterium glutamicum pta-ack* operon encoding phosphotransacetylase and acetate kinase. Microbiology *145*, 503-513.

Riedel, C., Rittmann, D., Dangel, P., Möckel, B., Sahm, H., and Eikmanns, B.J. (2001). Characterization, expression, and inactivation of the phosphoenolpyruvate carboxykinase gene from *Corynebacterium glutamicum* and significance of the enzyme for growth and amino acid production. J. Mol. Microbiol. Biotechnol. *3*, 573-583.

Rittmann, D., Schaffer, S., Wendisch, V.F., and Sahm, H. (2003). Fructose-1,6-bisphosphatase from *Corynebacterium glutamicum*: expression and deletion of the *fbp* gene and biochemical characterization of the enzyme. Arch. Microbiol. *180*, 285-292.

Saier, M.H. Jr. (1998). Multiple mechanisms controlling carbon metabolism in bacteria. Biotechnol. Bioeng. *58*, 170-174.

Sauer, U., and Eikmanns, B.J. (2005). The PEP-pyruvate-oxaloacetate node as the switch point for carbon flux distribution in bacteria. FEMS Microbiol. Rev. *29*, 765-794.

Schluesener, D., Fischer, F., Kruip, J., Rögner, M., and Poetsch, A. (2005). Mapping the membrane proteome of *Corynebacterium glutamicum*. Proteomics *5*, 1317-1330.

Schreiner, M., Fiur, D., Holátko, J., Pátek, M., and Eikmanns, B.J. (2005). E1 enzyme of the pyruvate dehydrogenase complex in *Corynebacterium glutamicum*: molecular analysis of the gene and phylogenetic aspects. J. Bacteriol. *187*, 6005-6018.

Schultz, J., Milpetz, F., Bork, P., and Ponting, C.P. (1998). SMART, a simple modular architecture research tool: identification of signaling domains. Proc. Natl. Acad. Sci. USA *95*, 5857-5864.

Schultz, C., Niebisch, A., Gebel, L., and Bott, M. (2007). Glutamate production by *Corynebacterium glutamicum*: dependence on the oxoglutarate dehydrogenase inhibitor protein OdhI and protein kinase PknG. Appl. Microbiol. Biotechnol. *76*, 691-700.

Seletzky, J.M., Noack, U., Fricke, J., Hahn, S., and Büchs, J. (2006). Metabolic activity of *Corynebacterium glutamicum* grown on L-lactic acid under stress. Appl. Microbiol. Biotechnol. *72*, 1297-1307.

Shiio, I., Momose, H., and Oyama, A. (1969). Genetic and biochemical studies on bacterial formation of L-glutamate. I. Relationship between isocitrate lyase, acetate kinase, and phosphate acetyltranferase levels and glutamate production in *Brevibacterium flavum*. J. Gen. Appl. Microbiol. *15*, 27-40.

Shiio, I., Ozaki, H., and Ujigawa, K. (1977). Regulation of citrate synthase in *Brevibacterium flavum*, a glutamate-producing bacterium. J. Biochem. *82*, 395-405.

Shirai, T., Nakato, A., Izutani, N., Nagahisa, K., Shioya, S., Kimura, E., Kawarabayasi, Y., Yamagishi, A., Gojobori, T., and Shimizu, H. (2005). Comparative study of flux redistribution of metabolic pathway in glutamate production by two coryneform bacteria. Metab. Eng. *7*, 59-69.

Stansen, C., Uy, D., Delaunay, S., Eggeling, L., Goergen, J.L., and Wendisch, V.F. (2005). Characterization of a *Corynebacterium glutamicum* lactate utilization operon induced during temperature-triggered glutamate production. Appl. Environ. Microbiol. *71*, 5920-5928.

Stülke, J., and Hillen, W. (1999). Carbon catabolite repression in bacteria. Curr. Opin. Microbiol. *2*, 195-201.

Stülke, J., and Hillen, W. (2000). Regulation of carbon catabolism in *Bacillus* species. Annu. Rev. Microbiol. *54*, 849-880.

Warner, J.B., and Lolkema, J.S. (2003). CcpA-dependent carbon catabolite repression in bacteria. Microbiol. Mol. Biol. Rev. *67*, 475-490.

Wendisch, V.F. (2006). Genetic regulation of *Corynebacterium glutamicum* metabolism. J. Microbiol. Biotechnol. *16*, 999-1009.

Wendisch, V.F., Spies, M., Reinscheid, D.J., Schnicke, S., Sahm, H., and Eikmanns, B.J. (1997). Regulation of acetate metabolism in *Corynebacterium glutamicum*: transcriptional control of the isocitrate lyase and malate synthase genes. Arch. Microbiol. *168*, 262-269.

Wendisch, V.F., de Graaf, A.A., Sahm, H., and Eikmanns, B.J. (2000). Quantitative determination of metabolic fluxes during coutilization of two carbon sources: comparative analyses with *Corynebacterium glutamicum* during growth on acetate and/or glucose. J. Bacteriol. *182*, 3088-3096.

Wittmann, C., and de Graaf, A.A. (2005). Metabolic flux analysis in Corynebacterium glutamicum. In Handbook of *Corynebacterium glutamicum*, L. Eggeling and M. Bott, eds. (Boca Raton, USA: CRC Press), pp. 277-304.

Yokota, A., and Lindley, N.D. (2005). Central metabolism: sugar uptake and conversion. In Handbook of *Corynebacterium glutamicum*, L. Eggeling and M. Bott, eds. (Boca Raton, USA: CRC Press), pp. 215-240.

From: Corynebacteria: Genomics and Molecular Biology. Edited by: Andreas Burkovski.

Molecular Mechanisms of Nitrogen Control in Corynebacteria

8

Eva Hänßler and Andreas Burkovski

Abstract
Nitrogen is one of the macro-elements necessary for cell maintenance and growth. As part of amino acids, nucleotides, and amino sugars nitrogen is an essential component of almost all macromolecules in a bacterial cell, e.g. proteins, nucleic acids, and the murein sacculus determining its shape. Besides for the standard model bacteria for Gram-negatives, *Escherichia coli*, and Gram-positives, *Bacillus subtilis*, nitrogen metabolism and nitrogen regulation is best understood in the mycolic acids-containing actinomycete *Corynebacterium glutamicum*. In this review, uptake and assimilation of nitrogen sources in *C. glutamicum* are shortly summarized, before the current knowledge on regulatory mechanisms including transcriptional control and post-transcriptional regulation is described. Furthermore, the available data on nitrogen metabolism and regulation in closely related actinomycetes are summarized.

1. Uptake and assimilation of nitrogen sources
Between 8 and 10 % of the *C. glutamicum* cell dry mass consists of nitrogen, which is after carbon (approximately 41 %) and oxygen (about 38 %) the third abundant element in this organism (Liebl, 2005). To satisfy its high nitrogen demand *C. glutamicum* is able to use a wide variety of nitrogen sources. As a first step of utilization all these compounds need to be taken up by the cell. This can either occur by passive diffusion or by active transport processes. So far transport systems as well as the respective assimilatory enzymes for ammonium, creatinine, glutamate, and urea have been characterized biochemically and on the genetic level (Börmann et al., 1992; Kronemeyer et al., 1995; Siewe et al., 1996; Jakoby et al., 1997; Beckers et al., 2001; Puskás et al., 2000; Meier-Wagner et al., 2001; Nolden et al., 2001a; Schulz et al., 2001; Beckers et al., 2004; Bendt et al., 2004). Further nitrogen sources used by *C. glutamicum* are glutamine, alanine, asparagine, serine, and threonine as well as different peptides. Uptake and assimilation of these substances is,

however, at least in respect to nitrogen supply, less investigated (for review, see Burkovski, 2003a; 2003b; 2005; 2007). Since glutamate and glutamine are the key donors of nitrogen, i. e. amino groups, in biochemical reactions, most nitrogen sources transported into the cell are decomposed in the cytoplasm to yield ammonium as a precursor for glutamate and glutamine synthesis.

Ammonium added to the medium serves as an excellent nitrogen source for *C. glutamicum.* In solution, the protonated ammonium ion (NH_4^+) is in equilibrium with uncharged ammonia (NH_3), which can enter the cell *via* diffusion across the cell membrane. In the presence of high ammonium concentrations as found in standard minimal media for *C. glutamicum*, ammonia diffusion is sufficient to promote cell growth and no energy for transport has to be spend. As soon as this passive process becomes limiting for nitrogen supply, two different ammonium transport systems, AmtA and AmtB, are recruited for uptake. These transporters belong to a familiy of ubiquitous membrane proteins found in bacteria, archaea, and eukaryotes (Huang and Peng, 2005). AmtA is a high affinity transport system, which works in a membrane potential-dependent manner (Siewe et al., 1996; Meier-Wagner et al., 2001; B. Walter, unpublished observation). Besides ammonium AmtA also accepts methylammonium as substrate, although this compound does not serve as nitrogen source in *C. glutamicum.* Determination of transport kinetics with the radioactively labeled substrate analog [^{14}C]methylammonium revealed a K_m of 44 ± 7 μM for methylammonium and maximal velocity of 20 ± 5 nmol min^{-1} (mg dry weight)$^{-1}$ (Siewe et al., 1996; Meier-Wagner et al., 2001). In contrast to AmtA, it was assumed previously that AmtB does not accept [^{14}C]methylammonium as substrate (Meier-Wagner et al., 2001), so that the characterization of this second transport system seemed more challenging. However, recent studies revealed that AmtB is a low affinity uptake system, which unlike AmtA, has channel-like properties. The rate of methylammonium transport was found to be strictly dependent on the substrate concentration and was not saturable up to a concentration of 5 mM (B. Walter, unpublished observation). This hints to a mechanism of transport by facilitated diffusion which has been proposed first for the *Escherichia coli* AmtB protein (Soupene et al., 1998; 2002; Zheng et al., 2004; Javelle et al., 2005; Winkler, 2006; for review of ammonium transport mechanisms in bacteria, see Javelle et al., 2007).

Once inside of the cell, ammonium can be assimilated *via* two different pathways. When it is present at high concentrations, ammonium is preferentially fixed by glutamate dehydrogenase (GDH). This enzyme catalyzes the formation of glutamate from ammonium and 2-oxoglutarate. During the reaction, one mol NADPH is oxidized per mol ammonium assimilated. Because of the low affinity to ammonium (K_m = 3.08 mM) a second pathway for ammonium assimilation is used at concentrations below 5 mM (Shiio and Ozaki, 1970; Tesch et al., 1998). This pathway consists of two reactions catalyzed by glutamate synthase (GOGAT) and glutamine synthetase (GS) encoded by *gltBD* and *glnA*, respectively (Jakoby et al., 1997; Beckers et al., 2001; Schulz et al., 2001). First, ammonium is coupled to glutamate to form glutamine in an ATP-dependent reaction by GS. In a subsequent step the amide group is then transferred reductively to 2-oxoglutarate by GOGAT, resulting in the net conversion of 2-oxoglutarate to glutamate. In contrast to GDH, the substrate affinity of GS is significantly higher so that under limiting ammonium concentrations this enzyme is preferred for ammonium fixation, despite a higher energy demand.

As mentioned above, *C. glutamicum* can also use other nitrogen sources than ammonium. For urea, glutamate, and creatinine transport systems have been identified in addition to the corresponding metabolizing enzymes. Just like ammonia, urea can easily pass the cell membrane by diffusion and is then hydrolyzed by the urease enzyme complex leading to two molecules of ammonium and one molecule of carbon dioxide. The genes encoding urease, *ureABCEFGD*, have been identified and an operon structure was proven (Nolden et al., 2000; Puskás et al., 2000; Beckers et al., 2004). At limiting nitrogen supply, urea can also enter the cell by an energy-dependent uptake system, an ABC transporter encoded by the *urtABCDE* operon (Beckers et al., 2004). Another alternative nitrogen source is creatinine, which is also taken up by a specific transport system, a permease encoded by *crnT*, and subsequently metabolized by creatinine deaminase encoded by the *codA* gene. The degradation of creatinine leads to the dead-end product N-methylhydantoin and one molecule ammonium (Bendt et al., 2004). In contrast to the two nitrogen sources just mentioned, glutamate can be directly incorporated without further metabolization. Uptake is mediated by a sodium-coupled permease, GltS, which is primarily active in complex media (Burkovski et al., 1996; Trötschel et al., 2003) and by an ABC transport system, encoded by the *gluABCD* genes (Kronemeyer et al., 1995). Uptake of glutamine was characterized biochemically. This amino acid is transported in symport with sodium ions into the cell (Siewe et al., 1995). Data on genes and regulation are not available.

Mechanisms, which are responsible for the regulation of processes like transport and assimilation of alternative nitrogen sources depending on the availability of ammonium, are summarized in the term nitrogen control. This includes sophisticated mechanisms of transcriptional regulation as well as fine-tuning of enzyme activity by post-translational control.

2. Transcriptional control

2.1 Isolation of the master regulator AmtR

First hints for a nitrogen-dependent expression control of ammonium transporters came from transport measurements. These revealed an induction of methylammonium uptake upon nitrogen deprivation (Siewe et al., 1996), which could be prevented by addition of chloramphenicol, suggesting regulation on the level of transcription. Later, transcription regulation of genes being part of the nitrogen starvation response was observed when *amtA* expression was analyzed in dependence on the availability of ammonium (Jakoby et al., 2000). RNA hybridization analyses showed a strong increase of *amtA* transcript in response to nitrogen starvation. The fact that additional plasmid encoded copies of the *amtA* gene resulted in deregulated methylammonium uptake, led to the hypothesis, that a transcriptional repressor might be involved and that this protein was titrated out by the additional *amtA* copies. The following screening procedure led to the identification of a TetR/ArcR-type regulator, which was designated AmtR (for ammonium transport repressor; Jakoby et al., 2000). The function of this newly identified protein was closer examined by construction of the respective deletion mutant and by binding studies. Indeed, *amtA* transcription and consequently the uptake rate of methylammonium were significantly increased even in the presence of nitrogen in the *amtR* deletion mutant compared to the wild type. Gel retardation experiments furthermore confirmed that AmtR binds to the upstream region of *amtA*, more precisely to the sequence ATCTATAGN_4ATAG. All together two palindromic sequences were found, which both serve as AmtR binding sites. The deduced consensus motif was then used to find additional binding sites upstream of other putative AmtR-regulated genes. By this approach a binding site upstream of the *amtB* gene was identified. As in many bacteria and archaea *C. glutamicum amtB* is located in the vincinity of *glnK* encoding a P_{II} type signal transduction protein; in actinomycetes these two proteins are additionally clustered with *glnD* encoding an uridylyltransferase/adenylytransferase (Jakoby et al., 2000; Thomas et al., 2000; Strösser et al., 2004). It was shown that *glnK* and *glnD* form an operon with *amtB* and are transcribed in an AmtR-dependent manner with high transcription levels during nitrogen deprivation (Jakoby et al., 2000; Nolden et al., 2001b). These transcription analyses suggested a function of AmtR as a global regulator of nitrogen control rather than being solely a regulator of ammonium transport.

2.2 Identification of the AmtR regulon

First studies on nitrogen transcription control focused on the characterization of distinct genes and operons; with the publication of the *C. glutamicum* genome sequence by two different industry-supported groups (Ikeda and Nakagawa, 2003; Kalinowski et al., 2003) global investigation approaches such as transcriptome and proteome analyses became available (Wendisch, 2003; Schaffer and Burkovski, 2005; see also elsewhere in this book) and were applied to elucidate nitrogen regulation in *C. glutamicum* (Silberbach and Burkovski, 2006).

To identify the AmtR regulon, a combination of bioinformatic tools, transcriptome and proteome analyses as well as DNA binding studies was used (Beckers et al., 2005). Based on the identified binding sequences upstream of *amtA* (Jakoby et al., 2000), the *amtB-glnK-glnD* operon (Jakoby et al., 2000), *glnA* (Nolden et al., 2001a), and the *gltBD* operon (Beckers et al., 2001), the genome was screened for further AmtR-target genes. The resulting genes were analyzed using RNA hybridization experiments, real time RT PCR, and gel retardation experiments (Beckers et al., 2005). Additionally, targets were verified by proteome analyses using two-dimensional gel electrophoresis followed by protein identification by mass spectrometry. As a consequence, it was also possible to define a more precise AmtR consensus motif. In addition to that, the transcription pattern of an *amtR* deletion mutant was compared to the wild type using DNA microarrays. By the combination of these techniques, it was revealed that at least 33 genes are directly regulated by AmtR, underlining its function as a global regulator of nitrogen metabolism. Genes controlled by AmtR can be grouped into different categories according to their function. These include genes encoding proteins involved in transport of ammonium (*amtA*, *amtB*) as well as transporters of alternative nitrogen sources (*crnT*, *urtABCDE, gluABCD*), enzymes of ammonium assimilation (*glnA*, *gltBD*) and utilization of alternative nitrogen sources like creatinine (*codA*) and urea (*ureABCEFGD*). Furthermore, the expression of proteins being part of signal transduction cascade (*glnK* and *glnD*) is AmtR-regulated. For the *vanABK* operon an indirect regulation was shown while the direct regulator remains unknown (Merkens et al., 2005).

The influence of long-term nitrogen limitation on the metabolism was studied by combining global approaches with cultivation in a chemostat (Silberbach et al., 2005b). The advantage of continuous cultivation lies in highly reproducible growth conditions over a period of days or weeks resulting in well adapted cells. Furthermore, this approach allows discriminating between general, for instance growth rate-dependent, stress response and specific adaptation to changes in environmental conditions. Ammonium limitation results in alterations in the expression of 285 genes in *C. glutamicum*. As expected, among those were genes of the AmtR regulon. Besides the activation of the nitrogen starvation response, changes in carbon metabolism and amino acid biosynthesis as well as a decrease in cell division were observed. Genes encoding proteins of L-aspartate, L-arginine, L-leucine, L-threonie, L-lysine, and aromatic amino acids biosynthesis were significantly less expressed, while genes encoding glycolytic enzymes and components of the respiratory chain exhibited an enhanced transcription (Silberbach et al., 2005a; 2005b). The latter results fit well with previous data, showing an increasing oxygen consumption rate of wild type cells in response to nitrogen deprivation (Schmid et al., 2000) and might reflect the enhanced ATP demand for ammonium assimilation *via* glutamine synthetase. The transcriptome analyses by DNA microarrays could be verified for several proteins by proteome analyses (Silberbach et al., 2005a).

2.3 Promoter structure

The identification of the AmtR regulon was a prerequisite for a closer, more detailed investigation of AmtR-mediated repression. As for instance described for NRI-P-controlled transcription of Ntr genes in *E. coli* (Atkinson et al., 2002), nitrogen-dependent gene expression in *C. glutamicum* is also marked by gradual expression of genes belonging to the AmtR regulon. While genes encoding for instance the ammonium uptake systems AmtA and AmtB are not transcribed under nitrogen surplus, the *glnA* gene is expressed at a low level in the presence of ammonium. This differential regulation of transcription allows the metabolism to fulfill the physiological need of glutamine synthesis by GS without turning on the nitrogen starvation response. The molecular

Table 8.1. Organisation of AmtR-controlled genes

Gene or operon	Protein(s)	AmtR binding site	Position relative to start codon	Position of transcription start relative to start codon	Reference
amtA-ocd-soxA	ammonium permease, putative ornithine decarboxylase, putative sarcosine oxidase	[2]TTTTTAC**CTAT**CGTTCT**A**T**AG**ATTTCTG [2]GTATTTT**CTAT**TCCGCT**A**T**AG**AT**A**AACC [1) 3)]AT**CTAT**AGAACG**A**T**AG** [1) 3)]AT**CTAT**AGCGGA**A**T**AG**	-103 to -76 -66 to -39 -97 to -86 -61 to -46	-56 and -53 (Jakoby et al., 2000)	Jakoby et al., 2000; Beckers et al., 2005
amtB-glnK-glnD	ammonium transporter, GlnK, GlnD	ATATTTT**CTAT**AGTTTA**A**C**AG**GT**A**ATTT GCTCTAA**CTAT**AGACCT**A**C**AG**AA**A**CTAA [1)] AT**CTAT**AGAATAG	-186 to -159 -118 to -91 -140 to -128	-57 (Nolden, unpublished observation)	Jakoby et al., 2000; Beckers et al., 2005
gluABCD	ABC transporter for the uptake of glutamate	GCAATAT**CTAT**CATGTG**A**T**AG**GT**A**AATT	-195 to -168	n. d.	Beckers et al., 2005
urtABCEFGD	ABC transporter for the uptake of urea	CAACTAC**CTAT**AGGCTG**A**C**AG**AA**A**CTCT CTAAAAA**CTAT**AGAGCT**A**T**AG**AA**A**CTTT [1)]**CTAT**AG [1)]**CTAT**AGAGCT**A**T**AG**A	-70 to -43 -44 to -17 -63 to -58 -37 to -23	n. d.	Beckers et al., 2004; Beckers et al., 2005
crnT	creatinine transporter	ATACTTT**CTAT**AGATTG**A**T**AG**AA**A**TGTG CCCTTTT**CTAT**CAATCT**A**T**AG**AA**A**CTTG [1)]AT**CTAT**AG	-81 to -54 -35 to -8 -68 to -61	n. d.	Bendt et al., 2004; Beckers et al., 2005

glnA	glutamine synthetase	AAAGTTTTG**AT**AGATCG**A**C**AG**GT**A**ATGC [1,3]AT**CTAT**	-151 to -124 -142 to -137	-109 (Schulz et al., 2001)	Nolden et al., 2001a; Beckers et al., 2005
gltBD	glutamate synthase	CGTTTTC**CTAT**AGGTTG**A**TC**G**AA**A**GTAA TTATTAT**C**G**A**ACGATTG**ATAG**AA**A**CAGG [1]AT**CTAT**AG	-117 to -90 -48 to -21 -102 to -109	-41 (Beckers et al., 2001)	Beckers et al., 2001; Beckers et al., 2005
gdh	glutamate dehydrogenase	[2]TACAAAT**T**GATAACCT**A**A**AG**AA**A**TTTT [1]AAC**CTA**A**AG**AA**A**TTTTCAAA [1]AAA**CTAT**TAACCGTT**AG**GTA	-370 to -343 -358 to -339 -208 to -189	-284 (Börmann et al., 1992)	Beckers et al., 2005; T. Müller unpublished results
ureABCEFGD	urease	AGTAGCG**CTA**CATCTGC**A**T**A**TCT**A**CCCC	-135 to -108	n. d.	Beckers et al., 2005
codA	creatinine deaminase	ATCTATT**CTAT**TGCTTG**A**C**AG**GT**A**TTAG	-86 to -59	n. d.	Beckers et al., 2005
amtR	AmtR repressor	[4] AT**C**G**AT**ACCAAATAG	-87 to -73	n. d.	Jakoby et al., 2000

Unless indicated otherwise binding sites were determined using Hidden-Markov models and subsequently verified by gel retardation assays.

[1] Determined using gel retardation assays
[2] Determined by Hidden Markov models
[3] Determined by deletion analyses
[4] Determined by sequence analysis

n. d. not determined

mechanism of this variation in AmtR stringency is not yet understood, but the bioinformatical and experimental identification of AmtR binding sites should help explaining mechanisms of AmtR repression more precisely. In Table 8.1 known AmtR binding sites are summarized and described in terms of their position relative to transcription and translation start site of the respective gene.

It is obvious that there are some major differences within the binding sites that might result in diverse affinity of AmtR to the promoter regions. Furthermore, positioning of the binding site with respect to the transcription start might also have an influence on stringency of repression. Third, it is likely that the number of AmtR binding sites upstream of a target gene might influence the stringency of repression. This was for example shown for the *E. coli* Lac repressor (Müller et al., 1996) and led to a concept of increased local concentrations of regulators at promoters (Müller-Hill, 1998; Dröge and Müller-Hill, 2001; Müller-Hill, 2006).

RNA hybridization experiments, real time RT PCR, and gelshift assays with upstream regions of AmtR-regulated genes already supported the idea that the affinity of the repressor to the respective upstream regions is not identical in respect to the investigated genes. Upstream of strongly repressed genes like *amtA* and *amtB,* canonic AmtR binding sites were identified (Jakoby et al., 2000), whereas the AmtR binding motif upstream of *glnA* differs significantly from the deduced consensus motif (Nolden et al., 2001a). In addition to that, the AmtR binding sites upstream of strictly regulated genes as *amtA* and the *gltBD* operon are located in the vicinity of the transcription start, so that transcription might be blocked in the presence of AmtR. The AmtR binding site of *glnA* is located upstream of the -10 region, that means the blocking effect might be diminished. In future, these assumption need to be verified by experimental approaches. First deletion studies within the *amtA* promoter region could already identify essential nucleotides involved in AmtR binding (Jakoby et al., 2000), so that based on mutational studies of either the repressor or the binding motif essential information can be provided.

2.4 Signal transduction

AmtR is a member of the TetR-family, the largest group of transcription regulator proteins in corynebacteria (Jakoby et al., 2000; Brune et al., 2005). Members of this family of transcription repressors typically interact with small molecules as it has been shown for TetR, the prototype of these proteins, which binds the antibiotics tetracycline, and QacR, which binds a number of cationic lipophilic drugs such as rhodamine G, crystal violet and ethidium (for review, see Ramos et al., 2005). In contrast to these Tet-type repressors, interaction of *C. glutamicum* AmtR with its target DNA does not depend on the presence or absence of small ligands, instead binding to promoters of nitrogen-dependent genes is regulated by specific protein-protein interaction. At decreasing ammonium concentrations AmtR is released from the DNA by interaction with the P_{II} type signal transduction protein GlnK

(Beckers et al., 2005). As a prerequisite for this interaction the trimeric GlnK needs to be present in its modified form. In response to nitrogen deprivation GlnK is adenylylated at tyrosyl residue 51 located in the T-loop of the protein. This modification is performed by GlnD. Initially, it was believed that GlnD acts as an uridylyltransferase due to a high homology to the *E. coli glnD* gene product (Nolden et al., 2001b), but eventually it was shown that for *C. glutamicum* GlnD this is not the case. Besides the adenylylation, the enzyme also catalyzes the demodification of GlnK depending on the nitrogen status of the cell. Studies with mutants expressing a truncated GlnD protein indicated that the modifying activity of this bifunctional enzyme is located at the N-terminal part, whereas the C-terminal part of the protein is crucial for its demodification activity (Strösser et al., 2004).

2.5 Metabolite effects

As described, *C. glutamicum* is able to respond to changes in the nitrogen supply. A prerequisite for a proper response is the perception of the cellular nitrogen status. Since various nitrogen sources such as ammonium, amino acids, creatinine, and urea can be metabolized by *C. glutamicum* (Burkovski, 2005; 2007) and contribute to nitrogen supply, it is most likely that not every single nitrogen source but a single solute or only a very limited number of key nitrogen metabolites are sensed. Deduced from other organisms and from their central position in metabolism ammonium, 2-oxoglutarate, and L-glutamine are likely candidates. L-glutamate, which is also a key metabolite of nitrogen assimilation, can be excluded to fulfil this function, since *C. glutamicum* accumulates up to 200 mM of this amino acid and the internal L-glutamate pool reacts only slowly to nitrogen starvation, while the response on transcription level is much faster (Nolden et al., 2001b; Müller et al., 2006). L-glutamate might function as a storage compound for nitrogen and/or carbon supply of *C. glutamicum* rather than reflecting the nitrogen status of the cell. This might also be correct for L-glutamine, since this amino acid can also be used as nitrogen and/or carbon source and its concentration in the cytoplasm is also very high compared to other bacteria. Additionally, the L-glutamine pool reacts as the glutamate pool only slowly to changes in the nitrogen supply compared to the transcription response (Nolden et al., 2001b; Müller et al., 2006). As a putative marker for good nitrogen supply, ammonium is discussed (Nolden et al., 2001b; Strösser et al., 2004; Müller et al., 2006). However, biochemical evidence for ammonium perception is missing.

A clue for a marker of low nitrogen supply came from mutant analyses of glutamate dehydrogenase-deficient *C. glutamicum* strains. Deletion of *gdh* has no obvious effect on growth and glutamate production in ammonium-rich standard minimal media (Börmann et al., 1992; Börmann-El Kholy et al., 1993). Additionally, Tesch and coworkers (Tesch et al., 1998) showed high GOGAT activity in a glutamate dehydrogenase-deficient strain even under nitrogen surplus. RNA hybridization experiments revealed that deletion of *gdh* leads to the deregulation of nitrogen control including deregulated expression of the *gltBD* operon (Müller et al., 2006). Metabolite analyses indicated that a high 2-oxoglutarate concentration found in the *gdh* mutant under any growth conditions and in the wild type under nitrogen starvation might be perceived as marker of a poor nitrogen nitrogen supply. The most likely sensor protein of 2-oxoglutarate might be signal transduction protein GlnK (for review, see Arcondeguy et al., 2001). Again, direct biochemical evidence for 2-oxoglutarate sensing by *C. glutamicum* is missing.

3. Post-transcriptional control

3.1 Regulatory localization changes and proteolysis of GlnK

As described above, adenylylation and deadenylyation of GlnK is crucial for nitrogen control in *C. glutamicum*. Interestingly, additional regulatory mechanisms are controlling GlnK function. First, depending on the nitrogen status of the cell, the GlnK protein changes its localization. During nitrogen starvation it is found exclusively in its adenylylated form in the cytoplasm (Strösser et al., 2004). In response to an improved nitrogen supply after a starvation period, the GlnK protein is sequestered to the cytoplasmic membrane. This localization change depends on the presence of the ammonium transporter AmtB (Strösser et al., 2004). Due to the conserved genetic organization of the corresponding genes within bacteria, interaction of GlnK and AmtB was suggested to be part of an ancient prokaryotic nitrogen control system (Javelle and Merrick, 2005).

Depending on membrane localization, in *C. glutamicum* the major amount of GlnK is degraded (for proteases involved, see chapter on general and regulatory proteolysis elsewhere in this book). Less than approximately 5 % of GlnK survives the first 30 min after an ammonium pulse and is protected from proteolysis by an unknown mechanism. The high redundancy of control mechanisms in respect to GlnK, i. e. transcription regulation, modification/demodification, change of localization, and proteolysis, hints to an essential function of this P_{II}-type signal transduction protein. However, in the laboratory, mutant strains deficient of *glnK*, *amtB*, or *glnD* are not impaired in growth. Obviously, control seems to be very robust and might be more important in the natural habitat of *C. glutamicum* rather than in the artificial environment of the laboratory.

3.2 Modification of glutamine synthetase

As already mentioned above glutamine synthetase is regulated on the level of transcription in an AmtR-dependent manner (Nolden et al., 2001a). However, since transcriptional control works in time ranges of several minutes, for a rapid adaptation to sudden changes in nitrogen supply, the enzyme is also regulated on the level of activity by post-translational modification (Jakoby et al., 1999). Measurements of GS activity confirmed that a decrease in nitrogen availability leads to an increase of enzyme activity, which in turn is reduced upon addition of ammonium. Since this effect was abolished in the presence of snake venome phosphodiesterase, it was postulated that GS is adenylylated in response to improved ammonium availability as it has been described in other microorganisms. Indeed, the adenylyltransferase GlnE was identified as the modifying enzyme, since a deletion of the respective gene led to a deregulation of GS activity in response to the nitrogen availability (Nolden et al., 2001a). Nucleotide sequence comparisons revealed a typical adenylylation site, a tyrosyl residue at amino acid position 405 (Jakoby et al., 1997), which was verified in later studies by mutant analyses (Jakoby et al., 1999). In analogy to the signal transfer shown in *E. coli* a signal pathway *via* GlnD, GlnK, and ATase to GS was suggested (Nolden et al., 2001b). Characterization of a *glnK* deletion strain showed later that ATase works independently from GlnK in *C. glutamicum* (Strösser et al., 2004). An GlnK-independent activity of GlnE was also shown for *Streptomyces coelicolor* (Hesketh et al., 2002). In *C. glutamicum* as in *S. coelicolor* the perception of the current nitrogen status by GlnE is not understood.

4. Related regulatory systems

Assimilation of nitrogen sources depends directly on carbon metabolism as donor of precursors and energy. Consequently, crosstalk of these metabolic networks is essential for an optimal adaptation to carbon and nitrogen availability. A direct link between ammonium assimilation and tricarboxylic acid (TCA) cycle is provided by the GDH reaction. At this branch point 2-oxoglutarate is either oxidized by 2-oxoglutarate dehydrogenase complex (ODHC) within the TCA cycle or glutamate is formed catalyzed by GDH. Alterations of *gdh* transcription result from both, changes in nitrogen supply and variation of carbon source. While nitrogen deprivation leads to an increase in transcript level, utilization of 2-oxoglutarate, glutamate, mannose, and fructose causes decreased expression of *gdh* compared to growth on glucose. Same results were obtained when a carbon source was lacking (Müller, unpublished results). This implies a so far not identified regulatory mechanism of *gdh* transcription in response to components of carbon metabolism.

More detailed knowledge is available on the control of the 2-oxoglutarate dehydrogenase complex. By interaction with OdhI, ODHC activity is inhibited. However, when OdhI is modified by protein kinase G (PknG) complex formation does not occur, resulting in the release of inhibition (Niebisch et al., 2006). Interestingly, this regulatory network responds to an intermediate of nitrogen metabolism. It has been proposed that phosphorylation activity of PknG is stimulated by glutamine (Niebisch et al., 2006; Schultz et al., 2007). Furthermore, it was demonstrated that the presence of an intact OdhI repressing ODHC is of crucial importance for glutamate production. The deletion of *pknG* leading to a decreased amount of phosphorylated OdhI had a positive influence on glutamate production depending on the cultivation condition (Schultz et al., 2007). By these results first insights about possible mechanisms of interaction between nitrogen and carbon metabolism were provided.

Studies on the AmtR regulon also gave some evidence that AmtR could indirectly control the expression of genes encoding vanillate demethylase (*vanAB*) and a protocatechuate transporter (*vanK*). Vanillate as well as protocatechuate can be used as sole carbon sources (Merkens et al., 2005). Even

Table 8.2. Distribution of genes involved in nitrogen uptake, assimilation, and control in sequenced *Corynebacterium* species

Gene	Function of the corresponding protein in *C. glutamicum*	Reference	*Corynebacterium*			
			diphtheriae	*efficiens*	*glutamicum*	*jeikeium*
amtA	ammonium uptake system	Siewe et al., 1996	-	+	+	-
amtB	ammonium uptake system	Meier-Wagner et al., 2001	-	+	+	+
amtR	TetR-type repressor of nitrogen-controlled genes	Jakoby et al., 2000[1]	+	+	+	-
codA	creatinine deaminase	Bendt et al., 2004	-	-	+	-
crnT	creatinine permease	Bendt et al., 2004	-	-	+	-
gdh	glutamate dehydrogenase	Börmann et al., 1992	+	+[2)]	+	+
glnA	glutamine synthetase, GS I-β type	Jakoby et al., 1997	-	+[2)]	+	+
glnA2	glutamine synthetase, GS I-α type	Nolden et al., 2001a	+	+	+	+
glnD	GlnK-regulating adenylyltransferase	Jakoby et al., 1999; Nolden et al., 2001a; Strösser et al., 2004[1]	+	+	+	-
glnE	GS-regulating adenylyltransferase	Nolden et al., 2001a	+	+	+	+
glnK	P_{II}-type nitrogen signal transduction protein	Jakoby et al., 1999; Nolden et al., 2001b	+	+	+	+
gltB	large glutamate synthase subunit	Beckers et al., 2001; Schulz et al., 2001	-	+	+	+
gltD	small glutamate synthase subunit	Beckers et al., 2001; Schulz et al., 2001	-	+	+	+
ureA	urease subunit	Nolden et al., 2000; Puskás et al., 2000	-	+	+	-

ureB	urease subunit	Nolden et al., 2000; Puskás et al., 2000	-	+	+	-
ureC	urease subunit	Nolden et al., 2000; Puskás et al., 2000	-	+	+	-
ureD	urease subunit	Nolden et al., 2000; Puskás et al., 2000	-	+	+	-
ureE	urease subunit	Nolden et al., 2000; Puskás et al., 2000	-	+	+	-
ureF	urease subunit	Nolden et al., 2000; Puskás et al., 2000	-	+	+	-
ureG	urease subunit	Nolden et al., 2000; Puskás et al., 2000	-	+	+	-
ureR	putative regulator of urease genes	Beckers et al., 2004	-	-	+	-
urtA	subunit of urease uptake system	Beckers et al., 2004	-	+	+	-
urtB	subunit of urease uptake system	Beckers et al., 2004	-	+	+	-
urtC	subunit of urease uptake system	Beckers et al., 2004	-	+	+	-
urtD	subunit of urease uptake system	Beckers et al., 2004	-	+	+	-
urtE	subunit of urease uptake system	Beckers et al., 2004	-	+	+	-

1) functional analysis in *C. diphtheriae* (Nolden et al., 2002)
2) two copies present in the genome

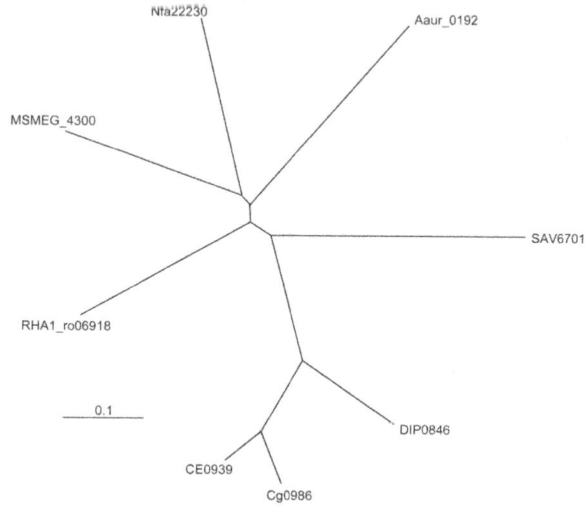

Figure 8.1. Phylogenetic tree of AmtR proteins in actinomycetes. The corresponding genes are annotated in the genomes of *C. efficiens* (CE0939), *C. glutamicum* (cg0986), *C. diphtheriae* (DIP0846), *Streptomyces avermitilis* (SAV6701), *Arthrobacter aurescens* (AAur0192), *Nocardia farcinica* (nfa22230), *Mycobacterium smegmatis* (MSMEG4303), and *Rhodococcus* strain A1 (RHA1 ro06918). *Corynebacterium* proteins are clustered in a defined sub-group, with *C. efficiens* and *C. glutamicum* proteins being closely related. Amino acid identities compared to the *C. glutamicum* AmtR protein range between 86 % for *C. efficiens* AmtR and 39 % for the *S. avermitilis* protein.

though it is not understood how AmtR is involved in regulation of gene expression, this might be another starting point for the investigation of the interplay between nitrogen and carbon metabolism

5. Nitrogen control in actinomycetes (corynebacteria)

As summarized above uptake and assimilation of nitrogen sources as well as nitrogen regulation is understood in great detail in *C. glutamicum.* Based on this knowledge genes and corresponding proteins involved in nitrogen assimilation and nitrogen regulation were predicted in other corynebacteria. Three other *Corynebacterium* species with available genome sequence, namely *Corynebacterium diphtheriae* (Cerdeno-Tarraga et al., 2003), *Corynebacterium efficiens* (Fudou et al., 2002), and *Corynebacterium jeikeium* (Tauch et al., 2005; for a global comparison of these genomes, see Kalinowski (2005) and the review by Tauch elsewhere in this book) were studied. Comparison of the different genome sequences revealed that nitrogen metabolism and control components are highly conserved among *Corynebacterium* species (Table 8.2; Walter et al., 2007). In *C. diphtheriae*, *C. efficiens*, and *C. glutamicum* nitrogen-dependent gene expression seems to be controlled by the master regulator AmtR, since a corresponding gene was annotated in these genomes. AmtR is not only found in corynebacteria but is widely distributed among actinomycetes (Figure 8.1). For *C. diphtheriae*, the function of AmtR in nitrogen control was already verified by mutant generation and RNA hybridization analyses (Nolden et al., 2002). Besides the *amtR* gene, also the DNA-binding motif of this global nitrogen regulator seems to be conserved (Walter et al., 2007). As described above, *C. glutamicum* AmtR is controlled by a signal cascade including GlnK and GlnD. Again, the corresponding genes are conserved in *C. diphtheriae* and *C. efficiens*. For *C. diphtheriae*, the importance of GlnD in nitrogen control was verified by the generation of a mutant strain and subsequent RNA hybridization analyses (Nolden et al., 2002). Modification of *C. efficiens* GlnK in response to nitrogen starvation was shown by biochemical analyses (Walter et al., 2007).

Besides these striking similarities, in frame of the genome screening approach species-specific differences were found as well. Compared to the pathogenic corynebacteria, the soil bacteria and amino acid producers *C. efficiens* and *C. glutamicum* have a broader set of genes encoding uptake systems for nitrogen sources, assimilatory enzymes, and regulatory proteins. On the other hand, a clear tendency towards gene decay was observed for *C. diphtheriae* and *C. jeikeium*. This includes ammonium and urea uptake systems, ammonium assimilation pathways, urea-cleaving enzymes, and regulatory proteins. A similar tendency towards loss of genes in pathogenic bacteria was observed for mycobacteria. Especially *Mycobacterium leprae* shows a drastic gene reduction and decay compared for example to *Mycobacterium smegmatis* (Cole et al., 2001; Vissa and Brennan, 2001).

6. Concluding remarks

Studies in the last ten to fifteen years have led to a detailed characterization of nitrogen control mechanisms in *C. glutamicum* and related species. Especially global analysis techniques such as transcriptome analyses, proteome profiling, and bioinformatic approaches led to new insights into nitrogen metabolism and its regulation. However, crucial questions have still to be answered. These include the sensor proteins involved, the signals perceived, and the interaction of nitrogen metabolism with other regulatory networks such as carbon regulation and others.

Acknowledgements

The authors wish to thank all former and current members of the laboratory for their contributions. Figure 8.1 was kindly provided by J. Amon (Erlangen). The continuous financial support of *Degussa GmbH*, the *Bundesministerium für Bildung und Forschung* and the *Deutsche Forschungsgemeinschaft* is gratefully acknowledged.

References

Arcondeguy, T., Jack, R., and Merrick, M. (2001). PII signal transduction proteins, pivotal players in microbial nitrogen control. Microbiol. Mol. Biol. Rev. *65*, 80-105.

Atkinson, M.R., Blauwkamp, T.A., Bondarenko, V., Studitsky, V., and Ninfa, A. (2002). Activation of the *glnA*, *glnK*, and *nac* promoters as *Escherichia coli* undergoes the transition from nitrogen excess growth to nitrogen starvation. J. Bacteriol. *184*, 5358-5363.

Beckers, G., Nolden, L., and Burkovski, A. (2001). Glutamate synthase of *Corynebacterium glutamicum* is not essential for glutamate synthesis and is regulated by the nitrogen status. Microbiology *147*, 2961-2970.

Beckers, G., Bendt, A.K., Krämer, R., and Burkovski, A. (2004). Molecular identification of the urea uptake system and transcriptional analysis of urea transporter- and urease-encoding genes in *Corynebacterium glutamicum*. J. Bacteriol. *186*, 7645-7652.

Beckers, G., Strösser, J., Hildebrandt, U., Kalinowski, J., Farwick, M., Krämer, R., and Burkovski, A. (2005). Regulation of AmtR-controlled gene expression in *Corynebacterium glutamicum*: mechanism and characterization of the AmtR regulon. Mol. Microbiol. *58*, 580-595.

Bendt, A.K., Becker, G., Silberbach, M., Wittmann, A., and Burkovski, A. (2004). Utilization of creatinine as an alternative nitrogen source in *Corynebacterium glutamicum*. Arch. Microbiol. *181*, 443-450.

Börmann, E.R., Eikmanns, B.J., and Sahm, H. (1992). Molecular analysis of the *Corynebacterium glutamicum gdh* gene encoding glutamate dehydrogenase. Mol. Microbiol. *6*, 317-326.

Börmann-El Kholy, E.R., Eikmanns, B.J., Gutmann, M., and Sahm, H. (1993). Glutamate dehydrogenase is not essential for glutamate formation in *Corynebacterium glutamicum*. Appl. Environ. Microbiol. *59*, 2329-2331.

Brune, I., Brinkrolf, K., Kalinowski, J., Pühler, A., and Tauch, A. (2005). The individual and common repertoire of DNA-binding transcriptional regulators of *Corynebacterium glutamicum*, *Corynebacterium efficiens*, *Corynebacterium diphtheriae* and *Corynebacterium jeikeium* deduced from the complete genome sequences. BMC Genomics *6*, 86.

Burkovski, A. (2003a). Ammonium assimilation and nitrogen control in *Corynebacterium glutamicum* and its relatives: an example for new regulatory mechanisms in actinomycetes. FEMS Microbiol. Rev. *27*, 617-628.

Burkovski, A. (2003b). I do it my way: Regulation of ammonium uptake and ammonium assimilation in *Corynebacterium glutamicum*. Arch. Microbiol. *179*, 83-88.

Burkovski, A. (2005). Nitrogen metabolism and its regulation. In: Handbook of *Corynebacterium glutamicum*, M. Bott, and L. Eggeling (eds.) (Boca Raton, FL., CRC Press LLC) p. 333-349.

Burkovski, A. (2007). Nitrogen control in *Corynebacterium glutamicum*: proteins, mechanisms, signals. J. Microbiol. Biotechnol. *17*, 187-194.

Burkovski, A., Weil, B., and Krämer, R. (1996). Characterization of a secondary uptake system for L-glutamate in *Corynebacterium glutamicum*. FEMS Microbiol. Lett. *136*, 169-173.

Cerdeno-Tarraga, A.M., Efstratiou, A., Dover, L.G., Holden, M.T.G., Pallen, M., Bentley, S.D., Besra, G.S., Churcher, C., James, K.D., De Zoysa, A. et al. (2003). The complete genome sequence and analysis of *Corynebacterium diphtheriae* NCTC13129. Nucleic Acids Res. *31*, 6516-6523.

Cole, S.T., Eiglmeier, K., Parkhill, J., James, K.D., Thomson, N.R., Wheeler, P.R., Honoré, N., Garnier, T., Churcher, C., Harris, D., et al. (2001). Massive gene decay in the leprosy bacillus. Nature *409*, 1007-1011.

Dröge, P., and Müller-Hill, B. (2001). High local protein concentrations at promoters: strategies in prokaryotic and eukaryotic cells. Bioessays *23*, 179-183.

Fudou, R., Jojima, Y., Seto, A., Yamada, K., Rimura, E., Nakamatsu, T., Hirashi, A., and Yamanaka, S. (2002). *Corynebacterium efficiens* sp. Nov., a glutamic-acid-producing species from soil and plant material. Int. J. Syst. Evol. Microbiol. *52*, 1127-1131.

Hesketh, A., Fink, D., Gust, B., Rexer, H.-U., Scheel, B., Chater, K., Wohlleben, W., and Engels, A. (2002). The GlnD and GlnK homologues of *Streptomyces coelicolor* A3(2) are functionally dissimilar to their nitrogen regulatory system counterparts from enteric bacteria. Mol. Microbiol. *46*, 319-330.

Huang, C.H., and Peng, J. (2005). Evolutionary conservation and diversification of Rh family genes and proteins. Proc. Natl. Acad. Sci. USA *102*, 15512-15517.

Ikeda, M., and Nakagawa, S. (2003). The *Corynebacterium glutamicum* genome: features and impacts on biotechnological processes. Appl. Microbiol. Biotechnol. *62*, 99-109.

Jakoby, M., Tesch, M., Sahm, H., Krämer, R., and Burkovski, A. (1997). Isolation of the *Corynebacterium glutamicum glnA* gene encoding glutamine synthetase I. FEMS Microbiol. Lett. *154*, 81-88.

Jakoby, M., Krämer, R., and Burkovski, A. (1999). Nitrogen regulation in *Corynebacterium glutamicum*: Isolation of genes involved and biochemical characterization of corresponding proteins. FEMS Microbiol. Lett. *173*, 303-310.

Jakoby, M., Nolden, L., Meier-Wagner, J., Krämer, R., and Burkovski, A. (2000). AmtR, a global repressor in the nitrogen regulation system of *Corynebacterium glutamicum*. Mol. Microbiol. *37*, 964-977.

Javelle, A., and Merrick, M. (2005). Complex formation between AmtB and GlnK: an ancestral role in prokaryotic nitrogen control. Biochem. Soc. Trans. *33*, 170-172.

Javelle, A., Thomas, G., Marini, A.M., Krämer, R., and Merrick, M. (2005). *In vivo* functional characterization of the *E. coli* ammonium channel AmtB: evidence for metabolic coupling to glutamine synthetase. Biochemical J. *390*, 215-222.

Javelle, A., Lupo, D., Li, X.-D., Merrick, M., Chami, M., Ripoche, P., and Winkler, F.K. (2007). Structural and mechanistic aspects of Amt/Rh proteins. J. Struct. Biol. *158*, 472-481.

Kalinowski, J. (2005). The genomes of amino acid-producing corynebacteria. In: Handbook of *Corynebacterium glutamicum*, M. Bott, and L. Eggeling (eds.) (Boca Raton, FL., CRC Press LLC) p. 37-56.

Kalinowski, J., Bathe, B., Bartels, D., Bischoff, N., Bott, M., Burkovski, A., Dusch, N., Eggeling, L., Eikmanns, B.J., Gaigalat, L., Goesmann, A., Hartmann, M., Huthmacher, K., Krämer, R., Linke, B., McHardy, A.C., Meyer, F., Möckel, B., Pfefferle, W., Pühler, A., Rey, D., Rückert, C., Rupp, O., Sahm, H., Wendisch, V.F., Wiegräbe, I., and Tauch, A. (2003). The complete *Corynebacterium glutamicum* ATCC 13032 genome sequence and its impact on the production of L-aspartate-derived amino acids and vitamins. J. Biotechnol. *104*, 5-25.

Kronemeyer, W., Peekhaus, N., Krämer, R., Eggeling, L., and Sahm, H. (1995). Structure of the *gluABCD* cluster encoding the glutamate uptake system of *Corynebacterium glutamicum*. J. Bacteriol. *177*, 1152-1158.

Liebl, W. (2005) *Corynebacterium* taxonomy. In: Bott, M. and L. Eggeling (eds.) Handbook of *Corynebacterium glutamicum*. (CRC Press LLC, Boca Raton, FL.), p. 9-34.

Meier-Wagner, J., Nolden, L., Jakoby, M., Siewe, R., Krämer, R., and Burkovski, A. (2001). Multiplicity of ammonium uptake systems in *Corynebacterium glutamicum*: Role of Amt and AmtB. Microbiology *147*, 135-143.

Merkens, H., Beckers, G., Wirtz, A., and Burkovski, A. (2005). Vanillate metabolism in *Corynebacterium glutamicum*. Curr. Microbiol. *51*, 59-65.

Müller, J., Oehler, S., and Müller-Hill, B. (1996). Repression of *lac* promoter as a function of distance, phase and quality of an auxiliary *lac* operator. J. Mol. Biol. *257*, 21-29.

Müller, T., Strösser, J., Buchinger, S., Nolden, L., Wirtz, A., Krämer, R., and Burkovski, A. (2006). Mutation-induced metabolite pool alterations in *Corynebacterium glutamicum*: towards the identification of nitrogen control signals. J. Biotechnol. *126*, 440-453.

Müller-Hill, B. (1998). The function of auxiliary operators. Mol. Microbiol. *29*, 13-18.

Müller-Hill, B. (2006). What is life? The paradigm of DNA and protein cooperation at high local concentrations. Mol. Microbiol. *60*, 253-255.

Niebisch, A., Kabus, A., Schultz, C., Weil, B. and Bott, M. (2006). Corynebacterial protein kinase G controls 2-oxoglutarate dehydrogenase activity via the phosphorylation status of the OdhI protein. J. Biol. Chem. *281*, 12300-12307.

Nolden, L., Beckers, G., Möckel, B., Pfefferle, W., Nampoothiri, K.M., Krämer, R., and Burkovski, A. (2000). Urease of *Corynebacterium glutamicum*: organization of corresponding genes and investigation of activity. FEMS Microbiol. Lett. *189*, 305-310.

Nolden, L., Farwick, M., Krämer, R., and Burkovski, A. (2001a). Glutamine synthetases of *Corynebacterium glutamicum*: transcriptional control and regulation of activity. FEMS Microbiol. Lett. *201*, 91-98.

Nolden, L., Ngouto-Nkili, C.-E., Bendt, A.K., Krämer, R., and Burkovski, A. (2001b). Sensing nitrogen limitation in *Corynebacterium glutamicum*: The role of *glnK* and *glnD*. Mol. Microbiol. *42*, 1281-1295.

Nolden, L., Beckers, G., and Burkovski, A. (2002). Nitrogen assimilation in *Corynebacterium diphtheriae*: pathways and regulatory cascades. FEMS Microbiol. Lett. *208*, 287-293.

Puskás, L.G., Inui, M., and Yukawa, H. (2000). Structure of the urease operon of *Corynebacterium glutamicum*. DNA Seq. 11, 383-394.

Ramos, J.L., Martinez-Bueno, M., Molina-Henares, A.J., Teran, W., Watanabe, K., Zhang, X., Gallegos, M.T., Brennan, R., and Tobes, R. (2005). The TetR family of transcriptional repressors. Microbiol. Mol. Biol. Rev. *69*, 326-356.

Schaffer, S., and Burkovski, A. (2005). Genome-based approaches: proteomics. In: Handbook of *Corynebacterium glutamicum*, M. Bott and L. Eggeling (eds.) (Boca Raton, FL, CRC Press LLC) pp. 99-118.

Schmid, R., Uhlemann, E.-M., Nolden, L., Wersch, G., Hecker, R., Hermann, T., Marx, A., and Burkovski, A. (2000). Response to nitrogen starvation in *Corynebacterium glutamicum*. FEMS Microbiol. Lett. *187*, 83-88.

Schultz, C., Niebisch, A., Gebel, L., and Bott, M. (2007). Glutamate production by *Corynebacterium glutamicum*: dependence on the oxoglutarate dehydrogenase inhibitor protein OdhI and protein kinase PknG. Appl. Microbiol. Biotechnol. *76*, 691-700.

Schulz, A.A., Collett, H.J., and Reid, S.J. (2001). Nitrogen and carbon regulation of glutamine synthetase and glutamate synthease in *Corynebacterium glutamicum* ATCC 13032. FEMS Microbiol. Lett. *205*, 361-367.

Shiio, I., and Ozaki, H. (1970). Regulation of nicotinamide adenine dinucleotide phosphate-specific glutamate dehydrogenase from *Brevibacterium flavum*, a glutamate-producing bacterium. J. Biochem. *68*, 633-647.

Siewe, R.M., Weil, B., and Krämer, R. (1995). Glutamine uptake by a sodium-dependent secondary transport system in *Corynebacterium glutamicum*. Arch. Microbiol. *164*, 98-103.

Siewe, R.M., Weil, B., Burkovski, A., Eikmanns, B.J., Eikmanns, M., and Krämer, R. (1996). Functional and genetic characterization of the (methyl)ammonium uptake carrier of *Corynebacterium glutamicum*. J. Biol. Chem. *271*, 5398-5403.

Silberbach, M., Hüser, A., Kalinowski, J., Pühler, A., Walter, B., Krämer, R., and Burkovski, A. (2005a). DNA microarray analysis of the nitrogen starvation response of *Corynebacterium glutamicum*. J. Biotechnol. *119*, 357-367.

Silberbach, M., Schäfer, M., Hüser, A., Kalinowski, J., Pühler, A., Krämer, R., and Burkovski, A. (2005b). Adaptation of *Corynebacterium glutamicum* to ammonium-limitation: a global analysis using transcriptome and proteome techniques. Appl. Environ. Microbiol. *71*, 2391-2402.

Silberbach, M., and Burkovski, A. (2006). Application of global analysis techniques to *Corynebacterium glutamicum*: new insights into nitrogen regulation. J. Biotechnol. *126*, 101-110.

Soupene, E., He, L., Yan, D., and Kustu, S. (1998). Ammonium acquisition in enteric bacteria: physiological role of the ammonium/methylammonium transport B (AmtB) protein. Proc. Natl. Acad. Sci. USA *95*, 7030-7034.

Soupene, E., Lee, H., and Kustu, S. (2002). Ammonium/methylammonium transport (Amt) proteins facilitate diffusion of NH_3 bidirectionally. Proc. Natl. Acad. Sci. USA *99*, 3926-3931.

Strösser, J., Lüdke, A., Schaffer, S., Krämer, R., and Burkovski, A. (2004). Regulation of GlnK activity: modification, membrane sequestration and proteolysis as regulatory principles in the network of nitrogen control in *Corynebacterium glutamicum*. Mol. Microbiol. *54*, 132-147.

Tauch, A., Kaiser, O., Hain, T., Goesmann, A., Weisshaar, B., Albersmeier, A., Bekel, T., Bischoff, N., Brune, I., Chakraborty, T., Kalinowski, J., Meyer, F., Rupp, O., Schneiker, S., Viehoever, P., and Pühler, A. (2005). Complete genome sequence and analysis of the multiresistant nosocomial pathogen *Corynebacterium jeikeium* K411, a lipid-requiring bacterium of the human skin flora. J. Bacteriol. *187*, 4671-4682.

Tesch, M., Eikmanns, B.J., de Graaf, A.A., and Sahm, H. (1998). Ammonia assimilation in *Corynebacterium glutamicum* and a glutamate dehydrogenase-deficient mutant. Biotechnol. Lett. *20*, 953-957.

Thomas, G., Coutts, G., and Merrick, M. (2000). The *glnKamtB* operon, a conserved gene pair in prokaryotes. Trends Genet. *16*, 11-14.

Trötschel, C., Kandirali, S., Diaz-Achirica, P., Meinhardt, A., Morbach, S., Krämer, R., and Burkovski, A. (2003). GltS, the sodium-coupled L-glutamate uptake system of *Corynebacterium glutamicum*: identification of the corresponding gene and impact on L-glutamate production. Appl. Microbiol. Biotechnol. *60*, 738-742.

Vissa, V.D., and Brennan, P.J. (2001). The genome of *Mycobacterium leprae*: a minimal mycobacterial gene set. Genome Biol. *2*, reviews1023.1-1023.8.

Walter, B., Hänßler, E., Kalinowski, J., and Burkovski, A. (2007). Nitrogen metabolism and nitrogen control in corynebacteria: variations of a common theme. J. Mol. Microbiol. Biotechnol. *12*, 131-138.

Wendisch, V.F. (2003). Genome-wide expression analysis in *Corynebacterium glutamicum* using DNA microarrays. J. Biotechnol. 104, 273-285.

Winkler, F.K. (2006). Amt/MEP/Rh proteins conduct ammonia. Pflugers Arch. *451*, 701-707.

Zheng, L., Kostrewa, D., Bernèche, S., Winkler, F.K., and Li, X.-D. (2004). The mechanism of ammonia transport based on the crystal structure of AmtB of *E. coli*. Proc. Natl. Acad. Sci. USA *101*, 17090-17095.

From: Corynebacteria: Genomics and Molecular Biology. Edited by: Andreas Burkovski.

Phosphorus Metabolism and its Regulation

9

Volker F. Wendisch and Michael Bott

Abstract

The phosphorus metabolism of *Corynebacterium glutamicum* and its regulation have been studied to some detail using a combination of genetic, biochemical and transcriptome analyses. *C. glutamicum* utilizes a number of inorganic and organic phosphates as sources of phosphorus. When orthophosphate (P_i), its preferred phosphorus source, is abundant, *C. glutamicum* accumulates polyphosphate. When P_i becomes scarce, a complex genetic response is elicited enabling *C. glutamicum* to scavenge low P_i concentrations, to take up transportable organophosphates and to access P_i from nontransportable P_i esters. The two-component signal transduction system PhoS-PhoR serves a role in the transcriptional regulation of the P_i starvation-inducible genes.

1. Sources of phosphorus for growth of *C. glutamicum*

Phosphorus makes up 1.5 to 2.1 % of the *C. glutamicum* cell dry weight (Liebl, 2005). Phosphorus is contained in the polymers DNA, RNA and inorganic polyphosphate (polyP). Phosphorous low-molecular weight compounds include inorganic orthophosphate (P_i), inorganic pyrophosphate, sugar phosphates, nucleotides, and phospholipids. *C. glutamicum* requires a source of phosphorus for growth and P_i was shown to sustain the highest growth rates, thus being the preferred source of phosphorus for growth. The Monod constant, i.e. the P_i concentration supporting growth with a half-maximal growth rate, is about 0.1 mM (Ishige et al., 2003). Besides P_i, *C. glutamicum* is able to use the following inorganic phosphates: pyrophosphate, trimetaphosphate and polyphosphate. The sugar phosphates glucose-6-phosphate, glucose-1-phosphate, fructose-6-phosphate, and ribose-5-phosphate, as well as phosphoenolpyruvate, glycerol-3-phosphate, 3-phosphoglycerate, phosphocholine, phosphoserine, AMP, UDP-glucose, and the systemic herbicide glyphosate (N-(phosphonomethyl) glycine) may serve as sources of phosphorus for *C. glutamicum* (Wendisch and Bott, 2005; Cuylen and Wendisch, unpublished). The phosphonates aminoethylphosphonate and ethylphosphonate did not serve as P sources.

2. Uptake of phosphorous compounds

According to the genome annotation (Kalinowski et al., 2003), three uptake systems for P_i exist in *C. glutamicum*, namely PitA, PstSCAB, and NptA. None of them has been analyzed biochemically yet. The protein encoded by *pitA* (cg0545) shows 28% sequence identity to PitA of *Escherichia coli* and therefore is proposed to be a low-affinity, high-velocity phosphate uptake system. *E. coli* PitA is the major P_i uptake system when P_i is present in excess (Rosenberg et al., 1977; Rosenberg et al., 1979). Divalent cations such as Mg^{2+} or Ca^{2+} are required for PitA activity and PitA presumably transports a soluble neutral metal phosphate complex ($MeHPO_4$) in symport with a proton (van Veen et al., 1994). The apparent K_m values of PitA for P_i transport vary between 25 μM (Rosenberg et al., 1977) and 38 μM (Willsky and Malamy, 1980) in whole cells and 11.9 μM in vesicles (van Veen et al., 1994) and V_{max} was determined to be 55 nmol P_i min^{-1} (mg protein)$^{-1}$ (Willsky and Malamy, 1980). The absence of PitA neither has an effect on growth with P_i nor on Pho regulon control in *E. coli* (Wanner, 1996). DNA microarray analysis indicates that in *C. glutamicum pitA* expression is reduced after a shift from P_i excess to P_i starvation (Schaaf and Bott, 2007).

The *C. glutamicum pstSCAB* operon (cg2846-2843) codes for an ABC-type transporter catalyzing ATP-driven high-affinity phosphate uptake. PstS is as an extracytoplasmic phosphate binding protein with a lipoprotein signal sequence (Ala-Leu-Val-Ala-Cys*-Ser) typical of Gram-positive bacteria (Sutcliffe and Harrington, 2002). *C. glutamicum* PstS shows 30 % amino acid sequence identity to *E. coli* PstS, for which a K_d value for P_i of ~1 μM was determined (Kubena et al., 1986). PstA and PstC are integral membrane proteins, and PstB contains the features of an ATP-binding protein. The *C. glutamicum* PstA, PstB and PstC proteins possess up to 56 % amino acid sequence identity to their *E. coli* orthologs. The apparent K_m value for P_i of the *E. coli* Pst system was reported to be around 0.2 μM (Rosenberg et al., 1977; Willsky and Malamy, 1980). As is true for many bacteria, the *pstSCAB* operon of *C. glutamicum* is strongly induced by phosphate starvation (Ishige et al., 2003)

In addition to PitA and PstSCAB, a third putative phosphate transporter exists in *C. glutamicum*. The protein encoded by cg3039 (*nptA*) shows 29 % sequence identity to the Na^+-phosphate symporter NptA of *Vibrio cholerae* (Lebens et al., 2002). For *V. cholerae* NptA, apparent K_m values of 300 μM for P_i and 75 mM for Na^+ and a V_{max} value with respect to P_i of 9 pmol min^{-1} (mg protein)$^{-1}$ were determined (Lebens et al., 2002). DNA microarray studies gave no evidence for a P_i-dependent regulation of *nptA* expression in *C. glutamicum* (Wendisch and Bott, unpublished data).

Many bacteria have uptake systems for organophosphates and these compounds can serve either as carbon source or as phosphorus source or both. In *C. glutamicum*, at least one such system is present. The *ugpAEBC* operon (cg1568-1571) presumably encodes an ABC transporter for the uptake of glycerol-3-phosphate. UgpB functions as an extracytoplasmic glycerol-3-phosphate binding protein and contains a characteristic lipoprotein signal sequence (Thr-Leu-Ala-Ala-Cys*-Ala), UgpA and UgpE are integral membrane proteins, and UgpC contains features of an ATP-binding protein. The corresponding proteins show up to 49 % (UgpC) amino acid sequence identity to the corresponding *E. coli* orthologs. The *E. coli* Ugp system has a K_m for glycerol-3-phosphate of ~1.5 μM and a V_{max} of ~3 nmol min^{-1} (mg protein)$^{-1}$ (Schweizer et al., 1982). Remarkably, the *E. coli* Ugp system cannot transport glycerol-3-phosphate for use as carbon source, presumably because high intracellular P_i inhibits uptake (Brzoska et al., 1994). The *C. glutamicum ugpAEBC* operon is induced under phosphate-limiting conditions (Ishige et al., 2003). Homologs of other organophosphate uptake systems such as GlpT for glycerol-3-phosphate, UhpT for hexose-6-phosphates, and PgtP for phosphoenolpyruvate, 2-phosphoglycerate, and 3-phosphoglycerate (Wanner, 1996) are not present in *C. glutamicum.* However, the *pctABCD* operon (*cg1652-1649*) encoding an ABC transporter was induced after P_i starvation (Ishige et al., 2003). Sequence comparisons revealed that the corresponding proteins have weak similarity to phosphonate ABC transporters. The periplasmic binding protein PctA contains a lipoprotein signal sequence (Ala-Leu-Val-Gly-Cys*-Ser). The PctABCD ABC transporter might be involved in the uptake of an as-yet-unidentified P compound.

3. Mobilization of nontransportable sources of phosphorus

C. glutamicum secretes few enzymes into the medium, which can be seen from its limited or absent amylolytic, cellulolytic, proteolytic, lipolytic, and nucleolytic activities (Yukawa et al., 2007). However, under phosphate-limiting conditions *C. glutamicum*, as many bacteria, secretes phosphatases which hydrolyze extracellular organophosphates so that the liberated free phosphate can subsequently be taken up by the P_i transport systems.

The *C. glutamicum* gene *ushA* (cg0397) was shown to encode a secreted enzyme which is active as 5′-nucleotidase (EC 3.1.3.5) and UDP-sugar hydrolase (EC 3.6.1.45) (Rittmann et al., 2005). The enzyme shows maximal activity at pH 5.0 with a K_m value of 0.18 mM for AMP. *C. glutamicum* UshA also accepts other (desoxy)nucleotides as substrates. Among the 5′-nucleoside monophosphates, GMP, IMP, and XMP, which contain 6-oxopurines, led to higher activities than the 6-aminopurine-containing AMP and the pyrimidine nucleoside monophosphates UMP and CMP. The specific activities observed for ADP, ATP, and dATP were higher for the di- and triphosphates than for the monophosphates. The cyclic mononucleotides 2′, 3′-cAMP and 3′, 5′-cAMP were not hydrolyzed by *C. glutamicum* 5′-nucleotidase. Thus, *C. glutamicum* 5′-nucleotidase exhibits broad substrate specificity, accepting 5′-nucleoside phosphates as well as 5′-deoxynucleoside phosphates, but not cyclic nucleotides. Transcription of *C. glutamicum ushA* starts at a cytosine 60 bp upstream of the translational start codon, which is preceded by a ribosome binding site (Kocan et al., 2006). The promoter of *ushA* contains a conserved –10 region but lacks a conserved –35 region. Upon phosphate starvation, *ushA* is induced at the transcript level almost 20-fold (Ishige et al.

2003) and the specific 5′-nucleotidase activity increased from <0.01 to 0.39 μmol min^{-1} mg^{-1} (Rittmann et al., 2005). UshA was shown to be required for growth of *C. glutamicum* on AMP and UDP-glucose as sole sources of phosphorus (Rittmann et al., 2005). In *E. coli ushA* is not phosphate starvation-inducible and dephosphorylation of 5′-AMP by the nucleotidase activity of UshA is part of the nucleotide salvage pathway and required for growth of *E. coli* on 5′-AMP as sole carbon source (Neuhard and Kelln, 1996; Zalkin and Nygaard, 1996). However, AMP is not a carbon source for *C. glutamicum* (Rittmann et al., 2005). Thus, in contrast to its ortholog from *E. coli*, UshA from *C. glutamicum* is an important component of the phosphate starvation response and necessary to access nucleotides and related compounds as sources of phosphorus.

The *nucH* gene (cg2668) encodes a protein which could function as nuclease, as it shows 31% sequence identity to NucH of *Aeromonas hydrophila* (Dodd and Pemberton, 1996). The predicted signal peptide and a single C-terminal transmembrane helix suggest that *C. glutamicum* NucH is a membrane protein. However, processing of the protein to yield a cell wall-associated or secreted protein cannot be ruled out. The *nucH* gene is strongly induced after P_i limitation (Ishige et al., 2003) and may serve a role in liberating P_i from extracellular nucleic acids.

The *phoC* gene (cg3393) encodes a protein containing a putative phosphoesterase domain (PF00149 in the PFAM database) in the carboxyterminal region. The three annotations available for the *C. glutamicum* genome predict three different start codons for *phoC*, leading to proteins of either 1461 amino acid residues (cg3393), or 1497 amino acid residues (NCgl2959), or 1344 amino acid residues (Cgl3064). The first one contains a signal peptide, whereas the latter two do not. At the carboxyterminal end, a putative transmembrane helix is present, in front of which a putative sortase motif (APGTG) was detected. Therefore, PhoC may function as a cell wall-associated phosphatase. Orthologous proteins are found in *Corynebacterium efficiens* (CE2911, 72 % amino acid sequence identity) and in *Bacillus subtilis* (YvnB, 33 % amino acid sequence identity), but like the *C. glutamicum* protein they have not been functionally characterized. The *phoC* gene showed a 12-fold increased mRNA level 120 min after a shift from P_i excess to P_i starvation, indicating that PhoC is involved in the adaptation process (Ishige et al., 2003).

Apart from PhoC, two additional secreted phosphatases are present in *C. glutamicum.* The protein encoded by the *phoB* gene (cg2700) shows 36 % amino acid sequence identity to the alkaline phosphatase PhoB of *B. subtilis* (Hulett, 2002) and the protein encoded by the *phoD* gene (cg2485) shows 38 % amino acid sequence identity to the alkaline phosphatase PhoD of *B. subtilis* (Eder et al., 1996) and 36 % identity to the PhoD ortholog from *Streptomyces griseus* (Moura et al., 2001). In contrast to *B. subtilis*, the *C. glutamicum phoB* and *phoD* genes were only slightly induced in some of the phosphate starvation experiments (*phoB*) or not at all (*phoD*) (Wendisch and Bott, unpublished results). Therefore, the function of PhoB and PhoD in *C. glutamicum* is unclear at present.

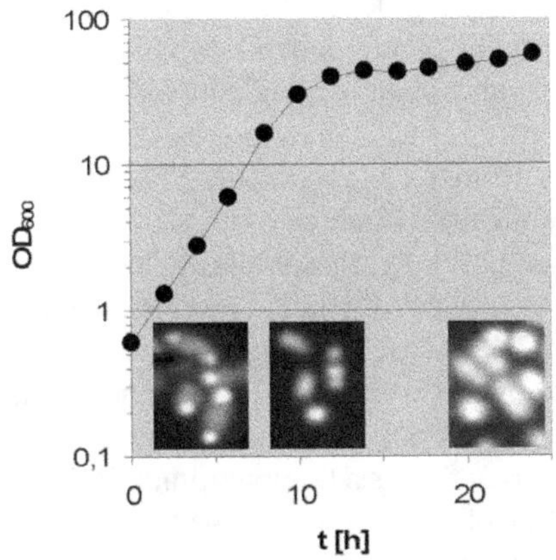

Figure 9. 1. Granular and cytoplasmic polyphosphate during growth of *C. glutamicum* in shake flasks. The growth curve is shown with insets showing the green channel of photographs of DAPI-stained *C. glutamicum* cells. By DAPI staining and ^{31}P NMR analysis it was shown that *C. glutamicum* ATCC 13032 primarily contains granular polyphosphate, which cannot be detected by ^{31}P NMR and which is located in 1 to 2 foci per cell, during the early exponential growth phase. Little polyphosphate was observed later in the exponential growth phase. In the stationary growth phase cells stained evenly with DAPI and cytoplasmic polyphosphate could be readily detected by ^{31}P NMR.

4. Enzymes of polyphosphate metabolism

C. glutamicum is able to accumulate up to 600 mM polyphosphate (polyP) intracellularly (Lambert et al., 2002), either as soluble polyphosphate or in the form of volutin granules (Pallerla et al., 2005). Volutin granules are primarily formed in the early exponential growth phase (Figure 9. 1) (Klauth et al., 2006) and may occupy up to 37% of the cell volume (Pallerla et al., 2005). Volutin granules have been purified and their composition has been analysed (Pallerla et al., 2005). They can be observed as metachromatic bodies after staining by toluidine blue or Neisser stains or they show green fluorescence after staining with DAPI. Electron dispersive X-rays (EDX) analysis revealed that volutin granules from *C. glutamicum* contain magnesium besides phosphorus. ^{31}P NMR spectroscopy showed that these volutin granules contain polyphosphate with chain lengths of 800-1000 (Lambert and Schoberth, unpublished observation). Volutin granules purified from *C. glutamicum* contain a number of different proteins, but only one abundant protein associated with volutin granules has been identified, the cg2091 gene product likely coding for polyP-dependent glucokinase (EC 2.7.1.63). The functional significance of this finding is not yet known (Pallerla et al., 2005). In *Agrobacterium tumefaciens* and *Rhodospirillum rubrum* volutin granula/acidocalcisomes were shown to contain pyrophosphatase (Seufferheld et al., 2004). The formation of volutin granules is stimulated by $MgCl_2$, but $MgCl_2$ did not significantly influence the expression of key genes involved in polyP metabolism (Pallerla et al., 2005). Similarly, phosphate starvation did not significantly influence the expression of these genes (Ishige et al., 2003). By inference, synthesis and degradation of polyP in *C. glutamicum* appears to be primarily regulated on the enzyme activity level rather than by gene expression control.

Inspection of the *C. glutamicum* genome sequence revealed several candidate genes encoding enzymes of polyP metabolism (Wendisch and Bott, 2005). According to sequence similarities to characterized enzymes, it was proposed that two polyphosphate kinases (EC 2.7.4.1) are encoded by *cg1046/ppk2A* and cg3007/*ppk2B*, two exopolyphosphatases (EC 3.6.1.11) by cg0488/*ppxI* and cg1115/*ppxII*, a polyphosphate-dependent glucokinase (EC 2.7.1.63) by cg2091/*ppgK* and a polyphosphate-dependent NAD kinase (EC 2.7.1.23) by cg1601/*ppnK*.

One of the polyphosphate kinases from *C. glutamicum*, PPK2B, has been purified and characterized biochemically (Lindner et al., 2007). This bacterium lacks the classical enzyme of polyP synthesis, class I polyphosphate kinase (PPK1), but it possesses two genes for putative class II polyphosphate kinases. Deletion of cg3007 decreased polyphosphate kinase activity and cellular polyP content, while overexpression of cg3007 increased both polyphosphate kinase activity and cellular polyP content (Lindner et al., 2007). Purified PPK2B of *C. glutamicum* is active as a homotetramer. PPK2B catalyzes both the synthesis and elongation of polyP requiring ATP or GTP as a polyphosphate kinase and the reverse reaction as a NDP kinase, i.e. polyP-dependent phosphorylation of ADP or GDP to yield ATP or GTP. PPK2B also formed polyP without the need of an added primer, i.e. only from ATP (Lindner et al., 2007). This reaction yielded polyP with an average chain length of about 125 as determined with ^{31}P NMR. The catalytic efficiency of *C. glutamicum* PPK2B was higher in the polyP forming direction than for NTP formation from polyP. PolyP chain elongation by PPK2B was inhibited by AMP, GMP and IMP with inhibition constants in the millimolar range (3.9 – 7.9 mM). Thus, at least to some extent polyP synthesis by PPK2B appears to be regulated by the 'energy charge' of the cell. AMP in the same concentration range (2 mM) activated *C. glutamicum* pyruvate kinase fourfold (Gubler et al., 1994), while the inhibition constant of AMP for *C. glutamicum* fructose-1,6-bisphosphatase was much lower (0.09 mM; Rittmann et al. 2003).

Currently, the role of the gene product of cg1046/*ppk2A* is not clear. Neither deletion nor overexpression of cg1046 changed the specific activity of polyphosphate kinase or cellular polyP content significantly (Lindner et al., 2007). Further studies are required to determine if the cg1046 gene product is active as polyP-dependent NDP kinase (i.e. catalyzing the reverse reaction of polyphosphate kinase) or whether it elongates polyP synthesized by PPK2B. cg0488/*ppxI* and cg1115/*ppxII* code for putative exopolyphosphatases. The gene product of *ppxI* has not been characterized yet, but genetic evidence suggests that it functions as an exopolyphosphatase, as overexpression of *ppxI* reduced the cellular polyP content (Knebel, Schoberth and Wendisch, unpublished results). Preliminary evidence suggests that purified enzyme encoded by *ppxII* hydrolyzes short-chain polyPs (Lindner, Schoberth and Wendisch, unpublished).

The gene product of cg2091/*ppgK* could be identified as protein associated with purified volutin granula (Pallerla et al., 2005). Purified His-tagged PpgK catalyzed polyP-dependent phosphorylation of glucose to glucose-6-phosphate (Lindner, Schoberth and Wendisch, unpublished results). However, its functional role has not yet been determined. This is of particular interest because *C. glutamicum* also possesses an ATP-dependent glucokinase encoded by cg2399/*glk* (Park et al., 2000). Expression of *glk* was shown to be constitutive and growth of *C. glutamicum* on complex media to be perturbed in its absence (Park et al., 2000). Although growth on minimal medium containing maltose, a source of intracellular free glucose, was impaired in a *glk* disruptant, some growth still occurred indicating a role of another glucokinase activity for growth on maltose.

C. glutamicum possesses no gene apparently encoding an ATP-dependent NAD kinase, an enzyme required for synthesis of NADP. However, cg1601/*ppnK* encodes a putative polyphosphate-dependent NAD kinase, which therefore might be the only enzyme capable of converting NAD to NADP.

5. Coping with P_i limitation: the role of polyP

When P_i, the preferred source of phosphorus for *C. glutamicum*, is abundant, *C. glutamicum* accumulates polyP (Klauth et al., 2006). Cells in the early exponential growth phase contain polyP in the form of volutin granula, while in the stationary growth phase soluble cytosolic polyP accumulates (Figure 9. 1) (Klauth et al., 2006). Under P_i limiting conditions, *C. glutamicum* is unable to accumulate high polyP concentrations (Klauth et al., 2006). It appears that the cells mobilize P_i from their endogenous stores of polyP, when *C. glutamicum* is shifted from P_i-abundant to P_i-limiting conditions (Lindner et al., 2007). When Δ*ppk2B* mutant cells that are unable to accumulate high intracellular polyP concentrations are shifted from P_i abundant to P_i-limiting conditions, these cells enter the stationary growth phase earlier than wild-type cells and form less biomass (Lindner et al., 2007). Mobilization of P_i from polyP stores is just one way to enable growth of *C. glutamicum* when P_i becomes scarce.

6. Coping with P_i limitation: phosphate starvation-inducible gene expression

Upon exhaustion of P_i supply, a genetic program is started and expression of a group of genes, the so-called phosphate starvation inducible (*psi*) genes, increases. Although intracellular polyP provides enough phosphorus to sustain growth for 4-6 generations in the absence of an external source of P_i, expression of the *psi* genes is induced almost immediately after depletion of extracellular P_i. The *pstSCAB* operon coding for the high-affinity ABC P_i uptake system is induced first, resulting in five-fold and 20-fold increased transcript levels within five and 30 minutes, respectively, after onset of P_i starvation in a medium transfer experiment (Ishige et al., 2003). With a delay of 5-10 minutes, transcript levels of the *ugpAEBC* operon encoding an ABC transporter for uptake of glycerol 3-phosphate and *glpQ1* encoding a glycerophosphoryl diester phosphodiesterase were increased (Ishige et al., 2003). Thirdly, expression of the following genes increased up to 30-fold: *ushA* encoding a secreted enzyme with UDP-sugar hydrolase and 5′-nucleotidase activity (Rittmann et al., 2005), *nucH* encoding a putative secreted nuclease, *phoC* which may encode a cell wall-associated phosphatase, *phoH1* encoding an ATPase of unknown function, and the *pctABCD* operon encoding an ABC transport system which might be involved in the uptake of a yet unknown phosphorus-containing compound (Wendisch and Bott, 2005). From the characterization of this timed transcriptional response, the strategy of *C. glutamicum* to cope with P_i limitation could be deduced. As soon as P_i becomes scarce, induction of the *pstSCAB* operon and synthesis of the high affinity P_i uptake system ensures uptake of low residual P_i from the environment. Secondly, increased expression of *ugpAEBC* and *glpQ1* results in the ability to efficiently liberate glycerol-3-phosphate from lipids and to import this organophosphate into the cell. Similarly, the *pctABCD* operon coding for an uptake system of unknown specificity was induced. Finally and with further delay as compared to induction of *ugpAEBC*, a number of secreted enzymes are produced to mobilize phosphate from nontransportable organosphosphates. The *nucH* gene product likely is a secreted nuclease for degradation of extracellular DNA and RNA to nucleotides. UshA was shown to be active as a 5′-nucleotidase as well as a UDP-sugar hydrolase (Rittmann et al., 2005), thus allowing access to phosphate present in nucleotides. Similarly, *phoC* (cg3393) encoding a putative cell-wall associated phosphoesterase was induced upon P_i limitation. P_i liberated by hydrolysis of nontransportable sources of phosphorus can be taken up into the cell by the P_i uptake systems.

7. Coping with P_i limitation: a role for the PhoS-PhoR two-component system

The *psi* genes are induced when phosphate becomes limiting, showing that the cell is able to sense phosphate availability. In bacteria where the phosphate starvation response has been studied, two-component signal transduction systems were found to play a key role. In their simplest form two-component systems consist of two modular proteins, a histidine kinase and a response regulator. The histidine kinase responds to a certain stimulus by autophosphorylation of a conserved histidine residue, and the phosphoryl group is subsequently transferred to a conserved aspartate residue of the response regulator, which then mediates changes in gene expression or cell behavior (for reviews see (Bourret et al., 1991; Mascher et al., 2006; Parkinson and Kofoid, 1992; Stock et al., 1989). The best-studied two-component systems involved in phosphate starvation are the PhoR-PhoB system of *E. coli* and the PhoR-PhoP system of *B. subtilis* (for reviews see (Wanner, 1996) and (Hulett, 2002)). *C. glutamicum* contains 13 two-component systems, whose genomic organisation and protein domain composition have recently been described (Kocan et al., 2006). Twelve of these are non-essential, as the corresponding genes could be deleted, whereas one appears to be essential (Kocan et al., 2006).

In order to test whether one of the twelve non-essential two-component systems of *C. glutamicum* is involved in the phosphate starvation response, the corresponding deletion mutants were cultivated in glucose minimal medium with different phosphate concentrations. One of the mutants, strain Δ*phoRS*, showed a growth defect under P_i limitation, but not under P_i excess. A second hint pointing towards a role of PhoRS in the P_i starvation response was the finding that *phoRS* were the only two-component genes whose expression was rapidly (within 10 minutes) and transiently induced after a shift to P_i starvation (Ishige et al., 2003). Transcriptome comparisons with DNA

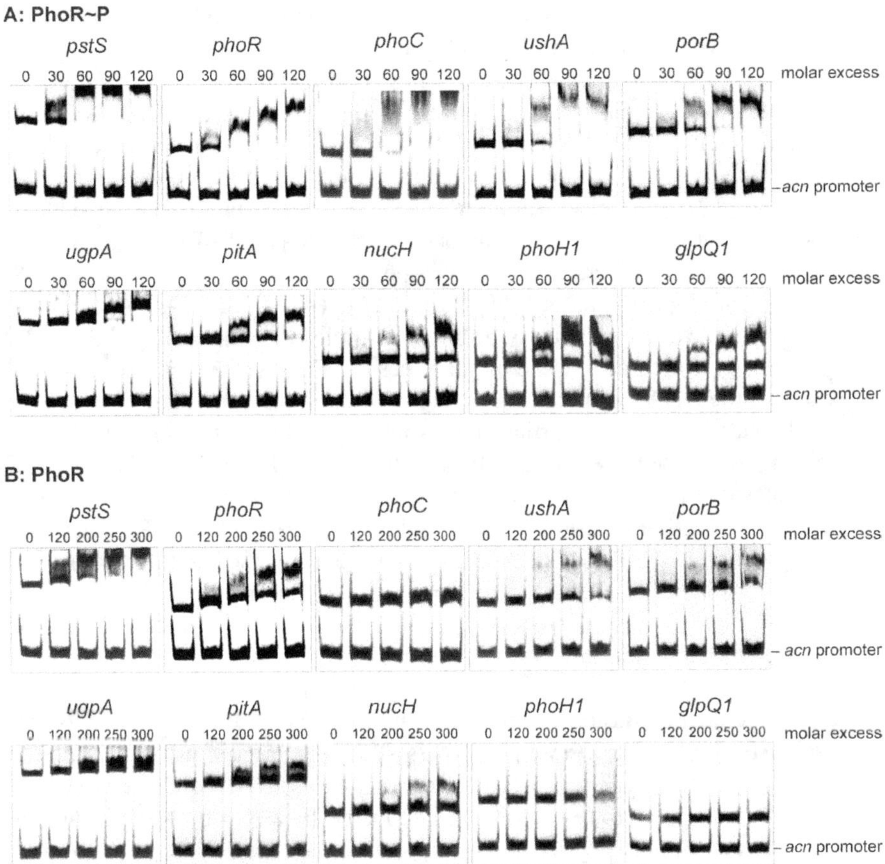

Figure 9. 2. Binding of PhoR~P and PhoR to target promoters. In (A) DNA fragments (14 – 38 nM) covering the promoter regions of *pstS*, *phoR*, *phoC*, *ushA*, *porB*, *ugpA*, *pitA*, *nucH*, *phoH1*, *glpQ1*, and *acn* (negative control) were incubated for 30 min at room temperature with a 0- to 120-fold molar excess of phosphorylated PhoR (PhoR~P). Subsequently, the reaction mixtures were separated by electrophoresis on a native 15% polyacrylamide gel and the gels were stained with SybrGreen. PhoR was phosphorylated by incubation for 60 minutes at room temperature with a two-fold lower concentration of MBP-PhoSΔ1-246 and 5 mM ATP. In (B) the same DNA fragments as used in A were incubated with a 0- to 300-fold molar excess of unphosphorylated PhoR and then treated as described above. Taken from Schaaf and Bott (2007) with permission.

microarrays showed that in the Δ*phoRS* mutant none of the characteristic *psi* genes except *pstSCAB* (e.g. *ugpAEBC*, *glpQ1*, *phoH1*, *ushA* or *nucH*) was induced within 60 minutes after a shift from P_i excess to P_i limitation (Kocan et al., 2006). In the case of the *pstSCAB* genes, induction was weaker in the Δ*phoRS* mutant than in the wild type. Primer extension experiments confirmed the microarray data for *pstSCAB*, *ugpAEBC*, *ushA*, *nucH*, and *phoRS* (Kocan et al., 2006). According to these results it was suggested that PhoS might function as a phosphate sensor that phosphorylates the response regulator PhoR when P_i becomes limiting. Phosphorylated PhoR in turn then should activate expression of the *psi* genes.

In order to test whether the *psi* genes are direct target genes of PhoR, *in vitro* studies with purified PhoR were performed (Schaaf and Bott, 2007). A fusion protein (MBP-PhoSΔ1-146) composed of the *E. coli* maltose binding protein and the kinase domain of PhoS, which exhibited constitutive auto-phosphorylation activity, was used as a tool for the rapid ATP-dependent phosphorylation of PhoR. Phosphorylated PhoR (PhoR~P) was shown to bind to eight promoters of *psi* genes/operons, i.e. *pstSCAB* (~35x), *phoRS* (~45x), *phoC* (~50x), *ushA* (~60x), *ugpAEBC* (~70x), *nucH*, *phoH1*, *glpQ1* (>120x), but with significantly different affinities as indicated by the values in brackets, which denote the molar excess of PhoR~P required for shifting ~50 % of the DNA fragment (Figure 9. 2). When unphosphorylated PhoR was used instead of (at least partially) phosphorylated PhoR, the binding affinity was always much lower, usually by a factor of five (Figure 9. 2). This clearly indicates that PhoR~P is the active form of the protein.

The induction kinetics of the individual *psi* genes after a P_i downshift (Ishige et al., 2003) may in part be determined by the binding affinity of their promoters for PhoR~P, as a correlation, although not strict, between these two parameters was observed: genes having a high affinity for PhoR are more rapidly induced than genes with a low affinity. The temporal pattern of induction of the different *psi* genes probably allows a highly efficient and economic adaptation to P_i limitation. The first response is the induction of *pstSCAB* and *phoRS*, whose promoters show the highest affinity to PhoR. The ABC transporter PstSCAB allows the cell to scavenge those low P_i concentrations in the environment that cannot be taken up by low-affinity P_i transporters, e.g. PitA. The *psi* genes that are induced later than *pstSCAB* and *phoRS* and whose promoters have a lower affinity for PhoR encode transporters for the uptake of organophosphates, such as glycerol-3-phosphate, or extracytoplasmic enzymes like PhoC, UshA (Rittmann et al., 2005) or NucH that cleave phosphate from organophosphates and thus make it accessible for import by the PstSCAB system.

Microarray data (Ishige et al., 2003), primer extension results (Kocan et al., 2006) and DNA binding studies (Schaaf and Bott, 2007) suggest that the *phoRS* genes are positively autoregulated. Elevated PhoR levels formed through autoinduction might be necessary to control all of the putative target promoters and at the same time could be responsible for a hierarchical organisation of the regulon. Whereas uninduced levels of PhoR~P could be sufficient to induce the high-affinity *pstSCAB* and *phoRS* promoters, elevated levels of PhoR~P might be required for induction of the lower affinity target promoters. Thus, the PhoS-PhoR system may function as a rheostat rather than a simple switch.

In addition to the promoter regions of the *psi* genes, PhoR~P also bound to the promoter regions of *pitA* and *porB* (Figure 9. 2). As outlined above, *pitA* encodes a secondary phosphate transporter and its mRNA level decreased after a shift to P_i limiting conditions. Therefore, PhoR~P might function as a repressor of *pitA*, in contrast to all other genes mentioned above,

No.	Gene	Sequence 5′ → 3′	Position
1	*pstS*	**CCTGTGA**GAATAAT**A**A**GAA**	-194 to -176
2	*phoR*	A**CTGTGAA**TGTA**CTGA**AA**G**	-43 to -61
3		**CCTGTGAA**aat**CCTGTGAA**	-

Figure 9. 3. Alignment of the PhoR binding sites in the promoter regions of *pstSCAB* and *phoRS* and of an artificial binding site (no. 3) with the currently highest affinity for PhoR~P. The latter one represents a perfect 8-bp direct repeat separated by 3-bp. The position is given with respect to the transcriptional start sites of *pstS* and *phoR*. Adapted from Schaaf and Bott (2007) with permission.

for which PhoR~P is assumed to act as an activator. The protein encoded by *porB* was described as an anion-specific channel or porin in the cell wall of *C. glutamicum* (Costa-Riu et al., 2003). As porins were recently reported to be required for P_i uptake in *Mycobacterium smegmatis* (Wolschendorf et al., 2007), PorB might have such a function in *C. glutamicum*. The mRNA level of *porB* was reduced in the Δ*phoRS* mutant not only under P_i limitation, but also under P_i excess, a result which is not yet understood.

Despite the knowledge of 10 promoter regions to which PhoR~P bound *in vitro*, the identification of the exact binding site proved to be very tedious. Recently, the high-affinity binding sites in the *pstSCAB* and the *phoRS* promoters were nailed down to 19-bp motifs which are centered at position -185 and -52 with respect to the transcription start sites of *pstS* and *phoR*, respectively (Figure 9. 3, no. 1 and 2). The *in vivo* relevance of these sites for PhoR-dependent transcriptional activation was supported by reporter gene fusion studies (Schaaf and Bott, 2007; Sorger-Hermann, 2006). An alignment of these two binding sites shows that in particular the 5′-terminal half is very similar (Figure 9. 3). As OmpR-type regulators, to which PhoR belongs, commonly bind to direct repeat sequences, it was suggested that the identified PhoR binding sites could represent two very loosely conserved 8-bp direct repeats separated by a 3-bp spacer (Schaaf and Bott, 2007). This assumption was supported by the finding that an artificial 19-bp binding site (Figure 9. 3, no. 3) composed of two perfect 8-bp direct repeats (identical with the 5′-terminal half of the *pstSCAB* binding site) showed a higher affinity to PhoR~P than the natural binding motif with the highest affinity, i.e. that of the *pst* promoter (Schaaf and Bott, 2007). Moreover, reduction of the spacer between the two repeats from three to two bp almost completely prevented PhoR~P binding (Schaaf and Bott, 2007). As the distance of 11 bp between the centres of the two repeats corresponds approximately to one full turn of the DNA helix, this spacing would allow the binding of two PhoR monomers in a side-by-side manner on the same face of the DNA helix. Whereas the position of the binding site in the *phoRS* promoter allows for a direct interaction of PhoR with the α-subunit of RNA polymerase or with the sigma factor, DNA bending or an interaction with additional regulatory proteins or (an) additional PhoR binding site(s) would be required for such an interaction in the *pstSCAB* promoter. A search for the occurrence of the artificial binding site with the highest affinity to PhoR in the promoter regions of the genes to which PhoR binds *in vitro* revealed that all contained one or two sequences with similarity to the query motif, but it was not yet tested experimentally whether these sequences indeed represent PhoR-binding sites (Schaaf and Bott, 2007).

8. Concluding remarks and outlook

The studies in recent years have shed some light on phosphate metabolism in *C. glutamicum*. Compounds that can serve as phosphorus sources were identified, the properties and biochemistry of polyphosphate were explored in several studies, the phosphate starvation stimulon was identified and the role of the PhoSR two-component system in the phosphate starvation response was characterized. However, the picture is far from being complete. The function(s) of polyphosphate need to be defined further, the functions of several proteins induced upon phosphate starvation have to be characterized biochemically, utilization of organophosphates as phosphorus sources has yet to be studied, the stimulus sensed by the histidine kinase PhoS is still unknown, and there is clear evidence for the involvement of additional transcriptional regulators besides PhoR in the phosphate starvation response that await characterization.

References

Bourret, R.B., Borkovich, K.A., and Simon, M.I. (1991). Signal transduction pathways involving protein phosphorylation in prokaryotes. Annu. Rev. Biochem. *60*, 401-441.

Brzoska, P., Rimmele, M., Brzostek, K., and Boos, W. (1994). The Pho regulon dependent Ugp uptake system for glycerol-3-phosphate in *Escherichia coli* is trans inhibited by P_i. J. Bacteriol. *176*, 15-20.

Costa-Riu, N., Maier, E., Burkovski, A., Krämer, R., Lottspeich, F., and Benz, R. (2003). Identification of an anion-specific channel in the cell wall of the Gram-positive bacterium *Corynebacterium glutamicum*. Mol. Microbiol. *50*, 1295-1308.

Dodd, H.N., and Pemberton, J.M. (1996). Cloning, sequencing, and characterization of the *nucH* gene encoding an extracellular nuclease from *Aeromonas hydrophila* JMP636. J. Bacteriol. *178*, 3926-3933.

Eder, S., Shi, L., Jensen, K., Yamane, K., and Hulett, F.M. (1996). A *Bacillus subtilis* secreted phosphodiesterase/alkaline phosphatase is the product of a Pho regulon gene, *phoD*. Microbiology *142*, 2041-2047.

Gubler, M., Jetten, M., Lee, S.-H., and Sinskey, A.J. (1994). Cloning of the pyruvate kinase gene (*pyk*) of *Corynebacterium glutamicum* and site-specific inactivation of *pyk* in a lysine-producing *Corynebacterium lactofermentum* strain. Appl. Environ. Microbiol. *60*, 2494-2500.

Hulett, F. M. (2002). The Pho regulon, In *Bacillus subtilis* and its closest relatives: from genes to cells, J.A. Sonenshein, and R.M. Losick, eds. (Washington, D.C.: ASM Press), pp. 193-201.

Ishige, T., Krause, M., Bott, M., Wendisch, V.F., and Sahm, H. (2003). The phosphate starvation stimulon of *Corynebacterium glutamicum* determined by DNA microarray analyses. J. Bacteriol. *185*, 4519-4529.

Kalinowski, J., Bathe, B., Bartels, D., Bischoff, N., Bott, M., Burkovski, A., Dusch, N., Eggeling, L., Eikmanns, B. J., Gaigalat, L., *et al.* (2003). The complete

Corynebacterium glutamicum ATCC 13032 genome sequence and its impact on the production of L-aspartate-derived amino acids and vitamins. J. Biotechnol. *104*, 5-25.

Klauth, P., Pallerla, S. R., Vidaurre, D., Ralfs, C., Wendisch, V.F., and Schoberth, S.M. (2006). Determination of soluble and granular inorganic polyphosphate in *Corynebacterium glutamicum.* Appl. Microbiol. Biotechnol. *72*, 1099-1106.

Kocan, M., Schaffer, S., Ishige, T., Sorger-Herrmann, U., Wendisch, V.F., and Bott, M. (2006). Two-component systems of *Corynebacterium glutamicum*: Deletion analysis and involvement of the PhoS-PhoR system in the phosphate starvation response. J. Bacteriol. *188*, 724-732.

Kubena, B. D., Luecke, H., Rosenberg, H., and Quiocho, F.A. (1986). Crystallization and X-ray diffraction studies of a phosphate-binding protein involved in active transport in *Escherichia coli.* J. Biol. Chem. *261*, 7995-7996.

Lambert, C., Weuster-Botz, D., Weichenhain, R., Kreutz, E.W., De Graaf, A.A., and Schoberth, S.M. (2002). Monitoring of inorganic polyphosphate dynamics in *Corynebacterium glutamicum* using a novel oxygen sparger for real time P-31 *in vivo* NMR. Acta Biotechnol. *22*, 245-260.

Lebens, M., Lundquist, P., Soderlund, L., Todorovic, M., and Carlin, N.I.A. (2002). The *nptA* gene of *Vibrio cholerae* encodes a functional sodium-dependent phosphate cotransporter homologous to the type II cotransporters of eukaryotes. J. Bacteriol. *184*, 4466-4474.

Liebl, W. (2005). *Corynebacterium* taxonomy, In Handbook of *Corynebacterium glutamicum*, L. Eggeling, and M. Bott, eds. (Boca Raton, Forida, USA: CRC Press, Taylor & Francis Group), pp. 9-34.

Lindner, S.N., Vidaurre, D., Willbold, S., Schoberth, S.M., and Wendisch, V.F. (2007). NCgl2620 encodes a class II polyphosphate kinase in *Corynebacterium glutamicum.* Appl. Environ. Microbiol. *73*, 5026-5033.

Mascher, T., Helmann, J.D., and Unden, G. (2006). Stimulus perception in bacterial signal-transducing histidine kinases. Microbiol. Mol. Biol. Rev. *70*, 910-938.

Moura, R.S., Martin, J.F., Martin, A., and Liras, P. (2001). Substrate analysis and molecular cloning of the extracellular alkaline phosphatase of *Streptomyces griseus.* Microbiology *147*, 1525-1533.

Neuhard, J., and Kelln, R.A. (1996). Biosynthesis and conversion of pyrimidines. *In* F.C. Neidhardt, R. Curtiss III, J.L. Ingraham, E.C.C. Lin, K.B. Low, B. Magasanik, W.S. Reznikoff, M. Riley, M. Schaechter, and H.E. Umbarger (ed), *Escherichia coli* and *Salmonella*: cellular and molecular biology, 2nd ed ASM Press, Washington, DC, 580-599.

Pallerla, S.R., Knebel, S., Polen, T., Klauth, P., Hollender, J., Wendisch, V. F., and Schoberth, S.M. (2005). Formation of volutin granules in *Corynebacterium glutamicum.* FEMS Microbiol. Lett. *243*, 133-140.

Park, S.Y., Kim, H.K., Yoo, S.K., Oh, T.K., and Lee, J.K. (2000). Characterization of *glk*, a gene coding for glucose kinase of *Corynebacterium glutamicum.* FEMS Microbiol. Lett. *188*, 209-215.

Parkinson, J.S., and Kofoid, E.C. (1992). Communication modules in bacterial signaling proteins. Annu. Rev. Genet. *26*, 71-112.

Rittmann, D., Schaffer, S., Wendisch, V.F., and Sahm, H. (2003). Fructose-1,6-bisphosphatase from *Corynebacterium glutamicum*: expression and deletion of the fbp gene and biochemical characterization of the enzyme. Arch. Microbiol. *180*, 285-292

Rittmann, D., Sorger-Hermann, U., and Wendisch, V.F. (2005). The phosphate starvation inducible gene *ushA* encodes a 5′-nucleotidase required for growth of *Corynebacterium glutamicum* on nucleotides as phosphorus source. Appl. Environ. Microbiol. *71*, 4339-4344.

Rosenberg, H., Gerdes, R. G., and Chegwidden, K. (1977). Two systems for uptake of phosphate in *Escherichia coli.* J. Bacteriol. *131*, 505-511.

Rosenberg, H., Gerdes, R.G., and Harold, F.M. (1979). Energy coupling to the transport of inorganic phosphate in *Escherichia coli* K12. Biochemical J. *178*, 133-137.

Schaaf, S., and Bott, M. (2007). Target genes and DNA-binding sites of the response regulator PhoR from *Corynebacterium glutamicum.* J. Bacteriol. *189*, 5002-5011.

Schweizer, H., Argast, M., and Boos, W. (1982). Characteristics of a binding protein-dependent transport system for sn-glycerol-3-phosphate in *Escherichia coli* that is part of the Pho regulon. J. Bacteriol. *150*, 1154-1163.

Seufferheld, M., Lea, C.R., Vieira, M., Oldfield, E., and Docampo, R. (2004). The H^+-pyrophosphatase of *Rhodospirillum rubrum* is predominantly located in polyphosphate-rich acidocalcisomes. J. Biol. Chem. *279*, 51193-51202.

Sorger-Hermann, U. (2006) Analyse des Mechanismus der Phosphatregulation in *Corynebacterium glutamicum*, PhD thesis, University of Düsseldorf, Düsseldorf.

Stock, J.B., Ninfa, A.J., and Stock, A.M. (1989) Protein phosphorylation and regulation of adaptive responses in bacteria. Microbiol. Rev. *53*, 450-490.

Sutcliffe, I. C., and Harrington, D. J. (2002). Pattern searches for the identification of putative lipoprotein genes in Gram-positive bacterial genomes. Microbiology *148*, 2065-2077.

van Veen, H.W., Abee, T., Kortstee, G.J.J., Konings, W.N., and Zehnder, A.J.B. (1994). Translocation of metal phosphate via the phosphate inorganic transport system of *Escherichia coli.* Biochemistry *33*, 1766-1770.

Wanner, B.L. (1996). Phosphorus assimilation and control of the phosphate regulon, *In* F.C. Neidhardt, R. Curtiss III, J.L. Ingraham, E.C.C. Lin, K.B. Low, B. Magasanik, W.S. Reznikoff, M. Riley, M. Schaechter, and H.E. Umbarger (ed), *Escherichia coli* and *Salmonella*: cellular and molecular biology, 2nd ed ASM Press, Washington, DC, 1357-1381.

Wendisch, V.F., and Bott, M. (2005). Phosphorus metabolism, In Handbook of *Corynebacterium glutamicum*, L. Eggeling, and M. Bott, eds. (Boca Raton, Florida, USA: CRC Press), pp. 377-396.

Willsky, G.R., and Malamy, M.H. (1980). Characterization of two genetically separable inorganic phosphate transport systems in *Escherichia coli.* J. Bacteriol. *144*, 356-365.

Wolschendorf, F., Mahfoud, M., and Niederweis, M. (2007). Porins are required for uptake of phosphates by *Mycobacterium smegmatis.* J. Bacteriol. *189*, 2435–2442

Yukawa, H., Inui, M., and Vertes, A.A. (2007). Genomes and Genome-Level Engineering of Amino Acid-Producing Bacteria. *In* Amino Acid Biosynthesis – Pathways, Regulation and Metabolic Engineering (Wendisch V.F., ed), Springer, Heidelberg, Germany.

Zalkin, H., and Nygaard, P. (1996). Biosynthesis of purine nucleotides. *In* F.C. Neidhardt, R. Curtiss III, J.L. Ingraham, E.C.C. Lin, K.B. Low, B. Magasanik, W.S. Reznikoff, M. Riley, M. Schaechter, and H.E. Umbarger (ed), *Escherichia coli* and *Salmonella*: cellular and molecular biology, 2nd ed ASM Press, Washington, DC, 561-579.

From: Corynebacteria: Genomics and Molecular Biology. Edited by: Andreas Burkovski.

Sulfur Metabolism in *Corynebacterium glutamicum* 10

Christian Rückert and Jörn Kalinowski

Abstract

Sulfur is an important element for life as it is a constituent of a number of essential organic molecules like cysteine, methionine, Coenzyme A, or iron sulfur clusters. In turn, these compounds are involved in a number of essential cellular processes like protein biosynthesis or the transfer of electrons and acyl groups. Yet, an external supply of these compounds is essential for many eucaryotes and, due to their scarcity in many foods and feeds, their biosynthesis is of great industrial interest. Therefore, the metabolism of sulfur in *Corynebacterium glutamicum* has been studied in more and more detail in recent years. Besides the pathways to obtain and utilize sulfur from the environment especially the reactions leading to and from the sulfur-containing amino acids cysteine and methionine have been analyzed in great detail, revealing a number of so far unique metabolic routes. In addition, the regulation of sulfur metabolism has been analyzed on the transcriptional as well as on the enzymatic level, revealing the presence of at least three transcriptional regulators and a high number of feed-back inhibitions of key enzymes. In this chapter, the current state of knowledge concerning the central pathways of sulfur metabolism (i.e. those consuming most of the cellular sulfur) is summarized and a short outlook on the perspectives to produce methionine and cysteine is given.

1. Introduction

Sulfur, which belongs to the chemical group of the chalcogens together with oxygen, selenium, and tellurium, is a common element contributing about 0.1 % to the mass of the earth crust (Brown, 1982). Reflecting the main oxidation states of sulfur, which are -2, 0, +4, and +6, inorganic sulfur can be found in form of sulfides (e.g. pryrite), as elemental sulfur (near volcanic regions), sulfites (e.g. in acid rain), and sulfates (like gypsum) while organic reservoirs include amino acids, sulfonate, and fossil fuels. For living organisms, one of the major functions of inorganic sulfur compounds is to provide energy, either as electron donor (e.g. as sulfide at volcanic vents in the abyssal oceanic regions) or as terminal electron acceptor in anaerobic habitats.

Table 10.1. *C. glutamicum* genes involved in sulfonate (ester) utilization and assimilatory sulfur reduction

CDS (CGB)	CDS (NCgl)	Name	regulated by	EC number	Function of encoded protein	Reference/Comment
Utilization of sulfonate (esters)						
cg0012	NCgl0009	*ssuR*	McbR		ROK-type transcriptional activator of sulfonate (ester) utilization	Koch et al., 2005a
cg1147	NCgl0968	*ssuI*	McbR, SsuR	1.5.1.29	NAD(P)H-dependent FMN reductase	Koch et al., 2005b
cg1151	NCgl0972	*seuA*	McbR, SsuR	1.14.14.-	$FMNH_2$-dep. monooxygenase, sulfonate ester degrading	Koch et al., 2005b
cg1152	NCgl0973	*seuB*	McbR, SsuR	1.14.14.-	$FMNH_2$-dep. monooxygenase, sulfonate ester degrading	Koch et al., 2005b
cg1153	NCgl0974	*seuC*	McbR, SsuR	1.14.14.-	Probable $FMNH_2$-dependent monooxygenase	Koch et al., 2005b
cg1156	NCgl0975	*ssuD2*	McbR, SsuR	1.14.14.5	$FMNH_2$-dep. aliphatic sulfonate monooxygenase	Koch et al., 2005b
cg1376	NCgl1173	*ssuD1*	McbR, SsuR	1.14.14.5	$FMNH_2$-dep. aliphatic sulfonate monooxygenase	Koch et al., 2005b
cg1377	NCgl1174	*ssuC*	McbR, SsuR		ABC-type aliphatic sulfonate transporter, permease subunit	Koch et al., 2005b
cg1379	NCgl1175	*ssuB*	McbR, SsuR		ABC-type aliphatic sulfonate transporter, ATPase subunit	Koch et al., 2005b
cg1380	NCgl1176	*ssuA*	McbR, SsuR		ABC-type aliphatic sulfonate transporter, binding lipoprotein	Koch et al., 2005b
Assimilatory reduction of sulfur						
cg0803	NCgl0671	*thtR*		2.8.1.1	Thiosulfate sulfurtransferase	based on similarity
cg2189	NCgl1923	*cysG*		4.2.1.75	Probable uroporphyrinogen III synthase	based on similarity

cg3049	NCgl2658	*fpr1*		1.18.1.2	Ferredoxin–NADP(+) reductase	Rückert et al., 2005
cg3112	NCgl2713	*cysZ*	McbR, CysR?		Probable sulfate transporter	Rückert et al., 2005
cg3113	NCgl2714	*cysY*	McbR, CysR?	4.99.1.4	Probable sirohydrochlorin ferrochelatase	Rückert et al., 2005
cg3114	NCgl2715	*cysN*	McbR, CysR?	2.7.7.4	Sulfate adenylyltransferase subunit 1	Rückert et al., 2005
cg3115	NCgl2716	*cysD*	McbR, CysR?	2.7.7.4	Sulfate adenylyltransferase subunit 2	Rückert et al., 2005
cg3116	NCgl2717	*cysH*	McbR, CysR?	1.8.4.10	Adenosine phosphosulfate reductase	Rückert et al., 2005
cg3117	–	*cysX*	McbR, CysR?		Ferredoxin-like protein	Rückert et al., 2005
cg3118	NCgl2718	*cysI*	McbR, CysR?	1.8.7.1	Ferredoxin–sulfite reductase	Rückert et al., 2005
cg3119	NCgl2719	*fpr2*	McbR, CysR?	1.18.1.2	Ferredoxin–NADP(+) reductase	Rückert et al., 2005

Still, the functions of sulfur performed in biomolecules is of even greater importance: Sulfur belongs to the five macro-elements found in all living organisms on earth, although it is the least common of the five, contributing between 0.5 - 1 % to the bacterial dry weight (Kertesz et al., 1999). Although not present in large amounts, sulfur is absolutely essential for life as it is part of a number of biomolecules like amino acids, prosthetic groups and co-factors. In most of these molecules, the sulfur moiety is the reactive part of the molecule, catalyzing vital reactions. This is due to its chemical property to be easily reduced and oxidized, making it an ideal candidate for electron transfer. The same property allows the cell to quickly react to changes of the cellular redox state by changing tertiary and quarternary protein structures through formation and breaking of disulfide bonds.

2. Sulfur assimilation

Baring the use of sulfur-containing amino acids, the first step of sulfur utilization for an aerobic soil bacterium is the uptake and usually reduction of sulfur from the environment. The source(s) of sulfur available depend(s) heavily on the habitat: While marine organisms in the water column encounter largely sulfate, microorganisms living in aerobic soils are faced with organically bound sulfur, in form of amino acids and peptides as well as sulfonates and sulfate esters (Autry and Fitzgerald, 1990). Of these compounds, the importance and the amount of sulfonates in the environment was underestimated until recently, although the most abundant source of sulfonates, plant sulfolipid, had been identified almost fifty years ago (Benson et al., 1959; Benson, 1963). Being an ubiquitous component of the thylakoid membrane, plant sulfolipid is present in all phototrophic organisms, from cyanobacteria to plants (Harwood, 1980). Other important sulfonates include the plant derived methanesulfonate, which arises indirectly from dimethylsulfide (Baker et al., 1991), 2-amino-ethanesulfonate (taurine) that can contribute up to 0.1 % of animal body mass (Huxtable, 1992), and xenobiotic compounds of human origin, that are used in high amounts as, e.g., detergents and dyes.

Another "non-native" sulfur source of industrial interest is the substance class of (di-)benzothiophene ((D)BT) derivatives. They occur in fossil fuels like coal and oil and are a major contributor to sulfur oxyacid (SO_X) pollution of the atmosphere. Therefore, microorganisms capable of selectively removing the sulfur moiety are of great interest and at least one *Corynebacterium* species has been described to be able to use DBT as sole source of sulfur (Omori et al., 1992).

2.1 Uptake and utilization of organic sulfur compounds

It could be demonstrated that *C. glutamicum* grows well on different inorganic sulfur sources, e.g. sulfate, sulfite, or sulfide, even at very low concentrations (100 μM). Likewise, amino acids like cysteine, cystathionine, or methionine can be utilized (Rückert, unpublished). Concerning the "classical" organic compounds, sulfonates and thiophenes, initial bioinformatic analyses indicated the presence of one pathway for the utilization of aliphatic sulfonates and one for the desulfurization of benzothiophene derivatives. Testing a large number of these substances, it was found that *C. glutamicum* can utilize almost all tested aliphatic sulfonates and sulfonate esters, but neither aromatic sulfonates nor benzothiophene derivatives (Koch et al., 2005b).

Although no enzymatic data is available for *C. glutamicum*, a detailed analysis of defined deletion mutants of two gene clusters possibly involved in the pathway for aliphatic sulfonate (ester) utilization, *cg1147-1156* and *cg1376-1380*, revealed a total of nine genes to be taking part in the pathway (Koch et al., 2005b), encoding known as well as novel functions (Table 10.1). The first step of sulfonate utilization, the uptake, occurs via a homologue of an ABC-type transporter, well known, e.g., from *Escherichia coli* (Kertesz, 1999). Encoded by *ssuCBA* (*cg1376-78*), this transporter seems to be involved in the uptake of all aliphatic sulfonates, although sulfonates with a carbon chain length of eight and more can be imported via an up to now unknown transporter (Koch et al., 2005b). Cleavage of the sulfonates is dependent on two sulfonatases, SsuD1 and SsuD2. Both are able to replace each other for most substrates, making *C. glutamicum* the first bacterium described to possess two broad-spectrum sulfonatases. Both are in turn dependent on SsuI, a protein weakly similar to oxidoreductases. SsuI was found to be essential for the utilization of all sulfonates and sulfonate esters tested, indicating that it is the functional replacement of the non-essential FMN reductase SsuE which is part of the *ssu* operon in *E. coli* and several *Pseudomonas* species (Kertesz, 1999).

An interesting finding, so far reported only for *C. glutamicum*, is the ability of the bacterium to utilize sulfonate esters. In addition to the sulfonatases SsuD1 and SsuD2, these compounds can be utilized via a third parallel pathway. It consists of at least two of the three proteins initially thought to be involved in the desulfurization of benzothiophene, Cg1151 and Cg1152 (now called SeuA and SeuB, for **s**ulfonate **e**ster **u**tilization; Koch et al., 2005b). Although the biological importance of sulfonate esters is not clear and the natural substrate(s) of SeuAB are still unknown, the presence of three systems to use them clearly indicates that at least one such compound must exist in the natural environment of *C. glutamicum.* As sulfonate esters were shown to be quite unstable (Koch et al., 2005b), one possible source of sulfonate esters might be as intermediates during the utilization of sulfur-containing aromatic compounds as carbon source: *C. glutamicum* has been shown to be able to use a wide variety of aromatic compounds (Shen et al., 2005) and it might therefore be hypothesized that intermediates resembling sulfonate esters occur during utilization of sulfur-substituted aromatics. The occurrence of a sultone, a cyclized sulfonate ester has been proposed by Oldfield et al. (1997) during desulfurization of dibenzothiophene. As SeuAB were found to be the proteins required to utilize butanesultone, these compounds appear to be the most likely targets of the SeuABC system (Figure 10. 1).

It can therefore be summarized that the utilization of organic sulfur compounds in *C. glutamicum* differs considerably from that described for the model organisms *E. coli* and *Bacillus subtilis* (Kertesz, 1999). Quite surpris-

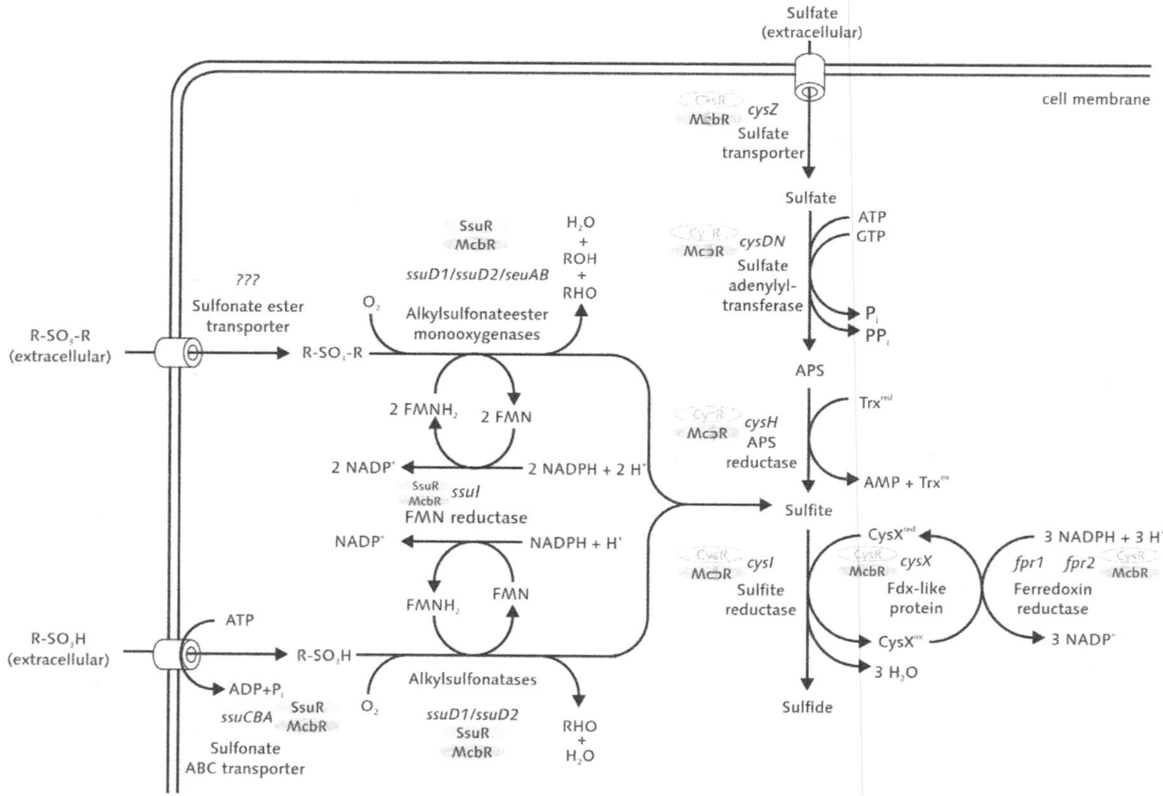

Figure 10.1. The current model of sulfonate utilization, assimilatory sulfate reduction, and its transcriptional regulation in *C. glutamicum*. Known transcriptional regulation is indicated by the name of the regulator in ellipses near the name of the regulated gene.

ingly, the *ssuIseuABCssuD2* cluster seems to be specific to *C. glutamicum*, even among the *Actinomycetales*: While the genes of the *ssuD1CBA* cluster can be detected in a number of species of this order, e.g. *Nocardia farcinica* and *Streptomyces coelicolor*, *ssuIseuABCssuD2* could not be found in any other completely sequenced actinomycete, not even in the closely related species *Corynebacterium efficiens*.

2.2 Transport and assimilation of inorganic sulfur

Whether using inorganic or organic sulfur compounds, microorganisms living under aerobic conditions need to reduce the sulfate or sulfite to sulfide before incorporating the latter into various biomolecules. The assimilatory sulfate reduction pathway has been studied in detail in the model organisms *E. coli* (reviewed by Kredich, 1996) and *B. subtilis*. But until recently, almost no data was available for the high-GC Gram-positives of the order *Actinomycetales* with the exception of the functional analysis of *cysH* of *Mycobacterium tuberculosis* (Williams et al., 2002). As for the utilization of organo-sulfur compounds, up to now no enzymatic data for the involved *C. glutamicum* proteins exists. Most of the participating genes have been proposed by Lee (2005), based on genome data. He noticed that several functions known from *E. coli* (Kredich, 1996) are missing, e.g. the flavoprotein subunit of sulfite reductase (CysJ) or a good candidate for a sulfate transporter. Recent studies on the genetic level by our group delivered a number of candidates to fill these gaps (Table 10.1; Rückert et al., 2005). Almost all of the genes thought to be directly involved in the reduction of sulfate and sulfite to sulfide are organized in one cluster, now called *fpr2 cysIXHDNYZ*. Based on the available data, the current model can be summarized as follows (Figure 10. 1).

Uptake of sulfate and sulfite in *C. glutamicum* is most likely mediated by the membrane protein CysZ. The gradual restoration of growth of the mutant strain with raising concentrations of sulfate leads to the hypothesis that CysZ might function as high-affinity sulfate transporter, analogous to the Pit-type transporter CysP in *B. subtilis* (Mansilla and de Mendoza, 2000) and Sbp CysTWA, an ABC-type transporter in *E. coli* (Sirko et al., 1990). The subsequent reduction of sulfate to sulfite seems to take place in only two steps in *C. glutamicum*, activation through adenylylation and direct reduction of APS. In contrast, in *E. coli* the intermediate APS must be phosphorylated, yielding PAPS, before sulfite can be liberated (Kredich, 1996). The same holds true for *B. subtilis*, although APS can also be used with greatly reduced catalytic efficiency (Berndt et al., 2004). In *C. glutamicum*, the gene encoding APS kinase, *cysC*, seems to be missing, as no candidate can be identified by similarity searches. Therefore, CysH is likely to act directly on APS, a hypothesis that is based on the presence of several sequence motifs (Kopriva et al., 2002) and the high degree of similarity to APS reductase from *M. tuberculosis* (Williams et al., 2002). Finally, the reduction of sulfite to sulfide shows the most marked differences to the two model organisms: *C. glutamicum* lacks the flavoprotein subunit known from *E. coli* (CysJ; Ostrowski et al., 1989) and *B. subtilis* (van der Ploeg et al., 2001), raising the question how reduction of sulfite takes place. As the putative hemoprotein subunit, CysI, shows a significantly higher degree of similarity to ferredoxin-dependent nitrite and sulfite reductases from plants than to CysI proteins from *E. coli* and *B. subtilis*, it can be hypothesized that the reduction of sulfite in *C. glutamicum* resembles that in plants. This hypothesis is supported by the presence of two potential NADP–ferredoxin reductases (Fpr1 and Fpr2), one of which is part of the studied cluster, as well as a small, ferredoxin-like protein (CysX). Taken together, these components might constitute the functional replacement of CysJ.

Interestingly, at least parts of the cluster are conserved among the completely sequenced members of the *Actinomycetales*. It can therefore be assumed that the pathway identified in *C. glutamicum* is archetypical for this bacterial order, setting it apart from the Gram-negative and low-GC Gram-positive bacteria like *E. coli* and *B. subtilis*, respectively.

Besides the genes of the aforementioned cluster, a number of additional genes might play a role in assimilatory sulfur reduction. Two, *cysG* and *cysY* encode proteins thought to be involved in siroheme biosynthesis (Rückert et al., 2005), an essential cofactor of sulfite reductase. In contrast to *E. coli* where all three reactions are performed by a multi-domain protein (Spencer et al., 1993), siroheme biosynthesis in *C. glutamicum* therefore is more like that encountered in *Bacillus megaterium* where the proteins are encoded by three separate proteins, *sirABC* (Raux et al., 2003). The third protein of interest is a putative thiosulfate sulfurtransferase (Cg0803), possibly involved in thiosulfate utilization. Although quite similar to ThtR from *M. tuberculosis*, the function of this protein in *C. glutamicum* is unclear. Growth on thiosulfate in this bacterium depends on sulfite reductase, and to a lesser extent, also on sulfate adenylyltransferase and APS reductase (Rückert et al., 2005). As *C. glutamicum* lacks the *O*-acetyl-serine sulfhydrylase B known from *E. coli* (Kredich, 1996), thiosulfate is most likely somehow broken down to sulfite and subsequently reduced to sulfide.

3. Metabolism of sulfur-containing amino acids

Perhaps the largest sink for sulfur in aerobic bacteria are sulfur-containing amino acids, by name the proteinogenic ones, cysteine and methionine, as well as intermediates like cystathionine and homocysteine. Still, their concentration as free amino acids in the *C. glutamicum* wild-type is usually below 0.5 μmol per gram cell dry weight (Krömer et al., 2005).

3.1 Cysteine

Of the two proteinogenic amino acids, cysteine is involved in the greater number of subsequent reactions, as it is used not only in proteins but also as the sulfur donor for, e.g., methionine, lipoic acid, thiamine, or mycothiol. Neither the uptake nor the export of cysteine has yet been studied in *C. glutamicum*, but the main pathway for the biosynthesis seems to be similar to that found in *E. coli* (Kredich, 1996). Serine is activated via acetylation by serine *O*-acetyltransferase, encoded by *cysE* (Wada et al., 2002), and cysteine

Table 10.2. *C. glutamicum* genes involved in cysteine and methionine metabolism

CDS (CGB)	CDS (NCgl)	Name	regulated by	EC/TC number	Function of encoded protein	Reference/Comment
Cysteine metabolism						
cg2344	NCgl2055				Putative cystathionine β-synthase	based on similarity
cg2833	NCgl2473	*cysK*	McbR	2.5.1.47	Cysteine synthase	Wada et al., 2004
cg2834	NCgl2474	*cysE*	McbR	2.3.1.30	Serine *O*-acetyltransferase	Wada et al., 2002
cg3086	NCgl2688			4.4.-	Putative cystathionine γ-lyase	based on similarity
Methionine metabolism						
cg0314	NCgl0254	*brnF*	Lrp		Branched chain amino acid exporter, large subunit	Trötschel et al., 2005
cg0315	NCgl0255	*brnE*	Lrp		Branched chain amino acid exporter, small subunit	Trötschel et al., 2005
cg0735	NCgl0608	*metI*	McbR	3.A.1.24.1	ABC-type methionine transporter, permease subunit	Trötschel et al., 2005
cg0736	NCgl0609	*metN*	McbR	3.A.1.24.1	ABC-type methionine transporter, ATPase subunit	Trötschel et al., 2005
cg0737	NCgl0610	*metQ*	McbR	3.A.1.24.1	ABC-type methionine transporter, substrate-binding lipoprotein	Trötschel et al., 2005
cg0754	NCgl0624	*metX*	McbR	2.3.1.31	Homoserine *O*-acetyltransferase	Park et al., 1998
cg0755	NCgl0625	*metY*	McbR	2.5.1.49	*O*-Acetylhomoserine sulfhydrylase	Hwang et al., 2002
cg0860	NCgl0719	*sahH*		3.3.1.1	Adenosylhomocysteinase	based on similarity
cg1290	NCgl1094	*metE*	McbR	2.1.1.14	Methionine synthase, vitamin B_{12}-independent	Rückert et al., 2003
cg1701	NCgl1450	*metH*	McbR	2.1.1.13	Methionine synthase, vitamin B_{12}-dependent	Rückert et al., 2003
cg1806	NCgl1541	*metK*	McbR	2.5.1.6	Methionine adenosyltransferase	Großmann et al., 2000
cg2383	NCgl2091	*metF*		1.5.1.20	N^5,N^{10}-Methylenetetrahydrofolate reductase (NAD(P)H)	Rückert et al., 2003
cg2536	NCgl2227	*aecD*		4.4.1.-	βC-S lyase, aminotransferase class I	Rossol and Pühler, 1992
cg2687	NCgl2360	*metB*	McbR	2.5.1.48	Cystathionine γ-synthase	Hwang et al., 1999
cg3253	NCgl2840	*mcbR*	McbR		TetR-family global transcriptional repressor of sulfur metabolism	Rey et al., 2003; 2005

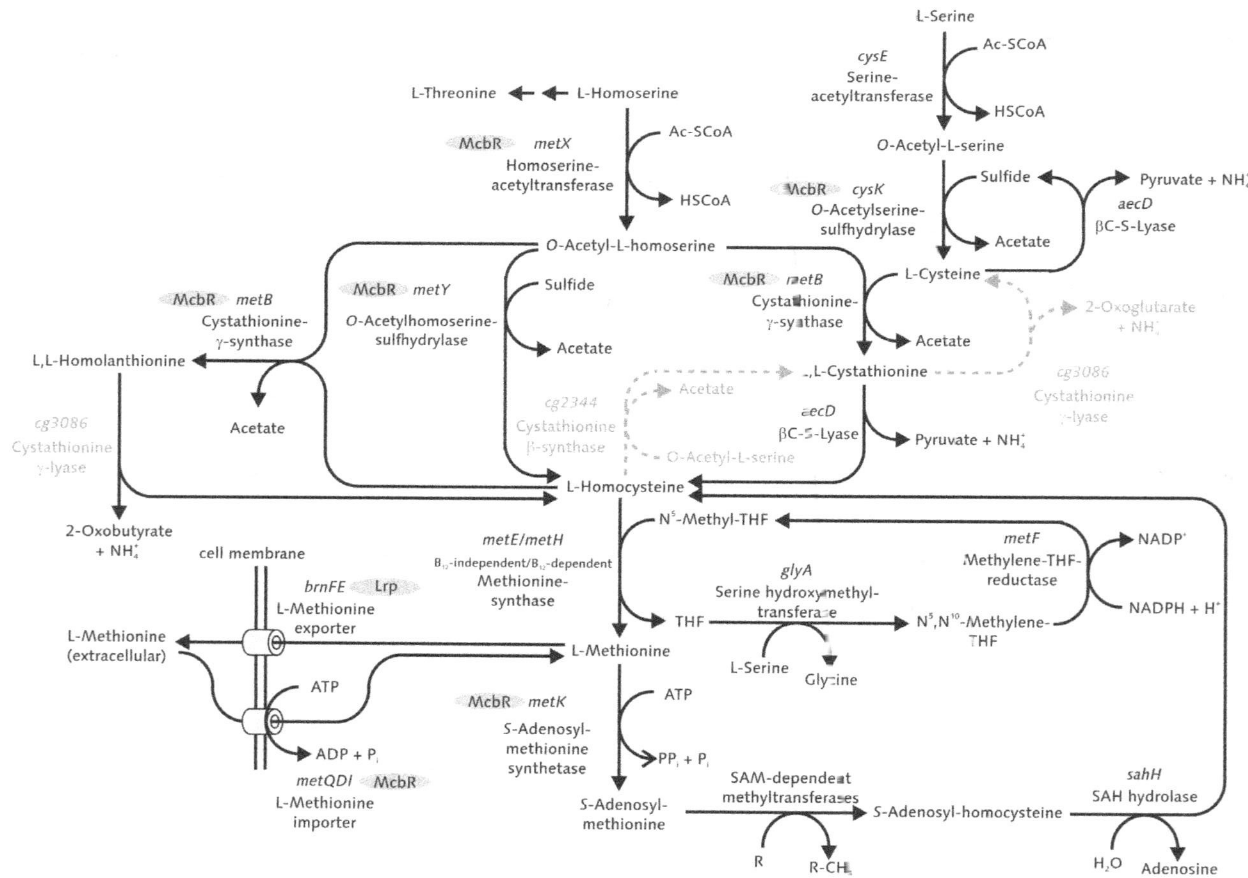

Figure 10.2. The current model of cysteine and methionine metabolism and its transcriptional regulation in *C. glutamicum*. Only genes are displayed where the biochemical reactions of the encoded proteins are known (solid black lines/type) or could be inferred from similarity data (dotted grey lines/type). Known transcriptional regulation is indicated by the name of the regulator in ellipses near the name of the regulated gene.

is subsequently formed by sulfhydrylation through *O*-acetyl-serine sulfhydrylase CysK (Wada et al., 2004). In contrast to *E. coli* that uses two isoforms (CysK and CysM; Kredich, 1996), only one active *O*-acetyl-serine sulfhydrylase seems to exist in *C. glutamicum*, as a deletion of *cysK* results in cysteine auxotrophy (Table 10.2; unpublished data). This is in accordance with the finding that growth on thiosulfate in *C. glutamicum* is dependent on at least sulfite reductase (Rückert et al., 2005).

As *C. glutamicum* can synthesize methionine directly from sulfide (Hwang et al., 2002) and can grow slowly on methionine as sole source of sulfur (Rückert, unpublished results), a pathway converting methionine to cysteine must exist. One possibility is the conversion of homocysteine to cysteine, accomplished by a possible cystathionine β-synthase (Cg2344) and a possible cystathionine γ-lyase, encoded by *cg3086* (Figure 10. 2). Another route has been proposed by Lee (2005), which starts with the cleavage of methionine, yielding methanethiol that is subsequently oxidized to methanesulfonate and then processed via the pathway for sulfonate utilization. A significant catabolic route for cysteine in *C. glutamicum* is realized via the βC-S lyase AecD. Besides acting as a cystathionine β-lyase (Kim et al., 2001; Rückert et al., 2003), AecD can also act on cysteine, relieving sulfide, ammonia, and pyruvate (Rossol and Pühler, 1992). Although the reaction rate with cysteine is about twentyfold lower compared to cystathionine, the presence of AecD inhibits cysteine accumulation in the cell (Wada et al., 2004). This enzyme therefore constitutes a major balancing point for sulfur distribution between cysteine and methionine.

3.2. Methionine

The second sulfur-containing proteinogenic amino acid, methionine also plays a vital role in cellular processes, most often as donor of methyl groups via the intermediate *S*-adenosyl-methionine (SAM). It belongs to the aspartate family of amino acids, sharing the precursor homoserine with threonine and isoleucine.

The first enzymes involved in conversion of homoserine to methionine were already characterized biochemically more than twenty five years ago (Kase and Nakayama, 1974a, 1974b; Ozaki and Shiio, 1982), due to the great interest in a methionine-producing strain. The corresponding genes were identified in recent years (Rossol and Pühler, 1992; Park et al., 1998; Hwang et al., 1999; Kim et al., 2001; Hwang et al., 2002), proving that the incorporation of the sulfur moiety takes place via a split pathway in *C. glutamicum* (Table 10.2 and Figure 10. 2). Unlike *E. coli* which uses succinylation (Greene, 1996), homoserine is activated by acetylation via homoserine *O*-acetyltransferase, encoded by *metX* (previously called *metA*; Park et al., 1998). The *O*-acetyl-homoserine formed can than be converted to homocysteine either directly through sulfhydrylation by the *metY* gene product, or by transsulfuration. In the latter case, *O*-acetylhomoserine and cysteine are converted to cystathionine by cystathionine β-synthase (MetB; Hwang et al., 1999) which is then cleaved by the βC-S lyase AecD (Rossol and Pühler, 1992; Hwang et al., 2002; Rückert et al., 2003). The role of this parallel pathway is not yet understood. Based on methionine production alone, direct sulfhydrylation would be preferable (Krömer et al., 2006b). But as high flux into the methionine pathway leads to homolanthionine accumulation which evidently cannot be removed efficiently in *C. glutamicum* (Krömer et al., 2006a), balancing the cysteine and methionine pools might be the reason for this metabolic setup.

For the last step of methionine biosynthesis, the incorporation of the methyl group, there are also two enzymes present which are highly similar to those known from *E. coli* (Greene, 1996): A B_{12}-dependent form, encoded by *metH*, and a B_{12}-independent form, encoded by *metE* (Rückert et al., 2003). The reason for this parallel reaction is unknown, but might be necessary due to the properties of the two enzymes. A loss of *metE* leads to methionine auxotrophy under high levels of oxygen unless B_{12} is supplied in the medium (Rückert et al., 2003). In *E. coli*, it has been demonstrated that MetH is at least a hundredfold more active than MetE (Greene, 1996), indicating that MetE might serve as a backup in the case that MetH activity becomes too low due to lack of unoxidized B_{12}.

In the cellular context, the methylation of homocysteine to form methionine is part of two interlocking cycles to transfer methyl groups from serine to a wide variety of targets. The first cycle, leading to the formation of methionine, starts with the transfer of a methylene group from serine to tetrahydrofolate (THF) by serine hydroxymethyltransferase (GlyA), resulting in the formation of glycine and N^5,N^{10}-methylene-THF. The latter is then reduced by N^5,N^{10}-methylene-THF reductase (MetF), providing the methyl group donor for MetE and MetH. In the second cycle, methionine is activated by adenylylation by methionine adenosyltransferase encoded by *metK* (Großmann et al., 2000), forming *S*-adenosyl-methionine (SAM). This compound is the central donor for methyl groups in the cell, e.g., for the methylation of DNA (Schäfer et al., 1997). All these reactions lead to the formation of *S*-adenosylhomocysteine (SAH), which plays a central role in the regulation as effector of the regulator McbR (see below). Conversion of SAH presumably takes place via the route described for, e.g., *Rhodobacter capsulatus* (Sganga et al., 1992), by adenosylhomocysteinase (SahH) mediated hydrolysis, relieving homocysteine for the next cycle.

Concerning the transport of methionine, only the export has been studied in detail. It was recently shown by Trötschel et al. (2005) that methionine excretion depends on branched-chain amino acid exporter, encoded by *brnFE* (Kennerknecht et al., 2002). For the import of methionine, a likely candidate is the ABC-type transporter encoded by *metQNI* (Trötschel et al., 2005). Not only are the proteins similar to those encoded by *metNIQ* in *E. coli* (Gál et al., 2002), but the genes are also controlled by the transcriptional repressor McbR (see below), another indication for an involvement in sulfur metabolism.

4. Metabolism of other sulfur-containing compounds

Among the cysteine-derived sulfur-containing compounds, mycothiol and iron-sulfur [Fe-S] clusters can be assumed to contribute a significant portion of cellular bound sulfur. For this and their essential cellular functions, they are beginning to emerge as future targets of interest.

4.1 Mycothiol

One of the sinks for cysteine in actinomycetes is the low molecular mass thiol mycothiol which is thought to act as the replacement of glutathione. Reviewed by Newton and Fahey (2002) in detail for mycobacteria, it was demonstrated to help maintain the cellular redox homeostasis as well as protectant against alkylating agents like certain antibiotics. Although not studied in detail in *C. glutamicum*, the biosynthesis can be assumed to be identical to that described for mycobacteria: *N*-acetylserine is linked to a pseudodisaccharide formed from D-glucosamine and *myo*-inositol in a four step pathway. In addition to the biosynthesis, two additional genes are thought to be involved in mycothiol metabolism: *mtr* encodes mycothiol reductase, needed to recover oxidized mycothiol disulfide, while the product of *mca*, mycothiol *S*-conjugate amidase, is needed to recycle mycothiol spent for detoxification of alkylating compounds.

For all genes known, possible orthologs can be identified in the *C. glutamicum* genome (Table 10.3) and the activities of MshB, MshC, and MshD have been verified experimentally by Feng et al. (2006). This group could also demonstrate a novel, mycothiol-dependent reaction needed for growth on aromatic compounds.

4.2. Iron-sulfur clusters

Iron-sulfur [Fe-S] clusters are among the most ubiquitous and versatile prosthetic groups, but their assembly in *C. glutamicum* has not been studied so far. Still, some predictions can be made based on the available genome data: Four gene clusters consisting of one (*cg1214*), one (*cg2214*), three (*cg1386-88*), and seven (*cg1765-59*) genes respectively (Table 10.3) can be assumed to play a role in [Fe-S] cluster formation and maintenance. Although their function has yet to be verified, strong circumstantial evidence indicates an involvement of at least *cg1765-59* and *cg1386-88*. The former cluster was found to be regulated in response to thiol oxidation stress (Nakunst et al., 2007) and is highly similar to the *nif-suf* cluster known, e.g., from *E. coli* (Johnson et al., 2005). But as other iron-sulfur cluster assembly systems seem to be absent in *C. glutamicum*, this cluster might constitute the essential system in this organism. The second cluster, *cg1386-88* (*fixAB nifS1*), displays some similarity to the *fixABCX* cluster from *E. coli*, but the strongest indication for an involvement in sulfur metabolism is the finding that this cluster is under McbR control (Rey et al., 2005). As [Fe-S] clusters can be assumed to play an important role in sulfur metabolism, these genes make for interesting targets for future analyses.

Table 10.3. *C. glutamicum* genes involved in the metabolism of mycothiol, iron-sulfur cluster maintenance and unknown functions

CDS (CGB)	CDS (NCgl)	Name	regulated by	EC number	Function of encoded protein	Reference/Comment
Metabolism of mycothiol						
cg0481	NCgl0389	*mshA*		2.4.1.-	Glycosyltransferase, probably involved in mycothiol synthesis	based on similarity
cg1127	NCgl0948	*mca*		3.5.1.-	Probable mycothiol *S*-conjugate amidase	based on similarity
cg1250	NCgl1055	*mshB*		3.5.1.-	*N*-acetyl-1-D-*myo*-inosityl-2-amino-2-deoxy-α-D-glucopyranoside deacetylase	Feng et al., 2006
cg1709	NCgl1457	*mshC*		6.3.2.-	1-D-*myo*-inosityl-2-amino-2-deoxy-α-D-glucopyranoside-cysteine ligase	Feng et al., 2006
cg2194	NCgl1928	*mtr*		1.-	Probable NADPH-dependent mycothiol reductase	based on similarity
cg2847	NCgl2487	*mshD*		2.3.1.-	De-acetyl-mycothiol *N*-acetyltransferase	Feng et al., 2006
Iron-sulfur cluster biosynhesis/maintenance						
cg1214	NCgl1022				Putative cysteine desulfurase / aminotransferase class IV	based on similarity
cg1386	NCgl1182	*fixA*	McbR		Putative electron transfer flavoprotein, beta subunit	based on similarity
cg1387	NCgl1183	*fixB*	McbR		Putative electron transfer flavoprotein, alpha subunit	based on similarity
cg1388	NCgl1184	*nifS1*	McbR	2.6.1.-	Putative, aminotransferase class IV / cysteine desulfinase	based on similarity
cg1759	NCgl1498		SufR, SigM		Putative metal-sulfur cluster biosynthetic enzyme	Nakunst et al., 2007

cg1760	NCgl1499		SufR, SigM		Protein involved in Fe-S cluster formation, NifU-family	Nakunst et al., 2007
cg1761	NCgl1500	*nifS2*	SufR, SigM	4.4.1.-	Cysteine desulfhydrase / aminotransferase class IV	Nakunst et al., 2007
cg1762	NCgl1501	*sufC*	SufR, SigM		Fe-S assembly ATPase, SufC-family	Nakunst et al., 2007
cg1763	NCgl1502	*sufD*	SufR, SigM		Fe-S assembly membrane protein, SufD-family	Nakunst et al., 2007
cg1764	NCgl1503	*sufB*	SufR, SigM		Fe-S assembly membrane protein, SufB-family	Nakunst et al., 2007
cg1765	NCgl1504	*sufR*	SufR, SigM		Transcriptional repressor of the *suf-nif* cluster	Nakunst et al., 2007
cg2214	NCgl1944				Putative Fe-S-cluster redox enzyme	based on similarity
cg3299	NCgl2874	*trxB1*	SigM		Thioredoxin (TRX)	Nakunst et al., 2007
cg3422	NCgl2984	*trxB*	SigM	1.6.4.5	Thioredoxin reductase	Nakunst et al., 2007
cg3423	NCgl2985	*trxC*	SigM		Thioredoxin	Nakunst et al., 2007
McbR-controlled genes of unknown function						
cg0156	NCgl0120	*cysR*	McbR		ROK-type transcriptional regulator, prob. controlling sulfur reduction	part of McbR regulon
cg2810	NCgl2463		McbR		Putative secondary H+/Na+ symporter	part of McbR regulon
cg3132	NCgl2731		McbR		Putative membrane protein	part of McbR regulon
cg3372	NCgl2940		McbR		Conserved hypothetical protein	part of McbR regulon
cg3399	NCgl2965		McbR		Permease, MFS-type	part of McbR regulon

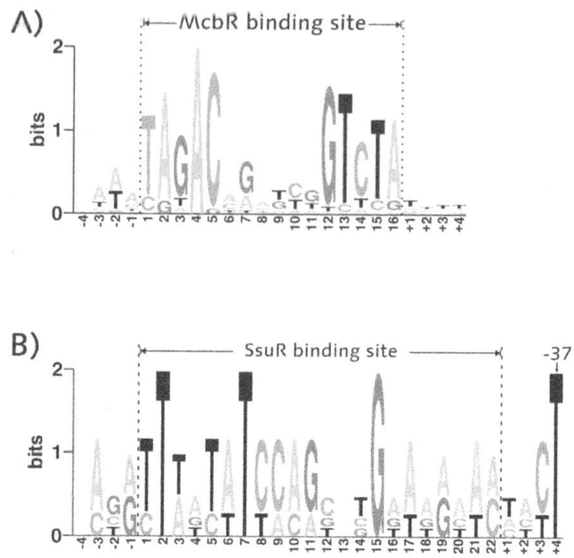

Figure 10.3 Sequence logo (Schneider and Stephens, 1990) of the (A) McbR and (B) SsuR binding sites. The sequence logo is based on the sequences of all putative McbR respectively SsuR binding sites identified in the *C. glutamicum* genome sequence, including 4 bp up- and downstream of each binding site.

5. Regulation of sulfur metabolism

As quite a number of intermediates and products are toxic at even low concentrations, a tight regulation of the various pathways is necessary for the cell to avoid the accumulation of toxic compounds above the necessary amount. The transcriptional regulation of the various pathways involved in sulfur metabolism in *C. glutamicum* has been studied in great detail in recent years, leading to the identification of two regulators and providing a third candidate.

5.1 The master regulator – McbR

Negative transcriptional regulation induced by methionine had been described for quite some time for several key genes: *hom* (encoding homoserine dehydrogenase; Folettie et al., 1988), *metB* (Kase and Nakayama, 1974b), *metX* (Kase and Nakayama, 1974a), and *metY* (Ozaki and Shiio, 1982). In order to identify the regulators responsible for the described transcriptional regulation, Rey et al. (2003) applied a DNA-affinity chromatography method to identify proteins binding specifically to the 5′-region of *metY*, finding the TetR-type transcriptional regulator Cg3253. A Δ*cg3253* mutant was demonstrated to abolish the repression caused by L-methionine in the wild-type, constitutively expressing a large number of genes involved in sulfur metabolism. Cg3253 was therefore named McbR, **m**ethionine and **c**ysteine **b**iosynthesis **r**epressor.

Subsequent microarray analyses by Rey et al (2005) revealed an increased transcription of 86 genes in the mutant strain, indicating that McbR is indeed the global regulator of sulfate metabolism in *C. glutamicum*. A combination of the results of the transcriptome studies with bioinformatical methods led to the identification of a 16 bp DNA sequence consisting of a highly conserved 5 bp inverted repeat separated by 6 variable base pairs (Figure 10. 3A). This motif is present upstream of 22 operons with increased expression in the Δ*mcbR* mutant strain, indicating a direct regulation of at least 45 genes by McbR.

In vitro tests by electrophoretic mobility shift assays (EMSA) and surface plasmon resonance (SPR) were used to identify the effector substance of McbR. Quite surprisingly, it was found to be S-adenosyl homocysteine (SAH) rather than methionine (Rey et al., 2005). This metabolite, the by-product of SAM-dependent transmethylation reactions, is assumed to be an indicator of cell growth: newly synthesized DNA, RNA, and proteins need to be methylated, increasing the SAH pool. This leads to an inactivation of McbR, increasing the synthesis of sulfur containing amino acids to provide the cell with steady supply of SAM during fast growth.

Besides identifying McbR as the "master regulator" of sulfur metabolism, the description of the McbR regulon proved to be of great value for the analysis of sulfur metabolism: As almost all genes known to be involved in the processing of this macro-element are under control of McbR, it can be hypothesized that those genes of the regulon with unknown function are also participating in sulfur turnover. This assumption could already been proven for *cg0012* (*ssuR*; Koch et al., 2005a), *cg1147* (*ssuI*; Koch et al., 2005b) as well as for *cg3112*, *cg3113*, and *cg3119* (*cysZ*, *cysY*, and *fpr2*, respectively; Rückert et al., 2005). Still, several genes of interest remain, as listed in Table 10.3. For one of these genes, *cg0156*, encoding a putative ROK-type transcriptional regulator, a function could recently be inferred from the phenotype of a deletion mutant. A Δ*cg0156* mutant becomes auxotrophic for cysteine respectively sulfide (Rückert, unpublished), indicating that it is involved in the control of the pathway for assimilatory sulfate reduction. Accordingly, *cg0156* was named *cysR*, in analogy to *cysB* of *E. coli*.

5.2 Utilization of alternatives – SsuR

As Koch et al. (2005b) had shown the gene clusters involved in sulfonate utilization to be tightly regulated, the presence of at least one other regulator controlling them was assumed. This regulator was identified to be the McbR-controlled activator Cg0012 (Koch et al., 2005a), classified as member of the ROK (repressor, ORF, sugar kinase) family (Brune et al., 2005). Comparison of the transcriptome of a deletion mutant to that of the wild-type revealed the mRNA levels of the *ssu* and *seu* genes to be reduced between 200 to 2,600 fold in the mutant strain. Therefore, *cg0012* was named *ssuR*, **s**ulfonate **s**ulfur **u**tilization **r**egulator.

Search for the binding site turned up a 22 bp motif upstream of the four SsuR-controlled promoters (Figure 10. 3B; Koch et al., 2005a). This motif displays similarity to the NagC binding sites described for *E. coli* (Plumbridge, 2001) and, based on a bioinformatical search and the transcriptomic data, only the four instances already identified seem to be present in the genome.

Concerning the effectors of SsuR, four compounds, mainly sulfate and to a lesser extent APS, sulfite, and sulfide were shown to act as inhibitors of the binding of SsuR to its target regions (Koch et al., 2005a). These four compounds are associated with the assimilatory reduction of sulfate, indicating that sulfonates are only utilized if the concentration of inorganic compounds is too low to provide the amount of sulfur needed. This assumption is backed by real-time RT-PCR data, showing that transcription of the *ssu* and *seu* genes is strongly reduced when the cells are grown in the presence of either sulfate or sulfite (Koch et al., 2005b).

This leads to a model of a hierarchical activation of sulfonate utilization: As expression of SsuR is also controlled by McbR, the *ssu* and *seu* genes are only activated when additional sulfur is required (McbR is inactivated) and the concentration of inorganic sulfur sources is low (SsuR is active).

5.3 The protein level – Feed-back Inhibition

In addition to regulation on the transcriptional level, several key enzymes in sulfur metabolism are regulated by feed-back inhibition. In the pathway leading to methionine, this starts with a cooperative inhibition of aspartate kinase, the entry point into the pathway leading to the aspartate derived amino acids, by threonine and lysine (Kalinowski et al., 1991). At the next branching point, homoserine dehydrogenase is controlled by threonine (Cremer et al., 1988), thereby controlling flux between lysine and the homoserine-derived amino acids. Downstream, in the pathway leading from homoserine to methionine, almost all enzymes were found to be controlled by one intermediate or the other. SAM was shown to inhibit homoserine *O*-acetyltransferase (MetX; Miyajima and Shiio, 1973) and cystathionine γ-synthase (MetB; Kase and Nakayama, 1974b), while methionine controls both MetX (Miyajima and Shiio, 1973) and *O*-acetylhomoserine sulfhydrylase (MetY; Ozaki and Shiio, 1982). This indicates that SAM rather than methionine might be considered the end point of the pathway.

Concerning cysteine metabolism, the second enzyme of the transsulfuration pathway, βC-S lyase (AecD), can again be considered as the link between cysteine and methionine: AecD is inhibited by cysteine as well as by glycine (Kase and Nakayama, 1974b), which might accumulate during methionine biosynthesis as a by-product of serine demethylation. Last but not least, also the first enzyme involved in cysteine biosynthesis, *O*-acetylserine sulfhydrylase (CysE), is thought to be inhibited by cysteine (Wada et al., 2002).

6. Towards biotechnological production of cysteine and methionine

The great scientific interest in the metabolism of sulfur in *C. glutamicum* is in part due to the commercial interest in the amino acids cysteine and especially methionine. Therefore, a significant number of studies to obtain a *C. glutamicum* strain producing methionine has been performed, as reviewed by Kumar and Gomes (2005). *In silico* studies by Krömer et al. (2006b) revealed that at least theoretically, methionine yields of 0.49 C-mol per C-mol glucose are possible with *C. glutamicum*. Current processes fall far short of this theoretical goal, with yields as high as 2.9 g/l (≈ 0.1 C-mol per C-mol glucose) using a feed-back resistant mutant (Park et al., 2007). This discrepancy is in part due to the fact that at least some of the carbon has to be used for biomass production but also to the presence of the various levels of transcriptional and post-translational regulation (see previous section). Another problem might be due to unwanted secondary reactions during high flux through the pathway, as shown with homolanthionine formation in a *mcbR* deletion mutant (Krömer et al., 2006a). The same holds true for cysteine, for which Wada et al. (2002) reported a yield of 0.29 g/l after 72 hours fermentation. In this case, the cysteine-degrading activity of βC-S lyase and feed-back inhibition of *O*-acetylserine sulfhydrylase significantly decrease cysteine yield. In summary, these low yields have so far precluded the implementation of commercial production processes for methionine and cysteine with *C. glutamicum*.

Still, the study by Krömer et al. (2006b) also provides first hints to overcome these difficulties. By switching to methanethiol as donor for the reduced sulfur moiety and the C_1 group, the theoretical yield can be increased as high as 0.91C-mol per C-mol glucose. Using such *in silico* methods by transforming the existing knowledge in a detailed model of sulfur metabolism can help rational design of production strains as well as optimization of the fermentation process. A recent example for this approach is a study by Kumar et al. (2005) who identified cysteine as a good donor for the sulfur moiety using a metabolic flux analysis. The subsequent supplementation of cysteine led to an increase in methionine production from 2.34 g/l to 3.39 g/l. Therefore, systems biology might provide the basis for commercially succesful methionine and cysteine production.

Acknowledgements

The authors thank Prof. Dr. Alfred Pühler for his continuos support and helpful discussions. The work was supported financially by the Bundesministerium für Forschung und Technologie within the GenoMik (grant 031U213D) and GenoMik-Plus (grant 0313805A) programs and by Degussa GmbH, Feed Additives.

References

Autry, A.R. and Fitzgerald, J.W. (1990). Sulfonate S: a major form of forest soil organic sulfur. Biol. Fertil. Soils *10*, 50–56.

Baker, S.C., Kelly, D.P., and Murrell, J.C. (1991). Microbial degradation of methanesulfonic acid - a missing link in the biogeochemical sulfur cycle. Nature *350*, 627–628.

Benson, A.A. (1963). The plant sulfolipid. Adv. Lipid Res. *1*, 387–394.

Benson, A.A., Daniel, H., and Wiser, R. (1959). A sulfolipid in plants. Proc. Natl. Acad. Sci. U.S.A. *45*, 1582–1587.

Berndt, C., Lillig, C.H., Wollenberg, M., Bill, E., Mansilla, M.C., de Mendoza, D., Seidler, A., and Schwenn, J.D. (2004). Characterization and reconstitution of a 4Fe-4S adenylyl / phospho-adenylyl sulfate reductase from *Bacillus subtilis*. J. Biol. Chem. *279*, 7850–7855.

Brown, E.A. (1982). Sulphur in the environment: a review. Environ. Pollut. Ser. B *3*, 671–673.

Brune, I., Brinkrolf, K., Kalinowski, J., Pühler, A., and Tauch, A. (2005). The individual and common repertoire of DNA-binding transcriptional regulators of *Corynebacterium glutamicum*, *Corynebacterium efficiens*, *Corynebacterium diphtheriae* and *Corynebacterium jeikeium* deduced from the complete genome sequences. BMC Genomics *6*, 86.

Cremer, J., Treptow, C., Eggeling, L., and Sahm, H. (1988). Regulation of enzymes of lysine biosynthesis in *Corynebacterium glutamicum*. J. Gen. Microbiol. *134*, 3221–3229.

Feng, J., Che, Y., Milse, J., Yin, Y.J., Liu, L., Rückert, C., Shen, X.H., Qi, S.W., Kalinowski, J., and Liu, S.J. (2006). The gene *ncgl2918* encodes a novel maleylpyruvate isomerase that needs mycothiol as cofactor and links mycothiol biosynthesis and gentisate assimilation in *Corynebacterium glutamicum*. J. Biol. Chem. *281*, 10778–10785.

Folettie, M.T., Shin, H.K., and Sinskey, A.J. (1988). Organization and regulation of the *Corynebacterium glutamicum hom-thrB* and *thrC* loci. Mol. Microbiol. *2*, 53–62.

Gál, J., Szvetnik, A., Schnell, R., and Kálmán, M. (2002). The *metD* D-methionine transporter locus of *Escherichia coli* is an ABC transporter gene cluster. J. Bacteriol. *184*, 4930–4932.

Greene, R.C. (1996). Biosynthesis of Methionine. In: *Escherichia coli* and *Salmonella*: Cellular and Molecular Biology, F.C. Neidhardt, R. Curtis III, J.L. Ingraham, E.C.C Lin, K.B. Low, B. Magasanik, W.S. Reznikoff, M. Riley, M. Schaechter, H.E. Umbarger, eds. (Washington D.C.: ASM Press), pp. 542–560.

Großmann, K., Herbster, K., and Mack, M. (2000). Rapid cloning of *metK* encoding methionine adenosyltransferase from *Corynebacterium glutamicum* by screening a genomic library on a high density colony-array. FEMS Microbiol. Lett. *193*, 99–103.

Harwood, J.L. (1980). Plant acyl lipids. In: The Biochemistry of Plants, P.K. Stumpf, E.E. Conn, eds., (New York: Academic Press), pp. 1–55.

Huxtable, R.J. (1992). Physiological actions of taurine. Physiol. Rev. *72*, 101–163.

Hwang, B.J., Kim, Y., Kim, H.B., Hwang, H.J., Kim, J.H., and Lee, H.S. (1999). Analysis of *Corynebacterium glutamicum* methionine biosynthetic pathway: Isolation and analysis of *metB* encoding cystathionine γ-synthase. Mol. Cells *9*, 300–308.

Hwang, B.J., Yeom, H.J., Kim, Y., and Lee, H.S. (2002). *Corynebacterium glutamicum* utilizes both transsulfuration and direct sulfhydrylation pathways for methionine biosynthesis. J. Bacteriol. *184*, 1277–1286.

Johnson, D.C., Dean, D.R., Smith, A.D., and Johnson, M.K. (2005). Structure, function, and formation of biological iron-sulfur clusters. Annu. Rev. Biochem. *74*, 247–281.

Kalinowski, J., Cremer, J., Bachmann, B., Eggeling, L., Sahm, H., and Pühler, A. (1991). Genetic and biochemical analysis of the aspartokinase from *Corynebacterium glutamicum*. Mol. Microbiol. *5*, 1197–1204.

Kase, H. and Nakayama, K. (1974a). Production of O-acetyl-L-homoserine by methionine analog-resistant mutants and regulation of homoserine-O-transacetylase in *Corynebacterium glutamicum*. Agr. Biol. Chem. *38*, 2021–2030.

Kase, H. and Nakayama, K. (1974b). The regulation of L-methionine synthesis and the properties of cystathionine-γ-synthase and β-cystathionase in *Corynebacterium glutamicum*. Agr. Biol. Chem. *38*, 2235–2242.

Kennerknecht, N., Sahm, H., Yen, M., Pátek, M., Saier Jr., M.H., and Eggeling, L. (2002). Export of L-isoleucine from *Corynebacterium glutamicum*: a two-gene-encoded member of a new translocator family. J. Bacteriol. *184*, 3947–3956.

Kertesz, M.A. (1999). Riding the sulfur cycle – metabolism of sulfonates and sulfate esters in gram-negative bacteria. FEMS Microbiol. Rev. *24*, 135–175.

Kim, J.W., Kim, H.J., Kim, Y., Lee, M.S., and Lee, H.S. (2001). Properties of the *Corynebacterium glutamicum metC* gene encoding cystathionine β-lyase. Mol. Cells *11*, 220–225.

Koch, D.J., Rückert, C., Albersmeier, A., Hüser, A.T., Tauch, A., Pühler, A., and Kalinowski, J. (2005a). The transcriptional regulator SsuR activates in the absence of sulphate the *Corynebacterium glutamicum ssu* and *seu* genes responsible for sulphonate utilization. Mol. Microbiol. *58*, 480–494.

Koch, D.J., Rückert, C., Rey, D.A., Mix, A., Pühler, A., and Kalinowski, J. (2005b). Role of the *ssu* and *seu* genes of *Corynebacterium glutamicum* ATCC 13032 in the utilization of sulfonates and sulfonate esters as sulfur sources. Appl. Environ. Microbiol. *71*, 6104–6114.

Kopriva, S., Büchert, T., Fritz, G., Suter, M., Benda, R., Schünemann, V., Koprivova, A., Schürmann, P., Trautwein, A.X., Kroneck, P.M.H., and Brunol, C. (2002). The presence of an iron-sulfur cluster in adenosine 5′-phosphosulfate reductase separates organisms utilizing adenosine 5′-phosphosulfate and phosphoadenosine 5′-phosphosulfate for sulfate assimilation. J. Biol. Chem. *277*, 21786–21791.

Kredich, N.M. (1996). Biosynthesis of Cysteine. In: *Escherichia coli* and *Salmonella*: Cellular and Molecular Biology, F.C. Neidhardt, R. Curtis III, J.L. Ingraham, E.C.C Lin, K.B. Low, B. Magasanik, W.S. Reznikoff, M. Riley, M. Schaechter, H.E. Umbarger, eds. (Washington D.C.: ASM Press), pp. 514–527.

Krömer, J.O., Fritz, M., Heinzle, E., and Wittmann, C. (2005). *In vivo* quantification of intracellular amino acids and intermediates of the methionine pathway in *Corynebacterium glutamicum*. Anal. Biochem. *340*, 171–173.

Krömer, J.O., Heinzle, E., Schröder, H., and Wittmann, C. (2006a). Accumulation of homolanthionine and activation of a novel pathway for isoleucine biosynthesis in *Corynebacterium glutamicum* McbR deletion strains. J. Bacteriol. *188*, 609–618.

Krömer, J.O., Wittmann, C., Schröder, H., and Heinzle, E. (2006b). Metabolic pathway analysis for rational design of l-methionine production by *Escherichia coli* and *Corynebacterium glutamicum*. Metab. Eng. *8*, 353–369.

Kumar, D., and Gomes, J. (2005). Methionine production by fermentation. Biotechnol. Adv. *23*, 41–61.

Kumar, D., Subramanian, K., Bisaria, V.S., Sreekrishnan, T.R., and Gomes, J. (2005). Effect of cysteine on methionine production by a regulatory mutant of *Corynebacterium lilium*. Bioresour. Technol. *96*, 287–294.

Lee, H.S. (2005). Sulfur Metabolism and its Regulation. In: Handbook of *Corynebacterium glutamicum*, L. Eggeling, M. Bott, eds., (Boca Raton, FL: Taylor and Francis Group), pp. 351–376.

Mansilla, M.C. and de Mendoza, D. (2000). The *Bacillus subtilis cysP* gene encodes a novel sulphate permease related to the inorganic phosphate transporter (Pit) family. Microbiology *146*, 815–821.

Miyajima, R. and Shiio, I. (1973). Regulation of aspartate family amino acid biosynthesis in *Brevibacterium flavum*: Properties of homoserine-O-transacetylase. J. Biochem. *73*, 1061–1068.

Nakunst, D., Larisch, C., Hüser, A.T., Tauch, A., Pühler, A., and Kalinowski, J. (2007). The extracytoplasmic function-type sigma factor SigM of *Corynebacterium glutamicum* ATCC 13032 is involved in transcription of disulfide stress-related genes. J. Bacteriol. *189*, 4696–4707.

Newton, G.L. and Fahey, R.C. (2002). Mycothiol biochemistry. Arch. Microbiol. *178*, 388–394.

Oldfield, C., Pogrebinsky, O., Simmonds, J., Olson, E.S., and Kulpa, C.F. (1997). Elucidation of the metabolic pathway for dibenzothiophene desulphurization by *Rhodococcus* sp. strain IGTS8 (ATCC 53968). Microbiology *143*, 2961–2973.

Omori, T., Monna, L., Saiki, Y., and Kodama, T. (1992). Desulfurization of dibenzothiophene by *Corynebacterium* sp. strain SY1. Appl. Environ. Microbiol. *58*, 911–915.

Ostrowski, J., Barber, M.J., Rueger, D.C., Miller, B.E., Siegel, L.M., and Kredich, N.M. (1989). Characterization of the flavoprotein moieties of NADPH-sulfite reductase from *Salmonella typhimurium* and *Escherichia coli*. Physicochemical and catalytic properties, amino acid sequence deduced from DNA sequence of *cysJ*, and comparison with NADPH-cytochrome P-450 reductase. J. Biol. Chem. *264*, 15796–15808.

Ozaki, H. and Shiio, I. (1982). Methionine biosynthesis in *Brevibacterium flavum*: Properties and essential role of O-acetylhomoserine sulfhydrylase. J. Biochem. *91*, 1163–1171.

Park, S.D., Lee, J.Y., Kim, Y., Kim, J.H., and Lee, H.S. (1998). Isolation and analysis of *metA*, a methionine biosynthetic gene encoding homoserine acetyltransferase in *C. glutamicum*. Mol. Cells *8*, 286–294.

Park, S.D., Lee, J.Y., Sim, S.Y., Kim, Y., and Lee, H.S. (2007). Characteristics of methionine production by an engineered *Corynebacterium glutamicum strain*. Metab. Eng. *9*, 327–336.

Plumbridge, J. (2001). DNA binding sites for the Mlc and NagC proteins: regulation of *nagE*, encoding the N-acetylglucosamine-specific transporter in *Escherichia coli*. Nucleic Acids Res. *29*, 506–514.

Raux, E., Leech, H.K., Beck, R., Schubert, H.L., Santander, P.J., Roessner, C.A., Scott, A.I., Martens, J.H., Jahn, D., Thermes, C., Rambach, A., and Warren, M.J. (2003). Identification and functional analysis of enzymes required for precorrin-2 dehydrogenation and metal ion insertion in the biosynthesis of sirohaem and cobalamin in *Bacillus megaterium*. Biochem. J. *370*, 505–516.

Rey, D.A., Pühler, A., and Kalinowski, J. (2003). The putative transcriptional repressor McbR, member of the TetR-family, is involved in the regulation of the metabolic network directing the synthesis of sulfur containing amino acids in *Corynebacterium glutamicum*. J. Biotechnol. *103*, 51–65.

Rey, D.A., Nentwich, S.S., Rückert, C., Koch, D.J., Pühler, A., Tauch, A., and Kalinowski, J. (2005). The McbR repressor modulated by the effector substance S-adenosylhomocysteine controls directly the transcription of a regulon involved in sulfur metabolism of *Corynebacterium glutamicum* ATCC 13032. Mol. Microbiol. *56*, 871–887.

Rossol, I. and Pühler, A. (1992). The *Corynebacterium glutamicum aecD* gene encodes a C-S lyase with a α,β-elimination activity that degrades aminoethylcysteine. J. Bacteriol. *174*, 2968–2977.

Rückert, C., Pühler, A., and Kalinowski, J. (2003). Genome-wide analysis of the L-methionine biosynthetic pathway in *Corynebacterium glutamicum* by targeted gene deletion and homologous complementation. J. Biotechnol. *104*, 213–228.

Rückert, C., Koch, D.J., Rey, D.A., Albersmeier, A., Mormann, S., Pühler, A., and Kalinowski, J. (2005). Functional genomics and expression analysis of the *Corynebacterium glutamicum fpr2-cysIXHDNYZ* gene cluster involved in assimilatory sulphate reduction. BMC Genomics *6*, 121.

Schäfer, A., Tauch, A., Droste, N., Pühler, A., and Kalinowski, J. (1997). The *Corynebacterium glutamicum cglIM* gene encoding a 5-cytosine methyltransferase enzyme confers a specific DNA methylation pattern in an McrBC-deficient *Escherichia coli* strain. Gene *203*, 95–101.

Schneider, T.D. and Stephens, R.M. (1990) Sequence Logos: A new way to display consensus sequences. Nucleic Acids Res. *18*, 6097-6100.

Sganga, M.W., Aksamit, R.R., Cantoni, G.L., and Bauer, C.E. (1992). Mutational and nucleotide sequence analysis of S-adenosyl-L-homocysteine hydrolase from *Rhodobacter capsulatus*. Proc. Natl. Acad. Sci. U. S. A. *89*, 6328–6332.

Shen, X.H., Huang, Y., and Liu, S.J. (2005). Genomic analysis and identification of catabolic pathways for aromatic compounds in *Corynebacterium glutamicum*. Microbes Environ. *20*, 160–167.

Sirko, A., Hryniewicz, M.M., Hulanicka, D., and Böck, A. (1990). Sulfate and thiosulfate transport in *Escherichia coli* K-12 nucleotide sequence and expression of the *cysTWAM* gene cluster. J. Bacteriol. *172*, 3351–3357.

Spencer, J.B., Stolowich, N.J., Roessner, C.A., and Scott, A.I. (1993). The *Escherichia coli cysG* gene encodes the multifunctional protein, siroheme synthase. FEBS Lett. *335*, 57–60.

Trötschel, C., Deutenberg, D., Bathe, B., Burkovski, A., and Krämer, R. (2005). Characterization of methionine export in *Corynebacterium glutamicum*. J. Bacteriol. *187*, 3786–3794.

van der Ploeg, J.R., Barone, M., and Leisinger, T. (2001). Functional analysis of the *Bacillus subtilis cysK* and *cysJI* genes. FEMS Microbiol. Lett. *201*, 29–35.

Wada, M., Awano, N., Haisa, K., Takagi, H., and Nakamori, S. (2002). Purification, characterization and identification of cysteine desulfhydrase of *Corynebacterium glutamicum*, and its relationship to cysteine production. FEMS Microbiol. Lett. *217*, 103–107.

Wada, M., Awano, N., Yamazawa, H., Hiroshi, T., and Nakamori, S. (2004). Purification and characterization of O-acetylserine sulfhydrylase of *Corynebacterium glutamicum*. Biosci. Biotechnol. Biochem. *68*, 1581–1583.

Williams, S.J., Senaratne, R.H., Mougous, J.D., Riley, L.W., and Bertozzi, C.R. (2002). 5′-adenosinephosphosulfate lies at a metabolic branch point in mycobacteria. J. Biol. Chem. *277*, 32606–32615.

From: Corynebacteria: Genomics and Molecular Biology. Edited by: Andreas Burkovski.

Regulation of Iron Homeostasis in *Corynebacterium glutamicum* 11

Julia Frunzke and Michael Bott

Abstract

The relevance of iron to corynebacteria was first noticed in the 1930s, when the synthesis of diphtheria toxin by *Corynebacterium diphtheriae* was shown to be dependent on the iron supply. Meanwhile, the DtxR protein responsible for this regulation has been intensively characterized biochemically. It functions as an intracellular Fe^{2+} sensor and, when its low-affinity iron-binding site is occupied, as a transcriptional regulator. In *C. diphtheriae*, several DtxR targets have been identified and thoroughly characterized, but no genome-wide studies were performed yet. In the non-pathogenic *Corynebacterium glutamicum*, on the other hand, the DtxR regulon has been determined by genome-wide studies including transcriptome comparisons of wild type and *dtxR* deletion mutants using DNA microarrays. By this approach, more than 50 genes were found to be repressed by DtxR under iron excess, most of which encode proteins involved in iron acquisition, notably several ABC transporters for siderophores. Moreover, genes which are presumably activated by DtxR have been identified, such as iron storage proteins and proteins responsible for iron-sulfur cluster assembly. Three of the target genes repressed by DtxR in *C. glutamicum* (*ripA*, *cgtR11*, *cg0527*) encode themselves transcriptional regulators. The AraC-type regulator RipA has been shown to repress a set of genes which encode prominent iron proteins of the cell, such as aconitase, succinate dehydrogenase or catalase. In this way, the iron demand of the cell is reduced, possibly allowing prolonged survival under iron limitation. In the case of the response regulator CgtR11, evidence for its involvement in heme utilization has been obtained, whereas the function of Cg0527 is not yet known. In summary, key players of a complex regulatory network controlling iron homeostasis in *C. glutamicum* have been elucidated in recent years.

1. Iron – essential but dangerous

Iron is an essential element for almost all organisms as constituent of proteins containing heme, iron-sulfur clusters, or mono- or binuclear iron species. Such proteins play important roles in many biological processes, such as electron transport, the tricarboxylic acid (TCA) cycle or gene regulation. Although iron is one of the most abundant elements in the earth crust, it is not readily available, since the predominant form of iron under aerobic conditions is the ferric state (Fe^{3+}) which has has been reported to have a solubility of 1.4 x 10^{-9} M at neutral pH (Chipperfield and Ratledge, 2000), in contrast to the ferrous state (0.1 M at pH 7.0). To cope with this problem, many microbes have evolved complex systems to solubilize iron from their environment. Besides being essential, iron can also be harmful to cells, mainly due to its reaction with hydrogen peroxide, a by-product of aerobic metabolism. In the Fenton reaction, Fe^{2+} reacts with H_2O_2 to yield Fe^{3+}, OH^- and the highly reactive and strongly damaging hydroxyl radical •OH (Pierre and Fontecave, 1999). Thus, cells have to cope with poor iron solubility on one hand and to avoid high levels of free intracellular Fe^{2+} on the other hand. As pointed out by Andrews et al. (2003), five major strategies are used by bacteria to deal with these problems: (i) high-affinity iron transport systems for scavenging iron from the environment; (ii) intracellular iron storage proteins providing an iron source when the external ones are transiently unavailable; (iii) redox stress resistance mechanisms such as degradation of reactive oxygen species and repair of damaged molecules; (iv) reduction of the cellular iron demand by partial repression of iron-containing proteins; and (v) a sophisticated regulatory system that controls all of the above-mentioned processes according to iron availability.

This chapter deals predominantly with iron homeostasis in the non-pathogenic Gram-positive soil bacterium *C. glutamicum*, but includes also relevant data from *C. diphtheriae*. First we will give an overview on the possibilities of *C. glutamicum* for iron acquisition and then we summarize current knowledge on a recently discovered complex regulatory network which controls iron homeostasis.

2. Iron acquisition by *Corynebacterium glutamicum*

2.1 Iron dependency of growth

Early studies showed that growth of *C. glutamicum* (formerly *Micrococcus glutamicus*) in synthetic medium depends on a sufficient supply of iron, which could be provided e.g. in form of ferrichrome (10 mg/l), or by high concentrations of iron salts (0.5 – 2.0 g/l). Moreover, a synergistic action between iron and several chelating agents on growth was demonstrated (Nakayama et al., 1964a; 1964b). Twenty-five years later, two studies addressed once more the problem of iron acquisition by *C. glutamicum* when a defined medium was required for metabolic studies. Von der Osten et al. (1989) showed that 0.1% (3.9 mM) citrate (provided as sodium salt) in combination with ≥20 mg/l $FeSO_4$ x 7 H_2O (72 μM) allowed good growth. Interestingly, citrate was consumed before glucose, but the depletion had no negative effect on the growth rate. The authors therefore suggested that iron is accumulated intracellularly in the early stages of cultivation. Liebl et al. (1989) confirmed the results of von der Osten et al., but noticed that the high concentrations of citrate required might be undesirable in metabolic studies. They could show that low concentrations (10 μM) of protocatechuate (3,4-dihydroxybenzoate) or catechol in combination with 72 μM $FeSO_4$ suffice to achieve the same growth as 0.1% citrate (Liebl et al., 1989). The standard minimal medium used in our and several other laboratories, called CGXII (Keilhauer et al., 1993), contains 30 mg/l (195 μM) protocatechuate. Like in the case of citrate (Polen et al., 2007), also protocatechuate is used as a carbon source by *C. glutamicum* (Merkens et al., 2005). Again, accumulation of iron in the initial growth phase might be responsible for maintaining a high growth rate when protocatechuate has been consumed. According to the studies described above, *C. glutamicum* is able to take up ferrichrome, ferric citrate and iron complexes of protocatechuate and catechol. The overall iron content of *C. glutamicum* cells has been determined to vary between 0.04 mg iron/g cell dry weight on CGIII complex medium and 0.3 - 0.5 mg iron/g cell dry weight on BMCG and CGXII minimal medium used for cultivation (Liebl, 2005).

Except for the studies described above and a very recent report (Dertz et al., 2006), no experimental data on iron transport in *C. glutamicum* appear to exist. Therefore, the current knowledge on iron acquisition is primarily based on the annotated genome sequence (Ikeda and Nakagawa, 2003; Kalinowski et al., 2003) and on DNA microarray studies in which global gene expression was analyzed in cells grown with different iron supplies or in mutant cells lacking regulators of iron metabolism (Brune et al., 2006; Krug et al., 2005; Wennerhold and Bott, 2006; Wennerhold et al., 2005).

2.2 *C. glutamicum* does not synthesize siderophores by itself

To overcome the poor solubility of ferric iron, many microorganisms synthesize and excrete low molecular mass (<1000 Da) compounds called siderophores that are characterized by their high specificity and affinity (K_d ~ 10^{-30} M^{-1}) towards ferric iron. More than 500 different siderophores are currently known and formation of iron complexes usually involves either hydroxamate groups, or catechol rings, or hydroxy acids. After import of iron siderophore complexes into the bacterial cytoplasm, the iron is released either by the reduction to Fe^{2+} catalyzed by ferri-siderophore reductases (Schröder et al., 2003) or by breaking down the siderophore structure as described for several siderophores, *e.g.* bacillibactin (Miethke et al., 2006) or enterobactin (Lin et al., 2005). In 1997, Budzikiewicz and coworkers (Budzikiewicz et al., 1997) reported the isolation of a cyclic catecholate siderophore from *C. glutamicum* strain ATCC14067 which they named corynebactin. A siderophore of identical structure was later isolated from *Bacillus subtilis* and designated bacillibactin (May et al., 2001). Moreover, the genes *dhbABCEF* responsible for bacillibactin synthesis were identified (May et al., 2001). Analysis of the genome sequence of *C. glutamicum* strain ATCC13032 revealed a gene homologous to *dhbC* (encoding isochorismate synthetase), but there were no genes present that are homologous to *dhbA*, *dhbB*, *dhbE* or *dhbF*. Moreover, no other genes obviously involved in siderophore biosynthesis were identified, leading to the conclusion that at least the type strain

Table 11.1. Transport systems possibly involved in iron acquisition in *Corynebacterium glutamicum* ATCC13032

Gene	Protein	Molecular mass, kDa	Function	DtxR binding site	Pfam domain and further information
Ferric-siderophore uptake systems					
cg0052		32.3	siderophore ABC transporter, permease protein	-	PF01032, FecCD transport family
cg0053		28.3	siderophore ABC transporter, ATP-binding protein		PF00005, ABC transporter
cg0054		31.0	cytoplasmic siderophore-interacting protein		PF04954, siderophore-interacting protein PF08021, siderophore-interacting FAD- binding domain
cg0405		34.0	secreted siderophore-binding lipoprotein	-	PF01497, periplasmic binding protein
cg0591		35.4	siderophore ABC transporter, permease protein	+	PF01032, FecCD transport family
cg0590		41.1	siderophore ABC transporter, permease protein		PF01032, FecCD transport family
cg0589		28.9	siderophore ABC transporter, ATP-binding protein		PF00005, ABC transporter
cg0748		35.1	secreted siderophore-binding lipoprotein	+	PF01497, periplasmic binding protein
cg0771		37.4	secreted siderophore-binding lipoprotein	+	PF01497, periplasmic binding protein
cg0770		37.8	siderophore ABC transporter, permease protein		PF01032, FecCD transport family
cg0769		35.3	siderophore ABC transporter, permease protein		PF01032, FecCD transport family
cg0768		29.1	siderophore ABC transporter, ATP-binding protein		PF00005, ABC transporter
cg0767		32.5	cytoplasmic siderophore-interacting protein		PF04954, siderophore-interacting protein PF08021, siderophore-interacting FAD- binding domain

cg0778	34.7	siderophore ABC transporter, permease protein	-	PF01032, FecCD transport family
cg0777	27.7	siderophore ABC transporter, ATP-binding protein		PF00005, ABC transporter
cg0776	33.8	secreted siderophore-binding lipoprotein		PF01497, periplasmic binding protein
cg0922	33	secreted siderophore-binding lipoprotein	+	PF01497, periplasmic binding protein
cg0921	30.8	cytoplasmic siderophore-interacting protein		PF04954, siderophore-interacting protein PF08021, siderophore-interacting FAD-binding domain
cg0924	35.7	secreted siderophore-binding lipoprotein	+	PF01497, periplasmic binding protein
cg0926	36.0	siderophore ABC transporter, permease protein	+	PF01032, FecCD transport family
cg0927	40.8	siderophore ABC transporter, permease protein		PF01032, FecCD transport family
cg0928	27.9	siderophore ABC transporter, ATP-binding protein		PF00005, ABC transporter
cg1405	31.2	cytoplasmic siderophore-interacting protein	+	PF04954, siderophore-interacting protein PF08021, siderophore-interacting FAD-binding domain
cg1418	35.9	secreted siderophore-binding lipoprotein	+	PF01497, periplasmic binding protein
cg1642	27.1	cytoplasmic siderophore-interacting protein	+	PF04954, siderophore-interacting protein PF08021, siderophore-interacting FAD-binding domain
cg2234	33.0	secreted siderophore-binding lipoprotein	+	PF01497, periplasmic binding protein
cg3404	36.6	secreted siderophore-binding lipoprotein	+	PF01497, periplasmic binding protein high similarity to FagD of *Corynebacterium pseudotuberculosis* (Billington et al. 2002)

Table 11.1 continued

Gene	*Protein*	*Molecular mass, kDa*	*Function*	*DtxR binding site*	*Pfam domain and further information*
Heme transport systems					
cg0466	HtaA	62.1	secreted heme-transport associated protein	+	PF04213, Htaa
cg0467	HmuT	37.3	hemin-binding periplasmic protein precursor		PF01497, periplasmic binding protein
cg0468	HmuU	37.1	hemin transport system, permease protein		PF01032, FecCD transport family
cg0469	HmuV	28.5	hemin transport system, ATP-binding protein		
cg0470		33.6	secreted heme-transport associated protein	+	PF04213, Htaa
cg0471		29	secreted heme-transport associated protein		PF04213, Htaa
cg3156		26.9	secreted heme-transport associated protein	+	PF04213, Htaa
Vitamin B12, heme or siderophore transport systems					
cg1832		37.3	permease of ABC transporter	-	PF01032, FecCD transport family
cg1833		33.5	secreted substrate-binding lipoprotein		PF01494, periplasmic binding protein
cg1834		27.7	ATP-binding protein of ABC transporter		PF00005, ABC transporter
cg2315		12.4	ATP-binding protein of ABC transporter	-	PF01497, periplasmic binding protein
cg2317		35.8	permease of ABC transporter		PF01032, FecCD transport family
cg2318		27.1	putative secreted vitamin B12-binding lipoprotein		PF00005, ABC transporter

Putative Fe^{2+}, Mn^{2+} or Zn^{2+} ABC transporter				
cg0043	23.7	ATP-binding protein of ABC transporter	+	PF00005, ABC transporter
cg0042	29.8	permease of ABC transporter		PF00950, ABC 3 transport family PF01032, FecCD transport family
cg0041	33.6	secreted Fe^{2+}, Mn^{2+} or Zn^{2+}-binding lipoprotein		PF01227, periplasmic solute binding protein family
Putative Fe^{3+} ABC transporter			+	
cg0508	37.9	secreted substrate-binding lipoprotein	-	PF01547, extracellular solute-binding protein
cg0507	55.6	permease of ABC transporter		PF00528, binding-protein-dependent transport system, inner membrane component
cg0506	37.8	ATP-binding protein of ABC transporter		PF00005, ABC transporter

ATCC13032 does not produce siderophores (Wennerhold and Bott, 2006). Recently, Raymond and coworkers reinvestigated corynebactin (alias bacillibactin) synthesis by *C. glutamicum* ATCC14067 and failed to detect the production of this siderophore (Dertz et al., 2006). The strain was also negative in the Chrome Azurol S (CAS) assay, which indicates the inability to synthesize siderophores that remove iron from the CAS chromophore. Finally, the strain did not take up ^{55}Fe-labeled bacillibactin, which would be expected for a strain that synthesizes and excretes this siderophore. Thus, *C. glutamicum* ATCC14067 like the type strain ATCC13032 does not produce siderophores (Dertz et al., 2006).

2.3 *C. glutamicum* has a large arsenal of genes for siderophore utilization

In contrast to the inability to synthesize siderophores by itself, *C. glutamicum* has a substantial arsenal of more than 25 genes annotated to be involved in the acquisition of iron from different siderophores. These include five ABC transporters, nine secreted siderophore-binding lipoproteins, and five cytoplasmic siderophore-interacting proteins (Table 11.1). Thus, it can be suggested that *C. glutamicum* is able to benefit in its natural habitat from siderophores produced by other organisms. However, as none of these transport components has been studied yet, their substrate specificity is still unknown. According to the results described in the preceding paragraph, ferrichrome and ferric citrate belong to the substrates accepted by these transporters.

Gram-negative bacteria have developed a sophisticated mechanism to transport ferric siderophores through the outer membrane. High-affinity receptor proteins located within the outer membrane (*e.g.* FhuA, FecA or FepA) are supplied with energy from the cytoplasmic membrane by the Ton complex (TonB, ExbB and ExbD, for review see (Postle and Kadner, 2003)), thereby allowing transport of the bound ligand. In the periplasm, the ferric siderophores are captured by soluble periplasmic binding proteins and subsequently transported through the cytoplasmic membrane (Faraldo-Gomez and Sansom, 2003). Especially the uptake systems for the siderophores enterobactin (FepABCD) and ferrichrome (FhuACDB) as well as ferric citrate (FecABCD) have been studied intensely in Gram-negative species. Homologs of FhuA, FecA, FepA, TonB, ExbB and ExbD are not present in corynebacteria. However, there is evidence that these bacteria and other members of the suborder *Corynebacterinae* possess a structure with functional similarity to the outer membrane of Gram-negative bacteria (Bayan et al., 2003; Puech et al., 2001). This structure is composed in particular of mycolic acids (2-alkyl-3-hydroxy acids of variable chain length), either linked to the cell wall sugar polymer arabinogalactan or esterified with trehalose. The question how siderophores and other iron complexes pass this hydrophobic layer has not been addressed yet. Porins may play a role in this process (Costa-Riu et al., 2003; Hünten et al., 2005).

2.4 *C. glutamicum* can utilize heme as iron source

In addition to siderophores many bacteria can also acquire iron from heme. Following the transport of heme into the cytoplasm *via* specific heme transporters, the iron is extracted by heme oxygenases that degrade the tetrapyrrole ring. Up to now heme uptake in Gram-positive organisms has rarely been studied, however, heme uptake systems of *C. diphtheriae* and *Corynebacterium ulcerans* and have recently been documented (Drazek et al., 2000; Schmitt and Drazek, 2001). The heme transporter of *C. diphtheriae* encoded by the *hmu* operon shows similarity to heme-dependent ABC transporters of Gram-negative bacteria (Drazek et al., 2000). The *hmu* operon consists of five genes including *htaA*, *hmuT*, *hmuU*, *hmuV* and an ORF of unknown function. The promoter-proximal *htaA* gene encodes a protein with a signal sequence and a putative membrane spanning region, suggesting that HtaA is secreted and possibly anchored to the cell surface. Its exact function is not yet known. The heme-dependent ABC transport system consists of the binding protein HmuT, the permease HmuU and the ATPase component HmuV (Drazek et al., 2000). Based on experimental data it has been proposed that HmuT is a heme- and hemoglobin-binding lipoprotein, sharing ~25% sequence identity to the periplasmic heme binding proteins (PBPs) from *Yersinia* species (Hornung et al., 1996; Wandersman and Stojiljkovic, 2000). The genome of *C. glutamicum* contains a putative operon (*cg0466-cg0469*) showing the same gene order as reported for the *hmu* operon of *C. diphtheriae*. The encoded proteins share about 50% sequence identity with the *C. diphtheriae* homologs. Additionally, the *C. glutamicum* genome contains three other genes predicted to encode proteins involved in heme transport, all of which contain a signal peptide and a putative C-terminal transmembrane helix (*cg0470*, *cg0471*, *cg3156*), and a heme oxygenase (*cg2445*) (Table 11.1). All these genes were recently shown to be repressed in *C. glutamicum* by the global iron regulator DtxR under conditions of sufficient iron supply (Brune et al., 2006; Wennerhold and Bott, 2006). The presence of the abovementioned genes and their iron-dependent regulation suggests that *C. glutamicum* is able to import heme and use it as an iron source. In fact, recent experiments in our laboratory have confirmed that *C. glutamicum* can grow with heme as sole iron source (Frunzke and Bott, unpublished results).

2.5 Survey of the *C. glutamicum* genome for the presence of putative ferrous and ferric iron transporters

Cells can acquire iron not only in form of iron complexes, but also in an uncomplexed state as ferric or ferrous ions. Analysis of the *C. glutamicum* genome revealed an apparent operon (*cg0508-cg0507-cg0506*) whose protein products share some similarity (25 – 30%) with the SfuABC system of *Serratia marcescens*, which has been described as the first siderophore-independent ferric iron ABC transporter (Angerer et al., 1990; Schryvers et al., 1998). Whether the proteins encoded by *cg0508-cg0507-cg0506* also catalyze ferric iron transport in *C. glutamicum* remains to be analyzed. As no DtxR binding site was identified in the promoter region of *cg0508*, these genes presumably do not belong to the DtxR regulon.

So far several systems have been described which mediate the uptake of ferrous iron and other divalent cations, like Mn^{2+}, Co^{2+}, Ni^{2+} or Mg^{2+} in various bacterial species (Koster, 2005). The Feo system of *Escherichia coli* was one of the first ferrous iron transport systems identified in bacteria (Cartron et al., 2006). Homologs of *feoB* and *feoA* are widely distributed in bacteria, but absent from the genome of *C. glutamicum*. MntH, an Nramp-like protein (natural resistance associated macrophage proteins) of *E. coli*, has also been described as divalent cation transporter, but with a preference for Mn^{2+}. Whereas the genome of *C. glutamicum* ATCC13032 does not encode proteins with significant identity to Nramp-like proteins, the genome of *C. glutamicum* strain R (Yukawa et al., 2007) encodes a protein (CgR_0158) which shares some sequence identity (34%) to *E. coli* MntH. In contrast to the Nramp proteins of various eukaryotes, the MntH proteins in bacteria seem to be of minor importance for the transport of ferrous iron under physiological conditions (Courville et al., 2006; Makui et al., 2000; Patzer and Hantke, 2001).

Several bacterial species use the Sit ABC transport system to mediate transport of Mn^{2+} and Fe^{2+}. In *C. glutamicum* the genes *cg0041* (secreted substrate-binding lipoprotein), *cg0042* (permease) and *cg0043* (ATP-binding protein) encode an ABC transporter whose proteins display some sequence identity (~30 %) with those encoded by *sitABCD* from *Enterobacterianeae* species (Janakiraman and Slauch, 2000). The *cg0043* gene is located upstream and divergent to *cg0042-cg0041*. A functional DtxR binding site has been identified in the intergenic region of *cg0042* and *cg0043*, but in DNA microarray studies no effect on the transcription of these genes has been observed in a *dtxR* deletion mutant (Wennerhold and Bott, 2006). However, transcriptome analysis of *C. glutamicum* wild type cells cultivated with different iron concentrations revealed a decreased mRNA level of this operon under iron limitation (Krug et al., 2005). As for the genes described above, further studies are required to determine the metal ion substrate(s) of this ABC transporter and its iron-dependent regulation.

3. Control of iron homeostasis in *Corynebacterium glutamicum*

3.1 DtxR, the repressor of the *Corynebacterium diphtheriae tox* gene

In the early 1930s it was demonstrated that high levels of iron in the growth medium repressed the production of diphtheria toxin by *C. diphtheriae* (Pappenheimer and Johnson, 1936). About 60 years later, this effect was shown to be mediated by the transcriptional regulator DtxR (diphtheria toxin regulator) (Boyd et al., 1990; Schmitt and Holmes, 1991). Under iron excess, DtxR in complex with Fe^{2+} represses transcription of the *tox* gene. When iron becomes limiting, Fe^{2+} dissociates from DtxR and apo-DtxR becomes inactive, causing derepression of the *tox* gene. In contrast to diphtheria toxin, which is encoded in the genome of some corynephages, DtxR is encoded

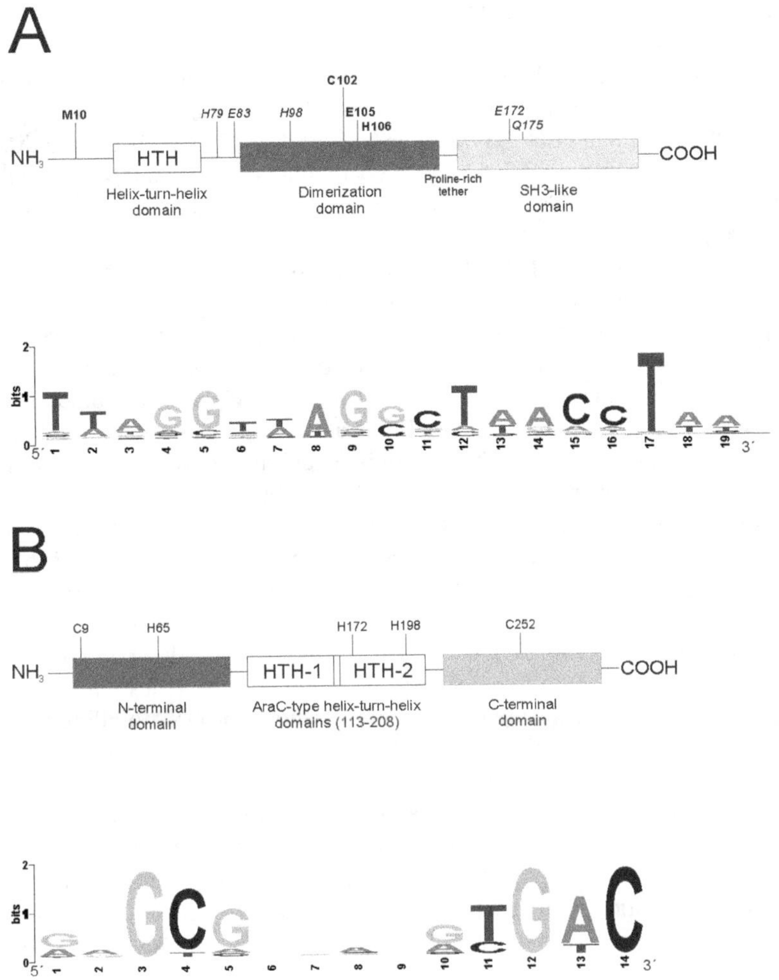

Figure 11.1. Domain structure and consensus DNA binding site of DtxR (A) and RipA (B) from *C. glutamicum*. In the structure of DtxR, amino acids printed in *italic* are ligands of the ancillary metal-binding site (site 1) and amino acids printed in **bold** constitute the primary binding site (site 2). The functional significance of the cysteine and histidine residues in the domain structure of RipA are not clear yet. They could be involved in binding of a metal ion. The presentation of the consensus binding sites was obtained using the WEBLOGO tool (Crooks et al., 2004). In the case of DtxR, the 52 binding sites shown in Table 3 of Wennerhold and Bott, 2006, were used to derive a consensus motif; in the case of RipA the 13 binding sites shown in Fig. 7 of (Wennerhold et al., 2005).

by the *C. diphtheriae* genome (locus tag DIP1414; (Cerdeno-Tarraga et al., 2003)). Meanwhile, it has been shown that DtxR is not a specific regulator of the *tox* gene, but controls many genes involved in iron acquisition in *C. diphtheriae* (Kunkle and Schmitt, 2003). Homologs of DtxR are widely distributed in bacteria and some have been characterized, *e.g.* the IdeR protein of *Mycobacterium tuberculosis* (Gold et al., 2001).

3.2 Metal ion activation of DtxR

The protein structures of DtxR from *C. diphtheriae* and IdeR from *M. tuberculosis* have been intensively investigated and provided important knowledge about the mechanism of metal ion activation of DtxR-like proteins (D'Aquino et al., 2005; Ranjan et al., 2006). The active form of DtxR is a homodimer, with each monomer consisting of two domains connected by a flexible tether of 23 amino acids that contains a proline-rich region (Qiu et al., 1995; Schiering et al., 1995). The N-terminal domain contains the helix-turn-helix motif responsible for DNA binding as well as two metal ion binding sites (Figure 11.1). The ancillary binding site (site 1) binds Fe^{2+} with high affinity ($K_d = 2 \times 10^{-7}$ M for Ni^{2+}, (D'Aquino et al., 2005)) and, when occupied, stabilizes the monomeric form of the repressor. When the iron concentration increases, the primary binding site (site 2), a low affinity Fe^{2+} binding site ($K_d = 6.3 \times 10^{-4}$ M for Ni^{2+}, (D'Aquino et al., 2005)), is occupied, leading to conformational changes in the N-terminal region and subsequently dimerization of the protein. Besides Fe^{2+}, several other divalent metal ions can activate the binding of DtxR to its target promoters *in vitro*, preferred in the following order $Fe^{2+} \sim Ni^{2+} > Co^{2+} >> Mn^{2+}$ (Spiering et al., 2003; Tao and Murphy, 1992). The C-terminal domain has been shown to be structurally similar to eukaryotic SH3 domains (Qiu et al., 1996). So far, the function of this domain is not completely understood. Although contributing two ligands (Glu170 and Gln173) to the ancillary ion binding site, studies with mutants lacking this domain indicate that it is not essential for DNA binding (Posey et al., 1999).

Crystal structures of the complex of metal-ion activated DtxR and a DNA fragment containing the *tox* operator have shown that two DtxR dimers interact with the target DNA at almost opposite sites of the DNA and do not interact with each other (Pohl et al., 1999; White et al., 1998). The minimal essential nucleotide sequence for DtxR binding is a 9-bp inverted repeat with an 1-bp gap (TWAGGTWAGSCTWACCTWA) (Tao and Murphy, 1994).

3.3 The DtxR regulon of *C. glutamicum*: genes repressed by DtxR

The genome of the *C. glutamicum* type strain ATCC13032 (Ikeda and Nakagawa, 2003; Kalinowski et al., 2003) contains a gene (*cg2103*) encoding a transcriptional regulator with 72% sequence identity to the diphtheria toxin repressor DtxR of *C. diphtheriae*. Therefore, it has been assumed that this regulator might carry out a similar regulatory function as DtxR from *C. diphtheriae*. In fact, the corresponding protein from *C. glutamicum* strain

Figure 11.2, opposite. The DtxR regulon of *Corynebacterium glutamicum*. Overview on those *C. glutamicum* genes or gene clusters that are preceded by a functional DtxR binding site and showed altered mRNA levels in a *dtxR* deletion mutant (Wennerhold and Bott, 2006). The black boxes indicate the position of the identified DtxR boxes. The mRNA ratios (Δ*dtxR* mutant *vs.* wild type) obtained from DNA microarray experiments are shown below the corresponding gene. The first value gives the ratio under iron excess (100 μM $FeSO_4$ in the growth medium), the second value

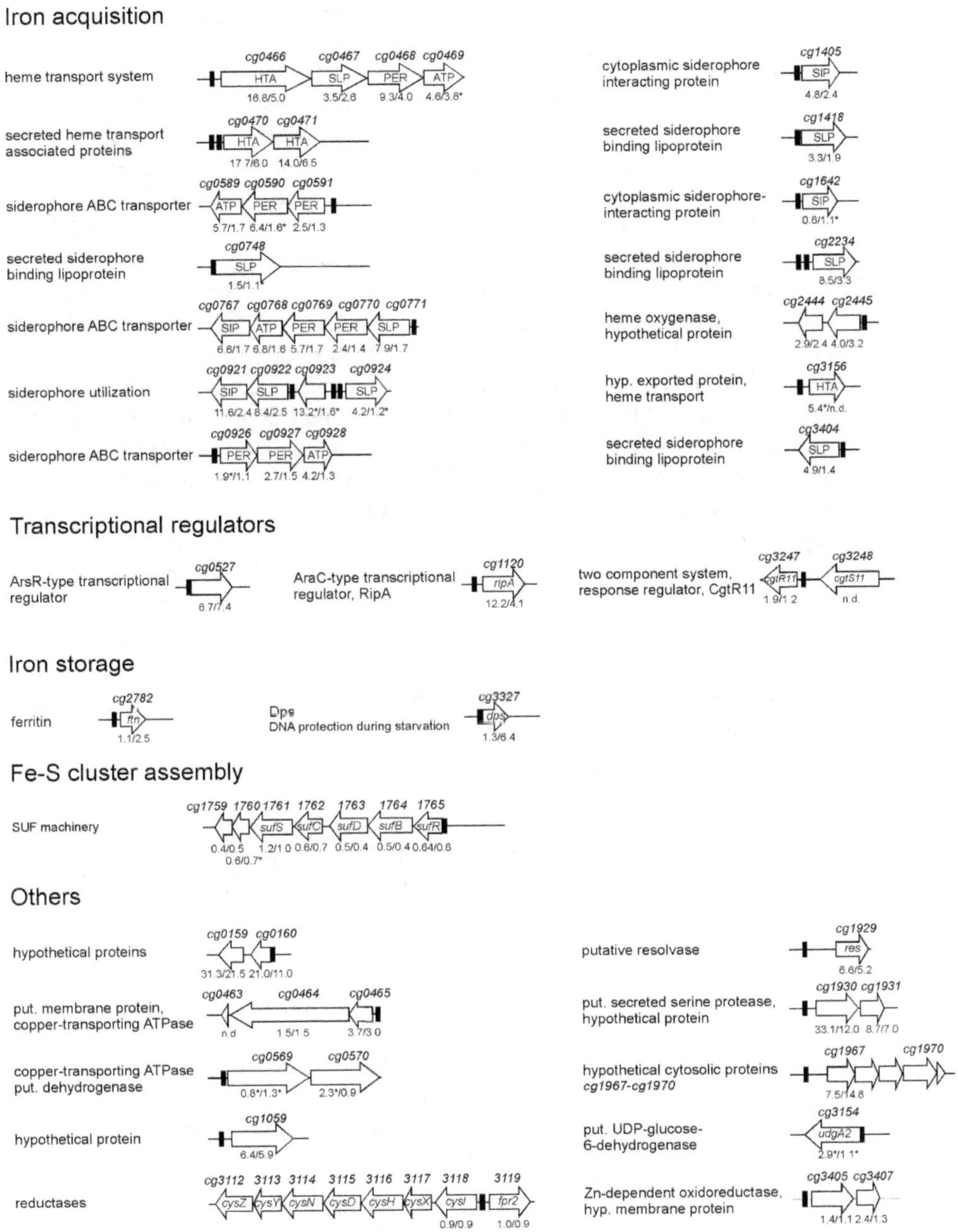

gives the ratio under iron limitation (1 μM $FeSO_4$). It has to be taken into account that for the majority of genes the indicated function was derived from the genome annotation rather than from experimental data. Additional genes which are preceded by a functional DtxR binding site have been omitted here, since their mRNA level was unchanged in a *dtxR* deletion mutant (Wennerhold and Bott, 2006). In the case of the genes *cg3115-cg3119*, increased mRNA levels in a Δ*dtxR* mutant were reported (Brune et al., 2006).

ATCC13869 (previously designated "*Brevibacterium lactofermentum*") was shown to bind the diphtheria *tox* promoter/operator in an iron-dependent manner (Oguiza et al., 1995). Recently, the DtxR regulon of *C. glutamicum* ATCC13032 has been determined in two independent studies using comparative transcriptome analysis of *dtxR* deletion mutants with the wild type in combination with computer-based approaches and *in vitro* binding studies in order to identify functional DtxR binding sites in the genome (Brune et al., 2006; Wennerhold and Bott, 2006). The results of these studies revealed that more than 50 genes are repressed by DtxR (Figure 11.2).

The majority of these genes encode proteins involved in iron acquisition, such as four ABC transporters (permease and ATPase) for siderophore uptake, seven secreted siderophore-binding lipoproteins, three cytoplasmic siderophore-interacting proteins, one ABC transporter for heme uptake (permease and ATPase), a secreted heme-binding lipoprotein, four proteins presumably associated in an unknown manner with heme transport, and heme oxygenase. Several genes repressed by DtxR encode proteins that are not obviously related to iron metabolism. These include two putative copper-transporting P-type ATPases, a putative resolvase, a putative secretory serine protease, a putative UDP-glucose-6-dehydrogenase, or a putative Zn-dependent oxidoreductase. The genes *cysI* (*cg3118*) and *cysJ* (*cg3119*) encode a putative sulfite reductase and a putative NADPH-ferredoxin reductase, respectively, and showed increased mRNA levels in one of the *dtxR* deletion mutants analyzed (Brune et al., 2006). Further studies are required to understand the increased expression of these genes under iron limitation. Interestingly, three genes repressed by DtxR encode themselves transcriptional regulators, adding a further level of complexity to the DtxR regulon. These are *ripA* (*cg1120*), *cg0527* encoding an ArsR-type regulator, and the response regulator *cgtR11* belonging to the CgtSR11 two-component signal transduction system (Kocan et al., 2006). Whereas the regulons of Cg0527 and CgtR11 are not yet known, that of RipA was identified recently and will be described in detail later in this chapter (Wennerhold et al., 2005).

3.4 The DtxR regulon of *C. glutamicum*: genes presumably activated by DtxR

The studies on the DtxR regulon revealed that this regulator can function not only as a transcriptional repressor but presumably also as an activator (Brune et al., 2006; Wennerhold and Bott, 2006). For example the genes *dps* (*cg3327*) and *ftn* (*cg2782*) had an about 10-fold decreased mRNA level in one of the *dtxR* deletion mutants (Brune et al., 2006). As the *C. glutamicum* Dps protein shows more than 50% sequence identity to *Mycobacterium smegmatis* Dps, which has been characterized structurally and functionally (Ceci et al., 2005; Gupta and Chatterji, 2003; Gupta et al., 2002; Roy et al., 2004), it presumably has a similar function and protects DNA from oxidative damage by sequestering intracellular Fe^{2+} and storing it in the form of Fe^{3+} oxyhydroxide mineral. Dps catalyzes the oxidation of Fe^{2+} by hydrogen peroxide: $2\ Fe^{2+} + H_2O_2 + 2\ H^+ \rightarrow 2\ Fe^{3+} + 2\ H_2O$. In this way, it prevents hydroxyl radical formation by the Fenton reaction and protects DNA from hydroxyl radical-mediated cleavage. Dps binds DNA with no apparent sequence specificity. In contrast to Dps, the sole function of ferritin is presumably restricted to iron storage. In *M. tuberculosis* it has been demonstrated that expression of the bacterioferritin-encoding gene *bfrA* and of the ferritin-encoding gene *bfrB* is activated by binding of the DtxR-homolog IdeR to a double iron box located 106 bp upstream of the transcriptional start site (Gold et al., 2001; Rodriguez and Smith, 2003). In *C. glutamicum* the DtxR binding sites in the promoter regions of *ftn* and *dps* were also located upstream of the -35 region, supporting an activating function of DtxR (Brune et al., 2006).

The synthesis of iron-sulfur clusters is usually also tightly regulated in response to iron availability and oxidative stress, since many iron-sulfur proteins play important roles in central metabolic pathways. A functional DtxR binding site has been identified in the upstream region of the *suf* operon in *C. glutamicum*. This operon consists of seven genes (*cg1765 – cg1759*) which showed an approx. twofold decreased mRNA level in one of the *dtxR* deletion mutants (Wennerhold and Bott, 2006). With the exception of the first gene, designated here as *sufR*, the proteins encoded by this operon are presumably involved in the assembly and repair of iron-sulfur clusters, constituting the so-called SUF operon in *C. glutamicum*. Recently, it has been shown that the SUF proteins of *M. tuberculosis* constitute the sole machinery for iron-sulfur cluster assembly in this organism (Huet et al., 2005). A similar situation can be assumed for *C. glutamicum*, since analysis of the genome sequence gave no evidence for alternative genes involved in iron-sulfur cluster assembly, e.g. genes encoding proteins of the Isc or Nif machinery (Johnson et al., 2005). Activation of the *suf* operon by DtxR might be explained by the fact that under conditions of sufficient iron supply, higher levels of several prominent iron-sulfur-cluster-containing proteins are formed, *e.g.* aconitase or succinate dehydrogenase, requiring higher levels of the SUF machinery. Curiously, the mRNA level of the *suf* genes was decreased in the Δ*dtxR* mutant not only under iron excess, but also under iron limitation (Wennerhold and Bott, 2006). Therefore, the role of DtxR in regulation of the *suf* operon is not completely clear. The first gene of the *C. glutamicum suf* operon encodes a transcriptional regulator of the DeoR family. It might regulate the expression of the *suf* operon in response to the cellular demand for iron-sulfur cluster assembly and repair, similar to the regulators IscR of *E. coli* (Giel et al., 2006; Schwartz et al., 2001; Yeo et al., 2006) or SufR of *Synechocystis* (Wang et al., 2004). Thus, expression of the *suf* operon of *C. glutamicum* is presumably regulated by two regulators, DtxR and SufR.

An interesting gene cluster with respect to iron-dependent regulation in *C. glutamicum* is the *sdhCAB* operon encoding succinate dehydrogenase. A functional DtxR binding site was identified 105 bp (with respect to the center of the 19-bp binding site) upstream of the transcriptional start site of *sdhC*, which suggests a transcriptional activation of *sdhCAB* by DtxR. The *sdhCAB* mRNA level was indeed about 5-fold reduced in the Δ*dtxR* mutant, however, this effect is most likely due to repression by RipA (see below) rather than to a missing activation by DtxR (Brune et al., 2006; Wennerhold and Bott, 2006). Further studies are required to elucidate the direct influence of DtxR on *sdhCAB* expression.

3.5 DtxR – the master regulator of iron metabolism in *C. glutamicum*

Based on the results described above, DtxR obviously is the master regulator of iron metabolism in *C. glutamicum*, controlling directly more than 60 genes (Figure 11.2). Its activity is determined by the occupation of a low affinity Fe^{2+} binding site and therefore by the cytoplasmic Fe^{2+} concentration, which thus seems to represent the decisive signal that reflects the cellular iron supply. Since at least three of the target genes that are repressed by DtxR function as transcriptional regulators (*ripA*, *cgtR11*, *cg0527*) by themselves, a set of further genes is controlled indirectly by DtxR.

3.6 RipA, a regulator of iron proteins

The *cg1120* gene of *C. glutamicum* encodes an AraC-type regulator (Figure 11.1) and was first noticed in DNA microarray experiments comparing cells grown under iron limitation and iron excess, where the *cg1120* mRNA level was strongly increased under iron limitation (Krug et al., 2005). At the same time, the mRNA of several genes encoding iron-containing proteins (*e.g* aconitase, succinate dehydrogenase and nitrate reductase) was decreased under iron limitation. Therefore, this regulator was considered as a candidate for repressing the iron protein-encoding genes and was subsequently named RipA for "regulator of iron proteins A". A *ripA* deletion mutant showed a growth defect under iron limitation, but not under iron excess (Wennerhold et al. 2005). Under iron limitation, the aconitase activity was ~2-fold higher in the Δ*ripA* mutant (~0.2 U (mg protein)$^{-1}$) than in the wild type (~0.1 U (mg protein)$^{-1}$), whereas it was similar under iron excess (~0.3 U (mg protein)$^{-1}$), supporting a repressor function of RipA. In a transcriptome comparison of Δ*ripA* mutant *vs.* wild type, about a dozen genes showed 2- to 4-fold increased mRNA levels in the mutant under iron limitation, but not under iron excess. This group of genes correlated well with those showing a decreased mRNA level in the wild type under iron limitation and included *acn*, *sdhCAB*, *narKGHJI* (nitrate/nitrite transporter and nitrate reductase), *leuCD* (isopropylmalate dehydratase), *catA* (catechol 1,2-dioxygenase), *pta* (phosphotransacetylase), and *katA* (catalase) (Figure 11.3). Except for the transporter NarK, phosphotransacetylase and acetate kinase (the *ackA* gene encoding the latter enzyme is co-transcribed with *pta* and presumably also co-regulated), the enzymes encoded by the other genes are known to contain iron. The function of RipA as a repressor of the abovementioned genes under iron limitation was supported by gel shift assays showing that purified RipA binds to the seven corresponding promoter upstream regions. In all cases, evidence for at least two RipA binding sites was obtained and from 13 experimentally verified binding sites the 14-bp consensus motif RRGCGNNNNRYGAC was derived (Figure 11.1). AraC-type regulators (Gallegos et al., 1997; Martin and Rosner, 2001) like RipA contain two adjacent helix-turn-helix (HTH)

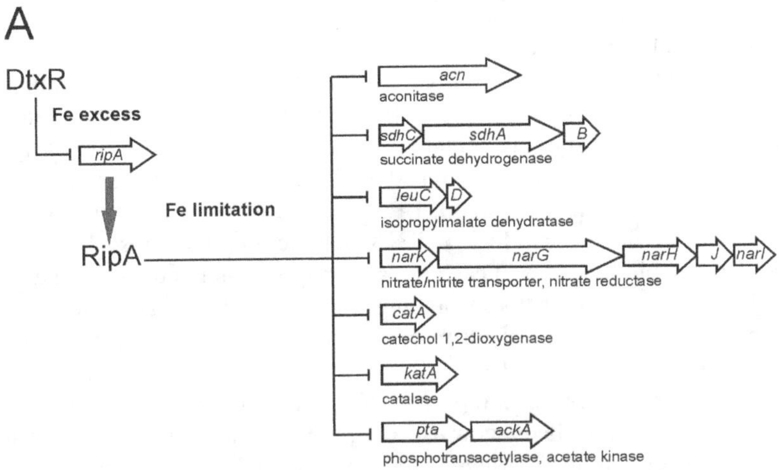

B

Iron cofactors of RipA targets

Aconitase	1x 4Fe-4S cluster
Succinate dehydrogenase	1x 4Fe-4S cluster 1x 3Fe-4S cluster 1x 2Fe-2S cluster 2x heme *b*
Nitrate reductase	3x 4Fe-4S cluster 1x 3Fe-4S cluster 2x heme *b*
Isopropylmalate dehydratase	1x 4Fe-4S cluster
Catechol 1,2-dioxygenase	1Fe per subunit
Catalase	1heme *b* per subunit
Phosphotransacetylase, Acetate kinase	/

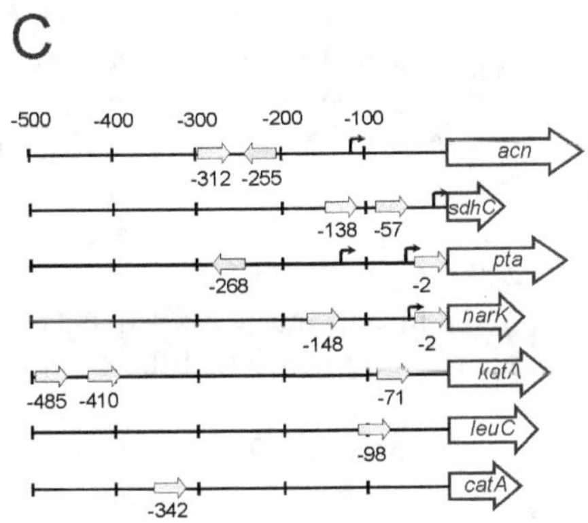

Figure 11.3. RipA-dependent regulation of prominent iron proteins in *C. glutamicum*. **(A)** Under iron excess, DtxR in complex with iron represses transcription of *ripA*. Under iron limitation, the AraC-type regulator RipA is synthesized and represses several genes encoding iron-containing proteins. In this way, RipA reduces the iron expenditure of the cell under iron limitation, which is probably beneficial for survival under this stress condition. **(B)** Iron cofactors of the proteins controlled by RipA. **(C)** Localization of the verified RipA binding sites in the promoter regions of its target genes. The orientation of the RipA binding motif within the promoter region is indicated by an arrow. The position beneath the arrows indicates the distance of the center of the binding site to the predicted start codon of the corresponding gene (position +1). For *acn*, *sdhC*, *pta* and *narK*, the transcriptional start sites have been determined experimentally and are indicated by black arrows.

motifs, which in the case of MarA insert in two adjacent segments of the major groove of the *mar* promoter (Rhee et al., 1998). Thus, one HTH region of RipA might interact with the RRGCG motif and the adjacent HTH region with the RYGAC motif. In Figure 11.3C, the location and orientation of the RipA binding sites in the seven target promoters is shown. In the case of the *acn* promoter, the experimentally verified RipA binding sites are centered at positions -146.5 and -203.5 with respect to the transcriptional start site, giving rise to the question how RipA exerts its repressing effect. A third putative RipA binding site which begins immediately downstream of the *acn* start codon might be relevant in this context, although a fragment containing only this site was not bound by RipA. In the case of the *sdhCAB* promoter, the two RipA binding sites are centered at -42.5 and -123.5 upstream of the transcription start site that was mapped 15 bp upstream of the *sdhC* start codon (Wennerhold and Bott, 2006; Brune et al. 2006). In this case, RipA might interfere with the binding of RNA polymerase. The presence of at least two DNA-binding sites in each RipA target promoter and the large and varying distances between these binding sites suggest that DNA looping might be involved in the mechanism of action of RipA, as reported e.g. for AraC (Lobell and Schleif, 1990).

The discovery of RipA as a repressor of iron proteins and its own repression by DtxR has unravelled a new aspect of the regulatory network controlling iron homeostasis in *C. glutamicum*. The reduced synthesis of several important iron proteins of the cell under iron deprivation may serve to reduce iron expenditure and to increase survival. Orthologs of RipA were identified in *Corynebacterium efficiens* (CE1047; 70% sequence identity) and *C. diphtheriae* (DIP0922; 52% sequence identity), but not in *Corynebacterium jeikeium* (Tauch et al., 2005) or *Mycobacterium* species. The *C. efficiens ripA* gene as originally annotated in the genome sequence (Nishio et al., 2003) encodes a protein of 400 amino acids (accession number BAC17857). According to the RipA sequences of *C. glutamicum* and *C. diphtheriae*, it seems more likely that the *C. efficiens* RipA contains only 332 amino acid residues (see accession number Q8FQS2). A well conserved DtxR binding site is centered 50 bp upstream of the ATG start codon of *C. efficiens ripA* (encoding the 332-amino-acid residue RipA protein), indicating repression by DtxR. In *C. diphtheriae*, the annotated genome sequence from strain NCTC13129 predicts that *ripA* (DIP0922) encodes a protein of 335 amino acid residues (Cerdeno-Tarraga et al., 2003). Inspection of the corresponding upstream sequence revealed a putative DtxR binding sites centered 128 bp (TAACCTTAGGCTTGCCTTT) upstream of the proposed start codon. Interestingly, Lee et al. (1997) previously identified a DtxR-regulated gene region designated IRP3 from *C. diphtheriae* strain C7 which is equivalent to the one described above. The DtxR binding site they identified experimentally by DNase I footprinting (TTAGGTGAGACGCACCCAT) is located upstream of an open reading frame encoding a 124-amino-acid polypeptide showing high identity to regions of *C. glutamicum* RipA (data not shown). The IRP3 DtxR binding site starts 267 bp downstream of the proposed start codon of DIP0922. Further studies are required to determine the relevance of the strain differences and the *in vivo* relevance of the DtxR binding sites. Due to the poor conservation of the RipA binding motif, it is difficult to predict RipA binding sites in *C. efficiens* and *C. diphtheriae* with confidence. Nevertheless, inspection of the promoter regions of potential RipA target genes revealed for example two putative RipA binding sites in front of the *acn* gene of *C. efficiens*, located at position -167 and -308 with respect to the start codon. Therefore, it can be speculated that all *ripA* genes are probably controlled by DtxR and that the RipA proteins presumably have an at least partially overlapping set of target genes.

A regulatory cascade with a similar function as that of DtxR and RipA is found in some Gram-negative bacteria, *e.g. E. coli* or *Pseudomonas aeruginosa* (Masse et al., 2007). In *E. coli* the ferric uptake regulator (Fur) in complex with iron represses a small (s)RNA named RhyB, which stimulates the degradation of mRNAs encoding iron-containing proteins like aconitase A (*acnA*), succinate dehydrogenase (*sdhCDAB*) and the iron storage proteins ferritin (*ftnA*) and bacterioferritin (*bfr*) (Masse et al., 2005).

3.7 Possible involvement of the *C. glutamicum* CgtSR11 two-component system in heme-dependent regulation

As described above, *C. glutamicum* possesses genes presumably involved in heme uptake as well as a gene encoding heme oxygenase (*hmuO*), all of which are repressed by DtxR under iron-replete conditions. Heme oxygenases catalyze the degradation of heme to α-biliverdin, accompanied by the liberation of the heme-bound iron. In *C. diphtheriae*, expression of *hmuO* was shown to be repressed by DtxR and induced by heme or hemoglobin (Schmitt, 1997). Subsequently, a two-component signal transduction system was identified that activated *hmuO* expression in a heme-dependent manner, designated ChrS-ChrA (Schmitt, 1999). However, a significant activation of heme-dependent *hmuO* expression was maintained in *C. diphtheriae* mutants lacking either *chrS* or *chrA*, suggesting the participation of an additional regulatory system in heme- or hemoglobin-dependent regulation (Bibb et al., 2005). Recent studies indeed showed that for full hemoglobin-dependent activation of *hmuO* an additional two-component system is required, designated as HrrS-HrrA, which shares significant sequence similarity to the ChrS-ChrA system (Bibb et al., 2007).

A BLAST search in the *C. glutamicum* genome revealed that both these systems show significant sequence similarity to the two-component system encoded by the *cgtS11-cgtR11* genes. The *cgtR11* gene has already been mentioned as being regulated by DtxR, which binds to a site located in non-coding region between *cgtS11* and *cgtR11*. This indicates the existence of a *cgtR11*-specific promoter. Based on the high sequence identity of CgtS11 to HrrS (56%) and ChrS (30%) and of CgtR11 to HrrA (86%) and ChrA (50%), a function of the *C. glutamicum* CgtS11-CgtR11 system in heme-dependent activation of *hmuO* transcription has been postulated (Wennerhold and Bott, 2006). Recent studies in our laboratory with a *cgtSR11* deletion mutant (Kocan et al., 2006) supported this suggestion (Frunzke and Bott, unpublished results). Therefore, *hmuO* expression in *C. glutamicum* also appears to be under dual regulation, repressed by DtxR in response to iron availability and activated by the two-component system CgtSR11 in response to heme availability.

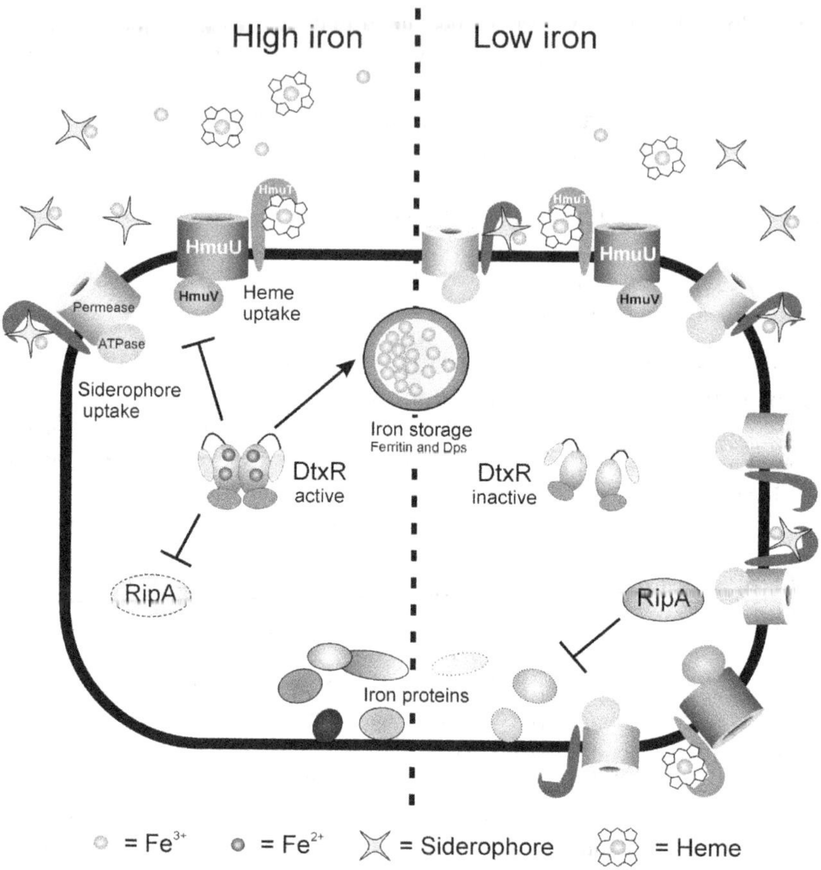

Figure 11.4. Model showing key features of iron dependent regulation in *C. glutamicum*. Under iron-replete conditions, DtxR binds iron, dimerizes and becomes active as a transcriptional regulator. It represses genes encoding high affinity iron uptake systems and activates the genes for the iron storage proteins Dps and ferritin. When intracellular iron becomes limiting, Fe^{2+} dissociates at least from the low-affinity primary binding site within DtxR, which then dissociates into its monomeric form and becomes inactive. As a consequence, a large set of proteins involved in iron acquisition is induced, in particular several ABC transporters for siderophores and an ABC transporter for heme, but also the AraC-type regulator RipA. RipA itself is a repressor of genes encoding several prominent iron proteins, *e.g.* aconitase or succinate dehydrogenase, and thereby reduces the iron demand of the cell under iron limitation (see Figure 11.3). Blunt-ended lines (–|) indicate transcriptional repression, arrows (→) stand for transcriptional activation.

4. Concluding remarks and outlook

The large number of *C. glutamicum* genes involved in iron acquisition (Table 11.1) impressively demonstrates the importance of this element for cellular functionality. Based on the genome sequence, comparative transcriptome analysis, and the definition of the DtxR and RipA regulons, key features of iron metabolism and its regulation in *C. glutamicum* have been elucidated in recent years. These are summarized in the model shown in Figure 11.4. However, there are obviously also many aspects that have not been addressed yet, such as the substrate spectrum of the different ABC transporters involved in siderophore transport, the release of iron from the siderophores, the fate of the siderophores after iron release, or the transport of free Fe^{3+} and Fe^{2+}. It is also still unclear how iron is taken up when *C. glutamicum* is grown in the laboratory in a minimal medium with an excess of iron, but without added siderophores, heme, citrate, protocatechuate or catechol. Thus, there is a lot of room for further studies.

References

Andrews, S.C., Robinson, A.K., and Rodriguez-Quinones, F. (2003). Bacterial iron homeostasis. FEMS Microbiol. Rev. *27*, 215-237.

Angerer, A., Gaisser, S., and Braun, V. (1990). Nucleotide sequences of the *sfuA, sfuB*, and *sfuC* genes of *Serratia marcescens* suggest a periplasmic-binding-protein-dependent iron transport mechanism. J. Bacteriol. *172*, 572-578.

Bayan, N., Houssin, C., Chami, M., and Leblon, G. (2003). Mycomembrane and S-layer: two important structures of *Corynebacterium glutamicum* cell envelope with promising biotechnology applications. J. Biotechnol. *104*, 55-67.

Bibb, L.A., King, N.D., Kunkle, C.A., and Schmitt, M.P. (2005). Analysis of a heme-dependent signal transduction system in *Corynebacterium diphtheriae*: deletion of the *chrAS* genes results in heme sensitivity and diminished heme-dependent activation of the *hmuO* promoter. Infect. Immun. *73*, 7406-7412.

Bibb, L.A., Kunkle, C.A., and Schmitt, M.P. (2007). The ChrA-ChrS and HrrA-HrrS signal transduction systems are required for activation of the *hmuO* promoter and repression of the *hemA* promoter in *Corynebacterium diphtheriae*. Infect. Immun. *75*, 2421-2431.

Billington, S.J., Esmay, P.A., Songer, J.G., and Jost, B.H. (2002). Identification and role in virulence of putative iron acquisition genes from *Corynebacterium pseudotuberculosis*. FEMS Microbiol. Lett. *208*,41-45.

Boyd, J., Oza, M.N., and Murphy, J.R. (1990). Molecular cloning and DNA sequence analysis of a diphtheria tox iron-dependent regulatory element (*dtxR*) from *Corynebacterium diphtheriae*. Proc. Natl. Acad. Sci. USA. *87*, 5968-5972.

Brune, I., Werner, H., Hüser, A.T., Kalinowski, J., Pühler, A., and Tauch, A. (2006). The DtxR protein acting as dual transcriptional regulator directs a global regulatory network involved in iron metabolism of *Corynebacterium glutamicum*. BMC Genomics *7*, 21.

Budzikiewicz, H., Bossenkamp, A., Taraz, K., Pandey, A., and Meyer, J.M. (1997). Corynebactin, a cyclic catecholate siderophore from *Corynebacterium glutamicum* ATCC 14067 (*Brevibacterium* sp. DSM 20411). Z Naturforsch C *52c*, 551-554.

Cartron, M.L., Maddocks, S., Gillingham, P., Craven, C.J., and Andrews, S.C. (2006). Feo - transport of ferrous iron into bacteria. Biometals *19*, 143-157.

Ceci, P., Ilari, A., Falvo, E., Giangiacomo, L., and Chiancone, E. (2005). Reassessment of protein stability, DNA binding, and protection of *Mycobacterium smegmatis* Dps. J. Biol. Chem. *280*, 34776-34785.

Cerdeno-Tarraga, A.M., Efstratiou, A., Dover, L.G., Holden, M.T.G., Pallen, M., Bentley, S.D., Besra, G.S., Churcher, C., James, K.D., De Zoysa, A., *et al.* (2003). The complete genome sequence and analysis of *Corynebacterium diphtheriae* NCTC13129. Nucleic Acids Res. *31*, 6516-6523.

Chipperfield, J.R., and Ratledge, C. (2000). Salicylic acid is not a bacterial siderophore: a theoretical study. Biometals *13*, 165-168.

Costa-Riu, N., Burkovski, A., Krämer, R., and Benz, R. (2003). PorA represents the major cell wall channel of the Gram-positive bacterium *Corynebacterium glutamicum*. J. Bacteriol. *185*, 4779-4786.

Courville, P., Chaloupka, R., and Cellier, M.F. (2006). Recent progress in structure-function analyses of Nramp proton-dependent metal-ion transporters. Biochem. Cell Biol. *84*, 960-978.

Crooks, G.E., Hon, G., Chandonia, J.M., and Brenner, S.E. (2004). WebLogo: A sequence logo generator. Genome Res. *14*, 1188-1190.

D'Aquino, J.A., Tetenbaum-Novatt, J., White, A., Berkovitch, F., and Ringe, D. (2005). Mechanism of metal ion activation of the diphtheria toxin repressor DtxR. Proc. Natl. Acad. Sci. USA. *102*, 18408-18413.

Dertz, E.A., Stintzi, A., and Raymond, K.N. (2006). Siderophore-mediated iron transport in *Bacillus subtilis* and *Corynebacterium glutamicum*. J. Biol. Inorg Chem. *11*, 1087-1097.

Drazek, E.S., Hammack, C.A., and Schmitt, M.P. (2000). *Corynebacterium diphtheriae* genes required for acquisition of iron from haemin and haemoglobin are homologous to ABC haemin transporters. Mol. Microbiol. *36*, 68-84.

Faraldo-Gomez, J.D., and Sansom, M.S. (2003). Acquisition of siderophores in gram-negative bacteria. Nat. Rev. Mol. Cell Biol. *4*, 105-116.

Gallegos, M.T., Schleif, R., Bairoch, A., Hofmann, K., and Ramos, J.L. (1997). AraC/XylS family of transcriptional regulators. Microbiol. Mol. Biol. Rev. *61*, 393-410.

Giel, J.L., Rodionov, D., Liu, M.Z., Blattner, F.R., and Kiley, P.J. (2006). IscR-dependent gene expression links iron-sulphur cluster assembly to the control of O_2-regulated genes in *Escherichia coli*. Mol. Microbiol. *60*, 1058-1075.

Gold, B., Rodriguez, G.M., Marras, S.A., Pentecost, M., and Smith, I. (2001). The *Mycobacterium tuberculosis* IdeR is a dual functional regulator that controls transcription of genes involved in iron acquisition, iron storage and survival in macrophages. Mol. Microbiol. *42*, 851-865.

Gupta, S., and Chatterji, D. (2003). Bimodal protection of DNA by *Mycobacterium smegmatis* DNA-binding protein from stationary phase cells. J. Biol. Chem. *278*, 5235-5241.

Gupta, S., Pandit, S.B., Srinivasan, N., and Chatterji, D. (2002). Proteomics analysis of carbon-starved *Mycobacterium smegmatis*: induction of Dps-like protein. Protein Eng. *15*, 503-511.

Hornung, J.M., Jones, H.A., and Perry, R.D. (1996). The *hmu* locus of *Yersinia pestis* is essential for utilization of free haemin and haem-protein complexes as iron sources. Mol. Microbiol. *20*, 725-739.

Huet, G., Daffe, M., and Saves, I. (2005). Identification of the *Mycobacterium tuberculosis* SUF machinery as the exclusive mycobacterial system of [Fe-S] cluster assembly: evidence for its implication in the pathogen's survival. J. Bacteriol. *187*, 6137-6146.

Hünten, P., Costa-Riu, N., Palm, D., Lottspeich, F., and Benz, R. (2005). Identification and characterization of PorH, a new cell wall channel of *Corynebacterium glutamicum*. Biochim Biophys Acta *1715*, 25-36.

Ikeda, M., and Nakagawa, S. (2003). The *Corynebacterium glutamicum* genome: features and impacts on biotechnological processes. Appl Microbiol. Biotechnol. *62*, 99-109.

Janakiraman, A., and Slauch, J.M. (2000). The putative iron transport system SitABCD encoded on SPI1 is required for full virulence of *Salmonella typhimurium*. Mol. Microbiol. *35*, 1146-1155.

Johnson, D.C., Dean, D.R., Smith, A.D., and Johnson, M.K. (2005). Structure, function, and formation of biological iron-sulfur clusters. Annu. Rev. Biochem. *74*, 247-281.

Kalinowski, J., Bathe, B., Bartels, D., Bischoff, N., Bott, M., Burkovski, A., Dusch, N., Eggeling, L., Eikmanns, B.J., Gaigalat, L., *et al.* (2003). The complete *Corynebacterium glutamicum* ATCC 13032 genome sequence and its impact on the production of L-aspartate-derived amino acids and vitamins. J. Biotechnol. *104*, 5-25.

Keilhauer, C., Eggeling, L., and Sahm, H. (1993). Isoleucine synthesis in *Corynebacterium glutamicum*: molecular analysis of the *ilvB-ilvN-ilvC* operon. J. Bacteriol. *175*, 5595-5603.

Kocan, M., Schaffer, S., Ishige, T., Sorger-Herrmann, U., Wendisch, V.F., and Bott, M. (2006). Two-component systems of *Corynebacterium glutamicum*: deletion analysis and involvement of the PhoS-PhoR system in the phosphate starvation response. J. Bacteriol. *188*, 734-732.

Koster, W. (2005). Cytoplasmic membrane iron permease systems in the bacterial cell envelope. Front Biosci *10*, 462-477.

Krug, A., Wendisch, V.F., and Bott, M. (2005). Identification of AcnR, a TetR-type repressor of the aconitase gene *acn* in *Corynebacterium glutamicum*. J. Biol. Chem. *280*, 585-595.

Kunkle, C.A., and Schmitt, M.P. (2003). Analysis of the *Corynebacterium diphtheriae* DtxR regulon: identification of a putative siderophore synthesis and transport system that is similar to the *Yersinia* high-pathogenicity island-encoded yersiniabactin synthesis and uptake system. J. Bacteriol. *185*, 6826-6840.

Lee, J.H., Wang, T., Ault, K., Liu, J., Schmitt, M.P., and Holmes, R.K. (1997). Identification and characterization of three new promoter/operators from *Corynebacterium diphtheriae* that are regulated by the diphtheria toxin repressor (DtxR) and iron. Infect. Immun. *65*, 4273-4280.

Liebl, W. (2005). *Corynebacterium* taxonomy. In Handbook of *Corynebacterium glutamicum*, L. Eggeling, and M. Bott, eds. (Boca Raton, Forida, USA, CRC Press, Taylor & Francis Group), pp. 9-34.

Liebl, W., Klamer, R., and Schleifer, K.H. (1989). Requirement of chelating compounds for the growth of *Corynebacterium glutamicum* in synthetic media. Appl Microbiol. Biotechnol. *32*, 205-210.

Lin, H., Fischbach, M.A., Liu, D.R., and Walsh, C.T. (2005). *In vitro* characterization of salmochelin and enterobactin trilactone hydrolases IroD, IroE, and Fes. J. Amer. Chem. Soc. *127*, 11075-11084.

Lobell, R.B., and Schleif, R.F. (1990). DNA looping and unlooping by AraC protein. Science *250*, 528-532.

Makui, H., Roig, E., Cole, S.T., Helmann, J.D., Gros, P., and Cellier, M.F. (2000). Identification of the *Escherichia coli* K-12 Nramp orthologue (MntH) as a selective divalent metal ion transporter. Mol. Microbiol. *35*, 1065-1078.

Martin, R.G., and Rosner, J.L. (2001). The AraC transcriptional activators. Curr. Opin. Microbiol. *4*, 132-137.

Masse, E., Salvail, H., Desnoyers, G., and Arguin, M. (2007). Small RNAs controlling iron metabolism. Curr. Opin. Microbiol. *10*, 140-145.

Masse, E., Vanderpool, C.K., and Gottesman, S. (2005). Effect of RyhB small RNA on global iron use in *Escherichia coli*. J. Bacteriol. *187*, 6962-6971.

May, J.J., Wendrich, T.M., and Marahiel, M.A. (2001). The *dhb* operon of *Bacillus subtilis* encodes the biosynthetic template for the catecholic siderophore 2,3-dihydroxybenzoate-glycine-threonine trimeric ester bacillibactin. J. Biol. Chem. *276*, 7209-7217.

Merkens, H., Beckers, G., Wirtz, A., and Burkovski, A. (2005). Vanillate metabolism in *Corynebacterium glutamicum*. Curr. Microbiol. *51*, 59-65.

Miethke, M., Klotz, O., Linne, U., May, J.J., Beckering, C.L., and Marahiel, M.A. (2006). Ferri-bacillibactin uptake and hydrolysis in *Bacillus subtilis*. Mol. Microbiol. *61*, 1413-1427.

Nakayama, K., Kinoshita, S., and Sato, Z. (1964a). Growth of glutamic acid producing bacterium and related bacteria. 1. Effect of iron salts, ferrichrome, amino acids and some other compounds. J. Gen. Appl Microbiol. *10*, 143-155.

Nakayama, K., Sato, Z., Kinoshita, S., and Tanaka, H. (1964b). Growth of glutamic acid producing bacterium and related bacteria. 2. Effect of chelating agent and its relation to inorganic salt. J. Gen. Appl Microbiol. *10*, 181-199.

Nishio, Y., Nakamura, Y., Kawarabayasi, Y., Usuda, Y., Kimura, E., Sugimoto, S., Matsui, K., Yamagishi, A., Kikuchi, H., Ikeo, K., *et al.* (2003). Comparative complete genome sequence analysis of the amino acid replacements responsible for the thermostability of *Corynebacterium efficiens*. Genome Res. *13*, 1572-1579.

Oguiza, J.A., Tao, X., Marcos, A.T., Martin, J.F., and Murphy, J.R. (1995). Molecular cloning, DNA sequence analysis, and characterization of the *Corynebacterium diphtheriae dtxR* homolog from *Brevibacterium lactofermentum*. J. Bacteriol. *177*, 465-467.

Pappenheimer, A.M., and Johnson, S.J. (1936). Studies in diphtheria toxin production. I : The effect of iron and copper. B J. Exp. Pathol. *17*, 335-341.

Patzer, S.I., and Hantke, K. (2001). Dual repression by Fe^{2+}-Fur and Mn^{2+}-MntR of the *mntH* gene, encoding an NRAMP-like Mn^{2+} transporter in *Escherichia coli*. J. Bacteriol. *183*, 4806-4813.

Pierre, J.L., and Fontecave, M. (1999). Iron and activated oxygen species in biology: the basic chemistry. Biometals *12*, 195-199.

Pohl, E., Holmes, R.K., and Hol, W.G. (1999). Crystal structure of a cobalt-activated diphtheria toxin repressor-DNA complex reveals a metal-binding SH3-like domain. J. Mol. Biol. *292*, 653-667.

Polen, T., Schluesener, D., Poetsch, A., Bott, M., and Wendisch, V.F. Characterization of citrate utilization in *Corynebacterium glutamicum* by transcriptome and proteome analysis. FEMS Microbiol. Lett. *273*, 109-119.

Posey, J.E., Hardham, J.M., Norris, S.J., and Gherardini, F.C. (1999). Characterization of a manganese-dependent regulatory protein, TroR, from *Treponema pallidum*. Proc. Natl. Acad. Sci. USA. *96*, 10887-10892.

Postle, K., and Kadner, R.J. (2003). Touch and go: tying TonB to transport. Mol. Microbiol. *49*, 869-882.

Puech, V., Chami, M., Lemassu, A., Laneelle, M.A., Schiffler, B., Gounon, P., Bayan, N., Benz, R., and Daffe, M. (2001). Structure of the cell envelope of corynebacteria: importance of the non-covalently bound lipids in the formation of the cell wall permeability barrier and fracture plane. Microbiology *147*, 1365-1382.

Qiu, X., Pohl, E., Holmes, R.K., and Hol, W.G.J. (1996). High-resolution structure of the diphtheria toxin repressor complexed with cobalt and manganese reveals an SH3-like third domain and suggests a possible role of phosphate as co-corepressor. Biochemistry *35*, 12292-12302.

Qiu, X., Verlinde, C.L.M.J., Zhang, S., Schmitt, M.P., Holmes, R.K., and Hol, W.G.J. (1995). Three-dimensional structure of the diphtheria toxin repressor in complex with divalent cation co-repressors. Structure *3*, 87-100.

Ranjan, S., Yellaboina, S., and Ranjan, A. (2006). IdeR in mycobacteria: from target recognition to physiological function. Crit. Rev. Microbiol. *32*, 69-75.

Rhee, S., Martin, R.G., Rosner, J.L., and Davies, D.R. (1998). A novel DNA-binding motif in MarA: the first structure for an AraC family transcriptional activator. Proc. Natl. Acad. Sci. USA. *95*, 10413-11048.

Rodriguez, G.M., and Smith, I. (2003). Mechanisms of iron regulation in mycobacteria: role in physiology and virulence. Mol. Microbiol. *47*, 1485-1494.

Roy, S., Gupta, S., Das, S., Sekar, K., Chatterji, D., and Vijayan, M. (2004). X-ray analysis of *Mycobacterium smegmatis* Dps and a comparative study involving other Dps and Dps-like molecules. J. Mol. Biol. *339*, 1103-1113.

Schiering, N., Tao, X., Zeng, H., Murphy, J.R., Petsko, G.A., and Ringe, D. (1995). Structures of the apo- and the metal ion-activated forms of the diphtheria *tox* repressor from *Corynebacterium diphtheriae*. Proc. Natl. Acad. Sci. USA. *92*, 9843-9850.

Schmitt, M.P. (1997). Utilization of host iron sources by *Corynebacterium diphtheriae*: identification of a gene whose product is homologous to eukaryotic heme oxygenases and is required for acquisition of iron from heme and hemoglobin. J. Bacteriol. *179*, 838-845.

Schmitt, M.P. (1999). Identification of a two-component signal transduction system from *Corynebacterium diphtheriae* that activates gene expression in response to the presence of heme and hemoglobin. J. Bacteriol. *181*, 5330-5340.

Schmitt, M.P., and Drazek, E.S. (2001). Construction and consequences of directed mutations affecting the hemin receptor in pathogenic *Corynebacterium* species. J. Bacteriol. *183*, 1476-1481.

Schmitt, M.P., and Holmes, R.K. (1991). Iron-dependent regulation of diphtheria toxin and siderophore expression by the cloned *Corynebacterium diphtheriae* repressor gene *dtxR* in *C. diphtheriae* C7 strains. Infect. Immun. *59*, 1899-1904.

Schröder, I., Johnson, E., and de Vries, S. (2003). Microbial ferric iron reductases. FEMS Microbiol. Rev. *27*, 427-447.

Schryvers, A.B., Bonnah, R., Yu, R.H., Wong, H., and Retzer, M. (1998). Bacterial lactoferrin receptors. Adv. Exp. Med. Biol. *443*, 123-133.

Schwartz, C.J., Giel, J.L., Patschkowski, T., Luther, C., Ruzicka, F.J., Beinert, H., and Kiley, P.J. (2001). IscR, an Fe-S cluster-containing transcription factor, represses expression of *Escherichia coli* genes encoding Fe-S cluster assembly proteins. Proc. Natl. Acad. Sci. USA. *98*, 14895-14900.

Spiering, M.M., Ringe, D., Murphy, J.R., and Marletta, M.A. (2003). Metal stoichiometry and functional studies of the diphtheria toxin repressor. Proc. Natl. Acad. Sci. USA. *100*, 3808-3813.

Tao, X., and Murphy, J.R. (1992). Binding of the metalloregulatory protein DtxR to the diphtheria *tox* operator requires a divalent heavy metal ion and protects the palindromic sequence from DNase I digestion. J. Biol. Chem. *267*, 21761-21764.

Tao, X., and Murphy, J.R. (1994). Determination of the minimal essential nucleotide sequence for diphtheria *tox* repressor binding by *in vitro* affinity selection. Proc. Natl. Acad. Sci. USA. *91*, 9646-9450.

Tauch, A., Kaiser, O., Hain, T., Goesmann, A., Weisshaar, B., Albersmeier, A., Bekel, T., Bischoff, N., Brune, I., Chakraborty, T., *et al.* (2005). Complete genome sequence and analysis of the multiresistant nosocomial pathogen

Corynebacterium jeikeium K411, a lipid-requiring bacterium of the human skin flora. J. Bacteriol. *187*, 4671-4682.

von der Osten, C.H., Gioannetti, C., and Sinskey, A.J. (1989). Design of a defined medium for growth of *Corynebacterium glutamicum* in which citrate facilitates iron uptake. Biotechnol. Lett. *11*, 11-16.

Wandersman, C., and Stojiljkovic, I. (2000). Bacterial heme sources: the role of heme, hemoprotein receptors and hemophores. Curr. Opin. Microbiol. *3*, 215-220.

Wang, T., Shen, G., Balasubramanian, R., McIntosh, L., Bryant, D.A., and Golbeck, J.H. (2004). The *sufR* gene (sll0088 in *Synechocystis* sp. strain PCC 6803) functions as a repressor of the *sufBCDS* operon in iron-sulfur cluster biogenesis in cyanobacteria. J. Bacteriol. *186*, 956-967.

Wennerhold, J., and Bott, M. (2006). The DtxR regulon of *Corynebacterium glutamicum*. J. Bacteriol. *188*, 2907-2918.

Wennerhold, J., Krug, A., and Bott, M. (2005). The AraC-type regulator RipA represses aconitase and other iron proteins from *Corynebacterium* under iron limitation and is itself repressed by DtxR. J. Biol. Chem. *280*, 40500-40508.

White, A., Ding, X., van der Spek, J.C., Murphy, J.R., and Ringe, D. (1998). Structure of the metal-ion-activated diphtheria toxin repressor/*tox* operator complex. Nature *394*, 502-506.

Yeo, W.S., Lee, J.H., Lee, K.C., and Roe, J.H. (2006). IscR acts as an activator in response to oxidative stress for the *suf* operon encoding Fe-S assembly proteins. Mol. Microbiol. *61*, 206-218.

Yukawa, H., Omumasaba, C.A., Nonaka, H., Kos, P., Okai, N., Suzuki, N., Suda, M., Tsuge, Y., Watanabe, J., Ikeda, Y., *et al.* (2007). Comparative analysis of the *Corynebacterium glutamicum* group and complete genome sequence of strain R. Microbiology *153*, 1042-1058.

From: Corynebacteria: Genomics and Molecular Biology. Edited by: Andreas Burkovski.

Structure and Synthesis of the Cell Wall

12

Lothar Eggeling, Gurdyal S. Besra and Luke Alderwick

Abstract

Apart from a general interest in understanding the living world, there are at least two other reasons for investigating the cell wall of *Corynebacterium glutamicum*. First of all, there is the influence of the cell-wall structure and its regulation on the amino acid efflux and, secondly, this knowledge would facilitate the understanding of the complex cell-wall structure of *Mycobacterium tuberculosis*. We summarize here what is known on the cell wall synthesis of *C. glutamicum* and describe how studies with this organism improved the understanding on the structure and synthesis of the polysaccharides arabinogalactan, lipomannan, and lipoarabinomannan in *Corynebacterianeae* encluding *M. tuberculosis*. While general reviews on cell wall synthesis and cell wall structure of *Corynebacterium* are available (Eggeling et al., 2001; Dover et al., 2004; Daffe, 2005), the present review focusses on the latest discoveries encluding the key functions of glycosyltransferases in building the cell wall of *C. glutamicum*.

1. Introduction

It is now well established that *C. glutamicum* and *M. tuberculosis* belong to a related group of bacteria, which is classified as the suborder *Corynebacterineae* encompassing families like *Corynebacteriaceae*, *Mycobacteriaceae*, and *Nocardiaceae* (Stackebrandt et al., 1997). These have been recognized to share common features like the presence of either mycolic acids or arabinose and galactose in sugar polymers, of their respective cell walls (Prauser and Falta 1968; Goodfellow and Minnikin, 1981). Prominent species among the *Corynebacteriaceae* are *C. glutamicum*, *Corynebacterium efficiens*, *Corynebacterium jeikeium*, *Corynebacterium diphtheriae* and *Corynebacterium ureolyticum*. Interestingly, the family *Corynebacteriaceae* encompasses both non-pathogenic and pathogenic bacteria, whereas the *Mycobacteriaceae* contains mostly pathogenic species. All these bacteria have basic features in common, like for instance, elements of cell wall synthesis and its structure. Among these bacteria, *C. glutamicum* represents a model organism, since it is genetically tractable and it contains the elementary structure of the cell wall within the *Corynebacterianeae*. It is considered to represent the archetype of the *Corynebacterianeae* and has a low frequency of structural alterations as manifested, for instance, in a low number of gene duplications (Nakamura et al., 2003). Thus, *Corynebacterium* species have only one *emb* gene, whereas *Mycobacterium* species have up to three. Moreover, it possesses the simplest mycolic acids, which contain 22-38 carbons, whereas those of the *Nocardiaceae* contain 44-60 and those of *Mycobacteriaceae* 60-90 (Barry et al., 1998). Indeed, elementary features of mycolic acid and cell wall synthesis have been discovered in *C. glutamicum*, making intensive use of gene and genome comparisons (Gande et al., 2004; Portevin et al., 2004; Alderwick et al., 2005; Tropis et al., 2005; Alderwick et al., 2006; Lea-Smith et al., 2007; Seidel et al., 2007a; 2007b; Tatituri et al., 2007a; 2007b; Mishra et al., 2007).

2. Overview on cell wall

The peptidoglycan (PG) of *Corynebacterium* is directly cross-linked of the type A1γ, where the cross-linkage of adjacent peptide chains is via positions 3 and 4, peptide bridges are absent, and *meso*-diaminopimelate is located at position 3 of the tetrapeptide subunits (Schleifer and Kandler, 1972). The cell wall of corynebacteria contains the heteropolysaccharide arabinogalactan (AG), which is linked to PG via its glycan part (Figure 12.1). Furthermore, AG is partially esterified via arabinan residues with mycolic acids, thus forming a large mycolyl-arabinogalactan-peptidoglycan (mAGP) complex (Brennan and Nikaido, 1995; Besra and Brennan, 1997). In addition the soluble polysaccharides lipoarabinomannan (LAM) and lipomannan (LM) are present. Furthermore, soluble lipids such as trehalose dimycolate (TDM) and trehalose

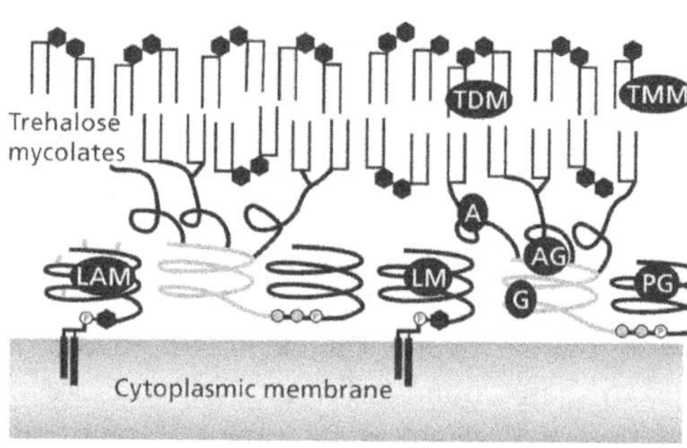

Figure 12.1. Overview on cell wall constitution. On top of the cytoplasmic membrane the peptidoglycan (PG) is located, and it may be arranged in parallel to the membrane or with glycan strands extending perpendicular to the cytoplasmic membrane as shown. To PG the glycan strands of arabinogalactan (AG), with its arabinan (A) and galactan (G) domain are linked (G is in grey). To the arabinan ends of AG the mycolic acids are esterified which together with soluble mycolic acids in the form of trehalose monomycolates (TMM) and trehalose dimycolates (TDM) form an outer membrane. Anchored in the cytoplasmis membrane are lipoarabinomannan (LAM) and lipomannan (LM), of which two types exist in *C. glutamicum* (see text). The grey short lines in LAM represent arabinose extensions.

monomycolate (TMM) are present in the cell envelope of *Corynebacterium* (Gebhardt et al., 2007). Their abundance, together with the covalently linked mycolates enables to form a lipid bilayer in *Corynebacterium* (Puech et al., 2001). Importantly, this hydrophobic layer has been shown in related genera to play an important role in drug and substrate transport due to its high impermeability (Jarlier and Nikaido, 1990; 1994). This outer lipid layer forms a structure reminiscent in architecture and function of the characteristic outer membrane of Gram-negative bacteria, although the molecular details are strikingly different. While the outer membrane in the envelope of Gram-negative bacteria is composed of phospholipids and lipopolysaccharides, the predominant constituents of the hydrophobic diffusion barrier in the corynebacterial cell envelope are the covalently linked mycolic acids, together with TMM, TDM, and phospholipids (Figure 12.1). Freeze-fracture electronmicroscopic analysis of mycolate-less *Corynebacterium amycolatum* or *Coynebacterium kroppenstedtii* are in agreement with an apparent lack of an outer membrane-like lipid bilayer (Puech et al., 2001). Mutants defective in protein components involved in extracytoplasmic lipid metabolism (mycolyl transferases) display a decreased mycolate content and an altered cell wall permeability which points to an important role in the outer membrane-like permeability barrier (Puech et al., 2000; 2001; Brand et al., 2003).

The overall three dimensional topology of the cell wall matrix is not very well understood. This is due to its complexity, lack of pure samples and absence of experimental data. The classical view of PG architecture is that the glycan strands are arranged parallel to the membrane. However, Dmitriev et al. (1999; 2000) suggested that peptidoglycan forms coils and runs perpendicular to the cell membrane, rather than being organized in a layered fashion. Also AG, LM and LAM might form coiled structures running parallel to the cell membrane, thus integrating with PG (Figure 12.1). More recently, a three dimensional solution structure of a 2-kDa synthetic peptidoglycan fragment of the cell wall was derived (Meroueh et al., 2006). This shows that the glycan backbone forms a right-handed helix with the first two amino acids of the pentapeptide adopting a limited number of conformations. Based on this structure the model derived ressembles that with the idea of a perpendicular arrangement of PG.

Regarding the outer lipid layer, freeze-etch electron microscopy techniques indicate that this hydrophobic barrier represents a true lipid bilayer (Puech et al., 2001). In accordance with an outer membrane bilayer, ion-permeable channels and pore-forming proteins, albeit different in monomer size and multimeric association from the trimeric porins of Gram-negative bacteria and also different from MspA of *M. smegmatis*, are present in different *Corynebacterium* species (Costa-Riu et al., 2003; Hünten et al., 2005a; 2005b; Schiffler et al., 2007)

3. Peptidoglycan

The peptidoglycan (PG) in *C. glutamicum* is of the A1γ type (Schleifer and Kandler, 1972). Its biosynthesis is well studied in *Escherichia coli* and an excellent recent review is available on this (Scheffers and Pinho, 2005). Due

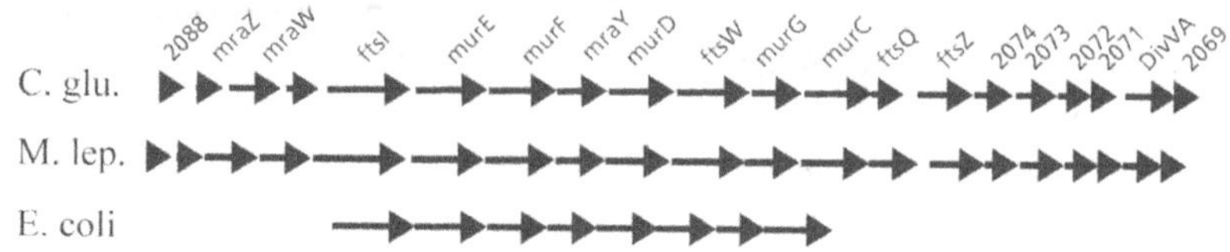

Figure 12.2. The eubacterial synthesis-division cluster. The functions of the genes are in part described in the text. Numbers are NCgl numbers. C. glu., *C. glutamicum*; M. lep., *Mycobacterium leprae*; E. coli; *Escherichia coli.*

to the similarity in structure, homology of genes and syntenic organization of cell-wall-genes (Figure 12.2) corynebacterial PG synthesis is assumed to be similar to that described for *E. coli.*

The biochemical reactions leading to PG can be divided into three stages. In the first stage the nucleotide sugars UDP-N-acetylglucosamine (UDP-GlcNAc) and UDP-N-acetylmuramic acid (UDP-MurNAc) are synthesized. UDP-MurNAc is generated from UDP-GlcNAc, and the proteins encoded by NCgl0386 (*murB*) and NCgl2470 (*murA*) might serve this function, as also concluded for *C. diphtheriae* (Dover et al., 2004). The first stage involves furthermore the synthesis of the nucleotide sugar-linked precursor UDP-N-acetylmuramyl-pentapeptide (UDP-MurNAc-pentapeptide). This occurs in the cytoplasm, and the genes *murE*, *murF*, *murD*, and *murC* are involved in these processes (Figure 12.2).

In the second stage, which is membrane-bound, the precursor lipid intermediates are synthesized. The phospho-MurNAc-pentapeptide moiety of UDP-MurNAc-pentapeptide is transferred to the membrane acceptor polyprenol phosphate, yielding lipid I (MurNAc-(pentapeptide)-pyrophosphoryl-polyprenol), and *mraY* encodes the corresponding transferase (Figure 12.2). Then GlcNAc from UDP-GlcNAc is added to lipid I by the action of the *murG* encoded transferase, yielding lipid II (GlcNAc-ß-(1,4)-MurNAc-(pentapeptide)-pyrophosphoryl-polyprenol) which is the substrate for the polymerization reactions resulting in cross-linked PG. The polyprenyl phosphate used in *C. glutamicum* may be undecaprenol, (C_{55}), which is demonstrated to be used for polyprenyl monophosphomannose synthesis in *C. glutamicum* (Gibson et al., 2003). The use of the lipid intermediate in PG synthesis serves to enable otherwise hydrophilic precursors to cross the lipid bilayer. For *E. coli* it is discussed that the structurally related FtsW and RodA proteins might be involved in this process, and orthologues are present in *C. glutamicum.* This is NCgl2079, *ftsW*, in the synthesis-division cluster (Figure 12.2) as well as NCgl0043, *rodA*, in a division-regulation cluster (Figure 12.3). Whereas the core of the synthesis-division cluster providing the substrate for PG synthesis is present in *E. coli*, the division-regulation cluster is specific to *Corynebacterianeae*.

In the third stage of PG synthesis, occuring at the periplasmic side, the disaccharide-peptide units are polymerized and incorporated into the growing PG. These transglycosylation and transpeptidation reactions are catalysed by the penicillin-binding proteins, PBPs, establishing the glycosidic bonds and

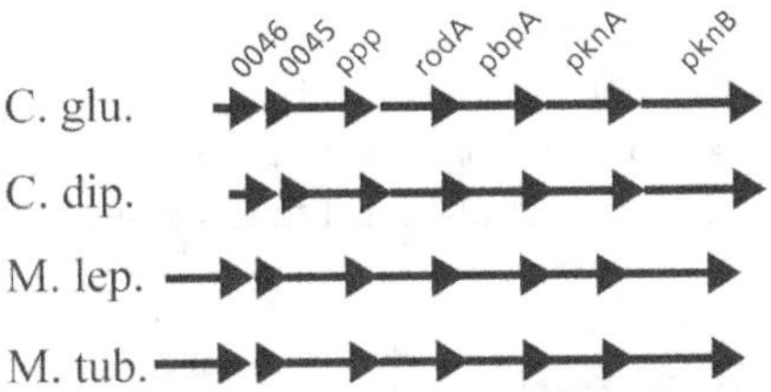

Figure 12.3. The corynebacterial division-regulation cluster. The cluster is specific to *Corynebacterianeae*. The functions of the genes are described in the text. Numbers are NCgl numbers. Abbreviations as above, and M. tub. *M. tuberculosis*.

peptide bonds of PG. *C. glutamicum* has the four penicillin-binding-proteins NCgl2178, NCgl2331, NCgl0042, and NCgl1933.

The synthesis-division cluster of *C. glutamicum* as shown in Figure 12.2, encompasses approx. 11 kb. It is retained in all *Corynebacterianeae*, even in *M. leprae* with its decayed genome, where only key genes prevail (Cole et al., 2001). Orthologues of *mraZ*, *mraW*, *ftsQ*, *ftsZ*, NCgl2074, NCgl2073 are present in *E. coli* but at other loci, with often poorly characterized function. In *C. glutamicum* a promoter is localized upstream of *mraZ* to derive a polycistronic message, at least including *mraZ* up to *murE*, with *ftsI* probably possessing an additional promoter (Valbuena et al., 2006). The genes *ftsI*, *murE* (Wijayarathna et al., 2001), as well as *ftsZ* (Honrubia et al., 2001; Ramos et al., 2005), and DivIVA (Ramos et al., 2004) are essential for *C. glutamicum* and the degree of expression alters cellular morphology. FtsI and FtsZ localise in *C. glutamicum* at a ring at mid-cell, whereas DivIVA does this at the cellular ends (Ramos et al., 2004). A tight genetic control exists, since depletion of *ftsI* in *C. glutamicum* results in an increased formation of the DivIVA protein (Valbuena et al., 2006). This structural and functional information confirms the view that cell division involves a number of essential proteins to synthesize and localize macromolecular elements in a spatial and timely ordered fashion (Errington et al., 2003).

Initiation of cell division and the hierarchy of the cytological events resulting in new cells obviously require sophisticated control at different levels in a well ordered time-frame, e.g. availability of building blocks, sensing integrity of the cell wall, proper location of enzyme complexes, or movement and location of macromolecules like DNA. It is thus not surprising that genes known to be involved in cell division are clustered with a number of regulatory genes.

A considerable accumulation of such genes is present in the division-regulation cluster where one of the four PBPs, *pbpA*, as well as *rodA* is present (Figure 12.3). Downstream of *pbpA* the two serine/threonine proteine kinase genes *pknA* and *pknB* are situated involved in signal transduction. It has been shown for *C. glutamicum* that the upstream located protein phosphatase, *ppp*, together with another serine/threonine proteine kinase, PknG, controls the phosphoryation status of OdhI which itself controls 2-oxoglutarate dehydrogenase activity (Niebisch et al., 2006; Schultz et al., 2007). This discovery clearly shows an interaction between cell wall synthesis and central metabolism and sheds glittering light on the mystery of L-glutamate excretion by *C. glutamicum*, which will be discussed below. The presence of *pknA*, *pknB*, *ppp*, and also the genes encoded by NCgl0045 and NCgl0046 which both are regulatory proteins containing a phosphopeptide recognition domain (FHA) illustrates the elaborated cellular control of cell wall synthesis and division. Also the PBP encoded by NCgl1933 is present in an operon containing an additional regulator (no figure shown), which is a signal transduction histidine kinase.

4. Lipomannan and lipoarabinomannan

As other *Corynebacterianeae C. glutamicum* also contains within its cell wall complex several structurally-related lipoglycans (Puech et al., 2001; Daffé, 2005). As shown for mycobacteria the various polysaccharide domains of this diverse range of molecules reside on the extracellular side of the membrane (Besra et al., 1997; Briken et al., 2004; Nigou et al., 2003; Morita et al., 2005). Three sub-families of phosphatidyl-innositol (PI) based glycolipids/lipoglycans are present in *C. glutamicum* (Figure 12.4), which are the PI-mannosides (PIMs), as well as lipomannan (LM) and lipoarabinomannan (LAM), (Besra et al., 1997). PI in *C. glutamicum* makes up 3.2 % of the total lipids and the PIMs 4.9 % (Hoischen and Krämer, 1990). The majority of membrane lipids (69.5 %) is contributed by phosphatidyl glycerol. Among the PIMs in *C. glutamicum* the dimannoside Ac(PIM$_2$) is dominant (Tatituri et al., 2007a), whereas in *M. tuberculosis* AcPIM$_2$ and the hexamannoside, AcPIM$_6$, are dominant PIM species (Besra et al. 1997; Morita et al., 2006). In addition to PI based LM, it was only recently discovered that in *C. glutamicum* another LM species is present, where 1,2-di-*O*-C16/C18:1-(α-D-mannopyranosyl)-(1,4)-(α-D-glucopyranosyluronic acid)-(1,3)-glycerol (ManGlcAGroAc2) serves as the lipid anchor (Figure 12.4B) (Tatituri et al., 2007b). Thus Cg-LM is two components, a dominant species based on the lipid anchor ManGlcAGroAc2 and a minor component akin to the characteristic mycobacterial PI-based LM.

In the biosynthesis pathway (Figure 12.4C), PI undergoes mannosylation at the 2-OH position of inositol through PimA (NCgl1603), which utilises GDP-Mannose (GDP-Man) as a high-energy sugar donor resulting in the formation of PIM$_1$ (Kordulakova et al., 2002). The further biosynthesis of the PIMs proceeds by the sequential addition of Man*p* residues to the 6-OH position of *myo*-inositol of PIM$_1$ or its acylated counterpart, AcPIM$_1$ (Figure 12.4C). It is still unclear whether the additional acyl residue present in AcPIM$_1$ (-R3 in Figure 12.4C) is added before further mannosylation of PIM$_1$ or after further elaboration of Man*p* residues at the 6-OH position of *myo*-inositol. However, in *M. tuberculosis* the orthologe of NCgl1604 has been identified as the acyltransferase responsible for the production of AcPIM$_1$ (Kordulakova et al., 2003). The genes encoding PimA, the acyltransferase and the PI synthase, *pgsA*, are adjacent and probably form an operon with at least three further genes and which all are organized in synteny in the *Corynebacterianeae* (not shown). PimA can not be inactivated in *C. glutamicum* (Krumbach and Eggeling, unpublished).

It is not yet clear how the further mannosylation of AcPIM$_1$ occurs. Since PIM$_4$ accumulates in *Mycobacterium smegmatis* (Kovacevic et al., 2006), and

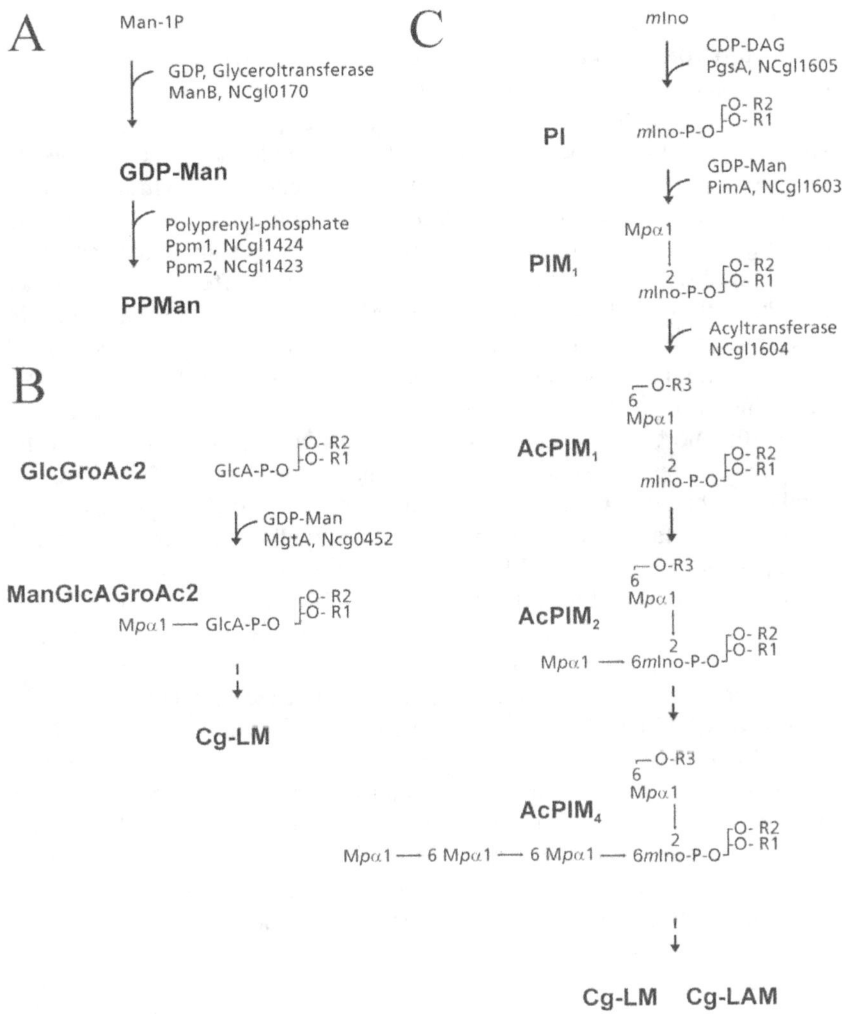

Figure 12.4. Overview on the initial steps of lipoglycan synthesis. In **A** is shown the mannose activation to the hydrophilic GDP-mannose (GDP-Man) and the lipophylic polyprenyl-P-mannose (PPMan). In **B** the key step of diacyl-mannopyranosyl-glucopyranosyluronic acid-glycerol (ManGlcAGroAc2) synthesis is shown leading to the majority of lipomannan (LM) in *C. glutamicum*, whereas in **C** the initial steps for the synthesis of phosphatidyl-inositol based lipomannan (LM) and lipoarabinomannan (LAM) synthesis are shown.

since PIMs, LM and LAM all share the mannosylation at the C-6 position of the *myo*-inositol it is likely that the synthesis pathway is in a linear manner from $PIM_1 \rightarrow PIM_2 \rightarrow PIM_3 \rightarrow PIM_4 \rightarrow$ LM $\rightarrow$ LAM as suggested for *M. tuberculosis* (Besra et al., 1997).

In *M. tuberculosis* PimB is thought to introduce the second mannose residue (Schaeffer et al., 1999; Jackson et al., 2000). However structural evidence suggests that the orthologe of *C. glutamicum*, which we termed

MgtA is a GDP-Man-dependent α-mannosyltransferase involved in ManGlcAGroAc2 synthesis (Tatituri et al., 2007b). Also PimC in *M. tuberculosis* is a candidate to be involved in the initial mannosylation steps of PI (Kremer et al., 2002). However, this gene is not present in *C. glutamicum* and even not all mycobacteria, suggesting some redundancy across species. PimE is a polyprenyl-phospho-mannose-dependent mannosyltransferase in *Mycobacterium*, which produces $AcPIM_5$ (Morita et al., 2006), and deletion of the gene encoding this enzyme in *M. smegmatis*, results in the accumulation of $AcPIM_4$, but leaving LM and LAM synthesis uneffected which further corroborates $AcPIM_4$ as an LM and LAM precurser, and indicates that in mycobacteria $AcPIM_5$ and PIM_6 are side-products. PimE is not present in *C. glutamicum*, but high identities of 39 % exist to a larger protein (NCgl2096) deserving further study.

Furthermore, it is also not yet clear at which step the switch from the cytosolic to the periplasmic steps in glycolipid synthesis occurs. The cytosolic synthesis is based on GDP-Mannose as a sugar donor and glycosyltransferases of the GT-A/B family, whereas the periplasmic synthesis requires polyprenyl monophosphomannose (PP-Man) as the sugar donor and membrane-bound glycosyltransferases of the GT-C family. These latter enzymes are well known in eukaryotes, but rare in prokaryotes except the suborder *Corynebacterineae* (Liu and Mushegian, 2003; Berg et al., 2007). Recent genetic and biochemical studies have recognized some of the GT-C candidates as involved in PIM, LM, LAM and AG synthesis. The fact that disruption of Ppm1 in *C. glutamicum* (Figure 12.4A) results in a lipoglycan-less mutant indicates that PP-Man as the mannose donor is essential for the formation of higher mannosylated lipoglycans (Gibson et al., 2003), and a reinvestigation of that mutant might help to define the step where the switch from the cytosolic to the periplasmic synthesis occurs. Since in *M. smegmatis* a mutant inactivated in *lpqW* accumulates PIM_4 , and LM and LAM synthesis is abolished, PIM_4 appears still to be synthesized in the cytosol, whereas higher mannose-containing derivatives are synthesized in the periplasm. Moreover, *lpqW* which is present in synteny in *Corynebacterineae*, encluding NCgl1054 in *C. glutamicum*, is probably a periplasmic binding protein involved in the transition step from cytosolic to periplasmic glycolipid synthesis.

As mentioned, it is only recently recognized that a number of GT-C glycosyltransferases are present in *Corynebacterineae*. They are characterized amongst others as membrane proteins and a glycosyltransferase motif similar to DXD, DDX, DEX, or DXE in a loop region directed towards the periplasm (Liu and Mushegian, 2003). In *C. glutamicum* the number of these specific glycosyltransferases likely all to be involved in non-cytoplasmic polysaccharide synthesis is currently 16 proteins, of which Emb, AftA and AftB are identified as arabinosyltransferases involved in AG synthesis, whereas NCgl0854 is a mannosyltransferase involved in glycosylation of a resuscitation factor (Brian et al., 2005; Mahne et al., 2006). Additional mannosyltransferases present in *C. glutamicum* have been identified which are MptA (Mishra et al., 2007), MgtA (Tatituri et al., 2007b), and NCgl2100, the orthologe of Rv2181 (Kaur et al., 2006).

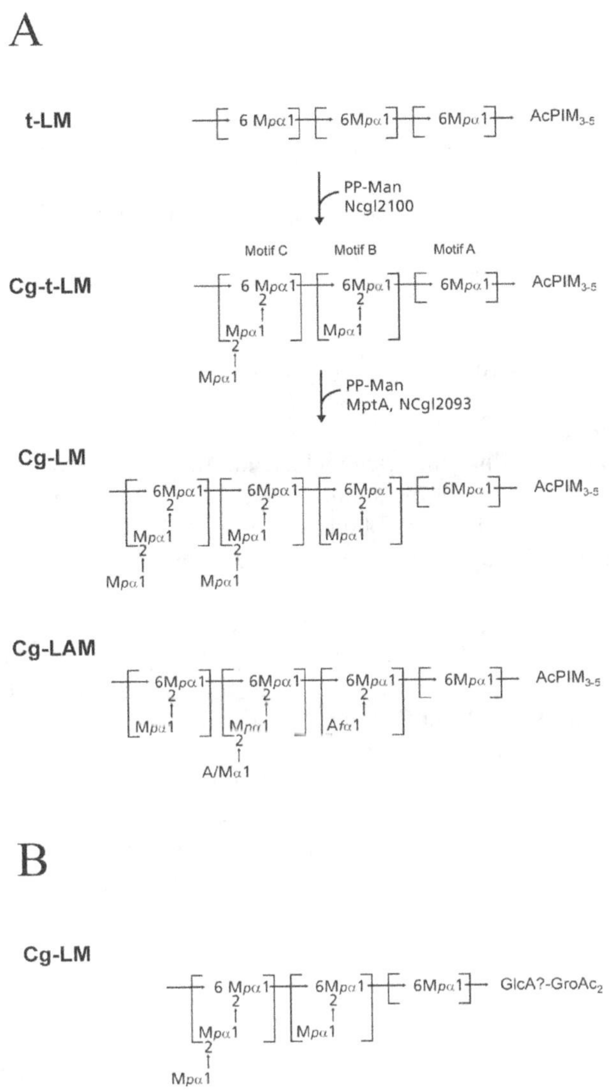

Figure 12.5. Overview on later steps of lipoglycan synthesis. In **A** is shown the addition of mannose residues from polyprenyl-P-mannose (PPMan) to result in Cg-t-LM. This is further mannosylated by the mannosyltransferase MptA. The further decoration with mannose and arabinose sugars by as yet undefined glycosyltransferases results in Cg-LM and Cg-LAM. In **B** the Cg-LM species anchored in the lipid bilayer via (ManGlcAGroAc2) is shown, and the mannosyltransferase MptA might be involved in this process, too.

The deletion of the Rv2131 orthologe of *M. tuberculosis* in *M. smegmatis* resulted in a glycolipid species with a preponderance of α(1→6) linked Man*p* residues. Therefore, Rv2131 and its corynebacterial orthologue NCgl2100, is probably the mannosyltransferase adding the branching α(1→2) linked Man*p* residues to a linear α(1→6) linked Man*p* chain as present in one of the higher mannosylated PIMs, like $AcPIM_4$ or $AcPIM_6$. The deletion of MptA in *C. glutamicum* resulted in the formation of a novel truncated lipomannan (t-LM) (Figure 12.5A) and a complete ablation of LM/LAM biosynthesis (Tatituri et al., 2007b), as also seen in *M. smegmatis* (Kaur et al., 2006). Purification and characterization of t-LM from *C. glutamicum* revealed an overall decrease in molecular mass, a reduction of α(1→6) and α(1→2) glycosidic linkages illustrating a reduced degree of branching compared to wild type LM. In an *in vitro* system, use of the donor β-D-mannopyranosyl-1-monophosphoryl-decaprenol together with the neoglycolipid acceptor α-D-Man*p*-(1→6)-α-D-Man*p*-*O*-C_8 as a substrate, confirmed NCgl2093 and Rv2174 as an α(1→6) mannopyranosyltransferase (MptA), involved in the later stages of the biosynthesis of the α(1→6) mannan core of LM.

5. Linker unit

The linker unit (LU) links the arabinoglactan to the peptidoglycan. In *M. tuberculosis* about 10-12 % of the muramic acid residues of the peptidoglycan make up the corresponding phosphodiester linkage (Lederer et al., 1975). It consists of the three sugars galactose, rhamnose and N-acetylglucosamine linked as Gal*f* (1→4)-Rha*p*-(1→3)-GlcNAc through a 1-*O*-phosphoryl bond of GlcNAc to the 6OH position of selected Mur residues as shown in Figure 12.6. Its composition in *M. tuberculosis* was reported by McNeil et al. (1990), after the early recognition that the linkage involves a phosphodiester bridge (Amar and Vilkas, 1973). In the first step the GlcNAc-P-P-polyprenyl is formed. This is catalysed by the decaprenol-mono-phosphate-α-*N*-acetylglucosamineyltransferase Rfe (NCgl1157), utilizing UDP-GlcNAc and polyprenylmonophosphate as substrates. The subsequent reaction is catalysed by the α-D-GlcNAc-α-1,3-L-rhamnosyltransferase WbbL (NCgl0709), which adds Rha*p* to GlcNAc-P-P-polyprenyl to form polyprenyl-P-P-GlcNAc-Rha (Mikusova et al., 1996; Mills et al., 2004).

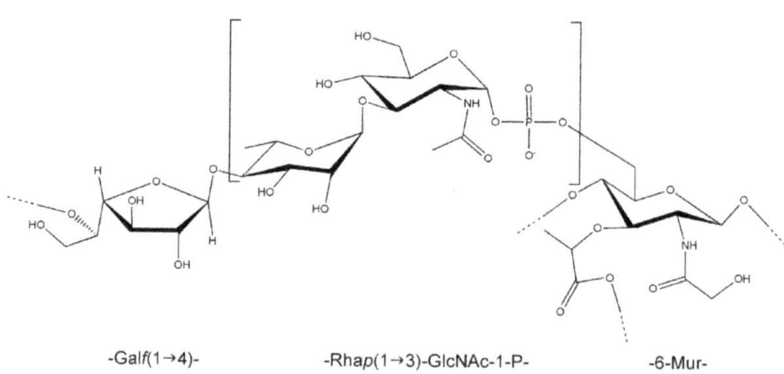

Figure 12.6. Molecular structure of the unique linkage unit (LU) which acts as a conduit between AG and PG. The Gal*f* represents the reducing end of the galactan domain of AG and the muramic acid residue represents the site of attachment to the PG. The disaccharide inbetween indicates the essential LU as determined by McNeil et al. (1990).

The Rha*p* added to GlcNAc-P-P-polyprenyl requires the activated sugar donor dTDP-Rha. As established for *M. tuberculosis*, this is synthesised *via* a short pathway involving four enzymes which utilise TTP and Glc-1-P as initial substrates. The first reaction involves the transfer of the thymidine moiety from TTP to glucose-1-phosphate by the enzyme α-D-glucose-1-phosphate thymidyltransferase RmlA (Ma et al., 2001). Sequential reorganisation of dTDP-Glc by the following enzymes dTDP-Glc dehydratase RmlB, dTDP-6-deoxy-D-*xylo*-4-hexulose epimerase RmlC (Stern et al., 1999) and finally dTDP-6-deoxy-D-*xylo*-4-hexulose reductase RmlD (Hoang et al., 1999) results in the formation of the high-energy sugar donor dTDP-Rha. Whereas in *M. tuberculosis* only the genes *rmlB* and *rmlC* are clustered, in *C. glutamicum*, and other *Corynebacterium* species as well, *rmlA*, *rmlD* and *rmlB* are adjacent probably forming an operon (not shown). An *rmlC* orthologue is not identifyable in *C. glutamicum*, whereas in *C. jekeium*, the genome organisation of *rmlC* and *rmlB* is syntenic to that in *M. tuberculosis*.

6. Arabinogalactan

Arabinogalactan (AG) is a complex macromolecule consisting of D-arabinose (Ara*f*) and D-galactose (Gal*f*) sugar residues present in their furanose form (Azuma et al., 1968; Kanetsuna, 1968; McNeil et al., 1987). However, beginning from the first understanding on the basic composition of this macromolecule and the identification of distinct structural motifs, it took about two decades until a more complete structural representation was elucidated (Daffé et al., 1990; Besra et al., 1997). Only recently, valuable

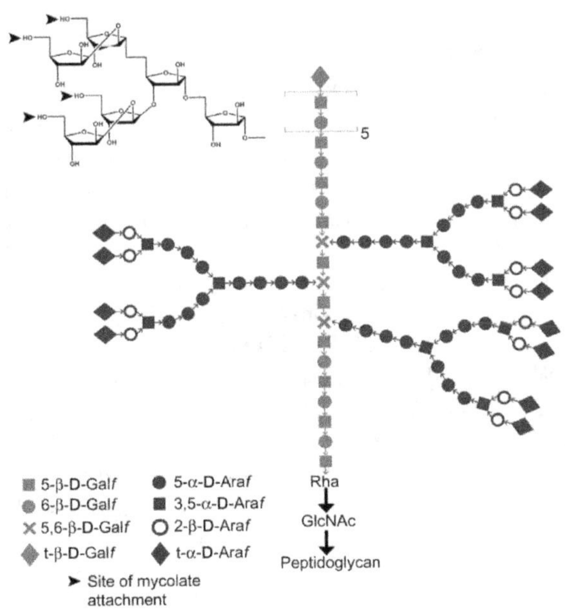

Figure 12.7. Molecular representation of arabinogalactan as integrated in the mycolyl-arabinogalactan-peptidoglycan complex. The model is derived from studies with *M. tuberculosis* and *C. glutamicum*.

insights into both structure and biosynthesis of AG came from the thorough analysis of defined deletion mutants of *C. glutamicum* (Alderwick et al., 2005; 2006; Seidel et al., 2007a; 2007b). The galactan is present in the form of alternating β(1→5) and β(1→6) linked galactofuranose (Gal*f*) residues extending up to 30 units in a linear chain (Figure 12.7). The reducing end of the galactan is covalently attached to peptidoglycan *via* the linker unit. The galactan is decorated with arabinan, branched from the C5 of a β(1→6) linked Gal*f* residue. Unlike the galactan domain of AG, the arabinan domain consists of several arabinose linkages and branching motifs. The majority of the arabinan consists of linear α(1→5) linked arabinofuranose (Ara*f*) residues with branching at the C3 hydroxyl forming a 3,5-Ara*f* linked fork. The arabinan then extends in a linear fashion before terminating in a unique hexaarabinofuranosyl (Ara_6) motif (McNeil et al., 1994). After a 3,5-Ara*f* branch point, the non-reducing termini consists of β(1→2) linked Ara*f* units and the entire Ara_6 motif consists of [Ara*f*-β(1→2)-Ara*f*-α(1-$]_2$→3,5-Ara*f*-α(1→5)-Ara*f*-α(1→ (McNeil et al., 1994). Whereas this model is derived for *M. tuberculosis*, and largely confirmed for *C. glutamicum* (Puech et al., 2001; Alderwick et al., 2005; Seidel et al., 2007a; 2007b), in the latter organism some of the terminal Ara*f* residues are linked to rhamnose (Alderwick et al., 2006). As revealed for *C. glutamicum*, there are three such arabinan domains attached per galactan domain at the 8th, 10th, and 12th galactofuranose (Gal*f*) residue (Alderwick et al., 2005) (Figure 12.7). The mycolic acids are esterified either to the 5OH-group of the penultimate β(1→2) linked Ara*f* residue, or the 5OH-group of the ultimate t-Ara*f* residue (McNeil et al., 1990; Seidel et al., 2007a).

7. Galactan precursor

The synthesis of this polysaccharide has been intensively studied in mycobacteria, and gene structures and genomic organisations suggests that it similar proceeds in all *Corynebacterianeae*. The Gal*f* residues present in the galactan domain of AG are supplied through the high-energy sugar nucleotide UDP-Gal*f*. This is synthesized by the UTP-glucose-1-phosphate uridylyltransferase GalU (NCgl0846) from glucose and UDP. Epimerisation via the uridine diphospho-glucose-4-epimerase GalE (NCgl0317) produces UDP-Gal*p* which is further converted to UDP-Gal*f* by UDP-galactopyranose mutase, Glf (NCgl2788). This latter reaction is characterised for *M. smegmatis* (Weston et al., 1997), as is the essentiallity of this gene (Pan et al., 2001). The gene encoding Glf is syntenic in *Corynebacterianeae* and marks the beginning of a prominent cell wall locus where the β-Gal*f*-transferase, a membrane-associated alkaline phosphatase, the prenyltransferase UbiA, mycolyltransferases, the arabinosyltransferase AftB (Seidel et al., 2007a), and genes of mycolic acid synthesis are located (Figure 12.8).

8. Galactan biosynthesis

As also derived from studies with mycobacteria, the prenylated linker unit polyprenyl-P-P-GlcNAc-Rha is extended at the C4 position of the Rha residue to result in a chain of Gal*f* residues linked in an alternating β(1→5)

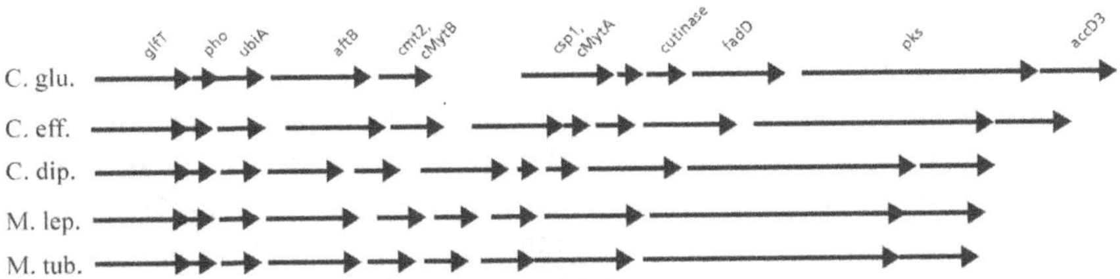

Figure 12.8. The corynebacterial cell-wall-synthesis cluster. This cluster comprises in *C. glutamicum* 20,952 kb. It contains genes essential for arabinogalactan synthesis, as well as for mycolic acid synthesis.

and β(1→6) fashion, up to a total of approximately 30 Gal*f* units. The bulk of galactan synthesis is by sequentially polymerising the galactan polysaccharide in an alternating β(1→5) and β(1→6) dependent manner utilising UDP-Gal*f* as substrate by the recently identified bi-functional galactofuranosyltransferase GlfT (Kremer et al., 2001). The amazing feature of GlfT to establish both β(1→5) and β(1→6) Gal*f* linkages was established in *in vitro* assays (Kremer et al. 2001, Mikusova et al., 2000), as well as only recently with the isolated enzyme (Rose et al. 2006). However, the kinetic parameters of the isolated enzyme, as well as earlier structural work suggests that the first and second, and possibly even the third Gal*f* residues linked to Rha*p* of the linker unit requires additional enzyme activities (Rose et al., 2006). Therefore, after the rhamnosyltransferase WbbL has attached a rhamnosyl residue to α-D-Glc*p*NAc-P-P-decaprenyl to generate polyprenyl-P-P-GlcNAc-Rha, a galactofuranosyltransferase other than GlfT must attach the first galactofuranosyl residue. This is reasonable because the acceptor is of an entirely different structure. As pointed out by Rose et al. (2006), it is furthermore highly unlikely that GlfT attaches also the second Gal*f* residue because of the structural features of the substrate. Therefore, two or three additional catalytic activities are required to build up the lipid bound growing galactan chain up to the point where GlfT begins to act and finish its extension. One gene product discussed as a candidate enzyme is RfbE (Rv3782, NCgl0195) (Mikusova et al., 2005), which according to own data obtained with neoglycolipids in an *in vitro* system adds the third Gal*f* residue (LG Dover, unpublished results).

9. Arabinan precursor

The arabinan precurser utilized by both *C. glutamicum* and *M. tuberculosis* is the lipid-linked sugar donor decaprenylmonophosphoryl-D-arabinose (DPA) (Wolucka et al., 1994; Lee et al., 1997a, 1998), and only recently more detailed knowledge on its unexpectedly membrane-linked synthesis was obtained (Huang et al., 2005). DPA originates from the pentose phosphate pathway (Scherman et al., 1995), and the initial reaction is the activation of ribose-5-phosphate by a pyrophosphate to result in 5-phospho-ribofuranose-pyrophosphate (pRpp). This reaction is catalysed by the ribose-phosphate pyrophosphokinase PrsA. Although, pRpp is also an intermediate of UMP synthesis, the gene *prsA* (NCgl0905) is in all *Corynebacterianeae* in an operon together with the glucosamine-1-phosphate acetyltransferase gene *glmU*, which might indicate its importance for cell wall synthesis, since the latter enzyme is known to synthesise components of cell surface polysaccharides in *Corynebacterium* species (Dawes et al., 1974; Mattos-Guaraldi et al., 1999). Interestingly, epimerisation of the ribose moiety occurs at the lipid-linked level where 5-phospho-ribofuranose-pyrophosphate (pRpp) is transferred to decaprenylmonophosphate forming decaprenylphosphoryl-5-phosphoribose (DPPR) by the 5-phospho-α-D-ribose-1-diphosphate: decaprenylphosphate-5-phosphoribosyltransferase UbiA (Huang et al., 2005) (Rv3806c and NCgl2789 in *M. tuberculosis* and *C. glutamicum*, respectively). An *ubiA*-disrupted mutant of *C. glutamicum* was obtained (Tatituri et al., 2007b). This produced an arabinan-deficient cell wall containing a core linear galactan polymer, thus confirming that the prenyltransferase function is essential to establish the AG polymer. It is hypothesized that DPPR then undergoes dephosphorylation and epimerization of the C2 anomeric hydroxyl of the ribose moiety by a two-enzyme complex encoded in *C. glutamicum* by NCgl0187 and NCgl0186 (Mikusova et al., 2005). The oxidation by NCgl0187 might form decaprenylphosphoryl-2-keto-β-D-erythro-pentofuranose as an intermediate, which is then reduced by NCgl0186 to form DPA. The phosphatase involved in dephosphorylation of DPPR has yet to be identified. A potential candidate is NCgl2782 which is directly upstream of *ubiA* in the prominent cell wall locus (Figure 12.8), where also mycolyltransferases, the arabinosyltransferase AftB (Seidel et al., 2007a), and genes of mycolic acid synthesis are located (De Sousa-D'Auria et al., 2003; Brand et al., 2003; Daffé, 2005). The phosphatase encoded by NCgl2782 is membrane embedded, as is the adjacent prenyltransferase UbiA, consistent with the lipid linked nature of its proposed substrate target.

10. Arabinan biosynthesis

The glycosidic linkage analysis of *C. glutamicum* AG shows that 2-Ara*f*, 5-Ara*f*, and 3,5-Ara*f* linkages are present in the arabinan domain (Puech et al., 2001; Alderwick et al., 2005), which are analogous to those found in the AG of *M. tuberculosis* (Besra et al., 1997). Furthermore, in *C. glutamicum* and *C. diphtheriae* 2,5-Ara*f* linkages are also evident (Dover et al., 2004), as are t-Rha*p* units capping the galactan chain as well as Ara*f* residues in the arabinan domain. In *M. tuberculosis* the arabinan domain terminates in a Ara_6 motif (Daffé et al., 1990), which might be similar present in *Corynebacterium* species. The formation of arabinan in AG occurs through the transfer of one Ara*f* unit to another using β-D-arabinofuranosyl-l-monophosphoryldecaprenol (DPA) as the high energy precursor. It is assumed that after maturation of AG, it is at some point mycolated (McNeil et al., 1987) and then ligated to the peptidoglycan by an as yet unidentified transglycosylase (Hancock et al., 2002; Yagi et al., 2003).

The understanding of arabinan synthesis is strongly connected to the use of the tuberculosis drug ethambutol (EMB). EMB treatment of *M. smegmatis* resulted in increased levels of DPA (Wolucka et al., 1994) and in EMB resistant mutants the polymerisation of Ara*f* into AG was strongly reduced (Mikusova et al., 1996). This was similarly observed for EMB treated *C. glutamicum*, where in agreement with the general structure of the mAGP complex also the mycolylation sites were lost (Radmacher et al., 2005a). Molecular access to the arabinan domain synthesis was not until Belanger et al. (1996) demonstrated that a 9.5 kb region from *Mycobacterium avium* encoding a region with three open reading frames *embR*, *embA* and *embB*, which, when over-expressed in *M. smegmatis*, produced an otherwise susceptible strain to become resistant to EMB. Interestingly, EMB resistance of *M. tuberculosis* correlates with mutations in *embB* (Shi et al., 2006). Also overexpression of the single *emb* gene of *C. glutamicum* results in increased EMB resistance (Radmacher et al., 2005a), and cell-free assay systems demonstrated that the glycosyltransferase encoded by *emb* utilises DPA (Belanger et al., 1996; Alderwick et al., 2005).

This was essentially the state of affairs until *C. glutamicum* was introduced as a model to study AG synthesis. Deletion of the single *emb* in *C. glutamicum* was possible and resulted in a viable yet slow growing strain with profound morphological changes (Alderwick et al., 2005). The sugar and glycosidic linkage analysis revealed an about 90% loss of cell wall arabinan, with all 5-Ara*f*, 3,5-Ara*f*, 2,5-Ara*f* residues (also the capping t-Rha*p* residues) being absent in the cell wall of the *C. glutamicum* Δ*emb* mutant, with the relative amounts of Gal*f* unchanged. The minor amounts of Ara*f* residues were present as t-Ara*f* units, implicating that they were directly attached to the galactan chain. This was somewhat surprising as it has long been thought that the gene products of *embA* and *embB* in mycobacteria are solely responsible for arabinan biosynthesis. Importantly, the occurrence of single Ara*f* residues at the galactan backbone without *emb* being present required a new arabinofuranosyltransferase activity which was identified as AftA, NCgl0185 (Alderwick et al., 2006). This DPA-dependent arabinofuranosyltransferase attaches the first "priming" Ara*f* residue onto the galactan chain for further elaboration by Emb. The gene *aftA* is located directly upstream of *emb*. Further upstream are the genes encoding the dehydrogenase and epimerase for the synthesis of arabinose (Figure 12.1 in Alderwick et al., 2006). The entire gene locus is retained in *Corynebacterianeae* indicative of an ancient core function of NCgl0187, NCgl0186, *aftA*, and *embC* involved in the synthesis of Ara*f* donors and their use to assemble a basic periplasmic arabinan domain which serves as a scaffold to tie mycolic acids. The existance of only t-Ara*f* residues in the *emb* deletion mutant of *C. glutamicum* resulted in a significant and concise fragmentation pattern by MALDI-TOF MS enable to conclude that the Ara domains in AG are located on the 8th, 10th and 12th Gal*f* residue.

Introduction of *emb* into a *C. glutamicum* Δ*emb* strain resulted in the full restoration of the wild type arabinan domain, encluding the Ara*f*-α(1→5) linked residues (Alderwick et al., 2006). However, it is not yet known whether Emb activity is restricted to this linkage type, and which chain length is established by Emb. The situation is even more complex, since individual *embA* and *embB* mutations in *M. smegmatis* led to an alteration of the terminal Ara_6

motif of AG but still produced a highly arabinosylated AG polymer (Escuyer et al., 2001). These studies relied on disruption mutants, and a reinvestigation with defined deletion mutants might be useful. However, a motif present in the Emb proteins equivalent to a proline-rich motif of several bacterial polysaccharide co-polymerases suggests the formation of a chain of Ara*f*-α(1→5) linked residues (Berg et al., 2005). Indeed, mutations introduced into the Pro-motif of the single *emb* of *C. glutamicum* at residues W627L, P667A, and Q674E resulted in a reduced Ara:Gal ratio, but all typical glycosidic linkages retained (Seidel et al., 2007a). This may imply that these residues in some way moderate the size of the stem of the arabinan chains in AG. This is further substantiated by the fact that, when all three residues are mutated together, a severe reduction in arabinose is observed, again with all glycosidic linkages retained.

In addition, the topology of the Emb proteins is slightly different than that of other DPA-dependent arabinosyltransferases, like AftA or AftB for instance. The Emb proteins are very large in size of 1094–1146 aa residues. As the detailed topological study on Emb of *C. glutamicum* has shown, Emb possesses 15 membrane embedded helices in the N terminal part, whereas the C terminal part of the protein extending from amino acid 754-1146 is located in the periplasm (Seidel et al., 2007b). The glycosyltransferase motif characteristic to the GT-C family (Liu et al., 2003) is present between helix III and IV, whereas the Pro-motif is present between helix VIII-IX (Berg et al., 2005; Seidel et al., 2007b). As already mentioned, the Pro-motif and possibly the long periplasmic extension might be related with the building of a longer polysaccharide chain consisting of Ara*f*-α(1→5) linked residues.

Although a number of further glycosyltransferases might be involved in the establishment of the full arabinan domain, only one further arabinosyltransferase is currently known, again from studies with *C. glutamicum* (Seidel et al., 2007b). This is AftB, whose gene is located in the prominent cell wall locus shown in Figure 12.8 where the genes coding for β-Gal*f*-transferase, a membrane-associated alkaline phosphatase, the prenyltransferase UbiA, mycolyltransferases, and genes of mycolic acid synthesis are clustered (Figure 12.8). Deletion of *aftB* (NCgl2780) revealed the complete absence of terminal β(1→2)-linked arabinofuranosyl residues. In addition, an *in vitro* assay utilizing the sugar donor DPA together with the neoglycolipid acceptor α-D-Araf-(1→5)-α-D-Ara*f*-O-C(8) as a substrate confirmed AftB as a terminal β(1→2) arabinofuranosyltransferase. Clearly additional arabinofuranosyltransferases still remain to be identified in *Corynebacterianeae*, in particular since initial Ara*f* residues at branching sites could require specialized arabinofuranosyltransferases as already observed for AftA (Alderwick et al., 2006). Indeed, in *C. glutamicum* the number of glycosyltransferases of the GT-C family likely all to be involved in non-cytoplasmic polysaccharide synthesis is currently as much as 16 (Liu et al., 2003). Of these, Emb, AftA and AftB are identified as DPA-dependent arabinosyltransferases involved in AG synthesis.

11. Fatty acids

The linear fatty acids are synthesized from malonyl-CoA which is made via carboxylation of acetyl-CoA. The reaction sequence involves two steps. In the first step the ATP-dependent biotin carboxylase catalyzes the transfer of a carboxyl group to biotin, which is linked to the biotin carboxyl carrier protein. In the second step, catalyzed by carboxyltransferase, the carboxyl group is transferred from carboxybiotin to acetyl-CoA to form malonyl-CoA (Cronan et al., 1996). In contrast to *E. coli*, where four proteins form the multienzyme complex catalyzing acetyl-CoA carboxylation, in *C. glutamicum* and other *Corynebacterianeae* domains are fused and two proteins carry the activities for acetyl-CoA carboxylation. The α-subunit possesses the biotin carboxylase and biotin carboxyl carrier protein function, and the ß-subunit the carboxyl-transferase activity. Furthermore, *in vivo* and *in vitro* studies have shown that in *C. glutamicum* the α-subunit, AccBC, is encoded by NCgl0670, and the ß-subunit, AccD1 by NCgl0678 (Gande et al., 2004). The acetyl-CoA carboxylase contains in addition a third small peptide, AccE, of 8.9 kDa. For the entire complex a mass of approximately 812,000 Da was determined which would agree with an $\alpha_6\beta_6\varepsilon_6$ structure (Gande et al., 2007).

The malonyl-CoA is utilized by the fatty acid synthase (FAS) of which *C. glutamicum* has two (Radmacher et al., 2005b). This is an exception among the *Corynebacterianeae*, since only *Corynebacterium ammoniagenes* has also two FAS, whereas *C. diphtheriae* or *C. jeikeium*, for instance, has only one FAS. FAS enzymes are large multienzyme complexes of more than 2 MDa, catalysing in successive cycles a multistep reaction to generate mature fatty acids (Cronan et al., 1996). In *Corynebacterianeae* the two FAS are of the eucaryotic type (FAS-I), where the individual domains are fused. Accordingly, the FAS-I encoding genes are very large, and the two enzymes of *C. glutamicum* are encoded by *fasA* (8907 kb; FAS-IA) and *fasB* (8988 kb; FAS-IB). *M. tuberculosis* has in addition FAS-II components where individual domains exist, and these are involved in elongation of the mero-chain of mycolic acids (Takayama et al., 2005). *C. glutamicum* also possesses three FAS-II modules (NCgl0281, NCgl0283 and NCgl0527), but their function is unknown.

Both FAS-I enzymes in *C. glutamicum* are functionally active, and *fasA* is essential but *fasB* is not. FAS-IB is primarily responsive for palmitate synthesis, as also shown with the isolated FAS-IB enzyme of *C. ammoniagenes* (Stuible et al., 1996; Stuible et al., 1997), which also has two FAS-I enzymes and is very closely related to *C. glutamicum* (Liebl, 2005). The isolated FAS-IB synthesizes stearate, but not oleate. In contrast, the isolated FAS-IA produces stearate and oleate, both representing the majority of fatty acids in the plasma membrane of *C. glutamicum* and explaining the essentiality of *fasA*. *fasB* contributes 32 % to both *fas* transcripts during growth on glucose and *fasA* is down regulated by 8.3-fold during growth on acetate as compared with glucose, thus showing a strong carbon-source dependent regulation of lipid synthesis genes, whose regulation is not yet understood (Radmacher et al., 2005b).

12. Mycolic acids

A unique feature of *Corynebacterianeae* is that they contain in addition to the linear fatty acids, mycolic acids, which are α-branched ß-hydroxy fatty acids. They form an outer lipid bilayer functionally comparable to the outer layer of Gram-negative bacteria. Cell-free extracts of *Corynebacterium* have been shown to utilize [^{14}C]palmitic acid (Lee et al., 1997b), with the newly synthesized mycolic acid exclusively labeled at C-1 and C-3. The Claisen condensation reaction was hypothesized to involve a carboxylation step since it is inhibited by avidin, an inhibitor of biotin-dependent enzymes in extracts of *Corynebacterium diphtheriae* that produces the putative precursor of mycolic acids, 2-tetradecyl-3-oxo-octadecanoic acid (Portevin et al., 2004). More detailed information was obtained only recently, again with *C. glutamicum*. Use of a *fasAB* double mutant supplied with uniformly labelled [^{13}C]oleate and analysis of isolated mycolic acids confirmed that mature mycolic acids consists exclusively of two fused [^{13}C]oleate molecules (Radmacher et al., 2005b).

However, enzymes involved in the carboxylation and condensation of the two fatty acids to afford mycolic acid synthesis were unknown, until the recent systematic study of the four *accD* paralogues in *C. glutamicum* with identities to β-subunits of acyl-CoA carboxylases (Gande et al., 2004). The mutational analysis revealed that *accD1* is required for malonyl-CoA synthesis, as already mentioned before, whereas surprisingly both *accD2* and *accD3* are required for mycolic acid synthesis. *accD4* can be inactivated without giving a distinct phenotype. The two carboxyltransferases *accD2* and *accD3* are retained in all *Corynebacterianeae*, including *M. leprae*, thus representing the key to mycolic acid synthesis by providing a carboxylated intermediate for condensation of the mero-chain with the α-branch (Portevin et al., 2005). Biochemical studies revealed that the carboxylase complex is made of the two highly similar ß-subunits AccD2 (NCgl0677) and AccD3 (NCgl2772) together with AccBC and AccE, the latter two identical to the subunits of the acetyl-CoA carboxylase complex (Gande et al., 2007). Structural considerations and crystal structures of the *M. tuberculosis* orthologue of AccD3 (Mt-AccD5), led us to suggest that the existence of two ß-subunits serves to lend specificity to the carboxylation multienzyme complex, which has a size of approximately the same than the acetyl-CoA carboxylase complex.

Directly adjacent to *accD3* the polyketide synthase gene *pks*, and the fatty acyl-ligase gene *fadD* is located (Figure 12.8). Deletion of the single *pks* in *C. glutamicum* led to the absence of mycolic acids (Portevin et al., 2004; Gande et al., 2004), as did the inactivation of *fadD* (Portevin et al., 2005). β-Ketoacyl synthases, like Pks, catalyze the formation of new carbon–carbon bonds by condensation of acyl chain precursors with an elongation substrate, usually malonyl or methylmalonyl residues that are covalently attached in a thioester linkage to an acyl-carrier protein (ACP) or a corresponding ACP domain. FadD, recently shown to constitute a new class of fatty acyl-AMP ligases, activates a linear fatty acid as acyl adenylate, which is probably then loaded onto the polyketide synthase Pks (Trivedi et al., 2004). The other fatty acid is carboxylated via the carboxylation complex consisting of AccD2, AccD3, AccBC, and AccE, and both fatty acids are loaded onto the polyketide synthase. This protein possesses a β-ketoacyl synthase domain with two ACP domains and a thioesterase domain. It can be assumed that the two phosphopantetheinyl arms of the ACP domains are present to bind at one site the acyl chain forming the mero-chain and at the other the acyl chain of the incoming α-branch. Further details of the interplay of a variety of domains required for activation, carboxylation, and condensation of fatty acids to produce mycolic acids are still to be solved. The product of the condensation forms a 2-alkyl-3-keto mycolate precursor and requires the subsequent reduction to result in mature mycolic acids. A recent bioinformatic comparison of the genomes of *Corynebacterianeae* identified among three candidates the gene encoded by NCgl2385, now termed *cmrA*, as the reductase (Lea-Smith et al., 2007). Deletion of the gene in *C. glutamicum* resulted in a slow growing strain that was deficient in arabinogalactan-linked mycolates. However, it synthesized abnormal forms of the mycolate-containing trehalose dicorynomycolate and trehalose mono-corynomycolate except containing an unreduced β-keto ester. The reduction step is apparently important for the attachment of mycolic acids to AG. The mycolates formed are thought to be passed to lipid carriers for transport across the plasma membrane (Besra et al., 1997).

All three known pathways for trehalose biosynthesis in bacteria, i.e. trehalose synthesis from UDP-glucose and glucose 6-phosphate (OtsA-OtsB pathway), from malto-oligosaccharides or alpha-1,4-glucans (TreY-TreZ pathway), or from maltose (TreS pathway) can operate in *C. glutamicum* (Tzvetkov et al., 2003; Wolf et al., 2003). Inactivation of only one of the three pathways did not have a severe impact on growth, while the simultaneous inactivation of the OtsA-OtsB and TreY-TreZ pathway or of all three pathways resulted in the inability of the corresponding mutants to synthesize trehalose and to grow efficiently on various sugar substrates in minimal media. This growth defect was largely reversed by the addition of trehalose to the culture broth. Further data show that trehalose availability is a prerequisite for the production of mycolates and the proper cell wall integrity. The mycoloyl residues synthesized in the absence of trehalose synthesis and grown on alpha-D-glucopyranosyl-containing oligosaccharides are transferred to free sugar acceptors, and it is suggested that also the transfer to native trehalose occurs outside the plasma membrane (Tropis et al., 2005). Furthermore, trehalose also operates as an osmoprotectant. Its synthesis is increased upon hyperosmotic conditions and this osmoregulated synthesis is mediated by the TreY-TreZ pathway (Wolf et al., 2003).

13. Glutamate export

Since *C. glutamicum* is used for amino acid production, the amino acids have to pass the cell membrane on their way into the medium, and the identification of specific export carriers was an impetus for amino acid research in general. The exporters currently identified are LysE exporting the basic amino acids L-lysine and L-arginine (Vrljic et al., 1996), BrnFE exporting the branched chain amino acids and L-methionine (Kennerknecht et al., 2002; Trötschel et al., 2005), and ThrE exporting L-threonine (Simic et al., 2001). These prior unknown carriers each represent a new type of transporter family which is in part widely spread among organisms, like ThrE present in bacteria, archaea and the fungal kingdom (Ming-Ren et al., 2001; Eggeling and Sahm, 2003). It should also be mentioned that the outer permeability layer consisting of mycolic acids may contribute to amino acid efflux. Indeed, under a number of conditions resulting in L-glutamate efflux, the amount of mycolic acids is reduced by up to approx. 40 %, and also the spectrum of mycolic acids usually consisting of 32, 30 and 34 carbons, is altered and mycolic acids with 22 and 24 carbons occur (Hashimoto et al., 2006). Moreover, when trehalose synthesis genes *treY* and *otsA* are inactivated, the L-glutamate accumulation rises from 40.2 to 45.6 g l^{-1} (Nakamura et al., 2001). Also, in the case of L-lysine, the mycolic acid layer contributes to the efflux properties. A mutant that lacks *treY* and *otsA* is devoid of mycolic acids, and excretes L-lysine at an increased rate accumulating 37% more lysine per cell mass than the original strain (Wolf, 2002; Gebhardt et al., 2007).

Whereas detailed knowledge on the export the amino acids L-lysine, L-threonine and branched-chain amino acids is now available, the situation with L-glutamate is different. Although L-glutamate was the very first amino acid produced by *C. glutamicum* (Kinoshita et al., 1957), and it is the amino acid with the highest production capacity of 1 800 000 tons per year (Shirai et al., 2005), the understanding of its excretion mechanism is still the least developed (Hoischen and Krämer, 1989; 1990). The first tantalizing fact is that always specific treatments are necessary, like growth under biotin limitation, addition of specific detergents, penicillin or ethambutol, or altered expression of lipid synthesis genes, like *cls* encoding the cardiolipin synthetase (Nampoothiri 2002), or *accD1* (Kimura et al., 1999), the subunit of the acetyl-CoA carboxylase (Gande et al., 2004). It is striking that all these different treatments or conditions have in common that they alter the cell envelope (Eggeling et al., 2001; Eggeling, 2005; Gebhardt et al., 2007). Obviously, the integrity of the cell envelope, encluding the plasma membrane, has to be taken into account in developing a model to explain L-glutamate excretion.

The second fact is the recent finding, that the 2-oxoglutarate dehydrogenase, which controls the flux of 2-oxoglutarate to either succinyl-CoA or L-glutamate is regulated involving a phosphorylation step via the Ser/Thr protein kinase PknG (Niebisch et al. 2006). The 2-oxoglutarate dehydrogenase is inhibited by the 15 kDa protein OdhI in its unphosphorylated form. Phosphorylation of OdhI is catalysed by PknG, whereas the dephosphorylation is catalysed by Ppp (Schultz et al., 2007). It is very likely that besides PknG one or more of the additional three Ser/Thr protein kinases present in *C. glutamicum* are involved in OdhI phosphorylation (Bott, 2007). Most importantly and as already mentioned, *pknB* as well as *ppp* are located within the conserved corynebacterial division-regulation cluster of cell wall-related genes (Figure 12.3). This strongly suggests a connection between the different cell wall treatments and L-glutamate efflux via a regulatory cascade.

The Ser/Thr protein kinases PknA, PknB, PknL are proteins spanning the membrane once. They transmit information from the extracytosolic side to the cytosol. Both PknB (NCgl0040) as well as PknA (NCgl2095) comprise an intracellular kinase domain and a reiterated extracellular PASTA (for "penicillin-binding protein and serine/threonine kinase associated") domain (Av-Gay and Everett, 2000). PASTA domains exhibit low affinity for beta-lactam antibiotics that are structurally similar to their likely normal ligands, which are the stem peptides of unlinked peptidoglycan. Thus a disturbed cell wall integrity might be sensed by the kinases. In addition, sensing of the lipid environment of the cytoplasmic membrane is plausible to ensure also the integrity of this component of the cell wall. Since the Ser/Thr protein kinases are membrane integrated proteins they may serve this function too, although there is so far no evidence for this. Consequently, a model to explain L-glutamate efflux which involves sensing of cell wall alterations and a regulatory cascade up to the central metabolism would elegantly combine the heterogeneity of the different triggers known to trigger L-glutamate efflux (Figure 12.9). The still speculative cascade involves sensing of cell wall integrity, and furthermore OdhI and the 2-oxoglutarate dehydrogenase as a major biosynthesis enzyme of relevance for L-glutamate production. It is known that the 2-oxoglutarate dehydrogenase activity is reduced under L-glutamate producing conditions (Kimura et al., 1999; Kimura, 2003).

However, the requested carrier to catalyse L-glutamate export of *C. glutamicum* via the plasma membrane is still poor defined. A great leap forward is the recent identification of NCgl1221 as to be involved in L-glutamate efflux (Nakamura et al., 2007) (Figure 12.9). Based on the observation that a *C. glutamicum* mutant with disrupted *odhA* (2-oxoglutarate dehydrogenase component) excreted L-glutamate at low level, but supressor mutants were

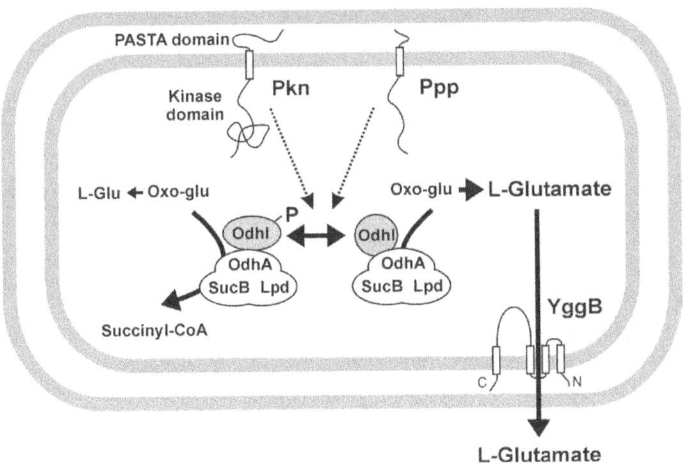

Figure 12. 9. Molecular components identified to be involved in L-glutamate efflux. The cascade triggering efflux involves cell wall alterations, their sensing by membrane bound serine/threonine kinases, inhibition of oxo-glutarate dehydrogenase activity by OdhI and the efflux carrier YggB.

readily obtained excreting L-glutamate at high level, NCgl1221 was cloned. Analysis of the suppressor mutants repeatedly identified this gene carrying amino acid substitutions or a Cys-Ser-Leu insertion. NCgl1221 encodes the YggB protein previously described as a putative mechanosensitive channel (Nottebrock et al. 2003). It is still a mystery why mutations result in increased efflux. However, even in the absence of an altered 2-oxoglutarate dehydrogenase, overexpression of NCgl1221 results in significantly increased glutamate excretion, and deletion of NCgl1221 in reduction of glutamate efflux. Having in mind that cell wall alterations are related with L-glutamate efflux (see above), and NCgl1221 bears similarities to a mechanosensitive channel, it is one possibility that L-glutamate excretion is a rather unspecific solute export related with osmotic stress due to disturbed cell wall integrity. Since a very high intracellular L-glutamate concentration in *C. glutamicum* of more than 100 mM prevails (Gutmann et al., 1992), it is questionable whether energy-driven export occurs as it is the case with LysE, Thre, and BrnFE, or indeed passive efflux as characteristic for a channel.

References

Alderwick, L.J., Radmacher, E., Seidel, M., Gande, R., Hitchen, P.G., Morris, H.R., Dell, A., Sahm, H., Eggeling, L., and Besra, G.S. (2005). Deletion of Cg-*emb* in corynebacterianeae leads to a novel truncated cell wall arabinogalactan, whereas inactivation of Cg-*ubiA* results in an arabinan-deficient mutant with a cell wall galactan core. J. Biol. Chem. *280*, 32362-32371.

Alderwick, L.J., Seidel, M., Sahm, H., Besra, G.S., and Eggeling L. (2006). Identification of a novel arabinofuranosyltransferase (AftA) involved in cell wall arabinan biosynthesis in *Mycobacterium tuberculosis*. J. Biol. Chem. *281*, 15653-15661.

Av-Gay, Y., and Everett, M. (2000). The eukaryotic-like Ser/Thr protein kinases of *Mycobacterium tuberculosis*. Trends Microbiol. *8*, 238-244.

Barry 3rd, C.E., Lee, R.E., Mdluli, K., Sampson, A.E., Schroeder, B.G., Slayden, R.A., and Yuan, Y. (1998). Mycolic acids: structure, biosynthesis and physiological functions. Prog. Lipid Res. *37*, 143-179.

Berg, S., Starbuck, J., Torrelles, J.B., Vissa, V., Crick, D.C., Chatterjee, D., and Brennan, P.J. (2005). Roles of conserved proline and glycosyltransferase motifs of EmbC in biosynthesis of lipoarabinomannan. J. Biol. Chem. *280*, 5651–5663.

Berg, S., Kaur, D., Jackson, M., and Brennan, P.J. (2007). The glycosyltransferases of *Mycobacterium tuberculosis* - roles in the synthesis of arabinogalactan, lipoarabinomannan, and other glycoconjugates. Glycobiology *17*, 35-56.

Bott, M. (2007). Offering surprises: TCA cycle regulation in *Corynebacterium glutamicum*. Trends Microbiol. *15*, 417-425.

Brand, S., Niehaus, K., Pühler, A., and Kalinowski, J. (2003). Identification and functional analysis of six mycolyltransferase genes of *Corynebacterium glutamicum* ATCC 13032: the genes *cop1*, *cmt1*, and *cmt2* can replace each other in the synthesis of trehalose dicorynomycolate, a component of the mycolic acid layer of the cell envelope. Arch. Microbiol. *180*, 33-44.

Brian, C., VanderVen, B., Harder, J.D., Crick, D.C., and Belisle, J.T. (2005). Export-mediated assembly of mycobacterial glycoproteins parallels eukaryotic pathways. Science *309*, 941-943.

Costa-Riu, N., Maier, E., Burkovski, A., Krämer, R., Lottspeich, F., and Benz, R. (2003). Identification of an anion-specific channel in the cell wall of the Gram-positive bacterium *Corynebacterium glutamicum*. Mol. Microbiol. *50,* 1295-1308.

Cronan, J.E., and Rock, C.O. (1996). Biosynthesis of membrane lipids. In Neidhardt, F.C., Curtiss III, R., Ingraham, J.L., Lin, E.C.C., Low, K.B., Magasanik, B., Reznikoff, W.S., Riley, M., Schaechter, M., and Umbarger, H.E., eds, *Escherichia coli* and *Salmonella*: cellular and molecular biology, 2nd ed. ASM Press, Washington, DC. pp 612-636

Daffé, M. (2005). The cell envelope of corynebacteria. In Handbook of *Corynebacterium glutamicum*, L. Eggeling; M. Bott, eds (CRC Tayler and Francis), pp. 121-148.

Dawes, J., Tuach, S.J., and McBride, W.H. (1974). Properties of an antigenic polysaccharide from *Corynebacterium parvum*. J. Bacteriol. *120,* 24-30.

De Sousa-D'Auria, C., Kacem, R., Puech, V., Tropis, M., Leblon, G., Houssin, C., and Daffé, M. (2003). New insights into the biogenesis of the cell envelope of corynebacteria: identification and functional characterization of five new mycoloyltransferase genes in *Corynebacterium glutamicum*. FEMS Microbiol. Lett. *224,* 35-44.

Dmitriev, B.A., Ehlers, S., and Rietschel, E.T. (1999). Layered murein revisited: a fundamentally new concept of bacterial cell wall structure, biogenesis and function. Med. Microbiol. Immunol. *187,* 173-181.

Dmitriev, B.A., Ehlers, S., Rietschel, E.T., and Brennan, P.J. (2000). Molecular mechanics of the mycobacterial cell wall: from horizontal layers to vertical scaffolds. Int. J. Med. Microbiol. *290,* 251-258.

Dover, L.G., Cerdeno-Tarraga, A.M., Pallen, M.J., Parkhill, J., and Besra, G.S. (2004). Comparative cell wall core biosynthesis in the mycolated pathogens, *Mycobacterium tuberculosis* and *Corynebacterium diphtheriae*. FEMS Microbiol. Rev. *28*, 225-250.

Eggeling, L. (2005). Export of amino acid and other solutes. In Handbook of *Corynebacterium glutamicum*, L. Eggeling; M. Bott, eds (CRC Tayler and Francis), pp. 187-214.

Eggeling, L., and Sahm, H. (2001). The cell wall barrier of *Corynebacterium glutamicum* and amino acid efflux. J. Biosc. Bioeng. *3*, 201-203.

Eggeling, L., Krumbach, K., and Sahm, H. (2001). L-Glutamate efflux with *Corynebacterium glutamicum*: Why is penicillin treatment or tween addition doing the same? J. Mol. Microbiol. Biotechnol. *3,* 67-68.

Eggeling, L., and Sahm, H. (2003). New ubiquitous translocators: Amino acid export by *Corynebacterium glutamicum* and *Escherichia coli*. Arch. Microbiol. *180,* 155-160.

Errington, J., Daniel, R.A., and Scheffers, D.J. (2003). Cytokinesis in bacteria. Microbiol. Mol. Biol. Rev. *67,* 52-65.

Gande, R., Dover, L.G., Krumbach, K., Besra, G.S., Sahm, H., Oikawa, T., and Eggeling, L. (2007). The two carboxylases of *Corynebacterium glutamicum* essential for fatty acid and mycolic acid synthesis. J. Bacteriol. *189,* 5257-5264.

Gibson, K.J., Eggeling, L., Maughan, W.N., Krumbach, K., Gurcha, S.S., Nigou, J., Puzo, G., Sahm, H., and Besra, G.S. (2003). Disruption of Cg-Ppm1, a polyprenyl monophosphomannose synthase, and the generation of lipoglycan-less mutants in *Corynebacterium glutamicum*. J. Biol. Chem. *278,* 40842-40850.

Gebhardt, H., Meniche, X., Tropis, M., Krämer, R., Daffé, M., and Morbach, S. (2007). The key role of the mycolic acid content in the functionality of the cell wall permeability barrier in *Corynebacterineae*. Microbiology *153*, 1424-1434.

Goodfellow, M., and Minnikin, D.E. (1981). The genera *Nocardia* and *Rhodococcus*. In The Prokaryotes, Starr, M.P., Stolp, H., Trüper, H.G., Balows, A., and Schlegel, G. eds (Springer-Verlag, Berlin), pp.2016-20127.

Gutmann, M., Hoischen, C., and Krämer, R. (1992). Carrier-mediated glutamate secretion by *Corynebacterium glutamicum* under biotin limitation. Biochim. Biophys. Acta *1112,* 115-123.

Hashimoto, K., Kawasaki, H., Akazawa, K., Nakamura, J., Asakura, Y., Kudo, T., Sakuradani, E., Shimizu, S., and Nakamatsu T. (2006). Changes in composition and content of mycolic acids in glutamate-overproducing *Corynebacterium glutamicum.* Biosci. Biotechnol. Biochem. *70,* 22-30.

Hoischen, C., and Krämer, R. (1989). Evidence for an efflux carrier system involved in the secretion of glutamate by *Corynebacterium glutamicum.* Arch. Microbiol. *151,* 342-347.

Hoischen, C., and Krämer, R. (1990) Membrane alteration is necessary but not sufficient for effective glutamate secretion in *Corynebacterium glutamicum.* J. Bacteriol. *172,* 3409-3416.

Honrubia, M.P., Ramos, A., and Gil, J.A. (2001). The cell division genes *ftsQ* and *ftsZ,* but not the three downstream open reading frames YFIH, ORF5 and ORF6, are essential for growth and viability in *Brevibacterium lactofermentum* ATCC 13869. Mol. Genet. Genomics *265,* 1022-1030.

Hünten, P., Costa-Riu, N., Palm, D., Lottspeich, F., and Benz, R. (2005a). Identification and characterization of PorH, a new cell wall channel of *Corynebacterium glutamicum.* Biochim. Biophys. Acta *1715,* 25-36.

Hünten, P., Schiffler, B., Lottspeich, F., and Benz, R. (2005b). PorH, a new channel-forming protein present in the cell wall of *Corynebacterium efficiens* and *Corynebacterium callunae.* Microbiology*151,* 2429-2438.

Jarlier, V., and Nikaido, H. (1990). Permeability barrier to hydrophilic solutes in *Mycobacterium chelonei.* J. Bacteriol. *172,* 1418-1423.

Jarlier, V., and Nikaido, H. (1994). Mycobacterial cell wall: Structure and role in natural resistance to antibiotics. FEMS Microbiol. Lett. *123,* 11-18.

Kaur, D., Berg, S., Dinadayala, P., Gicquel, B., Chatterjee, D., McNeil, M.R., Vissa, V.D., Crick, D.C., Jackson, M., and Brennan, P.J. (2006). Biosynthesis of mycobacterial lipoarabinomannan: Role of a branching mannosyltransferase. Proc. Natl. Acad. Sci. U S A *103,* 13664-13669.

Kennerknecht, N., Sahm, H., Yen, M.R., Patek, M., Saier, M.H. Jr, and Eggeling, L. (2002). Export of L-isoleucine from *Corynebacterium glutamicum*: a two-gene-encoded member of a new translocator family. J. Bacteriol. *184,* 3947-39456.

Kimura, E. (2003). Metabolic engineering of glutamate production. Adv. Biochem. Eng. Biotechnol. *79,* 37-57.

Kimura, E., Yagoshi, C., Kawahara, Y., Ohsumi, T., Nakamatsu, T., and Tokuda, H. (1999). Glutamate overproduction in *Corynebacterium glutamicum* triggered by a decrease in the level of a complex comprising DtsR and a biotin-containing subunit. Biosc. Biotech. Biochem. *63,* 1274-1278.

Kinoshita, S., Udaka, S., and Shimono, M. (1957). Studies on the amino acid fermentation. Production of L-glutamate by various microorganisms. J. Gen. Appl. Microbiol. *3,* 193-205.

Kovacevic, S., Anderson, D., Morita, Y.S., Patterson, J., Haites, R., McMillan, B.N., Coppel, R., McConville, M.J., and Billman-Jacobe, H. (2006). Identification of a novel protein with a role in lipoarabinomannan biosynthesis in mycobacteria. J. Biol. Chem. *281,* 9011-9017.

Kremer, L., Dover, L.G., Morehouse, C., Hitchin, P., Everett, M., Morris, H.R., Dell, A., Brennan, P.J., McNeil, M.R., Flaherty, C., Duncan, K., and Besra, G.S. (2001). Galactan biosynthesis in *Mycobacterium tuberculosis.* Identification of a bifunctional UDP-galactofuranosyltransferase. J. Biol. Chem. *276,* 26430-26440.

Kremer, L., Gurcha, S.S., Bifani, P., Hitchen, P.G., Baulard, A., Morris, H R., Dell, A., Brennan, P J., and Besra, G.S. (2002). Characterization of a putative alpha-mannosyltransferase involved in phosphatidylinositol trimannoside biosynthesis in *Mycobacterium tuberculosis*. Biochem. J. *363,* 437-447.

Lea-Smith, D.J., Pyke, J.S., Tull, D., McConville, M.J., Coppel, R.L., and Crellin, P.K. (2007). The reductase that catalyzes mycolic motif synthesis is required for efficient attachment of mycolic acids to arabinogalactan. J. Biol. Chem. *282,* 11000-11008.

Lee, R.E., Armour, J.W., Takayama, K., Brennan, P.J., and Besra, G.S. (1997a) Mycolic acid biosynthesis: definition and targeting of the Claisen condensation step. Biochim. Biophys. Acta *1346,* 275-284.

Liebl, W. (2005). *Corynebacterium* Taxonomy. In Handbook of *Corynebacterium glutamicum*, L. Eggeling; M. Bott, eds (CRC Tayler and Francis), pp. 9-36.

Liu, J., and Mushegian, A. (2003). Three monophyletic superfamilies account for the majority of the known glycosyltransferases. Protein. Sci. *12,* 1418-1431.

Mahne, M., Tauch, A., Pühler, A., and Kalinowski, J. (2006). The *Corynebacterium glutamicum* gene *pmt* encoding a glycosyltransferase related to eukaryotic protein-O-mannosyltransferases is essential for glycosylation of the resuscitation promoting factor (Rpf2) and other secreted proteins. FEMS Microbiol. Lett. *259,* 226-233.

Mattos-Guaraldi, A.L., Cappelli, E.A., Previato, J.O., Formiga, L.C., and Andrade, A.F. (1999). Characterization of surface saccharides in two *Corynebacterium diphtheriae* strains. FEMS Microbiol. Lett. *170,* 159-166.

Meroueh, S.O., Bencze, K.Z., Hesek, D., Lee, M., Fisher, J.F., Stemmler, T.L., and Mobashery, S. (2006). Three-dimensional structure of the bacterial cell wall peptidoglycan. Proc. Natl. Acad. Sci. U S A *103,* 4404-4409.

Mikusová, K., Huang, H., Yagi, T., Holsters, M., Vereecke, D., D'Haeze, W., Scherman, M.S., Brennan, P.J., McNeil, M.R., and Crick, D.C. (2005). Decaprenylphosphoryl arabinofuranose, the donor of the D-arabinofuranosyl residues of mycobacterial arabinan, is formed via a two-step epimerization of decaprenylphosphoryl ribose. J. Bacteriol. *187,* 8020-8025.

Ming-Ren, Y., Tseng, Y., Simic, P., Sahm, H., Eggeling, L., and Saier, M.H. Jr. (2001). The ubiquitous Thre family of putative transmembrane amino acid efflux transporters. Res. Microbiol. *153,* 19-25.

Mishra, A.K., Alderwick, L.J., Rittmann, D., Tatituri, R.V., Nigou, J., Gilleron, M., Eggeling, L., and Besra, G.S. (2007). Identification of an alpha(1->6) mannopyranosyltransferase (MptA), involved in *Corynebacterium glutamicum* lipomanann biosynthesis, and identification of its orthologue in *Mycobacterium tuberculosis*. Mol. Microbiol. *65,* 1503-1517.

Morita, Y.S., Velasquez, R., Taig, E., Waller, R.F., Patterson, J.H., Tull, D., Williams, S.J., Billman-Jacobe, H., and McConville, M.J. (2005). Compartmentalization of lipid biosynthesis in mycobacteria. J. Biol. Chem. *280,* 21645-21652.

Morita, Y.S., Sena, C.B., Waller, R.F., Kurokawa, K., Sernee, M.F., Nakatani, F., Haites, R.E., Billman-Jacobe, H., McConville, M.J., Maeda, Y., and Kinoshita, T. (2006). PimE is a polyprenol-phosphate-mannose-dependent mannosyltransferase that transfers the fifth mannose of phosphatidylinositol mannoside in mycobacteria. J. Biol. Chem. *281,* 25143-25155.

Nakamura, J., Izui, H., and Nakamatsu, T. (2001). Bacterium producing L-glutamic acid and method for producing L-glutamic acid. European Patent Application 1 174 508.

Nakamura, I., Hirano, S., Ito, H., and Wachi, M. (2007). Mutations of the *Corynebacterium glutamicum* NCgl1221 gene, encoding a mechanosensitive channel homolog, induce L-glutamic acid production. Appl. Environ. Microbiol. *73,* 4491-4498.

Nampoothiri, K.M., Hoischen, C., Bathe, B., Möckel, B., Pfefferle, W., Krumbach, K., Sahm, H., and Eggeling, L. (2002). Expression of genes of lipid synthesis and altered lipid composition modulates L-glutamate efflux of *Corynebacterium glutamicum.* Appl. Microbiol. Biotechnol. *58,* 89-96.

Niebisch, A., Kabus, A., Schultz, C., Weil, B., and Bott, M. (2006). Corynebacterial protein kinase G controls 2-oxoglutarate dehydrogenase activity via the phosphorylation status of the OdhI protein. J. Biol. Chem. *281,* 12300-12307.

Nottebrock, D., Meyer, U., Krämer, R., and Morbach, S. (2003). Molecular and biochemical characterization of mechanosensitive channels in *Corynebacterium glutamicum.* FEMS Microbiol. Lett. *218,* 305-309.

Portevin, D., De Sousa-D'Auria, C., Houssin, C., Grimaldi, C., Chami, M., Daffe, M., and Guilhot, C. (2004). A polyketide synthase catalyzes the last condensation step of mycolic acid biosynthesis in mycobacteria and related organisms. Proc. Natl. Acad. Sci. U S A *101,* 314-319.

Portevin, D., de Sousa-D'Auria, C., Montrozier, H., Houssin, C., Stella, A., Laneelle, M.A., Bardou, F., Guilhot, C., and Daffé, M. (2005). The acyl-AMP ligase FadD32 and AccD4-containing acyl-CoA carboxylase are required for the synthesis of mycolic acids and essential for mycobacterial growth: identification of the carboxylation product and determination of the acyl-CoA carboxylase components. J. Biol. Chem. *280,* 8862-88674.

Puech, V., Bayan, N., Salim, K., Leblon, G., and Daffé, M. (2000). Characterization of the *in vivo* acceptors of the mycoloyl residues transferred by the corynebacterial PS1 and the related mycobacterial antigens 85. Mol. Microbiol. *35,* 1026-1041.

Puech, V., Chami, M., Lemassu, A., Lanéelle, M.A., Schiffler, B., Gounon, P., Bayan, N., Benz, R, and Daffé, M. (2001). Structure of the cell envelope of corynebacteria: importance of the non-covalently bound lipids in the formation of the cell wall permeability barrier and fracture plane. Microbiology *147*, 1365-1382.

Radmacher, E., Stansen, K.C., Besra, G.S., Alderwick, L.J., Maughan, W.N., Hollweg, G., Sahm, H., Wendisch, V.F., and Eggeling, L. (2005a). Ethambutol, a cell wall inhibitor of *Mycobacterium tuberculosis*, elicits L-glutamate efflux of *Corynebacterium glutamicum.* Microbiology *151,* 1359-1368.

Radmacher E, Alderwick LJ, Besra GS, Brown AK, Gibson KJ, Sahm H, Eggeling L. (2005b). Two functional FAS-I type fatty acid synthases in *Corynebacterium glutamicum.*Microbiology *151*, 2421-2427.

Ramos, A., Honrubia, M.P., Vega, D., Ayala, J.A., Bouhss, A., Mengin-Lecreulx, D., and Gil, J.A. (2004). Characterization and chromosomal organization of the *murD-murC-ftsQ* region of *Corynebacterium glutamicum* ATCC 13869. Res. Microbiol. *155*, 174-184.

Ramos, A., Letek, M., Campelo, A.B., Vaquera, J., Mateos, L.M., and Gil, J.A. (2005). Altered morphology produced by *ftsZ* expression in *Corynebacterium glutamicum* ATCC 13869. Microbiology *151,* 2563-2572.

Schaeffer, M.L., Khoo, K.H., Besra, G.S., Chatterjee, D., Brennan, P.J., Belisle, J.T., and Inamine, J.M. (1999). The *pimB* gene of *Mycobacterium tuberculosis* encodes a mannosyltransferase involved in lipoarabinomannan biosynthesis J. Biol. Chem. *274,* 31625-31631.

Scheffers, D., and Pinho, M.G. (2005). Bacterial cell wall synthesis: new insights from localization studies. Microbiol. Molec. Biol. Rev. *69*, 585-607.

Schiffler, B., Barth, E., Daffé, M., and Benz, R. (2007). *Corynebacterium diphtheriae*: Identification and characterization of a channel-forming protein in the cell wall. J. Bacteriol. *189*, 7709-7719.

Schleifer, K.H., and Kandler, O. (1972). Peptidoglycan types of bacterial cell walls and their taxonomic implications. Bacteriol. Rev. *36*, 407-477.

Schultz, C., Niebisch, A., Gebel, L., and Bott, M. (2007). Glutamate production by *Corynebacterium glutamicum*: dependence on the oxoglutarate dehydrogenase inhibitor protein OdhI and protein kinase PknG. Appl. Microbiol. Biotechnol. *76,* 691-700.

Seidel, M., Alderwick, L.J., Birch, H.L., Sahm, H., Eggeling, .L, and Besra G.S. (2007a). Identification of a novel arabinofuranosyltransferase AftB involved in a terminal step of cell wall arabinan biosynthesis in *Corynebacterianeae*, such as *Corynebacterium glutamicum* and *Mycobacterium tuberculosis*. J. Biol. Chem. *282,* 14729-14740.

Seidel, M., Alderwick, L.J., Sahm, H., Besra, G.S., and Eggeling, L. (2007b). Topology and mutational analysis of the single Emb arabinofuranosyltransferase of *Corynebacterium glutamicum* as a model of Emb proteins of *Mycobacterium tuberculosis*. Glycobiology *17,* 210-219.

Shi, L., Berg, S., Lee, A., Spencer, J.S., Zhang, J., Vissa, V., McNeil, M.R., Khoo, K.H., and Chatterjee, D. (2006). The carboxy terminus of EmbC from *Mycobacterium smegmatis* mediates chain length extension of the arabinan in lipoarabinomannan. J. Biol. Chem. *281,* 19512-19526.

Shirai, T., Nakato, A., Izutani, N., Nagahisa, K., Shioya, S., Kimura, E., Kawarabayasi. Y., Yamagishi, A., Gojobori, T., and Shimizu, H. (2005). Comparative study of flux redistribution of metabolic pathway in glutamate production by two coryneform bacteria. Metab. Eng. *7,* 59-69.

Simic, P., Sahm, H., and Eggeling, L. (2001). L-Threonine export: Use of peptides to identify a new translocator from *Corynebacterium glutamicum*. J. Bacteriol. *183,* 5317-5324.

Stackebrandt, E., Rainey, F.A., and Ward-Rainey, N.L. (1997). Proposal for a new hierarchic classification system, Actinobacteria classis nov. Int. J. Syst. Bacteriol. *47,* 479-491.

Stuible, H., Wagner, C., Andreou, I., Huter, G., Haselmann, J., and Schweizer, E. (1996). Identification and functional differentiation of two type I fatty acid synthases in *Brevibacterium ammoniagenes*. J. Bacteriol. *178,* 4787-4793.

Stuible, H.P., Meurer, G., and Schweizer, E. (1997). Heterologous expression and biochemical characterization of two functionally different type I fatty acid synthases from *Brevibacterium ammoniagenes*. Eur. J. Biochem. *247,* 268-273.

Tatituri, R.V., Alderwick, L.J., Mishra, A.K., Nigou, J., Gilleron, M., Krumbach, K., Hitchen, P., Giordano, A., Morris, H.R., Dell, A., Eggeling, L., and Besra, G.S. (2007a). Structural characterization of a partially arabinosylated lipoarabinomannan variant isolated from a *Corynebacterium glutamicum ubiA* mutant. Microbiology *153*, 2621-2629.

Tatituri, R.V., Illarionov, P.A., Dover, L.G., Nigou, J., Gilleron, M., Hitchen, P., Krumbach, K., Morris, H.R., Spencer, N., Dell, A., Eggeling, L., and Besra, G.S. (2007b). Inactivation of *Corynebacterium glutamicum* NCgl0452 and the role of MgtA in the biosynthesis of a novel mannosylated glycolipid involved in lipomannan biosynthesis. J. Biol. Chem. *282,* 4561-4572.

Trivedi, O.A., Arora, P., Sridharan, V., Tickoo, R., Mohanty, D., and Gokhale, R.S. (2004). Enzymic activation and transfer of fatty acids as acyl-adenylates in mycobacteria. Nature *428,* 441-445.

Tropis, M., Meniche, X., Wolf, A., Gebhardt, H., Strelkov, S., Chami, M., Schomburg, D., Krämer, R., Morbach, S., and Daffé, M. (2005). The crucial role of trehalose and structurally related oligosaccharides in the biosynthesis and transfer of mycolic acids in Corynebacterineae. J. Biol. Chem. *280*, 26573-26585.

Trötschel, C., Deutenberg, D., Bathe, B., Burkovski, A., and Krämer, R. (2005). Characterization of methionine export in *Corynebacterium glutamicum*. J. Bacteriol. *187*, 3786-3794.

Tzvetkov, M., Klopprogge, C., Zelder, O., and Liebl, W. (2003). Genetic dissection of trehalose biosynthesis in *Corynebacterium glutamicum*: inactivation of trehalose production leads to impaired growth and an altered cell wall lipid composition. Microbiology *149*, 1659-1673.

Valbuena, N., Letek, M., Ramos, A., Ayala, J., Nakunst, D., Kalinowski, J., Mateos, L.M., and Gil, J.A. (2006). Morphological changes and proteome response of *Corynebacterium glutamicum* to a partial depletion of FtsI. Microbiology. *152*, 2491-2503.

Vrljić, M., Eggeling, L., and Sahm, H. (1996). A new type of transporter with a new type of cellular function: L-lysine export from *Corynebacterium glutamicum*. Mol. Microbiol. *22*, 815-826.

Vrljić, M., Garg, J., Bellmann, A., Wach, S., Freudl, R., Malecki, M. J., Sahm, H., Kozina, V.J., Eggeling, L., and Saier, M.H., Jr. (1999). The LysE superfamily: Topology of the lysine exporter LysE of *Corynebacterium glutamicum*, a paradyme for a novel superfamily of transmembrane solute translocators. J. Mol. Microbiol. Biotechnol. *1*, 327-336.

Wijayarathna, C.D., Wachi, M., and Nagai, K. (2001). Isolation of *ftsI* and *murE* genes involved in peptidoglycan synthesis from *Corynebacterium glutamicum*. Appl. Microbiol. Biotechnol. *55*, 466-470.

Wolf, A. (2002). Trehalosesynthese in *Corynebacterium glutamicum*. Dissertation University of Cologne.

Wolf, A., Krämer, R., and Morbach, S. (2003). Three pathways for trehalose metabolism in *Corynebacterium glutamicum* ATCC13032 and their significance in response to osmotic stress. Mol. Microbiol. *49*, 1119-1134.

From: Corynebacteria: Genomics and Molecular Biology. Edited by: Andreas Burkovski.

General and Regulatory Proteolysis in Corynebacteria

13

Johannes Amon, Alja Lüdke, Fritz Titgemeyer and Andreas Burkovski

Abstract

Proteases comprise a broad variety of functions in cells. They are involved in the degradation of polypeptides to supply amino acids for the synthesis of new proteins or for catabolism as carbon and energy source. Others remove misfolded or denatured proteins or are involved in the maturation of pre-proteins, the removal of signal peptides, and in the regulatory proteolysis of signal transduction proteins or transcriptional regulators. Furthermore, in pathogens, proteases are also important virulence factors. Reflecting this multiplicity of functions, in bacteria typically a broad set of proteases can be found. The composition of this set of proteolytic enzymes is specific for a single bacterium and depends on the ecological niche inhabited.

This review focuses on the role of proteases in corynebacteria. The proteolytic equipment of the sequenced *Corynebacterium* species is compared to the number of proteases in other important groups of Gram-positive bacteria, namely bacilli, streptomycetes, and mycobacteria. Recent data on the *Corynebacterium glutamicum* Clp and FtsH protease complexes are summarized and the putative role of secreted and surface-anchored proteases in pathogenic corynebacteria, especially in *Corynebacterium diphtheriae*, is discussed.

1. General functions of proteases

Proteolytic enzymes catalyze the cleavage of peptide bonds and represent approximately two percent of the total number of proteins in all types of organisms (Rao et al., 1998). This means that even in bacteria with strong reduction of genome size, such as in *Mycoplasma* species, several proteases can be found (a topical list of peptidases in various organisms is presented at the MEROPS database (Rawlings et al., 2006)). The multiplicity of these enzymes in cells is caused by the fact that proteases differ significantly in respect to their physiological function. For example, proteases can catalyze the total degradation of their substrate. Resulting amino acids or small peptides can subse-

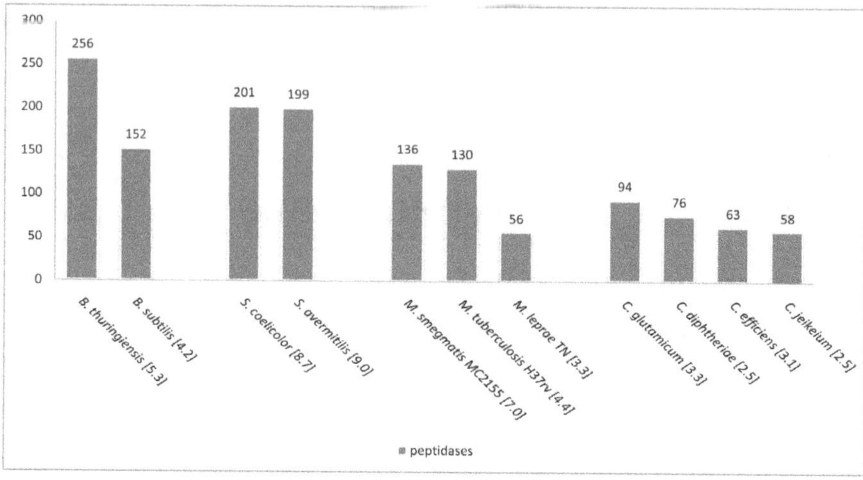

Figure 13.1. Peptidases and their homologues in selected Gram-positive bacteria. Shown are the total numbers of predicted putative peptidases according to the MEROPS database (Rawlings et al., 2006). Numbers in square brackets give the respective genome size in Mbp.

quently be used as building blocks for new proteins or as carbon and energy source. In this case, typically a set of rather unspecific endo- and exoproteases work together to ensure a fast and complete hydrolysis of their targets. Proteins used as growth substrates cannot be transported into a bacterial cell. Therefore, proteases are often transported across the cytoplasmic membrane and either bound to the surface of the cell or released into the surrounding medium. The resulting hydrolysis products are subsequently transported into the cell by specific amino acid or peptide uptake systems. Other proteases have a housekeeping function and degrade misfolded proteins or proteins damaged by heat, radiation or other detrimental factors (Wickner et al., 1999; Hengge and Bukau, 2003). Accessory subunits help to identify target proteins and to prevent hydrolysis of functional proteins in this case. Highly specific proteases are also involved in the maturation of preproteins (van Roosmalen et al., 2004) and in the regulatory proteolysis of signal transduction proteins or transcriptional regulators (Gottesman, 1999). Furthermore, in pathogens, proteases are involved in virulence and can damage host cells and tissues (Travis et al., 1995; Armstrong, 2006; Ribeiro-Guimaraes et al., 2007).

2. Distribution of proteases in bacilli, streptomycetes, and mycobacteria

Due to the multiple functions of proteases, all bacteria have a broad spectrum of these enzymes (for a topical list of peptidases in various species see MEROPS database (Rawlings et al., 2006)). Interestingly, the proteolytic capacity differs significantly between different bacteria. When major groups of Gram-positive bacteria are compared, namely bacilli, streptomycetes, mycobacteria, and corynebacteria (Figure 13.1), it becomes obvious that the spore-forming *Bacillus* species are proteolytically the most active group, especially when the number of proteases is correlated with genome size. Forty eight protease-encoding genes were annotated per Mbp of *Bacillus thuringiensis* chromosomal DNA and 36 genes per Mbp in the genome of *Bacillus subtilis*. The high number of proteins with putative proteolytic functions in *B. thuringiensis* might be the result of its pathogenic lifestyle. This bacterium kills insect larvae by a toxin, which destroys the gut epithel of the larvae, and uses the decaying bodies as nutrient source (Crickmore, 2005; Bravo et al., 2007).

Due to their high proteolysis activity and ability to secrete proteolytic enzymes into the surrounding medium, species like *B. subtilis, Bacillus licheniformis* or *Bacillus halodurans* are important industrial producers of alkaliphilic proteases as additives for laundry detergents, amongst various other secreted enzymes (Schallmey et al., 2004; Westers et al., 2004).

Despite their large genome sizes of about 9 Mbps, streptomycetes possess only a moderate number of proteases. In the *Streptomyces coelicolor* genome approximately 23 protease-encoding genes were found per Mbp chromosomal DNA, in case of *Streptomyces avermitilis* 22 genes. Taking the complex life cycle of streptomycetes and their various metabolic capabilities into account this is surprising. Polypeptides seem to be a less preferred growth substrate of these saprophytic soil bacteria, as streptomycetes can use various other growth substrates including the abundant polysaccharides chitin, xylan, and cellulose (Schrempf, 2001; Bertram et al., 2004).

Mycobacteria are also only moderately proteolytically active, especially *Mycobacterium smegmatis*, which features a comparatively small amount of proteins with peptidase function in respect to its genome size of about 7 Mbp. In this organism, only approximately 19 protease-encoding genes were annotated per Mbp of chromosomal DNA. An even more reduced number of proteases was found in *Mycobacterium leprae* with only 17 proteases encoded per Mbp. The reason might be the lifestyle of this obligate intracellular parasite resulting in severe gene decay termed *reductive evolution* (Eiglmeier et al., 2001). Nevertheless, some of the genes coding for proteins with putative proteolytic activity contribute to the process of human infection and leprosy skin lesions (Ribeiro-Guimaraes et al., 2007). This is also the case for the major human pathogen of the same genus, *Mycobacterium tuberculosis,* where proteases are studied as important pharmaceutical research targets in concerning infection and virulence (Ribeiro-Guimaraes and Pessolani, 2007). *M. tuberculosis* has the highest number of genes encoding proteases in this genus and approximately 30 of these genes are encoded per Mbp of chromosomal DNA (Rawlings et al., 2006), indicating a more important role of this enzyme class for survival of this bacterium.

3. The repertoire of protease-coding genes in corynebacteria

Based on published genome sequences of the *Corynebacterium* species *C. diphtheriae* (Cerdeno-Tarraga et al., 2003), *Corynebacterium efficiens* (Fudou et al., 2002), *Corynebacterium glutamicum* (Ikeda and Nakagawa,

Table 13.1. Protease-coding genes in sequenced genomes of corynebacteria. Amino acid sequence similarity searches were carried out using the BLAST algorithm (Altschul et al., 1990). The gene products were considered homologous when their identity was 50% or more and when the corresponding genes were situated in similar regions on the respective chromosomes. The putative proteases were validated based on protein domain searches according to the prediction programs in the Pfam (Bateman et al., 2004), MEROPS (Rawlings et al., 2006), and SMART (Letunic et al., 2006) databases. Non-classifiable secreted proteins with putative proteolytic functions are shown at the end of the table. Some of these proteins were already identified in proteomic studies (**bold**) but not yet further characterized. The obtained dataset was compared to proteome maps of the respective organism (Schaffer and Burkovski, 2005; Hansmeier et al., 2006a, 2006b, 2007; Lüdke et al., 2007; Lüdke, unpublished observations).

Protein name / family	*C. glutamicum* ATCC 13032 Bielefeld	*C. diphtheriae* NCTC13129	*C. efficiens* YS-314	*C. jeikeium* K411	Function
Signal peptidase I (SPase I, LepB)	*Cg2232*	*DIP1516*	*Ce1926*	*JK1183*	cleavage of hydrophobic, N-terminal signal or leader sequences from secreted proteins
Signal peptidase II (SPase II, LspA)	*Cg2347*	*DIP1584*	*Ce2033*	*JK0765*	release of signal peptides from bacterial membrane prolipoproteins including murein prolipoprotein.
Prepilin peptidase (PppA)	*Cg1830*	*DIP1346*	*Ce1744*	*JK1034*	processing of type 4 pilin precursor proteins to their mature forms by removal of leader peptides
	-	-	*Ce2562*	-	
Cell wall associated and secreted proteases (NlpC/ P60)	*Cg2401*	***DIP1621***	***Ce2080***	*JK0725*	probably peptidoglycan-lytic and/or invasion-associated
	Cg2402	*DIP1622*	*Ce2081*	*JK0726*	
	Cg1636	-	***Ce1582***	-	
	Cg0784	*DIP0640*	*Ce0701*	*JK1678*	
	Cg1735	***DIP1281***	***Ce1659***	***JK0967***	

SrtA	-	*DIP2012*	-	*JK1700*	anchoring of proteins (e.g. pili proteins) at the outer face of the cell wall
SrtB	-	*DIP0233*	*Ce2454*	-	
SrtC	-	***DIP0236***	*Ce2456*	-	
SrtD	-	*DIP2225*	*Ce2738*	-	
SrtE	-	*DIP2224*	*Ce2741*	-	
SrtF	*Cg3251*	*DIP2272*	*Ce2786*	*JK0103*	
FtsH	***Cg2984***	*DIP2002*	*Ce2542*	*JK0278*	for further details, see text
HtpX	-	-	-	*JK1570*	heat shock protein; clostridial origin
ClpX	*Cg2620*	*DIP1789*	*Ce2291*	*JK0546*	for further details, see text and Figure 13.2
ClpP2	***Cg2644***	*DIP1791*	***Ce2311***	***JK0545***	
ClpP1	***Cg2645***	*DIP1792*	***Ce2312***	*JK0544*	
ClpC	***Cg2963***	*DIP1983*	*Ce2529*	*JK0299*	
ClpB	***Cg3079***	*DIP2104*	***Ce2613***	***JK0201***	
HtrA	*Cg0998*	*DIP0856*	*Ce0950*	***JK1530***	heat shock protein
Aminopeptidase II	*Cg3325*	***DIP2346***	*Ce2832*	-	putative ATP-dependent Zn-protease (*Bacillus*-type)
M50 membrane-associated protease	*Cg2207*	*DIP1499*	*Ce1904*	*JK1166*	probably involved in the regulation of gene expression by proteolysis of transcription regulators
protease II	***Cg2873***	*DIP1926*	*Ce2486*	*JK0351*	prolyl oligopeptidase B
subtilisin-like serine protease	*Cg0665*	*DIP0554*	*Ce0578*	*JK1752*	similar to mycosins, cell wall-associated
prenyl protease	*Cg3209*	*DIP2248*	*Ce2734*	*JK0117*	unknown
putative metalloprotease	*Cg0029*	-	*Ce0316*	-	unknown, probable chaperone-function, membrane bound

Table 13.1 continued

Protein name / family	*C. glutamicum ATCC 13032 Bielefeld*	*C. diphtheriae NCTC13129*	*C. efficiens YS-314*	*C. jeikeium K411*	*Function*
chymotrypsin-like proteases, secreted and precursors	*Cg0356*	*DIP0307*	*Ce0292*	***JK1962***	
	-	***DIP0350***	-	-	
	-	*DIP0736*	-	*JK2027*	
	Cg0901	*DIP0743*	*Ce0808*	*JK1566*	
	Cg1243	***DIP0964***	*Ce1149*	-	
	-	-	*Ce1931*	-	
PepO	*Cg0193*	*DIP0154*	*Ce0152*	*JK0081*	exo-/endo-aminopeptidases
PepE	***Cg1681***	*DIP1238*	*Ce1613*	***JK1241***	
PepC	***Cg1693***	***DIP1250***	*Ce1628*	-	
PepQ	***Cg1826***	*DIP1341*	*Ce1738*	*JK1029*	
PepB	*Cg2419*	*DIP1637*	*Ce2096*	*JK0709*	
PepN	*Cg2662*	*DIP1798*	*Ce2320*	*JK0539*	
PepA	-	-	-	*JK1387*	
AmiA	*Cg0789*	*DIP0644*	*Ce0707*	*JK1763*	peptidoglycan hydrolases
AmiC	*Cg2285*	*DIP1551*	*Ce1986*	*JK0500*	
AmiB	***Cg3021***	*DIP2037*	*Ce2595*	*JK0802*	
MapA	*Cg0649*	*DIP0542*	*Ce0566*	*JK1764*	methionyl aminopeptidases
MapB	*Cg2198*	*DIP1496*	*Ce1901*	*JK1159*	
SppA	*Cg2043*	-	-	-	signal peptide peptidase homolog (prophage encoded)

Zn-dependent metallopeptidase	*Cg1842*	***DIP1354***	*Ce1753*	*JK1042*	
probable secreted proteases	*Cg0893*	*DIP0730*	*Ce0799*	*JK1605*	not further characterized ORFs with probable proteolytic functional domains
	Cg2902	-	***Ce2500***	-	
	Cg0349	*DIP0302*	*Ce0286*	*JK1973*	
	Cg0684	-	*Ce0596*	-	
	Cg1332	-	***Ce1275***	-	
	Cg3424	*DIP2375*	***Ce2935***	*JK2095*	
	Cg3131	*DIP0104*	-	-	
	Cg0409	*DIP0363*	*Ce0353*	*JK1938*	
	Cg0410	*DIP0364*	*Ce0354*	-	
	Cg0650	***DIP0544***	***Ce0567***	***JK0340***	
	Cg0611	-	***Ce0536***	-	
	-	-	***Ce2413***	-	
	Cg1577	-	***Ce1522***	-	
	Cg0980	*DIP0836*	***Ce0934***	***JK1559***	
	Cg0388	-	*Ce0331*	-	
	-	-	*Ce2742*	-	

2003; Kalinowski et al., 2003) and *Corynebacterium jeikeium* (Tauch et al., 2005), a comparative genomics analysis of genes encoding proteins with putative proteolytic functions was performed. A complete overview of protease genes found in each of the corynebacterial genomes is given in Table 13.1. A total number of 53, 51, 56, and 42 protease-encoding genes was predicted in the genomes of *C. glutamicum*, *C. diphtheriae*, *C. efficiens*, and *C. jeikeium*. Correlated to the genome size, *C. efficiens* has the lowest density of protease-encoding genes on the chromosome (less than 20 per Mbp), followed by *C. jeikeium* (approximately 23 genes per Mbp), *C. glutamicum* (about 29 genes per Mbp) and *C. diphtheriae* (more than 30 genes per Mbp). A core subset of 39 homologous proteases is common to all corynebacterial species. These findings are in good accordance with the numbers recently published for the closely related mycobacterial genus (Ribeiro-Guimaraes et al., 2007), comprising a similar set of highly conserved homologous proteases.

Two of the core proteases that are found in all corynebacteria are the signal peptidases I and II that cleave signal peptides from secreted (lipo-)proteins. Another conserved protein with similar function is the prepilin peptidase that processes type 4 pilin precursor proteins (prepilins) to their mature forms by removal of leader peptides. A gene encoding the prepilin peptidase (family A24 peptidase) is present in all corynebacterial genomes, while *C. efficiens* features an additional family A24 peptidase of unknown function (Ce2562) for which no homologs exist in other corynebacteria.

Members of the NlpC/P60 protein family (Anantharaman and Aravind, 2003), probably invasion-associated proteins, are highly conserved among non-pathogenic and pathogenic corynebacteria. Genes encoding these putative proteases are found in all corynebacterial genomes, where they are always located in a pair wise manner directly adjacent to another (*Cg2401/2402, DIP1621/1622, JK0725/0726, Ce2080/2081*) possibly as a result of gene duplication in an early corynebacterial ancestor cell. Other conserved NlpC/P60 family proteins found in all species are the uncharacterized Cg0784/DIP0640/Ce0701/JK1678 proteins as well as another hypothetical invasion-associated protein (Cg1735, DIP1281, JK0967, Ce1659). Additionally, both, *C. glutamicum* and *C. efficiens* feature an unknown NlpC/P60 protein with putative proteolytic function, Cg1636 and Ce1582, for which no homologs in other actinomycetes are found.

In all corynebacterial genomes a highly conserved gene coding for sortase F (*Cg3251, DIP2272, JK0103, Ce2786*) was observed. SrtF homologs seem to be the house-keeping sortases of corynebacteria and homologs are present in most actinomycetes. Sortases covalently attach a variety of proteins to the exterior of the bacterial cell wall and contribute to the diseases caused by the pathogenic bacteria. While Cg3251 is the only sortase-like protein in *C. glutamicum*, in *C. jeikeium* additionally a *srtA* homolog, *JK1700*, is found. An even more complex set of *srt* genes was annotated in the genome sequences of *C. diphtheriae*, with six and *C. efficiens* with five corresponding gene clusters (Scott and Zähner, 2006; Marraffini et al., 2006). These pili-specific sortases seem to form their own subfamily and the corresponding genes are

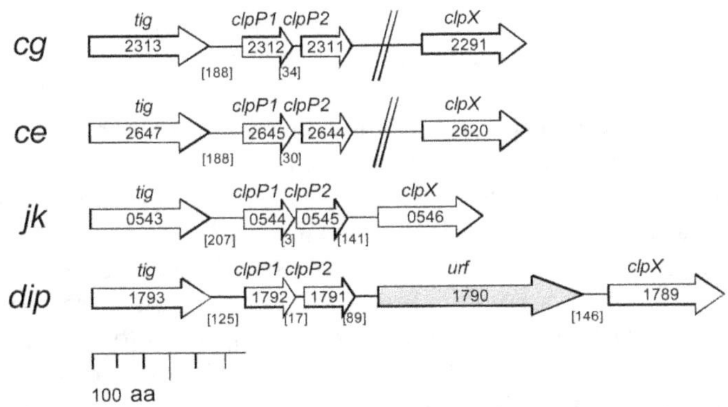

Figure 13.2. Genetic organization of *clp* gene clusters in corynebacteria. Arrows indicate the lengths and transcriptional orientations of annotated genes and predicted ORFs. Genes are shown by their annotation number, with the prefix of the respective organism. The gene names are assigned according to the annotations given by TIGR (http://www.tigr.org). Numbers in square brackets refer to the intergenic distance between two genes. Gene designations are as follows: *urf*, unknown reading frame; *tig*, trigger factor; *clpP1*, *clpP2*, proteolytic subunits; *clpX*, ATPase subunit; *cg*, *C. glutamicum*; *ce*, *C. efficiens*; *jk*, *C. jeikeium*; *dip*, *C. diphtheriae*.

located in clusters together with their respective pili-encoding genes (*spa*). Homologs of genes encoding the highly conserved membrane-bound FtsH protein are again found in all corynebacterial genomes (*Cg2984*, *Ce2542*, *DIP2002*, *JK0278*). The function of this AAA+ protein (ATPases associated with diverse cellular activities plus) in *C. glutamicum*, which is the only experimentally characterized FtsH protein in corynebacteria (Lüdke et al., 2007), will be discussed below. Also Clp protein complexes belong to the AAA+ class of proteases, which often have regulatory functions. These proteases are composed of accessory ATP-hydrolyzing subunits for recognition and unfolding of target proteins and proteins building a proteolytic chamber. The genes coding for ClpP1 and ClpP2 peptidase subunits are highly conserved in all available corynebacterial genomes. They are localized pair wise on the genome and are preceded by the trigger factor-encoding *tig* gene. A gene encoding the ATP-binding subunit ClpX is usually located in close neighbourhood on the chromosome (Figure 13.2), while the localization of the *clpC* gene, which encodes a second ATP-hydrolysing subunit, is less conserved. Details on targets and regulation of the Clp system are discussed below.

In the genus corynebacteria, only *C. jeikeium* features an ORF (*JK1570*) encoding a protein with the conserved catalytic domain of the *E. coli* HtpX endopeptidase that seems to be of clostridial origin and for which no other actinobacterial homologs are found. A protein with a putative role in heat shock response and probably involved in the degradation of mis- or unfolded proteins is encoded by the *htrA* homologs *Cg0998*, *DIP0856*, *Ce0950*, and *JK1530*.

Furthermore, an ORF encoding a putative M50 membrane-associated protease is found in all corynebacterial genomes. This not further characterized protease is highly conserved across all actinobacteria and seems to be crucial for this group of bacteria, deduced from the fact that a functional homolog is present even in *M. leprae*.

All corynebacterial genomes feature a diverse set of chymotrypsin-like proteases, of which some, like DIP0350 and Ce1931, seem to be unique to their respective organism. Most of them are synthesized as precursors and occur periplasmic or secreted with unknown functions. Common to all corynebacteria is also a more or less conserved set of various endo- and exo-aminopeptidases and carboxypeptidases that show a broad spectrum of putative biological functions, e. g. processing of proteins to mature forms, metabolism and lysis of bacterial cell walls, or assisting in the complete degradation of polypeptides to free amino acids.

Interestingly, the only corynebacterium to feature a signal peptide peptidase SppA homolog is *C. glutamicum*: *Cg2043* is encoding a unique putative protease IV that seems to be of archaeobacterial origin, and for which no homolog is present in other corynebacterial or actinobacterial genomes; homologs of the actinobacterial SppA (e.g. *M. tuberculosis* Rv0724) are not found in any corynebacterial genomes.

4. AAA+ proteases and the *C. glutamicum* Clp system

Since hydrolysis of a peptide bond is an exergonic reaction (ΔG approx. -10 kJ mol^{-1}), no energy-rich cofactors such as ATP are necessary for proteolysis (Goldberg and St. John, 1976). Nevertheless, some proteases are ATP-dependent. These typically oligomeric protease complexes are characterized by a proteolytic chamber with a narrow entrance suitable only for small (unfolded) proteins. The ATPase domain located at this entrance is responsible for substrate recognition, unfolding, and transport into the proteolytic core (Hlavacek and Vachova, 2002). All ATP-dependent proteases described so far belong to the AAA+ superfamily (Ogura and Wilkinson, 2001; Lupas and Martin, 2002). Five different AAA+ proteases were described in prokaryotes, namely, the proteasome, HslUV, ClpAP (and its homolog in Gram-positive bacteria, ClpCP), ClpXP, and FtsH. In corynebacteria, genes encoding the proteasome or HslUV were not detected; Clp and FtsH proteins are discussed in the following paragraphs.

In *C. glutamicum*, the Clp proteins are the best investigated proteolytic enzymes. Typically for actinomycetes, two genes, *clpP1* and *clpP2*, are located adjacent to each other (Figure 13.2), encoding the subunits of the proteolytic core. ClpC and ClpX are accessory ATPase subunits, which are members of the Clp/Hsp100 superfamily (Schirmer et al., 1996). Additionally, a *clpB* gene is annotated in the genome sequence. The corresponding protein has chaperone function and is not part of the Clp protease complexes.

A first hint for Clp function was found when the heat shock response of *C. glutamicum* was investigated. As shown by transcriptome analyses, expression of *clpP1*, *clpP2*, and *clpC* is increased in response to a heat shock (Muffler et al., 2002). Part of the underlying regulatory mechanism was already elucidated (Engels et al., 2004; 2005). Proteome analyses of

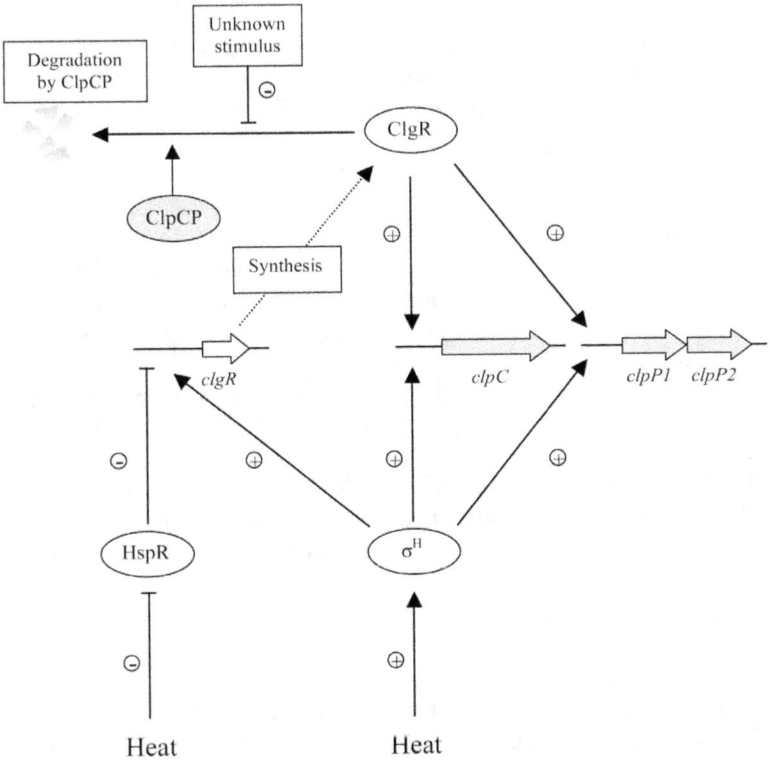

Figure 13.3. Regulation of *clpC* and *clpP1P2* transcription in *C. glutamicum*. Expression of *clpC* and *clpP1P2* genes is controlled by the transcriptional regulators HspR and ClgR. Additionally, expression is influenced by the ECF sigma factor σ^H. In response to heat shock, σ^H leads to activation of transcription *via* complex formation with RNA polymerase (RNAP). Besides direct induction of *clpC* and *clpP1P2* transcription by σ^H-RNAP, additionally σ^H-RNAP activates also the expression of the transcriptional activator ClgR. The resulting increase in ClgR concentration does only result in an increase of *clpC* and *clpP1P2* expression, when ClpC-dependent degradation of ClgR is blocked by an unknown stimulus. Furthermore, heat shock inactivates the transcriptional repressor HspR (Engels et al., 2004).

a *clpC* deletion strain in comparison to the wild type showed an approximately tenfold increased abundance of ClpP1 and ClpP2 in the mutant; the mRNA levels of the corresponding genes were increased by a factor of about five (Engels et al., 2004). DNA affinity purification experiments using the *clpP1P2* promoter region as bait resulted in the identification of ClgR. Further analyses showed that transcription of *clpP1*, *clpP2*, and *clpC* is controlled not only by this transcriptional regulator, but by a network of different regulatory proteins including ClgR, HspR, and the ECF (extracytoplasmic function) sigma factor σ^H (Figure 13.3). The regulon of ClgR was investigated in more detail by transcriptome profiling, primer extensions, and bioinformatic analyses, which revealed that at least five genes, namely *NCgl0240*, *NCgl0748*, *hflX*, *ptrB*, and *recR* are regulated by ClgR, while for about ten other genes a direct or indirect regulation by ClgR is discussed. When the (annotated) functions of the corresponding gene products are taken in consideration, it becomes obvious that this transcriptional activator controls expression of genes encoding proteins involved in proteolysis and DNA repair (Engels et al., 2005).

In an independent study it was shown that *C. glutamicum clpC* and *clpX* deletion mutations as well as a conditional *clpP1P2* mutant strain revealed an impaired degradation of the nitrogen signal transduction protein GlnK (Strösser et al., 2004). Since membrane contact is crucial for GlnK degradation, an indirect effect of the cytoplasmic Clp protease complexes on GlnK proteolysis was favoured. However, using a ClpP-specific polyclonal antiserum, also a minor fraction of Clp proteins was detectable at the membrane leaving the question of an involvement of Clp proteases in GlnK degradation open (Strösser et al., 2004).

5. The role of FtsH in *C. glutamicum*

FtsH proteins are membrane anchored proteases and were described as "machines for degrading membrane proteins" (Langer, 2000). They belong as Clp proteins to the AAA+ group of proteases, which often have regulatory functions. In fact, for the *C. glutamicum* FtsH an effect on the degradation of nitrogen signal control protein GlnK was reported (Strösser et al., 2004). The membrane localization of FtsH hints to a direct function on GlnK, since GlnK is only degraded, when it is sequestered to the membrane (Strösser et al., 2004). However, experimental evidence for an interaction of *C. glutamicum* FtsH and GlnK resulting in proteolysis of GlnK by FtsH is still missing.

Mutations of *ftsH* were described in different bacteria and were found to be remarkably species-specific. Effects of *ftsH* mutations range from drastic growth impairment in *Escherichia coli* (Begg et al., 1992) to effects on sporulation, development and stress response in *B. subtilis* (Deuerling et al., 1997; Lysenko et al., 1997) and *Caulobacter crescentus* (Fischer et al., 2002). In *C. glutamicum* cells, a deletion of the *ftsH* gene is well tolerated and has, in contrast to the situation in other bacteria, no strong effect on growth. Only a slightly increased doubling time of 2 h 38 min for the *ftsH* mutant compared to 2 h 18 min in the wild type was observed in minimal medium at 30 °C (Lüdke et al., 2007). This effect was more pronounced at 37 °C (Lüdke, unpublished observations), while no effect was found when cells were challenged by a sudden temperature increase to 39 °C or sudden or constant osmotic stress (Lüdke et al., 2007). Obviously, FtsH plays a less crucial role in *C. glutamicum* compared to other bacteria.

Besides by growth experiments, the influence of the membrane-bound AAA+ protease FtsH on membrane and cytoplasmic proteins of *C. glutamicum* was investigated in more detail. Proteome analyses using a combination of anion exchange chromatography and SDS-PAGE for the membrane fraction and standard two-dimensional gel electrophoresis for cytoplasmic proteins were carried out to identify FtsH targets in the cell (Lüdke et al., 2007). It was shown that deletion of the *ftsH* gene causes a strong increase of nine cytoplasmic and membrane proteins, namely biotin carboxylase/biotin carboxyl carrier protein, glyceraldehyde-3-phosphate dehydrogenase, homocysteine methyltransferase, malate synthase, isocitrate lyase, a conserved hypothetical protein (NCgl1985), succinate dehydrogenase A, succinate dehydrogenase B, and glutamate binding protein GluB. Thirty eight cytoplasmic and membrane-associated proteins showed a decreased abundance in an FtsH-lacking strain compared to the wild type. The decreasing amount of succinate dehydrogenase A in the cytoplasmic fraction of the *ftsH* mutant compared to the wild type and its increasing abundance in the membrane fraction indicates that FtsH might be involved in the cleavage of a membrane anchor of this membrane-associated protein and by this changes the localization of this enzyme. In summary, an astonishingly small amount of membrane and cytoplasmic proteins is affected by an *ftsH* deletion and the proteome analyses hint to a function of FtsH in regulation of energy and carbon metabolism as well as amino acid biosynthesis. A major role in stress response, as observed for other bacteria, was not found.

6. Proteases as virulence factors in corynebacteria

As mentioned above, proteases are not only involved in housekeeping functions and regulations, but also play an important role as virulence factors in pathogenic bacteria such as *Streptococcus pneumoniae*, *Pseudomonas aeruginosa*, and *Staphylococcus aureus* (Bethe et al., 2001; Malloy et al., 2005; Rigoulay et al., 2005; for review, see Amstrong, 2006). Often these host cell-damaging proteases are bound to the surface of the pathogen or secreted directly into the surrounding medium.

For *C. diphtheriae* and *C. jeikeium* detailed proteome analyses of surface-anchored and extracellular proteins were carried out recently (Hansmeier et al., 2006a; 2007). Interestingly, less proteases were identified in the nosocomial pathogen *C. jeikeium* compared to *C. diphtheriae*. The lipophilic lifestyle of *C. jeikeium* seems to reduce the need for a high proteolytic capacity. In contrast, several different proteases were identified that are located at the cell surface of *C. diphtheriae* or are secreted into the medium (Hansmeier et al., 2006a; see also Tab. 1). Initiated by these observations, the proteolytic capacity of *C. diphtheriae* was tested in respect to haemolysis of blood agar and halo generation on milk plates. Compared to *Bacillus cereus*, a negligible haemolysis and protease activity was observed (A. Wünsche, unpublished observation). In order to characterize major secreted proteases in more detail, mutants were generated, lacking the trypsin-like serine proteases DIP0350 and DIP0964, respectively. The gene encoding DIP0350 is localized on a genome island (see Genomics chapter) and was discussed to be involved in pathogenicity of *C. diphtheriae* (Hansmeier et al., 2006a). The two genes are expressed in a growth-dependent manner with the highest level of mRNA found in the exponential growth phase. Interestingly, loss of these proteases in corresponding mutant strains led to a significantly impaired ability to bind to host cells *in vitro* (A. Wünsche, unpublished observation). These observations indicate that proteolytic degradation of epithel tissue is not a major strategy of *C. diphtheriae* to colonize its host and proteases might rather be directly or indirectly involved in adhesion to epithelial cells.

The function of other proteolytic enzymes on the adherence of *C. diphtheriae* to host cells was demonstrated recently in a study addressing the function of sortases (Mandlik et al., 2007). As shown in Table 13.1, six different sortase-encoding genes are annotated in the *C. diphtheriae* genome sequence (Pallen et al., 2001; Cerdeno-Tarraga et al., 2003). The *srtA-E* genes are localized in three different clusters together with the pilus assembly genes *spaA-I*. While sortase SrtA is essential for the polymerization of SpaABC pili (Ton-That and Schneewind, 2003), SpaDEF pili require SrtB and SrtC (Gaspar and Ton-That, 2006), and the SrtD and SrtE sortases are required for the assembly of SpaHIG pili (Swierczynski and Ton-That, 2006). A *C. diphtheriae* mutant lacking all sortases, which is unable to assemble any proper pilus structures, showed a dramatically reduced binding to pharynx cells (Mandlik et al., 2007), emphasizing the importance of proteolysis in respect to pathogenicity in this organism.

7. Concluding remarks

Compared to other bacteria, corynebacteria seem to have only moderate proteolytic activity. Nevertheless, proteases obviously play a crucial role in many cellular processes. First investigations of *C. glutamicum* Clp protease complexes and FtsH hint to interesting regulatory functions of these AAA+ enzymes in this species. However, knowledge on this class of proteases is far from completeness and for example even the degradation signals are still unknown. The same is true for the function of putative degradation tags or amino acid residues influencing protein stability (Mogk et al., 2007). As discussed above, already the increased density of protease-encoding genes in the *C. diphtheriae* genome compared to other corynebacterial genomes and their partial localization on genomic islands suggest a role of these enzymes in pathogenicity. In fact, *C. diphtheriae* sortases were already shown to be crucial for pili assembly and adhesion to hosts cells, an effect also observed for two other secreted *C. diphtheriae* proteases. However, also in this case, only limited information is available. Future approaches might shed more light on the diverse function of known and uncharacterized proteases in different corynebacteria and hopefully elucidate the up to now completely unknown target recognition mechanisms of these enzymes.

References

Altschul, S.F., Gish, W., Miller, W., Myers, E.W., Lipman, D.J. (1990). Basic local alignment search tool. J. Mol. Biol. *215*, 403-410.

Anantharaman, V., and Aravind, L. (2003). Evolutionary history, structural features and biochemical diversity of the NlpC/P60 superfamily of enzymes. Genome Biol. *4*, R11.

Armstrong, P. (2006). Proteases and protease inhibitors: a balance of activities in host-pathogen interaction. Immunobiology *211*, 263-281.

Bateman, A., Coin, L., Durbin, R., Finn, R.D., Hollich, V., Griffiths-Jones, S., Khanna, A., Marshall, M., Moxon, S., Sonnhammer, E.L.L., Studholme, D.J., Yeats, C., and Eddy, S.R. (2004). The Pfam protein families database. Nucleic Acids Res. *32*, D138-D141.

Begg, K.J., Tomoyasu, T., Donachie, W.D., Khattar, M., Niki, H., Yamanaka, K., Hiraga, S., and Ogura, T. (1992). *Escherichia coli* mutantY16 is a double mutant carrying thermosensitive *ftsH* and *ftsI* mutations. J. Bacteriol. *174*, 2516-2417.

Bertram, R., Schlicht, M., Mahr, K., Nothaft, H., Saier, M.H., and Titgemeyer, F. (2004). *In silico* and transcriptional analysis of carbohydrate uptake systems of *Streptomyces coelicolor* A3(2). J. Bacteriol. *186*, 1362-1373.

Bethe, G., Nau, R., Wellmer, A., Hakenbeck, R., Reinert, R.R., Heinz, H.-P., and Zysk, G. (2001). The cell wall-associated serine protease PrtA: a highly conserved virulence factor of *Streptococcus pneumoniae*. FEMS Microbiol. Lett. *205*, 99-104.

Bravo, A., Gill, S.S., and Soberon, M. (2007). Mode of action of *Bacillus thuringiensis* Cry and Cyt toxins and their potential for insect control. Toxicon *49*, 423-435.

Cerdeno-Tarraga, A.M., Efstratiou, A., Dover, L.G., Holden, M.T.G., Pallen, M., Bentley, S.D., Besra, G.S., Churcher, C., James, K.D., De Zoysa, A., Chillingworth, T., Cronin, A., Dowd, L., Feltwell, T., Hamlin, N., Holroyd, S., Jagels, K., Moule, S., Quail, M.A., Rabbinowitch, E., Rutherford, K.M., Thomson, N.R., Unwin, L., Whitehead, S., Barrell, B.G., and Parkhill, J. (2003). The complete genome sequence and analysis of *Corynebacterium diphtheriae* NCTC13129. Nucleic Acids Res. *31*, 6516-6523.

Crickmore, N. (2005). Using worms to better understand how *Bacillus thuringiensis* kills insects. Trends Microbiol. *13*, 347-350.

Deuerling, E., Mogk, A., Richter, C., Purucker, M., and Schumann, W. (1997). The *ftsH* gene of *Bacillus subtilis* is involved in major cellular processes such as sporulation, stress adaptation and secretion. Mol. Microbiol. *23*, 921-933.

Eiglmeier, K., Parkhill, J., Honore, N., Garnier, T., Tekaia, F., Telenti, A., Klatser, P., James, K.D., Thomson, N.R., Wheeler, P.R., Churcher, C., Harris, D., Mungall, K., Barrell, B.G., and Cole, S.T. (2001). The decaying genome of *Mycobacterium leprae*. Lepr. Rev. *72*, 387-398.

Engels, S., Schweitzer, J.-E., Ludwig, C., Bott, M., and Schaffer, S. (2004). *clpC* and *clpP1P2* gene expression in *Corynebacterium glutamicum* is controlled by a regulatory network involving the transcriptional regulators ClgR and HspR as well as the ECF sigma factor σ^H. Mol. Microbiol. *52*, 285-302.

Engels, S., Ludwig, C., Schweitzer, J.-E., Mack, C., Bott, M., and Schaffer, S. (2005). The transcriptional regulator ClgR controls transcription of genes involved in proteolysis and DNA repair in *Corynebacterium glutamicum*. Mol. Microbiol. *57*, 576-591.

Fischer, B., Rummel, G., Aldrige, P., and Jenal, U. (2002). The FtsH protease is involved in development, stress response and heat shock control in *Caulobacter crescentus*. Mol. Microbiol. *44*, 461-478.

Fudou, R., Jojima, Y., Seto, A., Yamada, K., Rimura, E., Nakamatsu, T., Hirashi, A., and Yamanaka, S. (2002). *Corynebacterium efficiens* sp. Nov., a glutamic-acid-producing species from soil and plant material. Int. J. Syst. Evol. Microbiol. *52*, 1127-1131.

Gaspar, A.H., and Ton-That, H. (2006). Assembly of distinct pilus structures on the surface of *Corynebacterium diphtheriae*. J. Bacteriol. *188*, 1526-1533.

Goldberg, A.L., and St. John, A.C. (1976). Intracellular protein degradation in mammalian and bacterial cells – part 2. Ann. Rev. Biochem. *45*, 747-803.

Gottesman, S. (1999). Regulation by proteolysis: developmental switches. Curr. Opin. Microbiol. *2*, 142-147.

Hansmeier, N., Chao, T.C., Kalinowski, J., Pühler, A., and Tauch, A. (2006a). Mapping and comprehensive analysis of the extracellular and cell surface proteome of the human pathogen *Corynebacterium diphtheriae*. Proteomics *6*, 2465-2476.

Hansmeier, N., Chao, T.C., Pühler, A., and Tauch, A. (2006b). The cytosolic, cell surface and extracellular proteomes of the biotechnologically important soil bacterium *Corynebacterium efficiens* YS-314 in comparison to those of *Corynebacterium glutamicum* ATCC 13032. Proteomics *6*, 233-250.

Hansmeier, N., Chao, T.C., Daschkey, S., Müsken, M., Kalinowski, J., Pühler, A., and Tauch, A. (2007). A comprehensive proteome map of the lipid-requiring nosocomial pathogen *Corynebacterium jeikeium* K411. Proteomics *7*, 1076-1096.

Hengge, R., and Bukau, B. (2003). Proteolysis in prokaryotes: protein quality control and regulatory principles. Mol. Microbiol. *49*, 1451-1462.

Hlavacek, O., and Vachova, L. (2002). ATP-dependent proteinases in bacteria. Folia Microbiol. *47*, 203-212.

Ikeda, M., and Nakagawa, S. (2003). The *Corynebacterium glutamicum* genome: features and impacts on biotechnological processes. Appl. Microbiol. Biotechnol. *62*, 99-109.

Kalinowski, J., Bathe, B., Bischoff, N., Bott, M., Burkovski, A., Dusch, N., Eggeling, L., Eikmanns, B.J., Gaigalat, L., Goesmann, A., Hartmann, M., Huthmacher, K., Krämer, R., Linke, B., McHardy, A.C., Meyer, F., Möckel, B., Pfefferle, W., Pühler, A., Rey, D., Rückert, C., Sahm, H., Wendisch, V.F., Wiegräbe, I., and Tauch, A. (2003). The complete *Corynebacterium glutamicum* ATCC 13032 genome sequence and its impact on the production of L-aspartate-derived amino acids and vitamins. J. Biotechnol. *104*, 5-25.

Langer, T. (2000). AAA proteases: cellular machines for degrading membrane proteins. Trends Biol. Sci. *25*, 247-251.

Letunic, L., Copley, R.R., Pils, B., Pinkert, S., Schultz, J., and Bork, P. (2006). SMART 5: domains in the context of genomes and networks. Nucleic Acids Res. *34*, D257-D260.

Lüdke, A., Krämer, R., Burkovski, A., Schluesener, D., and Poetsch, A. (2007). A proteomic study of *Corynebacterium glutamicum* AAA+ protease FtsH. BMC Microbiol. *7*, 6.

Lupas, A.N., and Martin, J. (2002). AAA proteins. Curr. Opin. Struct. Biol. *12*, 746-753.

Lysenko, E., Ogura, T., and Cutting, S.M. (1997). Characterization of the *ftsH* gene of *Bacillus subtilis*. Microbiology *143*, 971-978.

Malloy, J.L., Veldhuizen, R.A.W., Thibodeaux, B.A., O'Callaghan, R.J., and Wright, J.R. (2005). *Pseudomonas aeruginosa* protease IV degrades surfactant proteins and inhibits surfactant host defense and biophysical functions. Am. J. Physiol. Lung Cell. Mol. Physiol. *288*, L409-L418.

Mandlik, A., Swierczynski, A., Das, A., and Ton-That, H. (2007). *Corynebacterium diphtheriae* employs specific minor pilins to target human pharyngeal epithelial cells. Mol. Microbiol. *64*, 111-124.

Marraffini, L.A., Dedent, A.C., Schneewind, O. (2006). Sortases and the art of anchoring proteins to the envelopes of gram-positive bacteria. Microbiol. Mol. Biol. Rev. *70*, 192-221.

Mogk, A., Schmidt, R., and Bukau, B. (2007). The N-end rule pathway for regulated proteolysis: prokaryotic and eukaryotic strategies. Trends Cell Biol. *17*, 165-172.

Muffler, A., Bettermann, S., Haushalter, M., Horlein, A., Neveling, U., Schramm, M., and Sorgenfrei, O. (2002). Genome-wide transcription profiling of

Corynebacterium glutamicum after heat shock and during growth on acetate and glucose. J. Biotechnol. *98*, 255-268.

Ogura, T., and Wilkinson, A.J. (2001). AAA+ superfamily ATPases: common structure – diverse function. Genes Cells *6*, 575-597.

Pallen, M.J., Lam, A.C., Antonio, M., and Dunbar, K. (2001). An embarrassment of sortases – a richness of substrates? Trends Microbiol. *9*, 97-102.

Rao, M.B., Tanksale, A.M., Ghatge, M.S.,and Deshpande, V.V. (1998). Molecular and biotechnological aspects of microbial proteases. Microbiol. Mol. Biol. Rev. *62*, 597-635.

Rawlings, N.D., Morton, F.R., and Barrett, A.J. (2006). MEROPS: the peptidase database. Nucleic Acids Res. *34*, D270-D272.

Ribeiro-Guimaraes, M.L., and Pessolani, M.C. (2007). Comparative genomics of mycobacterial proteases. Microb. Pathog., Epub ahead of print.

Ribeiro-Guimaraes, M.L., Tempone, A.J., Amaral, J.J., Nery, J.A., Antunes, S.L., and Pessolani, M.C. (2007). Expression analysis of proteases of *Mycobacterium leprae* in human skin lesions. Microb. Pathog., Epub ahead of print.

Rigoulay, C., Entenza, J.M., Halpern, D., Widmer, E., Moreillon, P., Poquet, I., and Gruss, A. (2005) Comparative analysis of the roles of HtrA-like surface proteases in two virulent *Staphylococcus aureus* strains. Infect. Immun. 73, 563-572.

Schaffer, S., and Burkovski, A. (2005). Genome-based approaches: proteomics. In: Handbook of *Corynebacterium glutamicum*. M. Bott, and L. Eggeling, eds. (CRC Press LLC, Boca Raton, FL.), pp. 99-118.

Schallmey, M., Singh, A., and Ward, O.P. (2004). Developments in the use of *Bacillus* species for industrial production. Can. J. Microbiol. *50*, 1-17.

Schirmer, E.C., Glover, J.R., Singer, M.A., and Lindquist, S. (1996). HSP100/Clp proteins: a common mechanism explains diverse functions. Trends Biochem. Sci. *21*, 289-296.

Schrempf, H. (2001). Recognition and degradation of chitin by streptomycetes. Antonie van Leuwenhook *79*, 285-289.

Scott, J.R., and Zähner, D. (2006). Pili with strong attachments: Gram-positive bacteria do it differently. Mol. Microbiol. *62*, 320-330.

Strösser, J., Lüdke, A., Schaffer, S., Krämer, R., and Burkovski, A. (2004). Regulation of GlnK activity: modification, membrane sequestration, and proteolysis as regulatory principles in the network of nitrogen control *in Corynebacterium glutamicum*. Mol. Microbiol. *54*, 132-147.

Swierczynski, A., and Ton-That, H. (2006). Type III pilus of corynebacteria: pilus length is determined by the level of its major pilin sububit. J. Bacteriol. *188*, 6318-6325.

Tauch, A., Kaiser, O., Hain, T., Goesmann, A., Weisshaar, B., Albersmeier, A., Bekel, T., Bischoff, N., Brune, I., Chakraborty, T., Kalinowski, J., Meyer, F., Rupp, O., Schneiker, S., Viehoever, P., and Pühler, A. (2005). Complete genome sequence and analysis of the multiresistant nosocomial pathogen *Corynebacterium jeikeium* K411, a lipid-requiring bacterium of the human skin flora. J. Bacteriol. *187*, 4671-4682.

Ton-That, H., and Schneewind (2003). Assembly of pili on the surface of *Corynebacterium diphtheriae*. Mol. Microbiol. *50*, 1429-1438.

Travis, J., Potempa, J., and Maeda, H. (1995). Are bacterial proteases pathogenic factors? Trends Microbiol. *3*, 405-407.

van Roosmalen, M.L., Geukens, N., Jongbloed, J.D.H., Tjalsma, H., Dubois, J.-Y.F., Bron, S., van Dijl, J.M., and Anné, J. (2004). Type I signal peptidases of Gram-positive bacteria. Biochim. Biophys. Acta *1694*, 279-297.

Westers, L., Westers, H. and Quax, W.J. (2004). *Bacillus subtilis* as cell factory for pharmaceutical proteins: a biotechnological approach to optimize the host organism. Biochim. Biophys. Acta *1694*, 299-310.

Wickner, S., Maurizi, M.R., Gottesman, S. (1999). Posttranslational quality control: folding, refolding, and degrading proteins. Sience *286*, 1888-1893.

From: Corynebacteria: Genomics and Molecular Biology. Edited by: Andreas Burkovski.

Environmental Stress Response of *Corynebacterium glutamicum* 14

Susanne Morbach and Reinhard Krämer

Abstract

Bacteria, like *Corynebacterium glutamicum*, are confronted with numerous stress situations. These can be caused by two principally different conditions, namely (i) by limitations in the supply of various nutrients, or (ii) by influences from the environment such as variations in temperature, osmolality, pH, and oxygen content, as well as increased heavy metal concentrations. In this chapter we will not describe adaptation processes related to nutrient limitation because this type of stress will be described elsewhere in this issue. Instead we will focus on external stress factors.

So far only data are available dealing with the response of *C. glutamicum* to influences of osmotic, heat and chill stress. These changes are often accompanied by the formation of reactive oxygen species, thus leading also to oxidative stress. In order to adapt to these unfavourable conditions, *C. glutamicum* has developed several adaptation strategies, which are acting at different levels of the regulatory network as for example by the transcriptional response of a whole sigma factor-dependent regulon or directly at the level of activity of an enzyme or a carrier. In this review we will summarize the knowledge about adaptation strategies of *C. glutamicum* to heat, oxidative, osmotic, and chill stress.

1. Heat stress

Under heat stress conditions the bacterial cell has to cope with negative effects of elevated temperature mainly on protein folding. These processes have to be strictly controlled by the bacterial cells, even under optimal growth conditions although to a lower extent. The conversion of an emerging linear peptide into the properly folded 3D-structure is one of the crucial steps during protein biosynthesis. Once the nascent peptide chain emerges from the ribosome it is in danger to be misfolded and to become aggregated. Although many proteins may fold autonomously, it was estimated that approx. 20-30% of all proteins require the assistance of molecular chaperones, like DnaK or GroEL, for the initial folding process (Bukau et al., 2000). It was even evaluated that a significant fraction of the newly synthesized peptide chains never reach the status of a properly folded protein (Ventura et al., 2006) and thus has to be degraded by proteases, e. g. by the Clp protease complex, to avoid the accumulation of aggregated proteins that are not accessible for refolding by chaperones.

Particular environmental conditions may shift the balance from a properly folded to a non-native protein pool, because they may provoke, in addition to the above described intrinsic impairment of the protein folding machinery, also a direct disintegration of the native 3-D protein structure. A prominent example for such a detrimental environmental condition is heat stress. The cellular response to heat stress is the transient induction of a relatively small set of genes, resulting in the synthesis of the members of the above mentioned protein quality control system of the cell, which are thus also called heat-shock proteins. Heat-shock proteins are classified into several families according to their molecular mass, namely Hsp100, Hsp90, Hsp70, Hsp60 and small Hsps (Bukau and Horwich, 1998; Narberhaus, 2002) as well as the newly identified Hsp33 family, which appears to be absent in high G+C Gram positive bacterial genomes (Ventura et al., 2006).

1.1 Protection mechanisms of *C. glutamicum* against heat stress

Recently, the genomic equipment of the G+C Gram-positive actinobacteria with regard to heat-shock proteins was reviewed in Ventura et al. (2006). **Table 14.1** summarizes the set of heat shock genes identified in the genomes of different corynebacteria. In *C. glutamicum* and *C. efficiens* two copies of DnaK and GroEL are present, which belong to the Hps70 and Hsp60 family, respectively. Both act as molecular chaperones and need ATP and co-chaperones for their activity. DnaJ (two copies) and GrpE (one copy) are part of the DnaK complex, and GroES (one copy) is the co-chaperone of GroEL. In addition, corynebacteria possess members of the Hsp100 family, which is characterized by ATP-depending proteases forming ring-shaped structures. One copy of *clpB*, *clpC* and *clpX* as well as two copies of *clpP* were identified in the genome, which are forming a bicistronic operon (Engels et al., 2004). ClpB seems to be involved in protein re-folding in cooperation with DnaK (Goloubinoff et al., 1999), whereas ClpC and ClpX represent the substrate recognition subunits of the two-component ClpCP and ClpXP protease complex (Goloubinoff et al., 1999).

1.2 Regulation of the heat stress response in *C. glutamicum*

1.2.1 Regulation at the transcriptional level

To date only a few analyses were performed investigating the adaptive response of *C. glutamicum* to heat stress. These investigations are not easy to compare since the heat stress conditions vary significantly in the range of the temperature shift and the time point of sample preparation (Muffler et al., 2002; Barreiro et al., 2004; Kim et al., 2005a, b; Engels et al., 2004).

Table 14.1. Chaperones and protease-encoding genes detected in the genomes of different *Corynebacterium species*.

Species	origin	Genome size (Mb)	*groEL*	*groES*	*grpE*	*dnaK*	*clpP*	*clpX*	*clpB*	*clpC*	*dnaJ1*	*dnaJ2*
C. diphtheriae	human	2.4	2	1	1	1	2	1	1	1	1	1
C. glutamicum	sewage (?)	3.3	2	1	1	2	2	1	1	1	1	1
C. efficiens	soil	3.1	2	1	1	2	2	1	2	1	1	1
C. jeikeium	human	2.4	2	1	1	1	2	1	1	1	1	1

Furthermore, only in one case a global transcriptome analysis was performed (Muffler et al., 2002). Thus, the results gained from these analyses shed light on different single aspects but do not yet provide a conclusive picture of the heat stress response of *C. glutamicum*. According to these data, the response of *C. glutamicum* can be separated in two steps. A primary response sufficient for adaptation to moderate heat stress through the synthesis of chaperones, and a drastic secondary response which is induced upon severe heat stress when protease synthesis might be essential to degrade protein aggregates, which cannot be refolded by chaperones (Barreiro et al., 2004; Barreiro et al., 2005; Engels et al., 2004).

Using RNA from cells, which were challenged in the exponential growth phase by a severe temperature shift from 30 to 50°C for 7 min, Muffler et al. (2002) showed in a DNA array, which was made by PCR fragments from a shotgun library of *C. glutamicum* that four different classes of genes had higher transcript levels, namely chaperone genes (*dnaK*, *grpE*, *dnaJ*, *groEL*, and *clpB*), protease genes (*clpP1*, *clpP2*, and *clpC*), and genes involved in the oxidative stress response (encoding methionine sulfoxide reductase, glutathion reductase, and thioredoxin), as well as *sigF*, a member of the sigma factor gene family. The increase was up to approx. 40-fold, but unfortunately DNA fragments were used, which covered in many cases more than one gene, thus the values represent mean values of the expression changes rather than the specific induction rate of individual genes. Taken together, *C. glutamicum* shows the typical induction of genes encoding heat shock proteins. In addition, transcript levels of the genes were increased which are involved in the oxidative stress response, indicating that under these conditions also reactive oxygen species were formed (Muffler et al., 2002).

In a different approach the transcriptional regulation of *groEL2*, *groES-groEL1* and the *dnaK* operon was characterized (Barreiro et al., 2004) after long-term adaptation to moderate heat stress of 40°C, which represents the upper limit for growth of *C. glutamicum* (Nishio et al., 2003). The gene *groEL2* and the operon *groES-groEL1* showed an identical induction pattern upon temperature increase. In both cases the same promoter was used under induced and non-induced conditions. The promoters of *groEL* and *groES-groEL* include a signature for a HrcA/CIRCE element, typically regulating chaperone expression under elevated temperatures (Zuber and Schumann, 1994). In contrast, the *dnaK* operon, composed of the genes *dnaK-grpE-dnaJ-hspR*, seems to be under the control of two promoters, namely P1 and P2 (Barreiro et al., 2004). P1 is proposed to be the house-keeping promoter which might be negatively regulated by HspR/HAIR (HspR-associated inverted repeat). HspR/HAIR binding motives were also identified in the promoters of *clpB*, *groEL2* (Barreiro et al., 2004), and *clgR* (Engels et al., 2004). In case of *Streptomyces coelicolor* the function of HspR as a repressor of the *dnaK* operon was experimentally verified (Bucca et al., 1997). Interestingly, the DnaK protein itself seems to be involved in the repression of its own operon by HspR. Bucca et al. (2000) proposed a novel model in which DnaK of *S. coelicolor*, besides being a chaperone, additionally functions as a transcriptional co-repressor by binding HspR and converting it into the active repressor. This interaction only takes place under growth conditions which do not promote protein denaturation, since otherwise DnaK becomes sequestered by misfolded proteins. The second promoter of the *dnaK* operon of *C. glutamicum* P2 might be controlled by the alternative ECF-type sigma factor SigH (Barreiro et al., 2004), because the sequence of the –10 and –35 regions of P2 is very similar to SigH-dependent promoter of *dnaK* of *Mycobacterium tuberculosis* (Raman et al, 2001; Manganelli et al., 2002). Taken together, evidence was provided that *Corynebacterium* species possess the typical transcriptional regulation circuits of the bacterial heat stress response involving SigH, HrcA/CIRCE and HspR/HAIR.

The analysis of *clpC* and *clpP1P2* expression regulation, encoding the protease complex ClpCP, revealed a complex regulation network allowing the integration of multiple environmental stimuli (Engels et al., 2004). At optimal growth temperatures the amount of the Clp complex is regulated via protein stability of ClpR. The concentration of this transcriptional activator, which binds to a palindromic binding motif upstream of *clpP1P2* and *clpC*, is titrated by ClpCP protease activity. Thus, in a *clpC* deletion mutant a higher concentration of ClpR was found which resulted in higher expression levels of *clpP1P2* and *clpC* (Engels et al. 2004). This mechanism is different to the regulation of Clp proteases in *Escherichia. coli* and *Bacillus subtilis,* although also in these cases the protein stability of the respective transcriptional regulators, namely σ^{38} and CtsR, is responsible for the actual cellular Clp concentration (Krüger and Hecker, 1998; Arsene et al., 2000). Interestingly, heat-triggered induction of *clpP1P2* and *clpC* did not depend on ClpR, but rather on the alternative ECF sigma factor SigH, indicating that at ambient temperatures *clp* transcription requires SigA and ClpR, whereas upon severe heat stress SigH is essential for induction of the *clp* protease genes.

1.2.2 Regulation at the translational level

Only one approach is published were the influence of heat stress on the proteome of *C. glutamicum* was investigated (Barreiro et al., 2005). In this study again a rather moderate temperature shift from 30 to 40°C for 60 min was performed. The amount of only six proteins was significantly changed after heat stress, namely the chaperones ClpB, DnaK, GroEL1, and GroEL2, the chaperone co-factor GrpE and the pyruvate quinone oxidoreductase PoxB (Barreiro et al., 2005). Depending on the protein a 3 to 40-fold increase was observed. Interestingly, the protein level of the peptidyl-propyl cis-trans isomerase, a foldase necessary for functional folding of proteins, a process which is hampered under conditions of heat stress, was decreased. Again no significant change in the synthesis of proteases was detected, which might explained by the hypothesis that a moderate temperature shift from 30°C to 40°C does not require the synthesis of proteases (Engels et al., 2005).

1.3 Interconnection of heat and oxidative stress of *C. glutamicum*: Function of SigH and SigM

Physiological analyses corroborated the observation that, as indicated by the DNA array analyses by Muffler et al. (2002), genes involved in the adaptation to oxidative stress are also important for the heat stress response. These environmental stress types seem to be connected *via* the expression regulation exerted by the alternative sigma factors SigH and SigM. By comparison of a *sigH* deletion mutant with wild type cells, Kim et al. (2005a) provided evidence that SigH is part of the adaptive response to heat and oxidative stress. Meanwhile an additional transcriptional regulator, WhcE, the gene of which is located downstream of *sigH*, was identified to be also part of the oxidative stress response. This regulator is carrying a helix-turn-helix motif as well as a redox sensing motif, which is typically found in redox-sensitive proteins. WhcE seems to be important for the stress-induced expression of *trxB*, encoding the thioredoxin reductase, whereby its own expression control is mediated by SigH (Kim et al., 2005b). Recently, by a detailed analysis of a SigM deletion mutant, Nakunst et al. (2007) showed that SigH is responsible for the expression of *sigM*, which codes like *sigH* also for an ECF-type sigma factor. SigM itself seems to directly control the expression of genes involved in the oxidative stress response, like *trxB* and *trxC*, and in heat stress adaptation, like the chaperone genes *groES*, *groEL* and *clpB*. Thus, in order to allow the adaptation at least to heat and oxidative stress a complex transcriptional regulatory network seems to be needed in *C. glutamicum.*

Taken together, the data obtained up to now, show that *C. glutamicum* adapts to heat stress conditions by activating the typical heat stress response including the synthesis of chaperones and proteases, which are needed to restore protein functionality or to degrade denatured proteins. In addition, the cells try to be prepared for the occurrence of reactive oxygen species, the concentration of which might be increased under several situations of environmental stress.

2. Osmotic Stress

The cytoplasmic membrane is freely permeable to water but forms an effective barrier for hydrophilic solutes, ions, and macromolecules like DNA or proteins. Bacteria actively regulate their cytoplasmic composition in order to keep the concentration of osmotically active solutes higher than in the environment, causing water to flow down its chemical potential into the cell. The resulting turgor pressure, which is exerted by the cytoplasmic membrane towards the cell wall, is considered necessary for enlargement of the cell envelope and thus for growth and cell division (Koch, 1983). In the soil, the natural environment of *C. glutamicum*, fluctuations of the external osmolarity are frequent as a result of sunshine and rainfall, respectively. Both conditions force *C. glutamicum* to adapt, either to avoid the dehydration of the cytoplasm which may result in growth stop and inactivation or misfolding of proteins or to prevent the danger of cell burst, respectively. Similar as in many other bacteria, two main strategies to cope with osmotic stress are

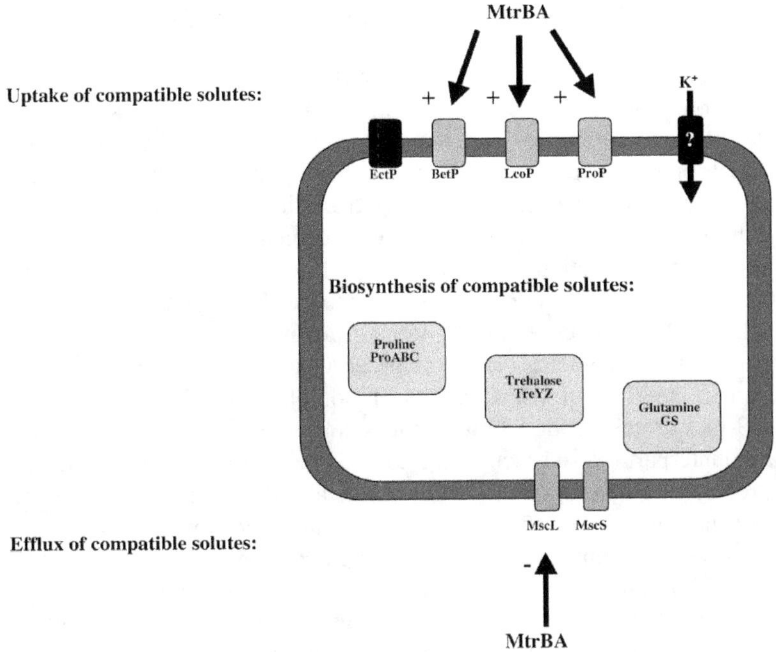

Figure 14.1. Systems of involved in the osmotic stress response of *C. glutamicum*. *C. glutamicum* has a set of different systems for the adaptation to hyperosmotic stress. If no compatible solutes are present in the environment the bacterium is able to the *de novo* synthesis of proline, trehalose or glutamine. If provided by the environment, the uptake of glycine betaine, proline or ectoine is preferred over *de novo* synthesis of compatible solutes. This uptake is mediated by the secondary uptake systems BetP, EctP and LcoP, belonging to the BCCT-family, and the MFS-transporter ProP. All four systems are efficiently regulated at the level of activity and besides EctP also at the level of transcription. The transcriptional control is mediated by the two-component system MtrB-MtrA. To reduce the driving force for water entry solutes can be released with high velocity from the cytoplasm via the mechanosensitive channels MscL and YggB under conditions of hypoosmotic stress.

observed in *C. glutamicum,* namely the release of solutes from the cytoplasm after an osmotic downshift and the accumulation of so-called compatible solutes under conditions of hypertonicity (Figure 14.1). Compatible solutes are mainly characterized by two fundamental properties. At the cellular level they do not interfere with the metabolism of the cell even in molar concentrations. Thus, they are ideally suited to increase the internal osmolality in order to redirect water fluxes. At the molecular level, they were shown to stabilize and protect enzyme proteins against denaturation because they lead to the preferential hydration of protein surfaces (Arakawa and Timasheff, 1985; Youxing et al., 1998).

2.1 Protection mechanisms of *C. glutamicum* against osmotic stress

In order to survive an osmotic downshift, *C. glutamicum* activates within milliseconds so-called mechanosensitive channels (Msc) (Ruffert et al., 1997). If the membrane is stretched by mechanical forces, mechanosensitive channels open large unspecific pores in the cytoplasmic membrane of bacteria, allowing a very fast release of solutes from the cytoplasm. As a result the driving force for water entry immediately decreases. In *C. glutamicum* two different channel activities were observed in patch clamp analyses (Ruffert et al., 1999), with a conductance similar to the *E. coli* channel activities of MscL and MscS (Berrier et al., 1996). Meanwhile the the genes of these channels were characterized (Nottebrock et al., 2003). Whereas a protein with high similarity to the *E. coli* MscL homolog was identified, the second MscS-like protein, called YggB, is less conserved when compared to its *E. coli* counterpart (Nottebrock et al., 2003). The *C. glutamicum* YggB protein is predicted to consist of 533 amino acids and it thus much larger than the *E. coli* homolog MscS (289 aa) (Levina et al., 1999), which was shown to form a homoheptamer with three transmembrane helices per subunit (Bass et al., 2002). Recently, Nakamura et al. (2008) related L-glutamate excretion in *C. glutamicum* to the presence of YggB. Strains carrying mutant forms of YggB excreted glutamate also under conditions in which wild type cells do not secrete this amino acid into the medium. Thus, it will be interesting to analyse the function of YggB of *C. glutamicum* in more detail, in order to decide whether it acts as a mechanosensitive channel and/or glutamate exporter. Besides, MscL and YggB, a third mechanosensitive channel lacking sequence identity to MscL and YggB, might be present in *C. glutamicum*, since a double *mscL/yggB* knock-out strain was still able to excrete betaine after an osmotic downshift with high efficiency (Nottebrock et al., 2003).

In contrast to the rapid response to an osmotic downshift by the activation of mechanosensitive channels, the adaptation to hyperosmotic stress is less fast and includes different strategies. After an osmotic upshift, *C. glutamicum* immediately accumulates compatible solutes (Kawahara et al., 1989; Farwick et al., 1995; Guillouet and Engasser, 1995a; 1995b; Skjerdal et al., 1996; Rönsch et al., 2003; Varela et al., 2003, Wolf et al., 2003; Ley, 2005) without showing a pronounced K^+/glutamate response typically for *E. coli* (Dinnbier et al., 1988; Record et al., 1998). The accumulation of compatible solutes is flexibly regulated in *C. glutamicum* since the species of compatible solute which is preferentially accumulated as well as the pathway used depends strictly on the external conditions and the genetic background of the cell (Skjerdal et al., 1996; Rönsch et al., 2003; Wolf et al., 2003; Ley, 2005). (1) Under conditions of nitrogen-surplus, which is typically the case in minimal media used for *C. glutamicum* fermentations, *de novo* synthesized proline is the preferred protecting substance (Kawahara et al., 1989; Farwick et al., 1995; Guillouet and Engasser, 1995a; 1995b; Skjerdal et al., 1996; Rönsch et al., 2003; Wolf et al., 2003; Ley, 2005). In *C. glutamicum* only one proline biosynthesis pathway is present, described by Ankri et al. (1996), which is responsible for both, the anabolic and stress-induced proline biosynthesis (Ley, 2005). (2) If cells cannot synthesize proline, as for example in a *proC* deletion strain, proline is replaced by glutamine when the supply for external proline is limited to 1 mM (Ley, 2005). (3) Under conditions of nitrogen limitation or starvation the biosynthesis of osmoprotecting solutes is shifted from amino acids to the nitrogen-free trehalose (Wolf et al., 2003). *C. glutamicum* possesses altogether three different pathways for the biosynthesis of trehalose, namely the OtsAB, the TreS, and the TreYZ pathway (Tzvetkov et al., 2003; Wolf et al., 2003), but the osmostress-induced trehalose biosynthesis was shown to depend almost completely on the TreYZ pathway using maltodextrins and glycogen as substrates for trehalose biosynthesis (Wolf et al., 2003). Besides the nitrogen supply, the kind of sugar used as carbon source also influences the ratio between proline and trehalose biosynthesis upon hyperosmotic conditions. When glucose or maltose were used instead of sucrose, *C. glutamicum* shifts the accumulation towards trehalose (Wolf et al., 2003), most likely because these sugars allow to build up higher internal glycogen contents (Seibold et al., 2007), which is the prerequisite for an active TreYZ-mediated trehalose biosynthesis. (4) The scenarios described up to now are restricted to conditions in which *C. glutamicum* is forced to synthesize compatible solutes *de novo*. Due to lower energy cost, *C. glutamicum* prefers the uptake of compatible solutes instead of their biosynthesis, if either glycine betaine, ectoine or proline are present in the environment (Skjerdal et al., 1996; Rönsch et al., 2003). *C. glutamicum* catalyses the uptake of compatible solutes by the secondary uptake systems BetP, EctP, ProP, and LcoP (Peter et al., 1996; 1998a; Steger et al., 2004). They are all characterized by a tight activity regulation allowing adapting their activity to the extent of the osmotic stress.

Taken together two main conclusions can be drawn. (i) The high redundancy of osmoregulated biosynthesis pathways and uptake carriers for compatible solutes indicates that an efficient adaptation to osmotic stress conditions is essential for the survival of *C. glutamicum* in its habitat. (ii) The accumulation of compatible solutes in *C. glutamicum* does not follow a rigid scheme but is flexibly adjusted to a variety of environmental parameters sensed by the cell in addition to the external osmolarity.

2.2 How is the complex network of osmostress-adaptation regulated?

2.2.1 Regulation at the transcriptional level

The transcriptional control of the systems in *C. glutamicum* involved in the osmostress response was investigated by several methods including RNA hybridisation, real time PCR, DNA micro array analysis, and transport measurements (Wolf et al., 2003; Steger et al., 2004; Möker et al., 2004; Weinand et al. 2007, O. Ley, unpublished). They all reveal that the transcriptional regulation of *C. glutamicum*, although the cells were challenged with severe osmotic stress, is characterized by moderate to low induction ratios compared to *B. subtilis* (Steil et al., 2003).

As mentioned above, the biosynthesis of proline upon a hyperosmotic upshift is mediated by the enzymes ProB, ProA, and ProC. The analysis how the cell regulates the internal proline level for needs of biosynthesis and osmotic response with one single biosynthesis pathway revealed that the control is mainly mediated via transcriptional regulation. Applying a severe osmotic upshift led to a 7-fold increase of the internal proline pool of up to 700 μmol/ g cell dry weight, which is accompanied by induction of the proline biosynthesis genes by a factor of up to 4 (Ley, 2005).

In contrast, transcriptional control seems to be of minor importance for the biosynthesis of trehalose. The genes of the TreYZ pathway, the major biosynthesis route of trehalose under hyperosmotic stress, were not found to be induced under these conditions (Wolf et al., 2003). Interestingly, the *otsA* gene was induced 5-fold, although the OtsAB pathway had only a minor importance for trehalose biosynthesis upon hypertonicity (Wolf et al., 2003). In this context it is important to note that these experiments were carried out under conditions of carbon excess allowing the formation of cytoplasmic glycogen, the substrate of the TreYZ pathway. Thus, it is tempting to speculate that under conditions of carbon limitation when the glycogen pool is consumed, the OtsAB pathway may become more important. This seems reasonable since the OtsAB pathway depends on the central metabolites UDP-glucose and glucose-6-phosphate, which, in contrast to maltodextrins, are likely to be present under carbon limitation, too.

As mentioned above, the secondary transporters for the uptake of compatible solutes are strictly regulated at level of activity; they are, however, also regulated at the transcriptional level. The corresponding genes, namely *ectP*, *betP*, *lcoP,* and *proP*, were induced by a factor of 2-10 after an osmolality shift from 0.3 to 1.8 osmol/kg (Möker et al., 2004; Weinand et al., 2007). The different induction levels resulted only in case of BetP, ProP and LcoP in an increase of protein concentration (Weinand et al., 2007). EctP seems to be the rescue system for *C. glutamicum*, since (i) the basal activity in non-shocked cells was rather high, (ii) the transport capacity was independent of the external osmolarity although the gene was induced two-fold, and (iii) the carrier shows the broadest substrate spectrum in *C. glutamicum* allowing the uptake of betaine, ectoine, or proline when they are provided in the environment. In contrast, the other three systems (BetP, ProP, and LcoP) represent adaptive systems, allowing the cell to optimally regulate the overall transporter capacity in dependence to the extent of the applied hyperosmotic stress (Weinand et al., 2007).

Concerning the molecular basis of regulation, i.e. the putative presence of transcriptional regulators of the above mentioned systems, little knowledge is available. Typical bacterial regulation circuits involved in stress regulation are two-component systems as well as alternative sigma factors. *C. glutamicum* possesses 13 different two-component systems (Möker et al., 2004; Kocan et al., 2006). A set of 12 strains was constructed in which the genes of one two-component systems were deleted at a time (Möker et al., 2004; Kocan et al., 2006). Upon a hyperosmotic shock none of the genes, encoding enzymes for the biosynthesis of compatible solutes, showed a changed expression pattern in 11 of the 12 mutant strains (Möker et al., 2004). Interestingly, the so-called MtrAB system was found to be involved in the transcriptional control of three of the genes coding for uptake systems for compatible solutes, namely the adaptive systems BetP, LcoP, and ProP. RNA hybridization experiments showed that after a hyperosmotic upshift the transcript levels of these genes were either dramatically decreased (*proP*) or transcription was almost completely abolished (*betP* and *lcoP*) in the mutant strain Δ*mtrAB* (Möker *et al.*, 2004). In addition, the expression of genes which are predicted to be involved in cell wall metabolism was elevated in the mutant strain. These observations led to the suggestion that the response regulator MtrA could act either a repressor or as an activator (Möker et al., 2004). Meanwhile, binding of MtrA to the promoter regions of *mepA*, *betP*, and *proP* was verified (Brocker and Bott, 2006), proving that the MtrB-MtrA system indeed directly regulates different cellular processes.

Stress-induced changes at the transcriptional level are frequently mediated by alternative sigma factors (Hecker and Völker, 2001; Helmann, 2002). In the *C. glutamicum* genome, besides the gene encoding the essential housekeeping factor *sigA*, six genes coding for alternative sigma factors were annotated (Ikeda and Nakagawa, 2003; Kalinowski et al., 2003). Future analyses have to address the question whether sigma factors are important for the osmostress regulation in *C. glutamicum*.

Taken together the analyses of the transcriptional regulation induced by hyperosmotic stress by applying different experimental techniques revealed that the expression changes in *C. glutamicum* are rather small ranging between 2-10-fold induction or repression (Möker et al., 2004; Steger et al., 2004; Weinand et al., 2007; Ley, 2005). This may be taken as an indication that *C. glutamicum* is under optimal growth conditions already pre-adapted to cope with hyperosmotic stress or, alternatively, this organism in general possesses a high resistance against environmental stress rendering extensive cellular remodeling by transcriptional regulation unnecessary.

2.2.2 Regulation at the activity level

To date the activity regulation of only a few enzymes or carriers involved in the osmotic stress response has been investigated in detail, including the osmoregulated transporter BetP, EctP LcoP and ProP as well as the histidine protein kinase MtrB (Peter et al., 1996; 1998a; Steger et al., 2004; Möker et al., 2007a, Möker et al., 2007b).

Although being different with respect to their substrate spectrum or affinity allowing to use a broad variety of compatible solutes from the environment (Morbach and Krämer, 2005), the activity regulation of these transporters is surprisingly similar (Peter et al., 1996; 1998a; Steger et al., 2004). These uptake systems are inactive at an osmolality below 0.4 osmol/kg. At higher osmolalities the carriers are activated in a dose-dependent manner with maximal activities found at approx. 1.3 osmol/kg (Morbach and Krämer, 2005). BetP, is the best characterized carrier of this kind in *C. glutamicum*

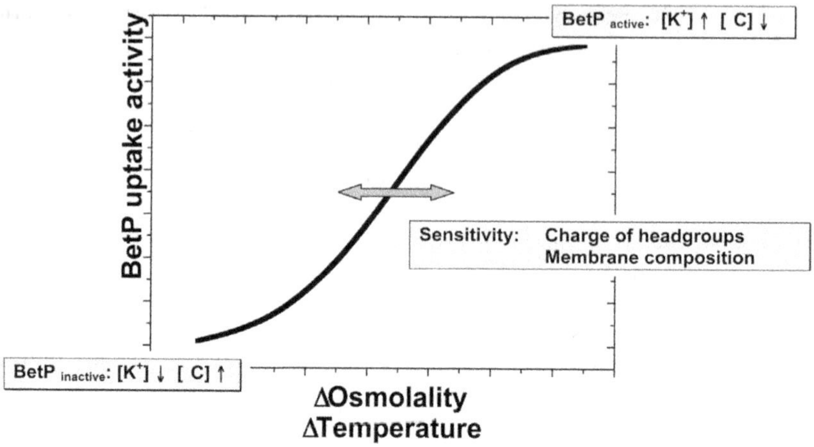

Figure 14.2. Modell of BetP regulation. The transporter BetP can be shifted into the active state by two independent triggers, namely by changes of the cytoplasmic K^+ concentration or by temperature changes, which are thought to be transduced via changes of the physical state of the membrane. Independent from the switch leading to the activation of BetP, its activity is further modulated by varying of the sensitivity of the system. The threshold for BetP activation depends on the fraction of charged phospholipids as well as on the membrane composition, including the fatty acid chain length, saturation as well as head group composition.

(Figure 14.2). It is predicted to possess 12 transmembrane segments as well as cytoplasmically located N- and C-terminal domains consisting of 62 and 55 amino acids, respectively. *In vivo* and *in vitro* experiments revealed that BetP itself harbors, besides its catalytical function, i. e. betaine uptake, the ability to sense osmotic stress and to adapt its activity level to the extent of the osmotic strength (Peter et al., 1996; 1998b, Rübenhagen et al., 2000). In other words BetP is both an osmosensor and an osmoregulator. Using the reduced complexity of an *in vitro* system, in which purified BetP was reconstituted in liposomes in its functionally active form, it was shown that BetP sensed the rise in internal potassium concentration as an indication for the presence of hyperosmotic stress conditions (Rübenhagen et al., 2001, Schiller et al., 2004a). The internal potassium concentration is in fact a reasonable stimulus for BetP activation, since the K^+ concentration immediately increases after a hyperosmotic upshift due to the induced water efflux from the cytoplasm. Meanwhile, the domain within BetP essential for K^+-sensing was identified (Schiller et al., 2004b; 2006). The putative K^+-sensor is located within a part of the C-terminal extension of BetP, which is predicted to form an α-helical structure (Schiller et al., 2006). Currently the question is addressed whether the K^+-sensor is exclusively located within the C-terminal domain, or whether other cytoplasmic parts of BetP (the N-terminal domain or internal loops), or the membrane is involved. In this context it is important to note that the membrane composition was shown to critically influence the sensitivity of the system. The higher the fraction of negatively charged phospholipids in the membrane the higher was the threshold for half-maximal activation (Schiller et al., 2006). Thus, future work will reveal whether the membrane acts as an interaction partner of the C-terminal extension or whether the phospholipid composition rather defines the "default" ratio between inactive and active BetP and thus the activation threshold of the system.

Recently, evidence was provided that the two-component system MtrB-MtrA is necessary for the activation of *betP*, *proP*, and *lcoP* expression (Möker et al., 2004). Since binding of MtrA to the promoters of *betP* and *proP* was shown (Brocker and Bott, 2006) and the induction level of the genes was found to depend on the osmotic strength applied (Möker et al., 2007a), it was suggested that MtrB, the histidine protein sensor kinase, acts as an osmosensor like BetP (Möker et al., 2007a). MtrB was successfully purified and actively reconstituted in liposomes, thus it was possible to investigate its osmosensing capabilities (Möker et al., 2007a). MtrB showed autophosphorylation as well as phosphatase activity towards its cognate response regulator MtrA *in vitro*. Thus, proteoliposomes are a suitable system to investigate which stimulus is used by MtrB for the detection of hyperosmotic stress (Möker et al., 2007b). In contrast to BetP, which showed a high specificity for the internal K^+ concentration, MtrB was efficiently activated by a broad variety of chemically diverse solutes, like amino acids, amino acid derivatives, sugars and phosphatidyl ethylene glycols. Since it seems rather unlikely that these solutes are bound to a specific binding pocket, it was concluded that the sensing event is likely based on a different mechanism, namely the change of the hydration state of MtrB thereby shifting the kinase into the active form (Möker et al., 2007b). Although to date the osmosensor domain within MtrB is not identified in detail, experimental evidence strongly suggested that cytoplasmic parts of the protein harbour the osmosensing function (Möker et al., 2007b). Another interesting aspect of MtrB dependent expression regulation is directed towards cell wall-related genes as mentioned above. The function of these gene products, however, was not investigated so far.

In many bacteria two-component systems are also involved in the response to envelope stress which involves participation of the cell wall (Raivio, 2005; Mascher, 2006). Thus, future investigations should address the question whether the MtrAB two-component system is, besides osmoregulation, rather important for cell proliferation or possibly also for the adaptation to envelope stress.

3. Chill stress

Lowering of the environmental temperature has in general an impact on several functions and structures in the bacterial cell. These are (i) rigidification of the plasma membrane, (ii) conformational changes of proteins, and/or (iii) conformational changes of nucleic acids. What are possible consequences for the cell? Membrane fluidity is reduced after an cold shock. This would result, depending on the lipid composition, in local and later in global phase transitions from a liquid-crystalline to a more rigid gel-like state. Such phase transitions hamper membrane-coupled processes mediated by membrane-embedded proteins, like transport and energy-converting processes. To restore membrane functionality at decreasing temperature, fatty acids (FA) with lower melting points are incorporated into the lipids in a species-specific mode. Melting points are lowered by FA acyl chain-length alteration, FA acyl chain branching and the introduction of *cis*-configured carbon double bonds (de Mendoza & Cronan, 1983; Russell 1984; de Mendoza et al. 1993). Protein conformation is in general changed at low temperature, and this severely may affect the structure of the ribosomes, leading to the accumulation of monosomes and 70S ribosomes. Together with an increased concentration of mRNA templates whose promoter regions are masked due to the formation of secondary structures, which are more stable at low temperatures, this results in a severely reduced transcription and translation efficiency. Bacteria respond with the synthesis of so-called cold-stress proteins (CSP), acting very likely as RNA chaperones and/or translation initiation-like factors, as well as with the production of other DNA or RNA modulating enzymes with the aim to restore transcription and translation. Furthermore, many bacterial species start to accumulate compatible solutes under chill stress conditions. Similar to the situation at osmotic or heat stress, protein function may be compromised due to conformational changes (Ventura et al., 2006). Thus, the accumulation of compatible solutes, which are known to protect the native structure and thus favor the active form of proteins, may counteract this process.

3.1 Protection mechanisms of *C. glutamicum* against chill stress

In contrast to *E. coli* and *B. subtilis* only few investigations have been carried out to characterize the cold shock response of *C. glutamicum.* These were mainly focused on temperature-dependent changes of the phospholipid composition as well as on effects on biosynthesis and uptake systems for compatible solutes, namely their transcriptional regulation and chill-dependent activity control (Özcan et al., 2005; Özcan et al., 2007).

C. glutamicum, which shows optimal growth at 30°C, seems to possess a higher tolerance against lower than against higher temperatures, since growth was detected down to a temperature of 2°C, but not above 40°C (Nishio et al., 2003; Özcan et al. 2005). In contrast, *Corynebacterium efficiens*, a close relative, is able to grow up to 50°C (Nishio et al. 2003), thus it might be interesting to investigate whether this species is still able to tolerate also low temperatures as observed for *C. glutamicum.* Interestingly, *C. glutamicum* cells, when shifted from the optimal growth temperature of 30°C to lower values, does not show any lag phase. In contrast to *E. coli* (Jones et al., 1987), cells keep growing with a decreased doubling time (Özcan, 2007). Similar to the response to hyperosmotic stress, under these conditions *C. glutamicum* starts to accumulate compatible solutes. In comparison to osmotic stress, lower concentrations of these protectants were sufficient for the adaptation of the cell. If no compatible solutes are available in the environment, proline is synthesized, but if present outside also in this case uptake of proline or betaine was preferred over *de novo* synthesis (Özcan, 2007). As described below in more detail, the transport systems for compatible solutes, BetP and LcoP, are responsible for this uptake (Özcan et al., 2005). The physiological advantage of this accumulation for *C. glutamicum* is still obscure, since it was not possible to show any positive effect on growth upon supplementation of the medium with glycine betaine or proline. In *B. subtilis* growth was accelerated by compatible solutes at temperatures closely above the minimum temperature allowing cell proliferation (Brigulla et al., 2003; Bremer personal communication). Thus, either 2°C does not represent the minimal growth temperature for *C. glutamicum* or the biosynthesis rate of proline under these conditions is still sufficient to promote proliferation of the non-supplemented control cells. Taken together it has to be stated that *C. glutamicum* seems to be very robust against environmental influences like chill or osmotic stress in general.

As mentioned above, bacteria respond to a decreased fluidity of the membrane caused by lower temperatures with the introduction of fatty acids with lower melting points. In a systematic analysis the influence of the growth temperature on the membrane composition of *C. glutamicum* was investigated (Özcan et al., 2007). The fatty acids of the corynebacterial plasma membrane consist to more than 90% of palmitic and oleic acid. At growth temperatures varying between 30 and 40°C an equal distribution between the two fatty acid species was found, which was shifted at the expense of pamitic acid to 70% oleic acid at 10°C (Özcan et al., 2007). Interestingly, not only the fatty acid species varied at different growth temperatures but also the phospholipid head group composition. At 30°C more or less equal fractions of phosphatidylglycerol, phosphatidylinositol, and cardiolipin were detected. Upon growth at low temperature, the fraction of phosphatidylglycerol was increased to approx. 60%, whereas at temperatures above 37°C cardiolipin was the predominant form (60%) (Özcan et al., 2007). This seems to our best knowledge to be the first description that a bacterium responds to varying temperatures with such dramatic changes within the phospholipid head group composition of the plasma membrane.

3.2. How is the network of chill stress-adaptation regulated?

3.2.1 Regulation at the transcriptional level

As mentioned above the two component system MtrB-MtrA is involved in the osmostress-dependent induction of the genes *betP*, *proP* and *lcoP* (Möker et al., 2004). Based on the observation that betaine and proline were accumulated in the cold, the transcriptional regulation of these transporter genes was investigated in the wild type and in Δ*mtrAB* cells upon a shift from 30°C to 15°C (N. Möker, unpublished; N. Özcan, unpublished). The comparison between the two strains revealed that (i) *betP*, *proP* and *lcoP* were induced significantly after applying chill stress and that (ii) the expression depended on the presence of the MtrBA system, indicating that it may has also a chill sensing function.

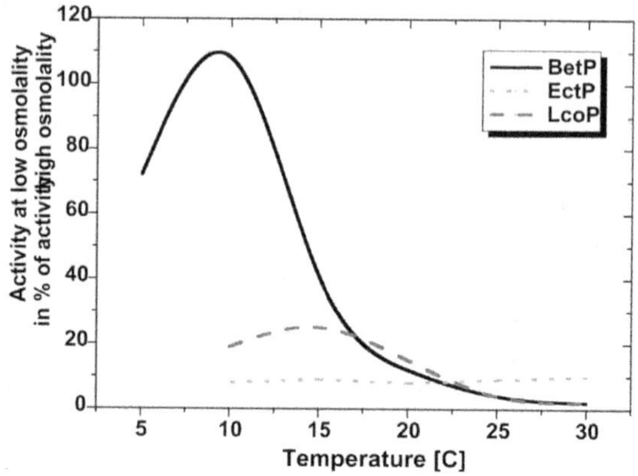

Figure 14.3. Chill activation of the transporters of the BCCT family. Comparison between activity regulation the BCCT family members BetP, EctP and LcoP in dependence of temperature. The figure shows the ratio between the activity at a low external osmolality of 0.2 osmol/kg versus the activity at high external osmolality of 1.25 osmol/kg. Whereas EctP is not stimulated by chill stress, LcoP and BetP respond to low temperatures. In case of BetP chill stress and osmotic stress seem to represent equivalent stress factors, since they lead to similar activation levels.

3.2.2 Regulation at the activity level

The uptake systems for compatible solutes, BetP, EctP, and LcoP, are immediately activated after shifting the cells to high external osmolality. Thus, it was interesting to investigate whether these transporters also sense temperature changes. Although the different carrier proteins showed almost identical osmostress-dependent activation profiles, their sensitivity towards chill stress was very different (Figure 14.3). At the absence of osmotic stress, EctP was not activated by chill and LcoP only to a very low extent, whereas BetP was efficiently stimulated at low temperatures with an activity optimum varying between 10 and 15°C (Özcan et al., 2005, Özcan et al., 2007). Interestingly, at these temperature the level of activation was as high as observed in response to stimulation by salt. This led to the interesting interpretation that BetP itself has, in addition to its osmosensing function, the capability to detect chill stress. It may be speculated that two individual signal input sites for these different stimuli are present, since chill activation of BetP was independent of the internal K^+ concentration, which was identified as being the stimulus for osmostress-dependent BetP activation (Özcan et al., 2005). In principle two different scenarios are reasonable. Either an additional (proteinaceous) factor acts as a primary chill sensor transducing this information to BetP, or the carrier protein detects temperature changes via alterations in the membrane state. The latter has been shown for DesK, the histidine kinase of a chill-activated two component system in *B. subtilis* (Aguilar et al., 2001). Unfortunately, it was not possible to successfully mimic the membrane composition of *C. glutamicum* in liposomes to date, and thus a functional *in vitro* system for studying chill stress conditions is not available. For this reason, it is currently not possible to discriminate between the two scenarios described above. Nevertheless, evidence has been provided that temperature changes are sensed at least partly *via* the membrane. The activation of BetP by both, osmotic and chill stress, is altered in *C. glutamicum* cells adapted to low temperature, which results in a significant rearrangement of the phospholipid composition as mentioned above. In cold-adapted cells, BetP has a decreased sensitivity to osmotic stress as well as to chill stress, indicating that a higher stimulus intensity is required to render BetP sensitive towards these environmental changes in cold-adapted cells (Özcan et al., 2007). Thus, at least the activation threshold of BetP depends on the membrane composition.

4. Concluding remarks

So far the presented data on the response of *C. glutamicum* to environmental stresses as heat, osmotic and chill stress revealed that this organisms seems to be very robust, since at least in case of severe osmotic and chill stress the adaptive response seems to take place mainly at the level of protein activity, whereas the transcriptional regulation is rather responsible for fine-tuning of the response but seems not to be essential. Transcriptional control seems to be more important for the adaptation to heat stress than to osmo- or chill stress conditions. Nevertheless, also in this case severe heat stress, which does not allow growth anymore, was necessary in order to induce the complete cell's quality control namely the induction of genes encoding chaperones as well as proteases. To date the different stress types were analyzed mainly individually. Future analyses will reveal, how far the stress responses are interconnected for examples by transcriptional control to guarantee an effective protection against more general effects provoked by different environmental stresses, like protein denaturation or the occurrence of reactive oxygen species.

Acknowledgement

We would like to thank the former and present members of the osmo- and chill stress group in Köln for their excellent work and the Deutsche Forschungsgemeinschaft for funding.

References

Aguilar, P.S., Hernandez-Arriaga, A.M., Cybulski, L.E., Erazo, A.C., and de Mendoza, D. (2001). Molecular basis of thermosensing: a two-component signal transduction thermometer in *Bacillus subtilis*. EMBO J. *20*, 1681-1691.

Ankri, S., Serebrijski, I., Reyes, O., and Leblon, G. (1996). Mutations in the *Corynebacterium glutamicum* proline biosynthetic pathway: a natural bypass of th *proA* step. J. Bacteriol. *178*, 4412-4419.

Arakawa, T., and Timasheff, S.N. (1985). The stabilization of proteins by osmolytes. Biophys. J. *47*, 4412-4419.

Arsene, F., Tomoyasu, T., and Bukau, B. (2000). The heat shock response of *Escherichia coli*. Int. J. Food Microbiol. *55*, 3-9.

Barreiro, C., Gonzalez-Lavado, E., Patek, M., and Martin, J.F. (2004). Transcriptional analysis of the *groES-groEL1*, *groEL2*, and *dnaK* genes in *Corynebacterium glutamicum*: characterization of heat shock-induced promoters. J. Bacteriol. *186*, 4813-4817.

Barreiro, C., Gonzalez-Lavado, E., Brand, S., Tauch, A., and Martin, J.F. (2005). Heat shock proteome analysis of wild-type *Corynebacterium glutamicum* ATCC 13032 and a spontaneous mutant lacking GroEL1, a dispensable chaperone. J. Bacteriol. *187*, 884-889.

Bass, R.B., Strop, P., Barclay, M., and Rees, D.C. (2002). Crystal structure of *Escherichia coli* MscS, a voltage-modulated and mechanosensitive channel. Science. *298*, 1582-1587.

Berrier, C., Besnard, M., Ajouz, B., Coulombe, A., and Gazi, A. (1996). Multiple mechanosensitive channels from *Escherichia coli*, activated at different thresholds of applied pressure. J. Membrane Biol. *151*, 175-187.

Brigulla, M., Hoffmann, T., Krisp, A., Völker, A., Bremer, E., and Völker, U. (2003). Chill induction of the SigB-dependent general stress response in *Bacillus subtilis* and its contribution to low-temperature adaptation. J. Bacteriol. *185*, 4305-4314.

Brocker, M., and Bott, M. (2006). Evidence for activator and repressor functions of the response regulator MtrA from *Corynebacterium glutamicum*. FEMS Microbiol. Lett. 264, 205-212.

Bucca, G., Hindle, Z., and Smith, C.P. (1997). Regulation of the *dnaK* operon of *Streptomyces coelicolor* A3(2) is governed by HspR, an autoregulatory repressor protein. J. Bacteriol. *179*, 5999-6004.

Bucca, G., Brassington, A.M., Schönfeld, H.J., and Smith, C.P. (2000). The HspR regulon of *Streptomyces coelicolor*: a role for the DnaK chaperone as a transcriptional co-repressordagger. Mol. Microbiol. *38*, 1093-1103.

Bukau, B., and Horwich, A.L. (1998). The Hsp70 and Hsp60 chaperone machines. Cell *92*, 351-366.

Bukau, B., Deuerling, E., Pfund, C., and Craig, E.A. (2000). Getting newly synthesized proteins into shape. Cell *101*, 119-122.

de Mendoza, D., and Cronan Jr., J.E. (1983). Thermal regulation of membrane lipid fluidity in bacteria. Trends Biochem. Sci. *8*, 49–52.

de Mendoza, D., Grau, R., and Cronan Jr., J.E. (1993). Biosynthesis and function of membrane lipids. In: *Bacillus subtilis* and other Gram-positive bacteria: physiology, biochemistry and molecular biology. Losick, R., Sonenshein, A.L., and Hoch, J.A. eds., (Washington, DC: American Society for Microbiology), pp. 411–425.

Dinnbier, U., Limpinsel, E., Schmid, R., and Bakker, E.P. (1988). Transient accumulation of potassium glutamate and its replacement by trehalose during adaptation of growing cells of *Escherichia coli* K-12 to elevated sodium chloride concentrations. Arch. Microbiol. *150*, 348-357.

Engels, S., Schweitzer, J.E., Ludwig, C., Bott, M., and Schaffer, S. (2004). *clpC* and *clpP1P2* gene expression in *Corynebacterium glutamicum* is controlled by a regulatory network involving the transcriptional regulators ClgR and HspR as well as the ECF sigma factor sigmaH. Mol. Microbiol. *52*, 285-302.

Farwick, M., Siewe, R.M., and Krämer, R. (1995). Glycine betaine uptake after hyperosmotic shift in *Corynebacterium glutamicum*. J. Bacteriol. *177*, 4690-4695.

Goloubinoff, P., Mogk, A., Zvi, A.P., Tomoyasu, T., and Bukau, B. (1999). Sequential mechanism of solubilization and refolding of stable protein aggregates by a bichaperone network. Proc. Natl. Acad. Sci. USA *96*, 13732-13737.

Guillouet, S., and Engasser, J.M. (1995a). Sodium and proline accumulation in *Corynebacterium glutamicum* as a response to an osmotic saline upshock. Appl. Microbiol. Biotechnol. *43*, 315-320.

Guillouet, S., and Engasser, J.M. (1995b). Growth of *Corynebacterium glutamicum* in glucose-limited continuous cultures under high osmotic pressure: Influence of growth rate on the intracellular accumulation of proline, glutamate and trehalose. Appl. Microbiol. Biotechnol. *44*, 496-500.

Hecker, M., and Völker, U. (2001). General stress response of *Bacillus subtilis* and other bacteria. Adv. Microb. Physiol. *44*, 35-91.

Helmann, J.D. (2002). The extracytoplasmic function (ECF) sigma factors. Adv. Microb. Physiol. *46*, 47-110.

Ikeda, M., and Nakagawa, S. (2003). The *Corynebacterium glutamicum* genome: features and impacts on biotechnological processes. Appl. Microbiol. Biotechnol. *62*, 99-109.

Jones, P.G., van Bogelen, R.A., and Neidhardt, F.C. (1987). Induction of proteins in response to low temperature in *Escherichia coli*. J. Bacteriol. *169*, 2092–2095.

Kalinowski, J., Bathe, B., Bartels, D., Bischoff, N., Bott, M., Burkovski, A., Dusch, N., Eggeling, L., Eikmanns, B.J., Gaigalat, L., Goesmann, A., et al. (2003). The complete *Corynebacterium glutamicum* ATCC 13032 genome sequence and its impact on the production of L-aspartate-derived amino acids and vitamins. J. Biotechnol. *104*, 5-25.

Kawahara, Y., Ohsumi, T., Yoshihara, Y., and Ikeda, S. (1989). Proline in the osmoregulation of *Brevibacterium lactofermentum*. Agric. Biol. Chem. *53*, 2475-2479.

Kim, T.H., Kim, H.J., Park, J.S., Kim, Y., Kim, P., and Lee, H.S. (2005a). Functional analysis of *sigH* expression in *Corynebacterium glutamicum*. Biochem. Biophys. Res. Commun. *331*, 1542-1547.

Kim, T.H., Park, J.S., Kim, H.J., Kim, Y., Kim, P., and Lee, H.S. (2005b). The *whcE* gene of *Corynebacterium glutamicum* is important for survival following heat and oxidative stress. Biochem. Biophys. Res. Commun. *337*, 757-764.

Kocan, M., Schaffer, S., Ishige, T., Sorger-Herrmann, U., Wendisch, V. F., and Bott, M. (2006). Two-component systems of *Corynebacterium glutamicum*: deletion analysis and involvement of the PhoS-PhoR system in the phosphate starvation response. J. Bacteriol. *188*, 724-732.

Koch, A. (1983). The surface stress theory of microbial morphogenesis. Adv. Microbiol. Physiol. *24*, 301-336.

Krüger, E., and Hecker, M. (1998). The first gene of the *Bacillus subtilis clpC* operon, *ctsR*, encodes a negative regulator of its own operon and other class III heat shock genes. J. Bacteriol. *180*, 6681-6688.

Levina, N., Tötemeyer, S., Stokes, N.R., Louis, P., Jones, M.A., and Booth, I.R. (1999). Protection of *Escherichia coli* cells against extreme turgor by activation of MscS and MscL mechanosensitive channels: identification of genes required for MscS activity. EMBO J. *18*, 1730-1737.

Ley, O. (2005). Charakterisierung der Regulation des Prolinbiosyntheseweges in *Corynebacterium glutamicum*. PhD thesis, Universität zu Köln, Germany.

Manganelli, R., Voskuil, M.I., Schoolnik, G.K., Dubnau, E., Gomez, M., and Smith, I. (2002). Role of the extracytoplasmic-function sigma factor σ^H in *Mycobacterium tuberculosis* global gene expression. Mol. Microbiol. *45*, 365-374.

Mascher, T. (2006). Intramembrane-sensing histidine kinases: a new family of cell envelope stress sensors in Firmicutes bacteria. FEMS Microbiol. Lett. *264*, 133-144.

Möker, N., Brocker, M., Schaffer, S., Krämer, R., Morbach, S., and Bott, M. (2004). Deletion of the genes encoding the MtrA-MtrB two-component system of *Corynebacterium glutamicum* has a strong influence on cell morphology, antibiotics susceptibility and expression of genes involved in osmoprotection. Mol. Microbiol. *54*, 420-438.

Möker, N., Krämer, J., Unden, G., Krämer, R., and Morbach, S. (2007a). *In vitro* analysis of the osmoregulated two-component system MtrB-MtrA from *Corynebacterium glutamicum*. J. Bacteriol. *189*, 3645-3649.

Möker, N., Reihlen, P., Krämer, R. and Morbach S. (2007b) Osmo sensing properties of the histidine protein kinase MtrB from *Corynebacterium glutamicum*. J. Biol. Chem. *282*, 27666-27677.

Morbach, S. and Krämer, R. (2005). Osmoregulation. In: Handbook of *Corynebacterium glutamicum*. Eggeling, L., and Bott, M. eds. (Boca Raton, USA: CRC Press, Taylor and Francis Group). pp. 417-435.

Muffler, A. Bettermann, S., Haushalter, M., Horlein, A., Neveling, U., Schramm, M., and Sorgenfrei, O. (2002). Genome-wide transcription profiling of *Corynebacterium glutamicum* after heat shock and during growth on acetate and glucose. J. Biotechnol. *98*, 255-268.

Nakamura, J., Hirano, S., Ito, H., and Watchi, M. (2008). Mutations of the *Corynebacterium glutamicum* NCgl1221 gene encoding a mechanosensitive channel homolog induce L-glutamic acid production. Appl. Environ. Microbiol., in press.

Nakunst, D., Larisch, C., Hüser, A.T., Tauch, A., Pühler, A., and Kalinowski, J. (2007). The extracytoplasmic function-type sigma factor SigM of *Corynebacterium glutamicum* ATCC 13032 is involved in transcription of disulfide stress related genes. J. Bacteriol. *189*, 4696-4707.

Narberhaus, F. (2002) Alpha-crystallin-type heat shock proteins: socializing minichaperones in the context of a multichaperone network. Microbiol. Mol. Biol. Rev. *66*, 64-93.

Nishio, Y., Nakamura, Y., Kawarabayasi, Y., Usuda, Y., Kimura, E., Sugimoto, S., Matsui, K., Yamagishi, A., Kikuchi, H., Ikeo, K., and Gojobori, T. (2003). Comparative complete genome sequence analysis of the amino acid replacements responsible for the thermostability of *Corynebacterium efficiens*. Genome Res. *13*, 1572-1579.

Nottebrock, D., Meyer, U., Krämer, R., and Morbach, S. (2003). Molecular and biochemical characterization of mechanosensitive channels in *Corynebacterium glutamicum*. FEMS Microbiol. Lett. *218*, 305-309.

Özcan, N., Krämer, R., and Morbach, S. (2005). Chill activation of compatible solute transporters in *Corynebacterium glutamicum* at the level of transport activity. J. Bacteriol. *187*, 4752-4759.

Özcan, N. (2007). Anpassung an Kältestress in *Corynebacterium glutamicum*. PhD thesis, Universitätzu Köln, Germany.

Özcan, N., Ejsing, C. S., Shevchenko, A., Lipski, A., Morbach, S. and Krämer, R. (2007) Osmolality, temperature and membrane lipid composition modulate the activity of betaine transporter BetP in *Corynebacterium glutamicum*. J. Bacteriol. *189*, 7485-7496.

Peter, H., Burkovski, A., and Krämer, R. (1996). Isolation, characterization, and expression of the *Corynebacterium glutamicum betP* gene, encoding the transport system for the compatible solute glycine betaine. J. Bacteriol. *178*, 5229-5234.

Peter, H., Weil, B., Burkovski, A., Krämer, R., and Morbach, S. (1998a). *Corynebacterium glutamicum* is equipped with four secondary carriers for compatible solutes: identification, sequencing, and characterization of the proline/ectoine uptake system, ProP, and the ectoine/proline/glycine betaine carrier, EctP. J. Bacteriol. *180*, 6005-6012.

Peter, H., Burkovski, A., and Krämer, R. (1998b). Osmosensing by N- and C-terminal extensions of the glycine betaine uptake system BetP of *Corynebacterium glutamicum*. J. Biol. Chem. *273*, 2567-2574.

Raman, S., Song, T., Puyang, X., Bardarov, S., Jacobs, W.R. Jr., and Husson, R.N. (2001). The alternative sigma factor SigH regulates major components of oxidative and heat stress responses in *Mycobacterium tuberculosis*. J. Bacteriol. *183*, 6119-6125.

Raivio, T.L. (2005). Envelope stress responses and Gram-negative bacterial pathogenesis. Mol. Microbiol. *56*, 1119-1128.

Record, M.T. Jr., Courtnay, E.S., Cayley, D.S., and Gutman, H.J. (1998). Responses of *E. coli* to osmotic stress: large changes in amounts of cytoplasmic solutes and water. Trends Biol. Sci. *24*, 143-148.

Rönsch H., Krämer, R., and Morbach, S. (2003). Impact of osmotic stress on volume regulation, cytoplasmic solute composition and lysine production in *Corynebacterium glutamicum* MH20-22B. J. Biotechnol. *104*, 87-97.

Rübenhagen, R., Jung, H., Krämer, R., and Morbach, S. (2000). Osmosensor and osmoregulator properties of the betaine carrier BetP from *Corynebacterium glutamicum* in proteoliposomes. J. Biol. Chem. *275*, 735-741.

Rübenhagen, R., Morbach, S., and Krämer, R. (2001). The osmoreactive betaine carrier BetP from *Corynebacterium glutamicum* is a sensor for cytoplasmic K^+. EMBO J. *20*, 5412-5420.

Ruffert, S., Lambert, C., Peter, H., Wendisch, V.F., and Krämer, R. (1997). Efflux of compatible solutes in *Corynebacterium glutamicum* mediated by osmoregulated channel activity. Eur. J. Biochem. *247*, 572-580.

Ruffert, S., Berrier, C., Krämer, R., and Ghazi, A. (1999). Identification of mechanosensitive ion channels in the cytoplasmic membrane of *Corynebacterium glutamicum*. J. Bacteriol. *181*, 1673-1676.

Russell, N.J. (1984). Mechanisms of thermal adaptation in bacteria: blueprints for survival. Trends Biochem. Sci. *3*, 108–112.

Skjerdal, O.T., Sletta, H., Flenstad, S.G., Josefsen, K.D., Levine, D.W., and Ellingsen, T.E. (1996). Changes of the intracellular composition in the response to hyperosmotic stress of NaCl, sucrose or glutamic acid in *Brevibacterium lactofermentum* and *Corynebacterium glutamicum*. Appl. Microbiol. Biotechnol. *44*, 635-642.

Schiller, D., Krämer, R., and Morbach, S. (2004a). Cation specificity of osmosensing by the betaine carrier BetP of *Corynebacterium glutamicum*. FEBS Lett. *563*, 108-112.

Schiller, D., Rübenhagen, R., Krämer, R., and Morbach, S. (2004b). The C-terminal domain of the betaine carrier BetP of *Corynebacterium glutamicum* is directly involved in sensing K^+ as an osmotic stimulus. Biochemistry *43*, 5583-5591.

Schiller, D., Ott, V., Krämer, R., and Morbach, S. (2006). Influence of membrane composition on osmosensing by the betaine carrier BetP from *Corynebacterium glutamicum*. J. Biol. Chem. *281*, 7737-7746.

Seibold, G., Dempf, S., Schreiner, J., and Eikmanns, B.J. (2007). Glycogen formation in *Corynebacterium glutamicum* and role of ADP-glucose pyrophosphorylase. Microbiology *153*, 1275-1285.

Steger, R., Weinand, M., Krämer, R., and Morbach, S. (2004). LcoP, an osmoregulated betaine/ectoine uptake system from *Corynebacterium glutamicum*. FEBS Lett. *573*, 155-160.

Steil, L., Hoffmann, T., Budde, I., Völker, U., and Bremer, E. (2003). Genome-wide transcriptional profiling analysis of adaptation of *Bacillus subtilis* to high salinity. J. Bacteriol. *185*, 6358-63570.

Tzvetkov, M., Klopprogge, C., Zelder, O., and Liebl, W. (2003). Genetic dissection of trehalose biosynthesis in *Corynebacterium glutamicum*: inactivation of trehalose production leads to impaired growth and an altered cell wall lipid composition. Microbiology *149*, 1659-1673.

Varela, C., Agosin, E., Baez, M., Klapa, M., and Stephanopoulos, G. (2003). Metabolic flux redistribution in *Corynebacterium glutamicum* in response to osmotic stress. Appl. Microbiol. Biotechnol. *60*, 547-555.

Ventura, M., Canchaya, C., Zhang, Z., Bernini, V., Fitzgerald, G.F., and van Sinderen, D. (2006). How high G+C Gram-positive bacteria and in particular bifidobacteria cope with heat stress: protein players and regulators. FEMS Microbiol. Rev. *30*, 734-759.

Weinand, M., Krämer, R., and Morbach, S. (2007). Characterization of compatible solute transporter multiplicity in *Corynebacterium glutamicum*. Appl. Microbiol. Biotechnol. *76*, 701-708.

Wolf, A., Krämer, R., and Morbach, S. (2003). Three pathways for trehalose metabolism in *Corynebacterium glutamicum* ATCC13032 and their significance in response to osmotic stress. Mol. Microbiol. *49*, 1119-1134.

Youxing, Q., Bolen, C.L., and Bolen, W. (1998). Osmolyte-driven concentration of a random coil protein. Proc. Natl. Acad. Sci. USA *95*, 9268-9273.

Zuber, U., and Schumann, W. (1994) CIRCE, a novel heat shock element involved in regulation of heat shock operon *dnaK* of *Bacillus subtilis*. J Bacteriol. *176*, 1359-1363.

From: Corynebacteria: Genomics and Molecular Biology. Edited by: Andreas Burkovski.

Index

A

AAA+ proteases 304
Acetate
  metabolism 165
  stimulon 34
AcnR 39, 155, 157, 161-162
Adenylylation191-193
Affimetrix gene-chips 34
AIEC/SDS-PAGE 65-66, 68-69
Alanine producer 4
Amino acid production. *See* Alanine producer, Cysteine production, Glutamate production, Methionine production, Valine production
Ammonium
  assimilation 184
  starvation stimulon 34
  uptake 184. *See also* Nitrogen uptake
AmtR 39, 185, 196
  binding sites 187-189
  promoters 187
  regulon 186
AmtR-dependent promoters 187-189
Anapleroptic fluxes 99. *See also* Pyruvate node
Arabinan
  biosynthesis 280
  precursor 279
Arabinogalactan 277

B

BAC/SDS-PAGE 64
Betaine. *See* Glycine betaine
BetP regulation 324. *See also* Chill stress, Osmotic stress

C

C. diphtheriae 9
  virulence factors 23, 307
  proteome 73
C. efficiens 9
  proteome 72
C. glutamicum
  metabolic network 80
  membrane proteins 69-72
  proteome 56
C. glutamicum R 8, 9
C. jeikeium 9
  lipid auxotrophy 22
  lipophilic corynebacteria 21-22
  lipophilism 22
  virulence factors 23, 307
  proteome 73
Carbon catabolite repression 156
  diauxic growth 156
  regulation of
    acetate metabolism 166
    ethanol metabolism 173
    sugar metabolism 160
Carbon flux distribution 92-94
Carbon metabolism 158, 168
  acetate
    metabolism 165
    stimulon 34
  citrate stimulon 34
  diauxic growth 156
  ethanol
    metabolism 173
    stimulon 34, 37
  inositol stimulon 34
  lactate
    metabolism 163
    stimulon 34
  myo-inositol stimulon 37
  phosphotransferase systems 157-158, 160
  propionate stimulon 34
  ribose stimulon 34
  serine stimulon 34
  sugar metabolism 157
Carbon sources 156. *See also* Carbon metabolism, Phosphotransferase systems

Cell wall synthesis 38, 270
 arabinan
  precursor 279
  biosynthesis 280
 arabinogalactan 277
 galactan
  biosynthesis 278
  precursor 278
 lipoarabinomannan 272
 lipomannan 272
 mycolic acids 284
 peptidoglycan 38, 269
 trehalose
  dimycolates 268
  monomycolates 268
Central metabolic network 80-81
Central metabolism 158, 168
Chaperones 314-315
Chill stress 325
ChIP-to-chip analysis 38
CitB 40
CitA-CitB 40
Citrate stimulon 34
ClgR 39, 305
Cloning vectors 122-124, 126
clp gene promoters 137
Clp protease 303-306. *See also* AAA+ proteases, FtsH, Proteases
Clustered regularly interspaced short palindromic repeats 25
ColE2-P9 115
Compatible solutes 38, 319
 glutamine 319
 glycine betaine 38, 319
 proline 38, 319
 trehalose 319
Core genes 26
Corynebacterium
 plasmids 114-116
 species 8
Creatinine
 uptake 185
 deaminase 185
CRISPRs 25
Cysteine 224
 metabolism 225-226
 production 234
Cytosolic proteome 57-58

D

2-D BAC 64
2-DE 55
Diauxic growth 156
Discovery of *Corynebacterium glutamicum* 1, 3
DNA affinity chromatography 38-39
DNA microarrays 34
 Affimetrix gene-chips 34
dnaK promoter 139
DtxR 39, 250-252
 binding sites 251
 regulon 252-255

E

ECF sigma factors 39, 137, 305
Episomes 10
Essential genes 26
Ethambutol 38, 41-42, 280-281, 286-287
Ethanol
 metabolism 173
 stimulon 34, 37
Expression vectors 127-128
Extra cytoplasmic function sigma factor. *See* ECF sigma factors

F

FarR 39
Fatty acids 22, 283
Ferric iron transport 249
Ferrous iron transport 249
Fimbriae 21
Flavor enhancer. *See* Kombu, MSG, Umami
Flux analysis 101. *See also* Isotope labelling
FtsH 306. *See also* AAA+ proteases, Clp proteases, Proteases

G

G+C content 9, 25
Galactan biosynthesis 278
Galactan precursor 278
GC skew 25
Genome
 size 9
 rearrangements 27
 sequencing strategies 8, 11, 13, 15, 20-21, 23-24
 stability 27
Genomic islands 12-16, 18-20, 22
Genomics of
 C. diphtheriae 17
 C. efficiens 15
 C. glutamicum 8
 C. jeikeium 21

C. pseudotuberculosis 24
C. urealyticum 23
GlnK proteolysis 192
Global expression analysis 41. *See also* Transcriptome analysis
Glutamate
excretion 3, 285
production 3-4, 6, 41
transport 185
Glutamine 319
GlxR 155, 157
Glycine betaine 38, 319. *See also* Compatible solutes
Glycosylation 59-60
GntR1 155, 157, 161
GntR2 155, 157, 161
GRAVY index 61
transmembrane helices 60

H

Heat stress 305, 313, 318
Heme 249
Heme-dependent regulation 259
High G+C 7
Homeostasis 34, 241
Housekeeping genes 133
HspR 305
Hypoosmotic shock 42, 320

I

incW 115
Inorganic sulfur compounds 223
Inositol stimulon 34
Iron
acquisition 242
heme 249
siderophores 243, 248
uptake systems 244-247
ferric iron transport 249
ferrous iron transport 249
stimulon 34
DtxR 39, 250-252
DtxR binding sites 251
DtxR regulon 252-255
RipA 39, 256
Iron-sulfur clusters 229
Isotope labelling 87. *See also* Flux analysis

K

α-Ketoglutarate 1-2, 157, 185, 191-193
Kombu 1

L

Lactate 36
metabolism 163
stimulon 34
LC-ESI 55
LC-MALDI 55
Lipid auxotrophy 22
Lipoarabinomannan 272
Lipomannan 272
Lipophilic corynebacteria 21-22
Lipophilism 22
LldR2 155, 157
LtbR 39
Lysine
producer 68
production 5-6, 43-44

M

Magnesium stimulon 34
McbR 39, 232-233
binding sites 232
Mass spectrometrie. *See* LC-ESI, LC-MALDI
Mechanosensitive channels 320
Membrane
proteome 61
C. glutamicum membrane proteins 69-72
GRAVY index 61
transmembrane helices 60
proteomics 60-64, 67
Metabolic
flux analysis 87
modelling concepts 85
sampling 89-90
network
analysis 86
description 86
pathway 168
Methionine 227
metabolism 225-226
production 102, 234
Modelling concepts 85
MSG 1
MtrA 39
MtrA-MtrB 38-39, 319
MudPIT 58, 65-66
Mutant mapping 39
Mycobacterium proteome 73-74
Mycolic acids 284
Mycothiol 229
myo-inositol stimulon 37

N

NADPH
  demand 97. *See also* Redox fluxes
  supply 43, 98. *See also* Redox fluxes
Nitrogen
  assimilation 194-195
    ammonium assimilation 184
    creatinine deaminase 185
    urease 185
  control 185. *See also* AmtR, AmtR regulon
  sources 183
  uptake 194-195.
    ammonium uptake 184
    creatinine uptake 185
    urea uptake 185

O

Organic sulfur compounds 220
Osmotic stress 318
  BetP regulation 324
  compatible solutes 38, 319
  glutamine 319
  glycine betaine 38, 319
  hypoosmotic shock 42, 320
  proline 38, 319
  trehalose 319
Oxidative stress 318
Oxygen deprivation 36
2-Oxoglutarate. *See* α-Ketoglutarate

P

pA501 115
pA505 115
pA3253 123
pAG1 114
pAG3 114
pAJ228 123
Pantothenate 45, 47
  producer 47
pAPE12 128
Pathogenicity 17, 22, 24
Pathway
  design 103
  fluxes 90-91
pB85776 115
pBHK18 122
pBKGEXm2 128
pBL1 114, 118, 122-123, 128, 130
pCC1 114, 119, 122
pCE2 114
pCE3 114
pCG1 114, 121-123, 128, 130-131
pCG2 114
pCG4 114
pCG11 123
pCGR1 9-10
pCJ84 114
pCR1 115
pCRA1 128
pCRB1 122
pCRYA4 115
pEBM2 122
pEC2 9-10
pEC3 9-10
pEC5 122
pEC901 128
pEC-K18*mob2* 122
pEC-XK99E 128
pECKA 122
pEGFP 130
pEK0 122
pEKEx1 128
pEKplCm 130
pEKpllacz 130
pEM1 123
pEM1dppc 123
pEMel-1 130
pEP2 122
Peptidoglycan 38, 269
pERP1 131
pET2 130
pGA1 114, 120, 122, 128
pGA2 114
pH adaptation 37
PhoR 39
  binding sites 213
PhoS-PhoR 39, 210
Phosphate
  limitation 209-210
  starvation 209
    stimulon 34
  compounds 204
    polyP 207, 209
    polyphosphate metabolism 207
  mobilization 205
  sources 203
  uptake 204
Phosphorylation 59
Phosphotransferase systems 157-158, 160
pIJ101 114
Pili 21
pJC1 122

pJUP05 130
pJV1 114
pK18*mob* 123
pK18*mob2* 123
pK18*mobsacB* 123
pK43 114
pK64 114
pkw0 122
pKW4 9-10, 115
pKX15 123
Plasmids 114-115. *See also* Expression vectors, Promoter-probe vectors, Shuttle vectors, terminator-probe vector
pLEW279a 115
pLEW279b 115
pMJ-A 122
pNG2 114-115, 121-122, 128
polyP 207, 209
Polyphosphate metabolism 207
Post-transcriptional regulation 175, 192
  adenylylation191-193
pPRE11 131
pPROBE17 130
pRAG5 131
Precursor demand 82-84
pRIM2 131
Production pathways 84
Proline 38, 319
Promoter activity 167
Promoter-probe vectors 127, 130-131
Prophages 12, 18-19. See also *tox* gene
Propionate stimulon 34
Proteases 296
  AAA+ proteases 304
  Clp protease 303-306
  FtsH 306
  GlnK proteolysis 192
Protein modification 58
  adenylylation191-193
  glycosylation 59-60
  phosphorylation 59
Proteome 55
  C. diphtheriae proteome 73
  C. efficiens proteome 72
  C. glutamicum proteome 56
  C. jeikeium proteome 72
  cytosolic proteome 57-58
  membrane proteome 61
  Mycobacterium proteome 73-74
  secreted proteome 58
Proteomics. *See* Proteome, Two-dimensional gel electrophoresis
pSCCD1 122
pSELF2000X 123
pSK1Cat 130
pSKFT2 123
pSRK21 122
pTET3 114
pTP10 115
PTS. *See* Phosphotransferase systems
pULMJ55 122
pULMJ88 130
pULMJ95 130
pUT2 130
pUT3 130
pVWEx1 128
pXK99E128
pXMJ19 128
pXZ608 114
pXZ1014 115
pXZ10145 115
pYM2 114
Pyruvate node 100. *See also* Anaplerotic fluxes
pZ8-1 128

R

RamA 155, 157, 169
RamB 155, 157, 169
Redox fluxes 97
Regulation of
  acetate metabolism 166
  ethanol metabolism 173
  sugar metabolism 160
Regulatory networks 169
Regulons 37-39
Ribose stimulon 34
RipA 39, 256

S

16S rDNA 7
Sampling 89-90
Secreted proteome 58
Serine stimulon 34
Shuttle vectors 128
Siderophores 243, 248. *See also* Iron acquisition
SigB 39, 136
SigB-dependent promoters 136
sigB gene promoters 139
SigH 318
SigM 39, 318

SigM-dependent promoters 138
Sigma factors 39, 132
  ECF sigma factors 39, 137, 305
  SigB 39, 136
  SigH 318
  SigM 39, 318
Sortases 21
SsuR 39, 232-234
SsuR binding sites 232
Stimulons 34-35
Stress response. *See* BetP regulation, Chaperones, Chill stress, Osmotic stress, Oxidative stress, Oxygen deprivation, pH adaptation
Sugar metabolism 157
SugR 39, 155, 157
Sulfonate utilization 218-219
Sulfur assimilation 218-220
Sulfur-containing amino acids 224
Sulfur uptake. *See* Uptake of inorganic sulfur sources, Uptake of organic sulfur sources
Suppressor mutant 40
Systems biology 101

T

TCA cycle 37, 69, 175
  acetate metabolism 165
  acetate stimulon 34
  AcnR 39, 155, 157, 161-162
  central metabolic network 80-81
  citrate stimulon 34
  DtxR 39, 250-252
  ethanol metabolism 173
  pyruvate node 100
  RamA 155, 157, 169
  RamB 155, 157, 169
  RipA 39, 256
Terminator-probe vector 130
Thermotolerance 16
tox gene 17, 20, 250
Transcriptional regulators
  AcnR 39, 155, 157, 161-162
  AmtR 39, 185, 196
  CitB 40
  ClgR 39, 305
  DtxR 39, 250-252
  GlxR 155, 157
  GntR1 155, 157, 161
  GntR2 155, 157, 161
  LtbR 39
  LldR2 155, 157
  McbR 39, 232-233
  MtrA 39
  PhoR 39
  RamA 155, 157, 169
  RamB 155, 157, 169
  RipA 39, 256
  SugR 39, 155, 157
  SsuR 39, 232-234
Transcriptome analysis. *See* Affimetrix gene-chips, DNA microarrays, Global expression analysis, Mutant mapping, Stimulons
Transmembrane helices 60
  GRAVY index 61
  membrane proteome 61
Trehalose 319
  dimycolates 268
  monomycolates 268
Two-component systems 38-40, 210, 259
  CitA-CitB 40
  MtrA-MtrB 38-39, 319
  PhoS-PhoR 39, 210
Two-dimensional gel electrophoresis 55
  AIEC/SDS-PAGE 65-66, 68-69
  BAC/SDS-PAGE 64
  2-D BAC 64
  2-DE 55

U

Umami 1
Uptake of
  inorganic sulphur sources 223
  organic sulfur sources 220
Urea uptake 185
Urease 185
Urinary tract infections 23

V

Valine
  production 4, 45-47
  stimulon 34
Virulence factors 23, 307